COMPATIBLE FOREST MANAGEMENT

Managing Forest Ecosystems

Volume 8

Series Editors:

Klaus von Gadow

Georg-August-University,
Göttingen, Germany

Timo Pukkala

University of Joensuu,
Joensuu, Finland

and

Margarida Tomé

Instituto Superior de Agronomía,
Lisbon, Portugal

Aims & Scope:

Well-managed forests and woodlands are a renewable resource, producing essential raw material with minimum waste and energy use. Rich in habitat and species diversity, forests may contribute to increased ecosystem stability. They can absorb the effects of unwanted deposition and other disturbances and protect neighbouring ecosystems by maintaining stable nutrient and energy cycles and by preventing soil degradation and erosion. They provide much-needed recreation and their continued existence contributes to stabilizing rural communities.

Forests are managed for timber production and species, habitat and process conservation. A subtle shift from *multiple-use management* to *ecosystems management* is being observed and the new ecological perspective of *multi-functional forest management* is based on the principles of ecosystem diversity, stability and elasticity, and the dynamic equilibrium of primary and secondary production.

Making full use of new technology is one of the challenges facing forest management today. Resource information must be obtained with a limited budget. This requires better timing of resource assessment activities and improved use of multiple data sources. Sound ecosystems management, like any other management activity, relies on effective forecasting and operational control.

The aim of the book series ***Managing Forest Ecosystems*** is to present state-of-the-art research results relating to the practice of forest management. Contributions are solicited from prominent authors. Each reference book, monograph or proceedings volume will be focused to deal with a specific context. Typical issues of the series are: resource assessment techniques, evaluating sustainability for even-aged and uneven-aged forests, multi-objective management, predicting forest development, optimizing forest management, biodiversity management and monitoring, risk assessment and economic analysis.

Compatible Forest Management

Edited by

Robert A. Monserud

USDA Forest Service Pacific Northwest Research,
Portland, Oregon, U.S.A.

Richard W. Haynes

USDA Forest Service Pacific Northwest Research,
Portland, Oregon, U.S.A.

and

Adelaide C. Johnson

USDA Forest Service Pacific Northwest Research,
Juneau, AK, U.S.A.

KLUWER ACADEMIC PUBLISHERS

DORDRECHT / BOSTON / LONDON

A C.I.P. Catalogue record for this book is available from the Library of Congress.

ISBN 1-4020-1536-4

Published by Kluwer Academic Publishers,
P.O. Box 17, 3300 AA Dordrecht, The Netherlands.

Sold and distributed in North, Central and South America
by Kluwer Academic Publishers,
101 Philip Drive, Norwell, MA 02061, U.S.A.

In all other countries, sold and distributed
by Kluwer Academic Publishers,
P.O. Box 322, 3300 AH Dordrecht, The Netherlands.

Printed on acid-free paper

Printed in the Netherlands.

Contents

Contributors

Darius M. Adams
Department of Forest Resources, College of Forestry,
Oregon State University
Corvallis, OR 97331, USA
darius.adams@orst.edu

Susan J. Alexander
USDA Forest Service, Pacific Northwest Research Station
3200 SW Jefferson Way, Corvallis, OR 97331, USA
salexander@fs.fed.us

Heidi Ballard
Department of Environmental Science, Policy and Management,
University of California,
151 Hilgard Hall #3110, Berkeley, CA 94720, USA
hballard@nature.berkeley.edu

R. James Barbour
USDA Forest Service, Pacific Northwest Research Station
620 SW Main St., Suite 400, Portland, OR 97205, USA
jbarbour01@fs.fed.us

John Bliss
Department of Forest Resources, College of Forestry,
Oregon State University
Corvallis, OR 97331, USA
john.bliss@orst.edu

Mason D. Bryant
USDA Forest Service, Pacific Northwest Research Station
2770 Sherwood Lane, Juneau, AK 99801, USA
mdbryant@fs.fed.us

Andrew B. Carey
USDA Forest Service, Pacific Northwest Research Station
3625 93rd Ave., Olympia, WA 98512, USA
acarey@fs.fed.us

John H. Cissel
USDI Bureau of Land Management
3200 SW Jefferson Way, Corvallis, OR 97331, USA
jcissel@or.blm.gov

Roger N. Clark
USDA Forest Service, Pacific Northwest Research Station
400 N 34th St. Suite 201, Seattle, WA 98103, USA
rnclark@fs.fed.us

David V. D'Amore
USDA Forest Service, Pacific Northwest Research Station
2770 Sherwood Lane, Juneau, AK 99801, USA
ddamore@fs.fed.us

Robert L. Deal
USDA Forest Service, Pacific Northwest Research Station
620 SW Main St., Suite 400, Portland, OR 97205, USA
rdeal@fs.fed.us

Toni L. De Santo
USDA Forest Service, Pacific Northwest Research Station
2770 Sherwood Lane, Juneau, AK 99801, USA
tdesanto@fs.fed.us

Ellen M. Donoghue
USDA Forest Service, Pacific Northwest Research Station
620 SW Main St., Suite 400, Portland, OR 97205, USA
edonoghue@fs.fed.us

Richard T. Edwards
USDA Forest Service, Pacific Northwest Research Station
2770 Sherwood Lane, Juneau, AK 99801, USA
rtedwards@fs.fed.us

Roger D. Fight
USDA Forest Service, Pacific Northwest Research Station
620 SW Main St., Suite 400, Portland, OR 97205, USA
rfight@fs.fed.us

Takashi Gomi
Department of Geography, University of British Columbia
1984 West Mall, Vancouver, British Columbia VV6T 1Z2, Canada
gomi@interchange.ubc.ca

Richard W. Haynes
USDA Forest Service, Pacific Northwest Research Station
620 SW Main St., Suite 400, Portland, OR 97205, USA
rhaynes@fs.fed.us

Paul E. Hennon
USDA Forest Service, Pacific Northwest Research Station
2770 Sherwood Lane, Juneau, AK 99801, USA
phennon@fs.fed.us

Susan Stevens Hummel
USDA Forest Service, Pacific Northwest Research Station
620 SW Main St., Suite 400, Portland, OR 97205, USA
shummel@fs.fed.us

Adelaide C. Johnson
USDA Forest Service, Pacific Northwest Research Station
2770 Sherwood Lane, Juneau, AK 99801, USA
ajohnson03@fs.fed.us

K. Norman Johnson
Department of Forest Resources, College of Forestry,
Oregon State University
Corvallis, OR 97331, USA
norm.johnson@orst.edu

Becky K. Kerns
USDA Forest Service, Pacific Northwest Research Station
3200 SW Jefferson Way, Corvallis, OR 97331, USA
bkerns@fs.fed.us

Ryan Kimbirauskus
Department of Entomology, Michigan State University
E. Lansing, MI 48824, USA
kimbira1@pilot.msu.edu

Christian LeSage
Department of Entomology, Michigan State University
E. Lansing, MI 48824, USA
lesagec@michigan.gov

Eini C. Lowell
USDA Forest Service, Pacific Northwest Research Station
620 SW Main St., Suite 400, Portland, OR 97205, USA
elowell@fs.fed.us

David D. Marshall
USDA Forest Service, Pacific Northwest Research Station
3625 93rd Ave., Olympia, WA 98512, USA
dmarshall@fs.fed.us

John R. Mills
USDA Forest Service, Pacific Northwest Research Station
620 SW Main St., Suite 400, Portland, OR 97205, USA
jmills@fs.fed.us

Robert A. Monserud
USDA Forest Service, Pacific Northwest Research Station
620 SW Main St., Suite 400, Portland, OR 97205, USA
rmonserud@fs.fed.us

Claire A. Montgomery
Department of Forest Resources, College of Forestry,
Oregon State University
Corvallis, OR 97331, USA
claire.montgomery@orst.edu

Ewa H. Orlikowska
USDA Forest Service, Pacific Northwest Research Station,
2770 Sherwood Lane, Juneau, AK 99801, USA
eorlikowska@fs.fed.us

David Pilz
Department of Forest Science, College of Forestry
Oregon State University,
Corvallis, OR 97331, USA,
david.pilz@orst.edu

Allison Reger
USDA Forest Service, Willamette National Forest
Box 10607, Eugene, OR 97440, USA
areger@fs.fed.us

Mark E. Schultz
USDA Forest Service, Pacific Northwest Research Station
2770 Sherwood Lane, Juneau, AK 99801, USA
mschultz01@fs.fed.us

Thomas A. Spies
USDA Forest Service, Pacific Northwest Research Station
3200 SW Jefferson Way, Corvallis, OR 97331, USA
tspies@fs.fed.us

George H. Stankey
USDA Forest Service, Pacific Northwest Research Station
3200 SW Jefferson Way, Corvallis, OR 97331, USA
ghstankey@fs.fed.us

Frederick J. Swanson
USDA Forest Service, Pacific Northwest Research Station
3200 SW Jefferson Way, Corvallis, OR 97331, USA
fswanson@fs.fed.us

Mark S. Wipfli
USDA Forest Service, Pacific Northwest Research Station
1133 N. Western Ave., Wenatchee, WA 98801, USA
mwipfli@fs.fed.us

Foreword

The context for compatible management is established by three converging factors: developing public recognition of the array of goods and services produced by forests, the growing debate about sustainable forest management, and recognition within the scientific community of the connection between processes and outputs. These factors shape the emerging recognition that we need approaches to forest management that emphasize the compatibility among various goods and services and avoid counterproductive arguments about conflicting tradeoffs between biophysical and socioeconomic components of ecosystems.

The Pacific Northwest has a long history of deliberate forest management debates that have contributed considerable scientific information on the compatibilities and tradeoffs among commodity production and the other values wanted by the public from these forests. Forest management has been frequently portrayed as producing either wood commodities or environmental values; this portrayal has focused much discussion on direct tradeoffs rather than on opportunities for compatible outputs of different management strategies. The discussion has become increasingly value laden as people attempt to describe the links between management practices and sustainable forest production. It has led to an effort at the Pacific Northwest Research Station (U.S. Department of Agriculture, Forest Service) to develop a greater understanding of potential compatibility among commodities and ecological, social, and cultural values.

This book summarizes various aspects of compatibility: the importance of considering multiple production opportunities; persistent emphasis on multiple-use forestry; respect for science-based management and an understanding of the incremental nature of scientific information; the power of markets to provide incentives for forest management; the importance of spatial and temporal scales; and the provisional nature of social values and acceptability. Forest managers can use this information to integrate key scientific findings with their stewardship of forest lands. A key point is that scientific information can be used to increase opportunities for producing compatible goods and services in a manner that is socially acceptable and economically viable. These goods and services may include wood, wildlife habitat, scenery, recreation, water quality (including water as a commodity), and riparian habitat.

This outcome is consistent with the emerging emphasis we see on sustainable forest management and concerns about meeting rising demands for goods and services from the forest in an environmentally acceptable manner.

While we have attempted to provide an integrative approach, we acknowledge that as a forest management text, some aspects of the ecosystem are not treated in the detail generally considered by forest managers. Notably, this detail is lacking in the aquatic and geologic areas. We are beginning to appreciate the dynamic nature of natural disturbances, such as floods, which play a key role in changing landscapes. Within stream channels, disturbance alters accumulation patterns of sediment and woody debris in riparian and upland forests; disturbance also alters trajectories of forest development. Because disturbance does not occur uniformly over space or time, irregular patterns are introduced and pose challenges to forest managers.

Finally, no book springs to life without the hard work of dedicated individuals. We acknowledge the assistance and creative accomplishments of the 37 authors who contributed to the various chapters. We appreciate the support of the former station director, Thomas Mills, who was a tireless advocate of compatible forestry. We acknowledge our anonymous reviewers who challenged us for greater clarity. We greatly appreciate the hard work and high standards of Judy Mikowski who managed the production, Rhonda Mazza who edited, Jenny Beranek who did the page layout, Margaret Hamilton and Lynn Starr who proofread the text, and Valerie Rapp who wrote the section introductions. Without their diligence this book would not have been possible.

Finally, we appreciate the use of a map (Figure 1, Chapter 1) of the Temperate Rainforests of the Pacific Northwest by Ecotrust.

11 May 2003

Robert A. Monserud
Richard W. Haynes
Adelaide C. Johnson

Unit Conversion Table

When you know:	Multiply by:	To find:
Meters (m)	3.28	Feet (ft)
Cubic meters (m^3)	35.3	Cubic feet (ft^3)
Square meters (m^2)	10.8	Square feet (ft^2)
Centimeters (cm)	2.54	Inches (in)
Hectares (ha)	2.47	Acres (ac)
Kilometers (km)	0.62	Miles (mi)
Kilogram (kg)	2.205	Pounds (lbs)
Metric ton (t)	1.1	Short tons (t)
	Product conversions	
Logs:		
Cubic meters	5.66	Thousand board feet (log scale)
Lumber:		
Cubic meters	2.36	Thousand board feet (full sawn)
Panel products:		
Cubic meters	1.13	Thousand square feet (3/8" basis)

Color Plates

1-1 Figure 1. The coastal temperate rainforest of North America. Map reprinted by permission from Ecotrust www.ecotrust.org.

3-5 Figure 5. Biologically simplified, 40-yr-old, even-aged conifer ecosystem (A) and a mixed red alder-conifer ecosystem of the same age (B) showing the more diverse stand structure and abundant understory.

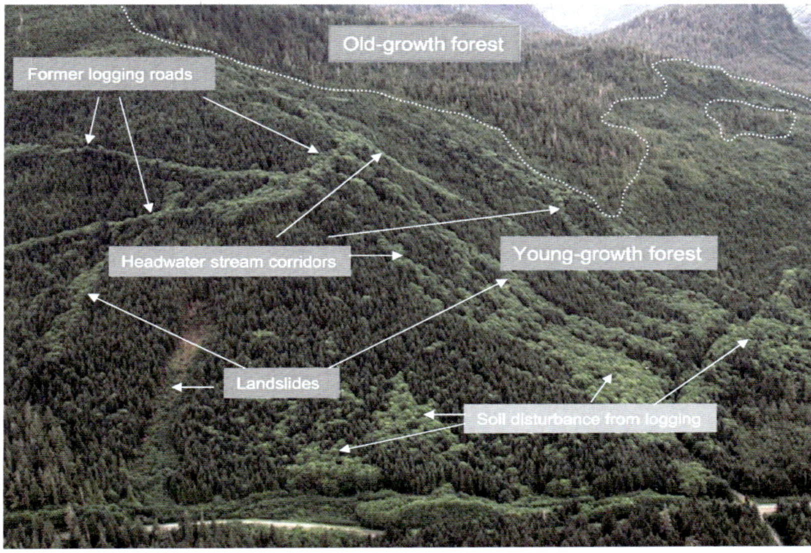

3-6 Figure 6. Distribution and occurrence of red alder in a section of the Maybeso drainage, Prince of Wales Island, southeastern Alaska.

3-7 Figure 7. Headwater stream draining an upland forest where it has already transitioned into a lower gradient, fish-bearing reach.

3-10 Figure 10. Red alder and conifer woody debris provides both physical and biological functions in streams and their riparian zones.

7-1a&b Figure 1. EnVision (McGaughey 2001) illustrates pre- treatment (1a) and post-treatment (1b) fire risk by depicting dead trees, accumulated understory vegetation, and down fuels.

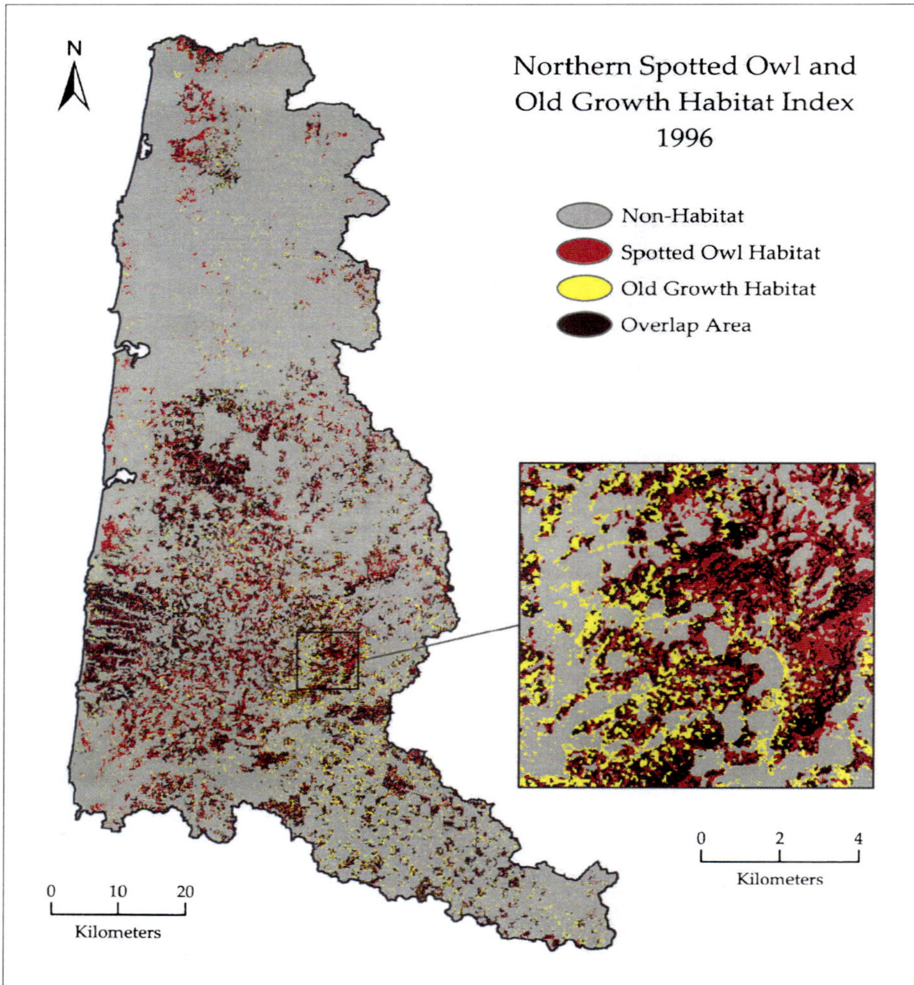

8-3 Figure 3. Area of overlap between suitable habitat for northern spotted owls (McComb et al. 2002) and moderate to high old-growth development, and areas where owl and old-growth habitat do not overlap in the central Oregon Coast Range.

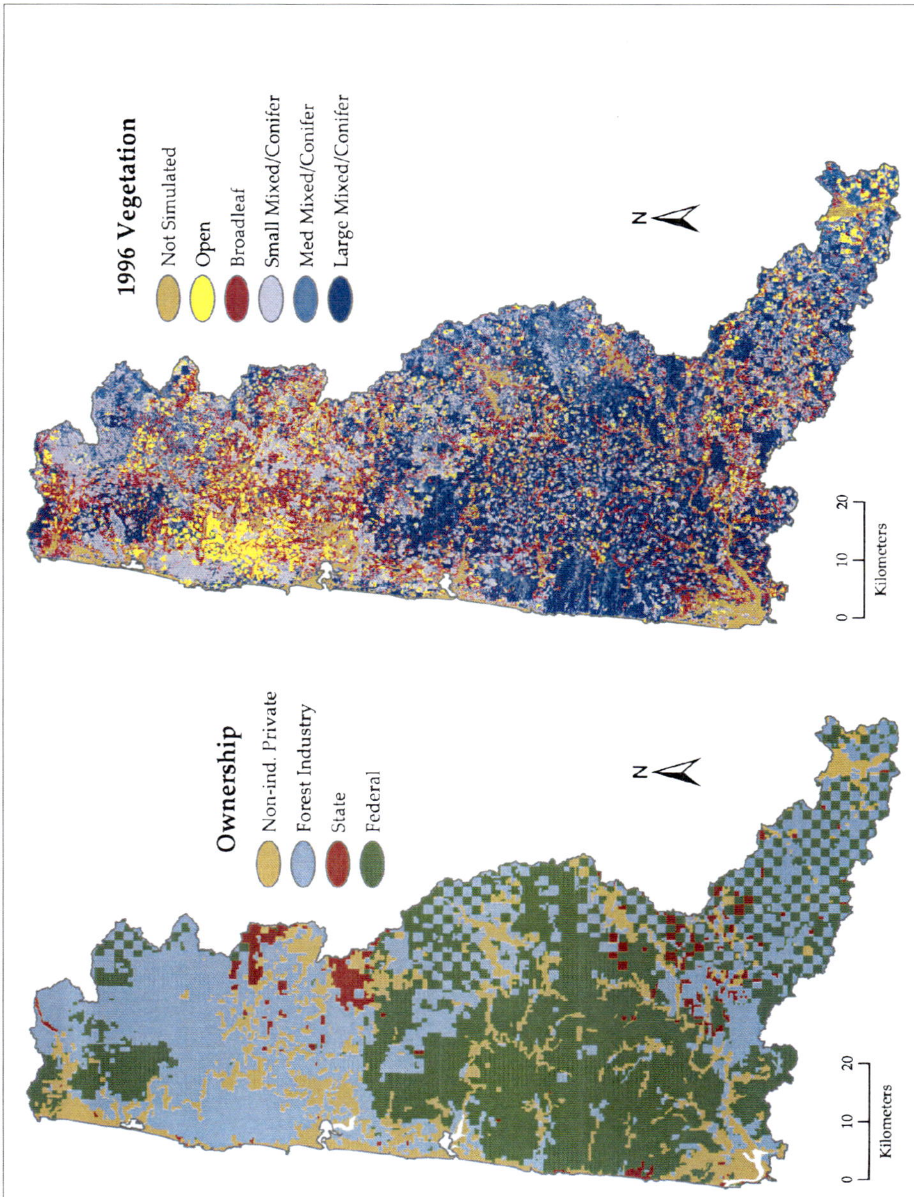

8-5 Figure 5. Vegetation and ownership patterns in 1996 across multiple ownerships in the central Oregon Coast Range (from Ohmann and Gregory 2002).

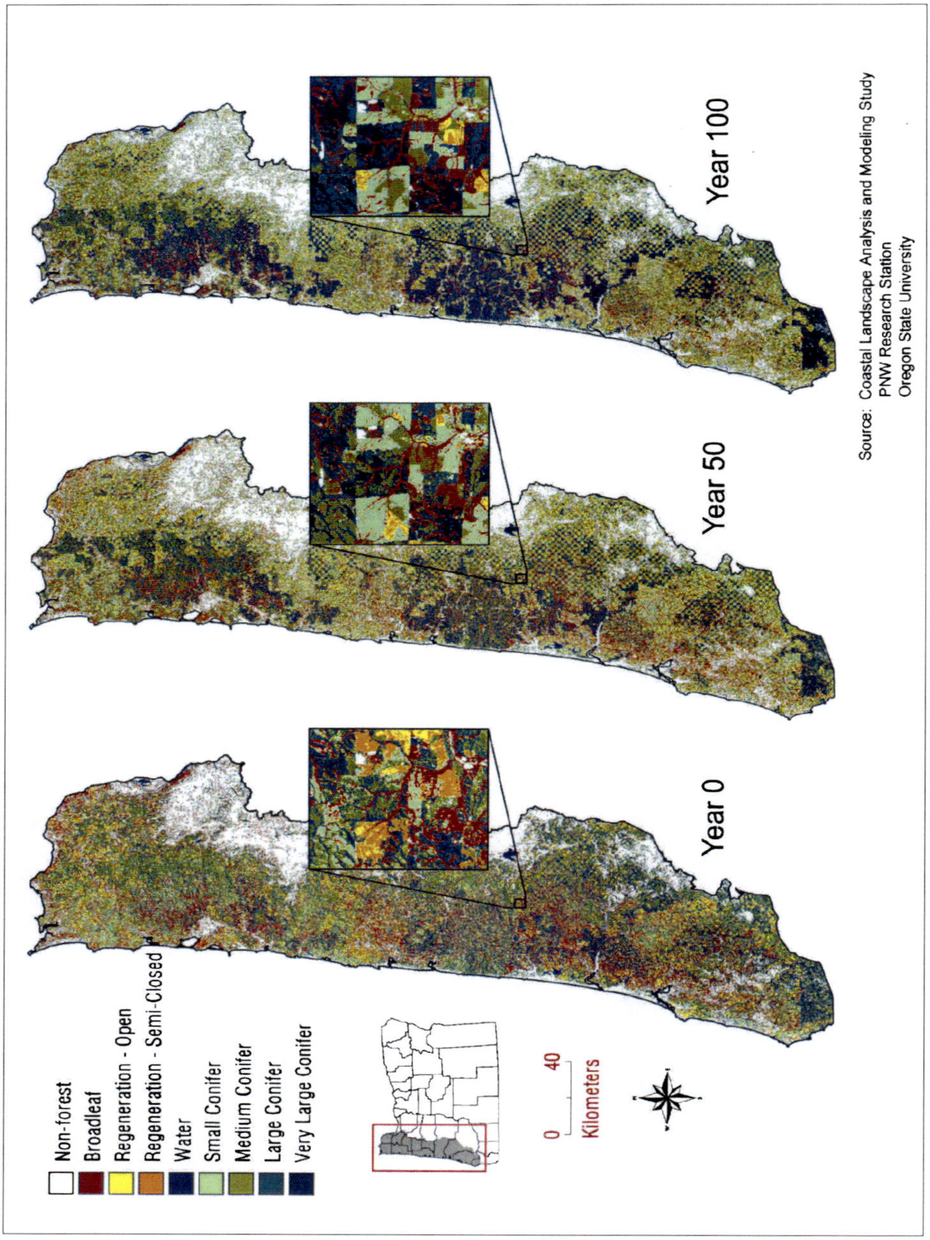

8-7 Figure 7. Simulated changes in forest structure and composition classes for 100 years under current policies (Spies et al. 2002a).

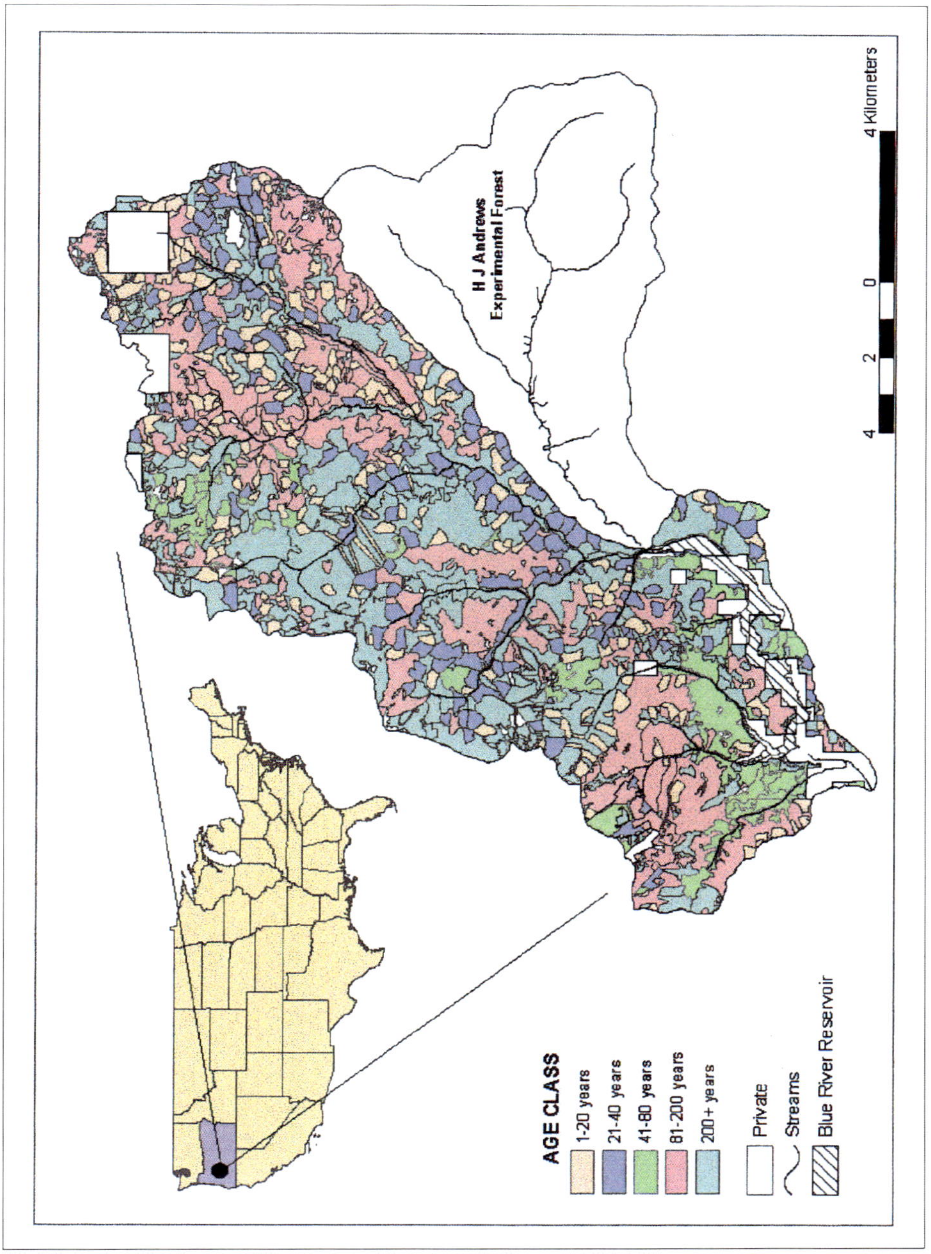

9-1 Figure 1. Location of the Blue River, Oregon, landscape study area and current vegetation age classes

11-1a&b Figure 1. Tree exhibiting multiple flushing (a) and ramicorn branching (b).

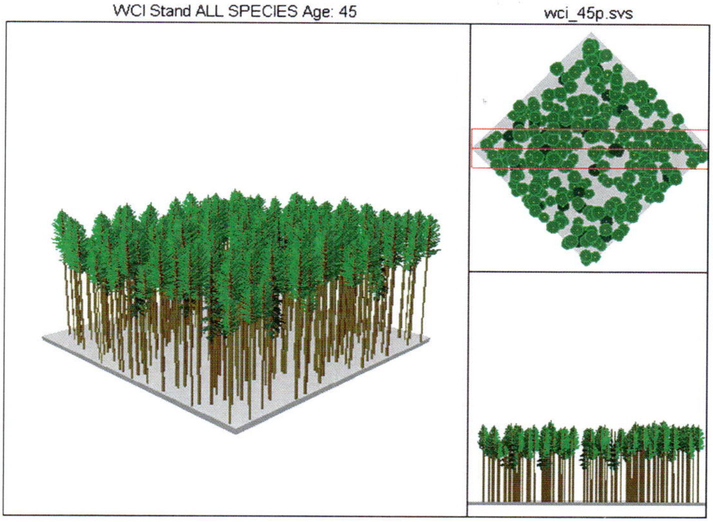

11-4 Figure 4. Simulated stand for the Timber Production regime at rotation age 45 years from the Stand Visualization System (McGaughy 1997).

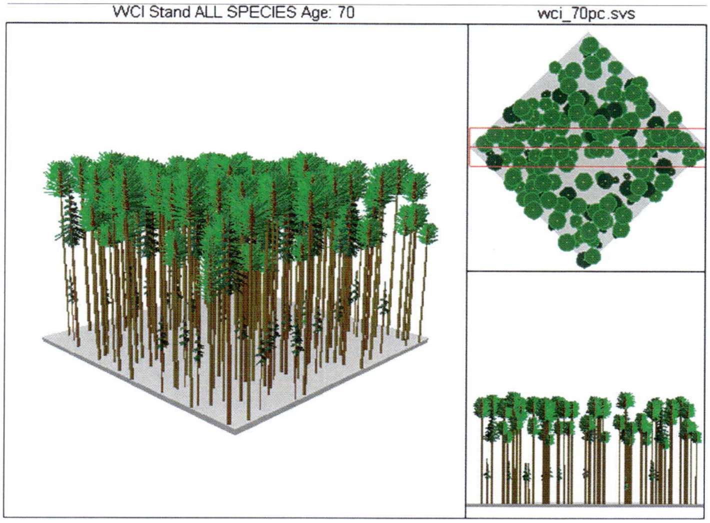

11-5 Figure 5. Simulated stand for the Extended Rotation regime at rotation age 70 years from the Stand Visualization System (McGaughy 1997).

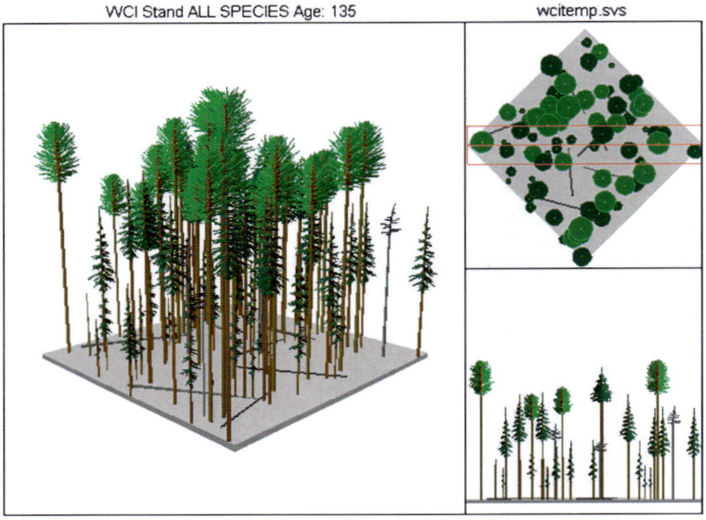

11-6 Figure 6. Simulated stand for the Biodiversity regime at rotation age 135 years from the Stand Visualization System (McGaughy 1997).

14-1 Figure 1. A keystone complex: spotted owl-flying squirrel-truffle-Douglas-fir.

Introduction

Forest science is always rooted in a place—a specific forest, with its unique mix of climate, trees, other plants, wildlife, and human uses. In this book, scientists discuss their research on compatible forest management in the moist maritime forests of the Pacific Northwest. They focus on the zone northward from the California-Oregon border to southeastern Alaska. Although the findings are rooted in these forests, they offer much that is relevant to scientists and managers of other forests.

Compatible forest management is an approach that simultaneously produces multiple products of value without decreasing other values, all in a socially acceptable manner. The central idea is to expand the framework of forest management from an "either-or" bipolar framework to a framework with many options. Perhaps options exist where change can have positive results for several values.

Chapter 1

Compatible Forest Management: Background and Context

Richard W. Haynes
Robert A. Monserud
USDA Forest Service, Pacific Northwest Research Station,
620 SW Main St., Suite 400, Portland, OR 97205, USA
Adelaide C. Johnson
USDA Forest Service, Pacific Northwest Research Station,
2770 Sherwood Lane, Juneau, AK 99801, USA

1. Introduction

Let us begin by defining our area of interest: the moist maritime forests of the Pacific Northwest (PNW). Geographically (as shown in Figure 1), this includes western Oregon and Washington (the west side of the Cascade Range, the Coast Range, and the Olympic Mountains), coastal British Columbia (Coast Mountains), and island-dominated southeastern Alaska (as far north as the Kenai Peninsula). Collectively, this is the world's northernmost temperate rainforest (Walter 1985). Although this coastal rainforest zone stretches southward to San Francisco, California, we focus our research northward from the California-Oregon border. The east-west extent of this coastal forest varies as a function of climate and elevation; the zone ranges from a few kilometers at the northern and southern extremes to several hundred kilometers in the middle of its range. Much of the northern half of this range (British Columbia north of Vancouver Island to southeastern Alaska) is in a relatively undisturbed, natural state. The forest from Vancouver Island, British Columbia to northern California contains some of the world's most valuable and productive commercial timberlands. An important subset of the PNW is the Douglas-fir region (western Washington and Oregon), which is dominated by the fast-growing coastal Douglas-fir (*Pseudotsuga menziesii* (Mirb.) Franco).

R.A. Monserud, R.W. Haynes and A.C. Johnson (eds.)
Compatible Forest Management, 3-32

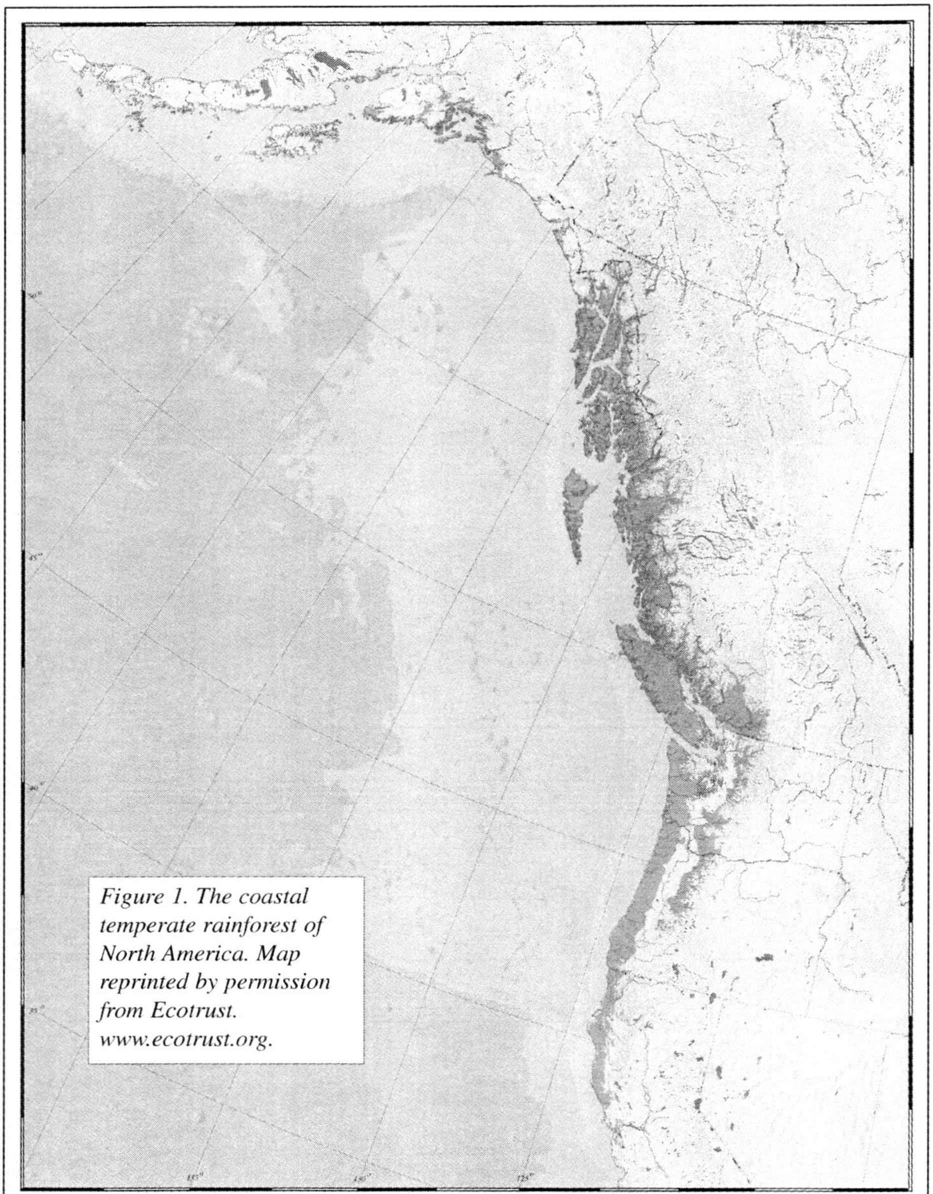

Figure 1. The coastal temperate rainforest of North America. Map reprinted by permission from Ecotrust. www.ecotrust.org.

The Pacific Northwest has a relatively long tradition of science-based forest management (Meidinger and Pojar 1991, Kimmins 1992, Peterson et al. 1997, Duncan 2000). Since the mid-1980s there has been an evolving interest in forest management that is compatible with commodity production and ecological, social, and cultural values (Peterson and Monserud 2002).

Compatible management focuses on outcomes that can be managed for simultaneously. In the simplest sense, compatible forest management simultaneously produces multiple products of value without decreasing other values, all in a socially acceptable manner. Hereafter, we refer to this as *compatible forest management.*

Interest in compatible forest management regimes has been stimulated by the public debate over forest management, which often portrays management choices as tradeoffs between biophysical and socioeconomic components of ecosystems (Peterson and Monserud 2002). This portrayal often leads to a focus on dichotomous tradeoffs rather than on the opportunities for compatible changes in outputs (either goods or services) that may exist among alternative management strategies. Here we focus on management strategies and emphasize broad goals such as wood production, biodiversity, and habitat conservation while maintaining other values from forestlands desired by the public. We show how different management approaches can offer alternatives to the divisive arena of "either-or" choices.

In general, our purpose is to explore the proposition: "*Commodity production (timber and nontimber forest products) and the other forest values (biodiversity, fish and wildlife habitat) can be simultaneously produced from the same area in a socially acceptable manner.*" This proposition will be expanded to consider natural and managed systems across a range of scales, uniform and non-uniform landscapes, and dynamic and static systems. We will examine the proposition from the point of view of several disciplines (e.g., wildlife, sociology, and economics). Inferences will be drawn about the nature and extent of joint production at different geographic and temporal scales. Inferences will also be drawn about how variation in joint production capability is related to the inherent productivity of the land (stand, watershed, province), the legacies of past land management, and differences due to ownership itself.

Much of the work discussed in this book is a product of recent research in the Pacific Northwest that addresses two aspects of the compatibility issue. First, how are various forest management practices related to an array of associated goods and services? Second, how do different approaches to forest management affect relatively large and complex ecosystems? Much of this work was organized as a research initiative (the Wood Compatibility Initiative) at the Pacific Northwest Research Station (Haynes and Monserud 2002, Johnson et al. 2002, Peterson and Monserud 2002). Its purpose was to expand information on the options that may increase the compatibility between wood production and other societal values derived from forestlands. We generally take a broad view of values related to forests (e.g., Donoghue 2003). These include life support, economic, recreation, aesthetic, wildlife, biotic diversity, scientific, natural history, cultural, spiritual, and intrinsic values (Rolston and Coufal 1991).

To set the context for the rest of the book, we summarize the management history in the Pacific Northwest (with a description of the forests), the forest sector, and the conceptual model underlying discussions of compatibility. We also provide a brief discussion of the policy issues motivating our interest in compatible forest management, and a brief introduction to the following chapters.

2. Management History

European settlement of the Pacific Northwest began in the mid-1800s (Alaska was settled earlier by the Russians). At that time, forests were seen as unlimited or as impediments to settlement. By the early 1900s, however, efforts were underway in the United States to establish federal forest reserves to protect forest resources, water quality and wildlife habitat. Most timber production during this period came from private timberlands and early forest management problems dealt with fire protection, regeneration, growth and yield studies, and harvest practices. Until World War II, most management on federal timberlands was focused on conserving forest resources. Following World War II there was a deliberate decision to increase the level of forest management on public timberlands to help meet the growing demand for forest products (e.g., homes) from a booming population and economy.

From the 1940s until the late 1960s in the United States, there was general agreement among both federal and private land managers that timber production was the primary objective in management of most forest land (Curtis et al. 1998, Peterson and Monserud 2002). Basic assumptions were that wood production in old-growth stands was essentially static (no net growth), and that insects and disease were diminishing the amount of usable wood in these stands. It seemed desirable, therefore, to replace old-growth forests with young, rapidly growing stands (USDA FS 1963, Curtis et al. 1998). Furthermore, in the Douglas-fir region, clearcut logging and broadcast burning were justified as mimicking the catastrophic, stand-replacing fires typical of the region before fire suppression (Halpern 1995). Management practices during this period attempted to meet increasing wood demands by relying on the economic efficiencies of clearcutting and plantation management. At the same time, public concerns about fire protection and, later, restocking of cutover timberland led the states of Oregon, Washington, California, and Alaska to adopt forest practice acts in 1971, 1974, 1974, and 1978, respectively. These rely on a combination of best management practices, logger and landowner education, and enforcement activities by state agencies.

Over the years, conflicts among demands for different forest values have intensified (Cissel et al. 1999, Peterson and Monserud 2002). The public has become increasingly aware that forests can produce more than wood (Behan 1990, Beese and Phillips 1997). Current public debate over forest management of public lands centers on interactions between wood production and the needs of wildlife, aquatic resources, biodiversity, and social acceptance (Peterson and Monserud 2002). These include most of the major factors setting the public agenda on the future of forest management in the Pacific Northwest. Given that the public has strongly held values and opinions regarding the appropriate use of forest land (especially public forestland), the process of deciding how best to manage it has become increasingly polarized. Simply put, public opinion has often been at odds with forest management goals, especially on public land.

As a consequence, the focus of management in the Pacific Northwest during the 1990s shifted from producing timber to maintaining sustainable forest ecosystems (Behan 1990). This shift culminated in the Northwest Forest Plan (USDA and USDI 1994a,b) for western Oregon and Washington, the new Forest Practices Code of British Columbia (1994), and the Tongass Land Management Plan (USDA 1997a,b,c) for the Tongass National Forest in southeastern Alaska. Instead of the traditional goal of economically efficient wood production, which relied on even-aged plantations, the focus in these recent efforts is toward "old-growth" and multi-resource ecosystem management, which attempt to protect the habitat of endangered species and fish habitat, and promote biodiversity (FEMAT 1993, Clayoquot Scientific Panel 1995). There is interest and need for science-based silvicultural practices and management regimes that will reduce conflicts among user groups while producing the many values associated with forest lands on a biologically and economically sustainable basis (Curtis et al. 1998, Committee of Scientists 1999). In this changing context, silviculture still plays a key management role in the sense that various treatments may be applied to forest stands to maintain and enhance their utility for a given purpose (Smith's (1986, p.1) definition of silviculture).

Although forest management is science-based, much of that science is not based on scientific experimentation (Franklin et al. 1999). It is rare that resource management approaches are formulated from a systematic scientific approach that is tested before widespread application (Franklin et al. 1999, Monserud 2003). For example, dispersed patch clearcutting was widely adopted on national forest land after World War II without experimental testing, largely with the support of Isaac (1943) and other experts. Alternative silvicultural approaches were rejected (Isaac 1956) based on a few ill-chosen case studies rather than a scientific test of alternative methods (Curtis 1998). The consequent lack of research into alternatives to clearcutting has severely handicapped current efforts to meet changing objectives and public concerns (Curtis 1998). Other harvest experiments were implemented at a very small scale and generally lacked statistical rigor and replication (Tappeiner et al. 1997, Franklin et al. 1999). Since then, methods for regenerating vigorous young stands of primary timber species following clearcut logging have been extensively researched and tested throughout the Pacific Northwest (Loucks et al. 1996, Smith et al. 1997). Foresters have accumulated an enormous amount of information and experience on even-age silviculture and plantation management in the past 50 years (Curtis et al. 1998), but this scientific basis did not exist when widespread implementation of modern plantation management began decades earlier in the Douglas-fir region. After a half century of implementation, it was a short step to the belief that this intensive plantation management was the only silvicultural regime that worked in the Pacific Northwest (Peterson and Monserud 2002).

The management history of the coastal forests of British Columbia and southeastern Alaska is similar to that in the Douglas-fir region—a strong reliance on clearcut logging (especially in old growth) followed by even-age plantation management (Beese and Bryant 1999, McClellan et al. 2000). Except for experiments with shelterwood cutting in mature and old-growth stands (e.g., Williamson 1973), well-documented comparative trials of other possible silvicultural systems in the Pacific Northwest were conspicuously lacking before the mid-1990s (Curtis 1996).

Recently, several new large-scale, multidisciplinary silvicultural experiments from the Pacific Northwest have been initiated (e.g., Curtis 1996, Carey et al. 1999, Halpern and Raphael 1999, McClellan et al. 2000). All are implementing silvicultural alternatives (e.g., variable retention, and variable density thinning) to the widely used plantation management of the previous 50 years (Monserud 2002, 2003). Some are attempting to hasten the approach to old-growth structure and composition. All examine the joint production of wood and some other forest value (e.g., biodiversity).

2.1. Forests of the Pacific Northwest

We focus our research on the temperate rainforests on the northwest coast of North America. This zone includes the Douglas-fir region of western Oregon and Washington north to the vast coastal Sitka spruce (*Picea sitchensis* (Bong.) Carr.)/western hemlock (*Tsuga heterophylla* (Raf.) Sarg.) forests of British Columbia and southeastern Alaska. These coniferous forests contain the highest quality wood-producing lands on the continent, and exhibit some of the greatest biomass accumulations and highest productivity levels of any in the world, temperate or tropical (Franklin and Dyrness 1973, Fujimori et al. 1976, Franklin and Waring 1981, Walter 1985, Franklin 1988). These forests are valued for their scenery, recreational opportunities, watershed protection, and fish and wildlife habitat (Peterson and Monserud 2002). The northern extent of this rainforest (coastal British Columbia and southeastern Alaska) is relatively unaltered (Everest et al. 1997).

Moist maritime conditions characterize the region, producing expanses of forests dominated by massive evergreen conifers, including Douglas-fir, western hemlock, western redcedar, (*Thuja plicata* Donn ex D. Don), Sitka spruce, Pacific silver fir (*Abies amabilis* Dougl. ex Forbes), and noble fir (*A. procera* Rehd.). A mediterranean climate of mild winters and relatively dry summers prevails (Walter 1985), although summer dryness decreases markedly to the north. This climate favors needle-leaved conifers by permitting extensive photosynthesis outside the growing season and reducing transpiration losses during the summer months (Waring and Franklin 1979). The summer climate is controlled by a large, semi-permanent, high-pressure center in the Pacific Ocean, which greatly reduces the frequency and intensity of Pacific storms (Meidinger and Pojar 1991). Infrequent catastrophic events, such as wildfires and hurricanes at intervals of several hundred years, allow for the genetic expression of enormous biomass and extremely tall and long-lived forest trees (Franklin 1988). Such favorable conditions for forest growth result in dominance by coniferous species that are the tallest, largest, and oldest of their genera (Franklin 1988). Unlike the other temperate rainforests of the world (see Walter 1985), this is the only one dominated by conifers rather than hardwoods (Franklin 1988).

The vegetation zone classification of Franklin (1988) is used, which is based on the system of Franklin and Dyrness (1973) but has larger geographic coverage. Most of the studies covered in this book lie within three forest vegetation zones: the broad Douglas-fir-western hemlock zone, the coastal Sitka spruce-western hemlock zone, and the montane Pacific silver fir-western hemlock zone (Franklin and Dyrness 1973, Franklin 1988). However, as Appendix 1 illustrates, there are many ways to describe forest developmental stages and their associated characteristics (Appendix 1, developed by B. Kerns; see list of contributors).

2.2. Timber Markets

Current forest conditions in the southern part of the Pacific Northwest are a function of the markets for various forest products. This area is one of the major timber producing regions in the world and is recognized for its productive timberlands, forest management institutions, well-organized markets, and large-scale timber processing industries. Indeed, many of the concerns about compatible forest management arise from the opportunities to derive valuable products from the forest. Here we briefly discuss the importance of various aspects of timber markets in Oregon and Washington, both in terms of incentives for forest management and for maintaining a vigorous processing sector that offers opportunities to profitably use an array of timber species and sizes.

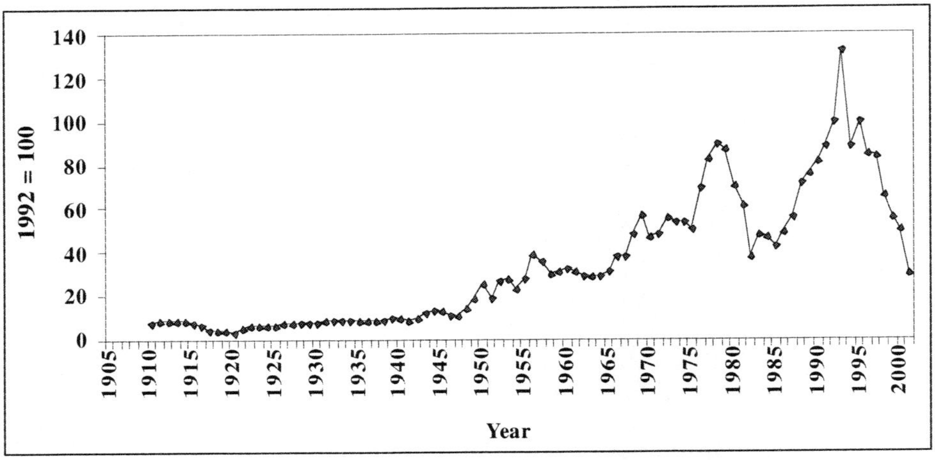

Figure 2. Oregon and Washington stumpage price index, 1909-2001.

Figure 2 illustrates the growth and the volatility of stumpage prices in Oregon and Washington. It provides evidence of the returns landowners have seen from investments in forest management. It also illustrates how different events have influenced timber prices and shaped the perceptions of land managers. For example, before World War II there were efforts to limit harvests from public timberlands. The goal was to provide incentives for private land managers to improve forest stewardship. The elimination of World War II price controls, rapid economic adjustment, and expansion in demand led to increases in prices. By the late 1950s, the increases in public harvests and slowdowns in economic growth were limiting the growth in stumpage prices. In the next three decades there were increased harvests, decrease in available timber, structural shifts in the economy, rapid inflation in commodity prices, and a severe recession in the early 1980s. All of these factors contributed to the price volatility illustrated in Figure 2. In the 1990s, the reductions in federal harvest, loss of both lumber and manufacturing productive capacity, and the reductions in exports following the economic slowdown in Asian counties all acted to reduce stumpage prices.

Stumpage prices (Figure 2) have increased in real terms, offering an incentive to apply forest management regimes that improve stand conditions. More recently, the growing importance and value of non-timber forest products (NTFP) is being recognized and addressed in management regimes (Alexander et al. 2001).

Tables 1 and 2 reveal that the timber industry in Oregon and Washington (the data for both states are further split between the Douglas-fir region and the ponderosa pine region—the area east of the Cascade Mountains) has declined for much of the past decade. In terms of importance in the United States, it peaked in the 1960-70s when it supplied roughly 33% of the U.S. softwood

Table 1. Lumber and plywood production for Oregon and Washington and United States, 1952-97, with projections to 2050.

	Historical							Projections				
Product and region	1952	1962	1970	1976	1986	1990	1996	2010	2020	2030	2040	2050
						Million cubic meters						
Softwood lumber:												
Douglas-fir region	24.3	20.3	17.9	20.1	21.9	21.2	17.5	19.6	22.7	23.4	26.7	30.2
Ponderosa pine region	5.4	5.7	5.7	6.4	6.6	6.8	3.5	3.3	4.2	5.4	6.6	8.0
Oregon and Washington	29.7	26.0	23.6	26.4	28.6	28.1	21.0	22.9	26.9	28.8	33.3	38.2
Total U.S.	71.3	63.2	64.9	70.1	83.3	84.5	78.6	76.7	85.2	88.3	93.2	102.2
Plywood:												
Douglas-fir region	2.4	7.0	7.5	8.1	7.7	4.3	2.2	1.5	0.8	0.7	0.7	0.7
Ponderosa pine region	0.0	.2	.7	.9	.9	.9	1.0	1.0	1.0	1.0	1.0	1.0
Oregon and Washington	2.4	7.2	8.2	8.9	8.6	5.2	3.2	2.5	1.8	1.7	1.7	1.7
Total U.S.	2.7	8.4	12.7	18.9	21.6	18.3	16.5	12.7	11.7	11.3	11.3	11.3

Table 2. Softwood timber removals, harvest, growth, and inventory in Oregon and Washington and United States, 1952-97, with projections to 2050.

	Historical							Projections				
Item	**1952**	**1962**	**1970**	**1976**	**1986**	**1991**	**1997**	**2010**	**2020**	**2030**	**2040**	**2050**
						Million cubic meters						
Douglas-fir region:												
Softwoods--												
Removals	56	55	68	70	67	50	34	43	45	44	47	51
Harvest	59	58	65	68	72	59	33	44	46	46	48	52
Net annual growth	28	34	40	44	61	54	70	77	79	81	84	82
Inventory	3110	2955	2806	2651	2363	2233	2791	3209	3569	3944	4353	4690
Ponderosa pine region:												
Softwoods--												
Removals	11	14	17	18	21	18	13	9	10	12	13	14
Harvest	10	13	16	17	19	17	15	9	11	12	14	15
Net annual growth	14	17	18	17	18	18	19	23	22	22	22	22
Inventory	1123	1149	1165	1099	921	971	1056	1252	1381	1493	1590	1670
United States:												
Softwoods--												
Removals	220	215	260	283	322	307	286	273	305	333	355	378
Harvest	213	207	246	269	319	311	286	279	312	340	363	387
Net annual growth	219	272	321	354	354	339	384	446	460	469	474	464
Inventory	12 220	12 728	13 018	13 215	12 817	12 732	13 693	15 536	17 216	18 705	20 019	21 001

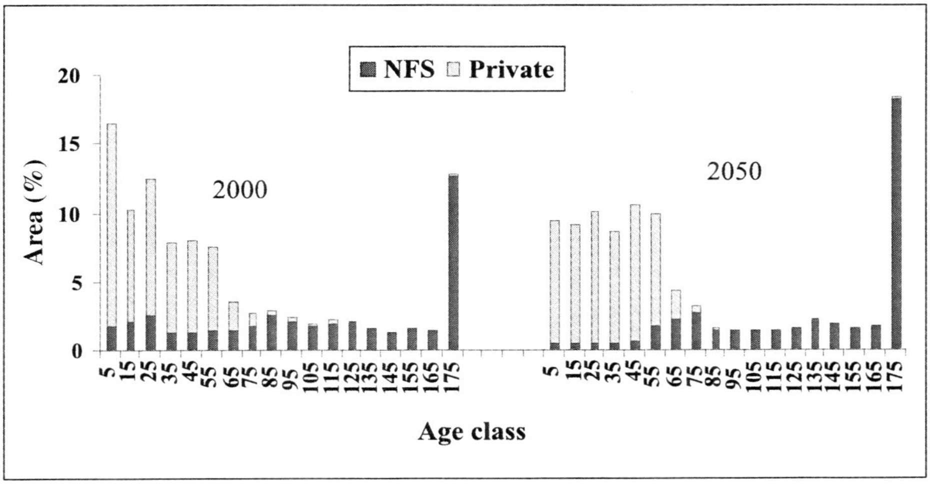

Figure 3. Age-class distribution by ownership for softwood forest types on timberland in the Douglas-fir region for 2000 and projected for 2050.
Note: NFS = National Forest System.

harvest. In 1997, these two states accounted for 27% of the softwood lumber, 19% of the softwood plywood, 17% of the softwood harvest, and 28% of the softwood timber inventories of the United States (Haynes 2003). The steep declines that began in the early 1990s followed reductions in federal harvest (Table 3) associated with the adoption of various habitat conservation strategies, such as the Northwest Forest Plan (USDA and USDI 1994a,b). Projections (shown in Table 1) reveal a more stable role for Oregon and Washington. Lumber is expected to increase after 2020 as private timber inventories improve.

The projected 60% increase in inventory volume for the Douglas-fir region (Table 2) reflects an improving timber resource situation. Most of this increase in timber inventories is on public timberlands as a result of habitat conservation strategies, and a sharp reduction in harvesting. The most dramatic changes in the timber inventories are shown in Figure 3 (also see Haynes et al. 2003). The age-class structure of the inventories is expected to shift from the unimodal distribution shown on the left to the more bimodal distribution shown on the right. The later type of age-class distribution suggests that there will be fewer mid-seral forests by 2050. In Oregon and Washington, most of the younger stands will be on private timberlands, whereas the older forest will be on public lands. This changing mix of demographics may alter our perceptions of forest health so that we perceive the situation as a public land management problem in dealing with a gradually aging inventory that may (for some types) exacerbate an already severe overstocking problem.

Table 3. Softwood harvest in Oregon and Washington and United States, by ownership, 1952-97, with projections to 2050.

Item	Historical							Projections				
	1952	1962	1970	1976	1986	1991	1997	2010	2020	2030	2040	2050
	Million cubic meters											
Douglas-fir region:												
National forest	10	17	14	14	19	8	2	1	2	2	2	3
Other public	4	8	10	12	12	7	5	5	5	5	5	5
Forest industry	35	28	35	36	35	31	18	28	29	27	27	29
Nonindustrial private forest	9	6	7	6	7	12	8	9	11	12	14	15
Total	59	58	65	68	72	59	33	44	46	46	48	52
Ponderosa pine region:												
National forest	3	7	8	8	11	10	2	2	2	3	3	4
Other public	1	2	3	3	2	2	2	2	2	2	2	2
Forest industry	3	3	3	4	5	3	5	3	3	3	3	4
Nonindustrial private forest	3	2	1	2	2	2	6	3	4	5	5	6
Total	10	13	16	17	19	17	15	9	11	12	14	15
United States:												
National forest	27	46	54	53	61	51	20	15	19	21	22	23
Other public	11	16	20	23	23	22	16	16	16	16	16	16
Forest industry	79	65	82	97	119	117	103	113	136	152	148	148
Nonindustrial private forest	95	80	90	96	116	121	147	134	140	151	176	200
Total	213	207	246	269	319	311	286	279	312	340	363	387

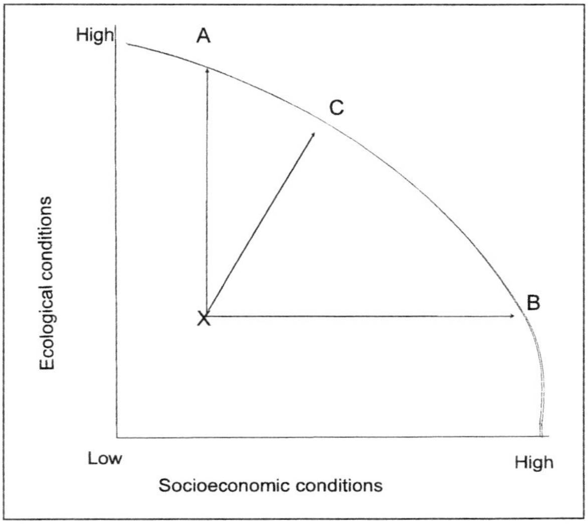

Figure 4. Hypothetical joint production function between ecological and socioeconomic conditions showing opportunities for compatible changes of both.

Timber markets have influenced the goals for land management and the evolution and application of various management regimes. How then can markets be harnessed to assist in the development of strategies that emphasize compatible forest management? First, what are the management regimes that produce a reliable supply of sawtimber? How can landowners be assured a fair return for implementing these management practices? Second, what are the realistic opportunities for smaller scale enterprises using less popular species and smaller sizes of timber? These opportunities have to be attractive in capital markets and offer a fair wage to workers. Third, what is the public's willingness to accept and pay for forestry restoration? Is it possible to combine some traditional activities (like thinning and under-burning) with efforts to restore selected ecological conditions?

3. The Conceptual Model

Notions of compatible forest management rest on two conceptual models. The first (Figure 4) represents an abstract representation of the challenge facing land managers, and provides a framework for considering the nature of the various relations involved in the compatibility arguments. The second (Figure 5) represents the more traditional view of the challenge facing land managers. It also provides a framework for looking at the links among the components of the land management problem.

Figure 4 illustrates the challenge facing land managers who are trying to manage for ecological and socioeconomic well-being (the general forest management problem has been described by Gregory 1972). The curve represents the production possibility frontier. This is the set of all combinations of ecological and socioeconomic conditions with no waste or inputs left over from which more of one output could be achieved without giving up some of the other (see Montgomery 2003). If, for example, our current position is point X, society will theoretically be better off if we move closer to the production possibility frontier in any positive direction. However, people who place high value on socioeconomic conditions are concerned that improvements in ecological conditions will likely mean a move to the left of point A, at which point socioeconomic conditions will be reduced. Similarly, people who place high value on ecological conditions are concerned that improvements in socioeconomic conditions will likely mean a move below point B, at which point ecological conditions will suffer. Resistance to change means we forgo opportunities to move toward C, the point where both ecological and socioeconomic conditions improve and everybody is better off. This move toward C—in which nobody is worse off and at least someone is better off—is a move closer to Pareto optimality, a useful concept that does not require the marketplace to determine value. In this simple two-dimensional example, all points bounded by A-X-B are desirable; either condition is at least as good as at point X, the status quo. The challenge is to identify point C—and the path to reach it—in a dynamic world with multiple inputs and multiple desired outputs.

Figure 5 illustrates the basic interactions among multiple values, ideas, actions, and outcomes that provide the context for research. Social values influence institutional policy that in turn affects managerial decisions and actions, resulting in a change in forest resource components and the associated mix of outcomes. Those decisions and proposed actions are evaluated—and often challenged—by society before they are implemented. Note that social concerns are not just at the top of this cycle in constructing policy and goals. Social actions are woven through dimensions such as water quality, biodiversity, and economics. Thus, we need to distinguish social activity and public use from the outcome of social acceptability. Once a management action is executed, it is evaluated in relation to the desired mix of measurable outcomes (often stated in the management prescription). The evaluation is complicated because many of the values are realized in different areas and over varying lengths of time after the management action. This also suggests that much of the research information, given its uncertain nature, should be amenable to socioeconomic evaluation of risks and consequences.

Figure 4 largely serves as a guide in discussions of broadscale policy and science issues. Some land managers and scientists see it as too simple. They argue, for example, that ecological integrity should have two or three axes,

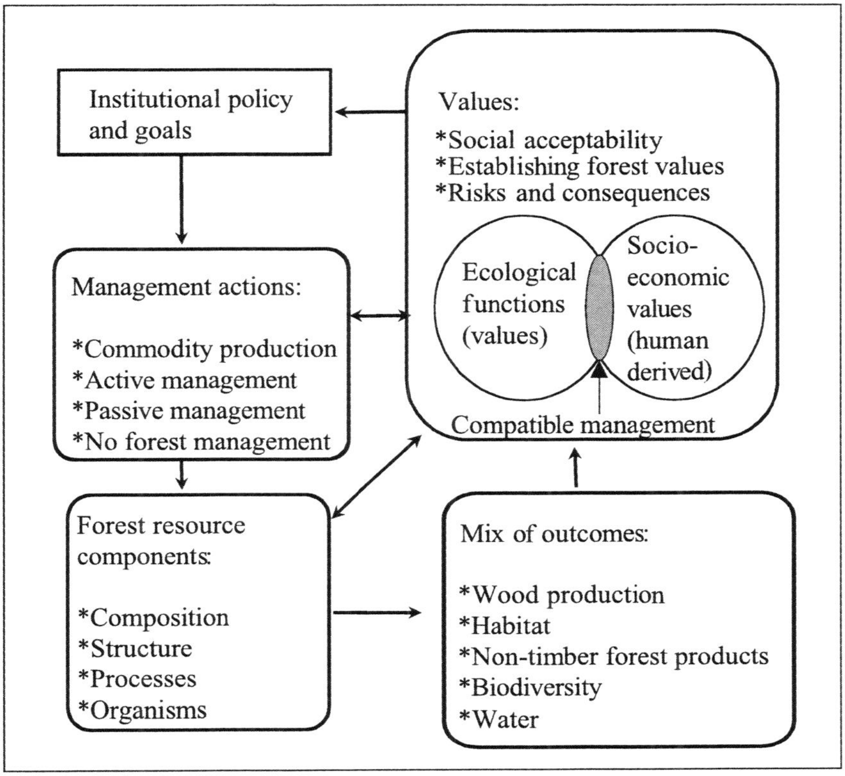

Figure 5. A general conceptual model for the Wood Compatibility Initiative.

given the complexity and sometimes competing or contradictory dimensions to the problem. Though this would add more dimensions, the essential topology of policy and science issues would be much the same.

The second conceptual model (Figure 5) provides a practical framework for the various components and links that can be used at multiple spatial scales. It illustrates the key role that social values play in shaping institutional context, management actions, and various forest resource components. It also illustrates the role that a mix of outputs has in shaping social values. This conceptual model specifies the management regimes and considers the multiple ownerships making up different broad-scale landscapes. Thus, questions regarding the management of public lands are examined within the context of broader, spatially complex landscapes.

These models are applied at three different spatial scales in the development of compatible forest management. The first scale is the traditional stand scale of forest management. Compatibility, at this scale, usually depends on management actions that meet relatively specific land management objectives.

The emphasis may be on a selected stand, but there is also interest in the cumulative effects of actions across neighboring stands and the entire watershed.

The second scale is composed of broader landscapes that include multiple watersheds, sub-basins, or counties. The importance of this larger scale is that it sets the context for finer scales where management (and much research) is actually implemented. In this case, most tradeoffs still involve choices among management actions. Notions of public tradeoffs in terms of social acceptability regarding specific land management actions are important at this scale.

The third scale is the ecoregion. This corresponds, for example, to the Douglas-fir region. At this scale, tradeoffs involve choices among different mosaics of fine and mid-scale management objectives. Broad-scale measures are often developed from relatively coarse data to describe conditions on relatively large areas such as a subregion (see Haynes 2003). Planning tools such as ATLAS (Mills and Kincaid 1992) provide analytical models that can be used to examine tradeoffs for western Oregon and Washington, for example, in the context of changes in the U.S. forest sector. Such planning tools can also be used to explore the links among different land management-owner strategies, broad-scale resource conditions, and various outputs at the regional scale (see Haynes et al. 2003, Spies and Johnson 2003).

4. Policy Context

The context for research on compatible management is set by three converging policy interests: shifting public recognition of the array of goods and services produced by forests; the growing debate about sustainable forest management; and recognition within the scientific community of the connectiveness among processes and outputs. These interests all shape the emerging recognition that we need approaches to forest management that emphasize the compatibility among various goods and services so we can avoid counterproductive arguments about tradeoffs.

The Douglas-fir region has a long history of very deliberate forest management debates (e.g., Behan 1990, Curtis et al. 1998, Peterson and Monserud 2002). These debates have contributed to considerable scientific information on the compatibilities and tradeoffs between commodity production and the other values (e.g., nontimber forest products, fish, and wildlife habitat) that the public desires from these forests (Johnson et al. 2002).

By using this information, we can integrate key scientific findings to enhance good stewardship of our forestlands, both public and private. Key in this is the role that scientific information can be used to increase opportunities for producing compatible bundles of goods and services. These goods and services include wood, wildlife habitat, scenery, recreation, water quality (including water as a commodity), and riparian habitat provided in a manner that is socially acceptable and economically viable. This is consistent with the emerging emphasis we see on sustainable forest management and concerns about ways to meet rising demands for goods and services from the forest in an environmentally acceptable manner.

4.1. The Relation between Sustainable and Compatible Forest Management

Sustainable forest management is an enduring issue that in the past decade has taken on increased importance out of concerns for resource overexploitation (Powers 2001), sustainable development (World Commission on Environment and Development 1987), and possible climate change effects (Watson et al. 1995, Schwalm and Ek 2001). During its 200-year history, the concept of sustainable forest ecosystem management has been the focus of scientific and political discussion, with varying degrees of intensity—it is promoted with vehement fervor during periods of social or economic crisis and less intensely during periods of stability (von Gadow et al. 2000).

Sustainable forestry is difficult to define and quantify (Amaranthus 1997, Schlaepfer and Elliott 2000). In its broadest sense, sustainability is a deliberate management goal that implies an ability to maintain the productivity and ecological integrity of the forest in perpetuity (Monserud, in press). This statement includes the interactions among both temporal and spatial scales by expressing the need to consider the timeframe and the spatial extent of management goals. Clearly, measures of sustainability become increasingly complex as the scale increases from stand to landscape to region, and on to national and global scales. Almost all of the work with criteria and indicators of sustainability is at the national scale. Forest management, however, is implemented at the stand scale, which introduces a profound scale effect (Hall 2000).

Before the 1990s, most countries managed their forests under the principle of *sustained yield*, with a nearly exclusive focus on timber yield of forest stands or contiguous groups of stands (Tittler et al. 2001). The selection of sustainable forestry practices depends on what one is trying to sustain, a choice often driven by utilitarian principles (Amaranthus 1997). In response to rising social pressure for a larger variety of goods and services from the forest, the concept of sustainable forestry has expanded to include much larger areas and forest uses. Wilson and Wang (1999) define *sustainable forestry* as comprising a host of management regimes to maintain and enhance the long-term health and integrity of forest ecosystems and forest-dependent communities, while providing ecological, economic, social, and cultural opportunities for the benefit of present and future generations. This is a multi-dimensional definition including biological, sociological, political, and economic factors (Perry and Amaranthus 1997, Wilson and Wang, 1999). Although an all-encompassing definition might seem appealing, it does not lend itself easily to quantifiable science, especially at small scales.

The shift in attitude from sustained yield to sustainable forestry was triggered internationally by the Brundtland Report on sustainable development in 1987 (World Commission on Environment and Development 1987) and later solidified at the Earth Summit in Rio de Janeiro in 1992, with the adoption of the Forest Principles (Tittler et al. 2001). In response, several initiatives and international agreements have attempted to quantify broad-scale sustainability (Mendoza and Prabhu 2000, Tittler et al. 2001), such as the Montreal Accord (Mihajlovich 2001). The emphasis is on criteria and indicators for judging if sustainable forestry is occurring over broad areas, usually at the national scale (Mendoza and Prabhu 2000).

The development of forest certification programs by nongovernment groups such as the Forest Stewardship Council or the American Forest and Paper Association is another indicator of the shift in public attitudes and the need for landowners to demonstrate their commitment to responsible forest management. These programs develop principles and objectives or criteria for sustainable forest management that can be applied to a participant's forests. Both emphasize the use of forest planning, best management practices, and logger and landowner education to achieve sustainable forest management (see AF&PA 1999).

In this book we examine forest management as a set of compatible production opportunities at the stand and landscape levels, and from the landowners' and managers' perspectives. These opportunities simultaneously seek to produce multiple forest goods and services, maintain the ecological integrity of the forest resource, and reduce social conflict regarding management. Clearly, this goal is complementary to (and likely a subset of) sustainable forest management. The emphasis in this book is on management approaches that can increase joint production in a socially acceptable manner. We refer to this as compatible forest management. In our work, we look across scales (stand to regional) to understand how compatibility can be increased. We seek solutions that reduce conflict and eschew dichotomous tradeoffs.

5. Research Summary

This book is structured around 17 chapters that provide information on options that may increase compatibility between wood production and other societal values derived from forestlands. Our goal is to emphasize a more balanced view of forest management rather than the often diametrically opposed views that pit environmental issues against economic interests with little regard to the equity of those positions.

The following chapters summarize research findings and the results of formal syntheses. Potential management opportunities are presented and illustrated for a range of case studies (e.g., chapters 2, 3, 8, 14). These case studies are drawn from an area that has a rich history of forest management based both on market incentives as well as public regulations. They serve as models for practicing forest managers elsewhere.

Chapters 2 and 3 give examples of research associated with compatible management. Chapter 2 reviews the production possibility approach to compatibility research. Forest policymakers may find this approach helpful when they evaluate competing forest uses and identify situations where opportunity to improve current management exists. Chapter 3 is a case study from southeastern Alaska where the current objective is to manage upland forests for wood products, wildlife, and fish. In this ecosystem, the presence of red alder (*Alnus rubra* Bong.) may help alleviate some of the problems associated with fish and wildlife habitat that develop in the dense conifer-dominated young-growth forests that typically regenerate following clearcutting. This is also an ecosystem where compatibility and tradeoffs among resources are clearly evident.

Chapters 4 through 7 deal with silviculture, experimental approaches to joint forest production, stand modeling, and landscape management issues. The discussion on silviculture in Chapter 4 begins with a review of silvicultural methods in the Pacific Northwest. New silvicultural systems are discussed in relation to traditional methods and new management goals, such as biodiversity. Chapter 5 examines several large-scale management experiments initiated in the Pacific Northwest in the 1990s, partially in response to contentious resource management debates. All of these experiments focus on joint production. They examine alternative silvicultural treatments aimed at enhancing wildlife habitat, biodiversity, or the conservation of aquatic resources in a manner that is socially acceptable. We examine several of these experiments for their utility in providing an experimental basis for compatible forest management. Chapters 6 and 7 highlight the importance of simulating stand growth and landscape dynamics under heterogeneous structures and mixed-species compositions to evaluate the compatibility of management alternatives. The adaptability and utility of forest growth simulation models is assessed for alternative silvicultural systems under experimentation in the Pacific Northwest.

Chapters 8 through 10 address scale. For instance, Chapter 8 questions whether alternative policies, examined at broader scales, provide greater overall compatibility among wood production and ecological integrity than current policies. This initial work reveals that management changes that seem drastic at stand or landscape levels may have small or slow effects on ecological conditions at larger spatial scales. Chapter 9 expands this discussion to include problems encountered when scaling from the stand to landscape in the Cascade Range of western Oregon. Here, both broad time and space contexts are considered, and multiple goods and services from these natural resource systems are examined. Chapter 10 focuses on contemporary management regimes in the Douglas-fir region where the diversity of timberland ownerships has led to a patchwork mosaic. Management regimes have evolved to take advantage of the inherent biophysical productivity, while being robust in a highly volatile economic and (more recently) regulatory climate.

Chapters 11 through 14 deal with different aspects of compatible management. First, the implications for timber quality associated with these management regimes are assessed, including prospective issues in marketing larger trees and the types of timber that might be available from various habitat conservation strategies used in the Douglas-fir region. Chapter 12 describes the opportunities for managing for nontimber forest products. It links understory development and relations with overstory variables and provides insights to help guide compatible forest management. For example, the authors assess silvicultural treatments that may affect commercial understory species supply, product attributes, species, and site-specific responses. Chapter 13 describes access to nontimber forest products. Emphasis is placed on user groups, current regulations, and opportunities for compatible management. Chapter 14 investigates different approaches for managing for wildlife where opportunities exist for a simultaneous increase in both wildlife habitat and wood production. Examples include management options developed for different landowners in western Washington.

Chapter 15 suggests that the relation between social values and forests must be understood to determine if compatible management exists. Furthermore, this knowledge of forest values needs to be integrated into forest management strategies. Emphasis is placed on the role of values in shaping change, assessing social values, and participatory approaches to management. The need to devise management strategies that are supported by the public is outlined in Chapter 16. Specifically, it is critical to understand how the public forms opinions and judges the acceptability of forest management practices and the decisionmaking processes. This chapter summarizes key findings from several studies that examine the acceptability of a range of management practices.

The concluding chapter summarizes our progress toward developing compatible management regimes. These regimes can help us meet the increasing demands for wood and other timberland goods and services in the Pacific Northwest. They require both ecologists and economists to seek compatible approaches to forest management. These approaches will need to improve environmental conditions while accommodating growth in use. The recent emphasis on sustainable forest management expands this management quest to include concerns about social equity. In this evolving context, there are emergent lessons from the Pacific Northwest that can apply to forest managers elsewhere.

6. References

American Forest & Paper Association [AF&PA]. 1999. Sustainable forestry initiatives standard: principles and objectives. Washington, DC: American Forest & Paper Association. 5 p.

Alexander, S.J., McLain, R.J., and Blatner, K.A. 2001. Socio-economic research on non-timber forest products in the Pacific Northwest. *Journal of Sustainable Forestry*, 13(3/4): 95-103.

Amaranthus, M.R. 1997. Forest sustainability: an approach to definition and assessment. General Technical Report PNW-GTR-416. Portland OR: U.S. Department of Agriculture, Forest Service, Pacific Northwest Research Station. 14 p.

Beese, W.J., and Bryant, A.A. 1999. Effect of alternative silvicultural systems on vegetation and bird communities in coastal montane forests of British Columbia, Canada. *Forest Ecology and Management*, 115: 231-242.

Beese, W.J., and Phillips, E.J. 1997. Harvesting costs and impacts of alternative silviculture prescriptions in old growth. In: Forest management into the next century: What will make it work? (pp. 102-110). Publication 7276. Madison, WI: Forest Products Society.

Behan, R. 1990. Multiresource forest management: a paradigmatic challenge to professional forestry. *Journal of Forestry*, 88(4): 12-18.

Bormann, F.H., and Likens, G.E. 1979. Pattern and process in a forested ecosystem. New York: Springer-Verlag. 253 p.

Carey, A.B., and Curtis, R.O. 1996. Conservation of biodiversity: a useful paradigm for forest ecosystem management. *Wildlife Society Bulletin*, 24: 610-620.

Carey, A.B., Thysell, D.R., and Brodie, A.W. 1999. The forest ecosystem study: background, rationale, implementation, baseline conditions, and silvicultural assessment. General Technical Report PNW-GTR-457. Portland, OR: U.S. Department of Agriculture, Forest Service, Pacific Northwest Research Station. 129 p.

Cissel, J.H., Swanson, F.J., and Weisberg, P.J. 1999. Landscape management using historical fire regimes: Blue River, Oregon. *Ecological Applications*, 9(4): 1217-1231.

Clayoquot Scientific Panel. 1995. Scientific panel for sustainable forest practices in Clayoquot Sound, Report 5, sustainable ecosystem management in Clayoquot Sound: planning and practices, Victoria, BC. 296 p.

Committee of Scientists. 1999. Sustaining the people's lands: recommendations for stewardship of the national forests and grasslands into the next century. Washington, DC: U.S. Department of Agriculture. 235 p. Retrieved April 1999 from http://www.fs.fed.us/news/science/.

Curtis, R.O. 1996. Silvicultural options for harvesting young-growth production forests. Prospectus for DNR/PNW Cooperative Project. 5 p. Unpublished document. On file with: Forestry Sciences Laboratory, 3625 93rd Avenue SW, Olympia, WA 98512-9193.

Curtis, R.O. 1998. "Selective cutting" in Douglas-fir: history revisited. *Journal of Forestry*, 96(7): 40-46.

Curtis, R.O., DeBell, D.S., Harrington, C.A., Lavender, D.P., St. Clair, J.B., Tappeiner, J.C., and Walstad, J.D. 1998. Silviculture for multiple objectives in the Douglas-fir region. General Technical Report PNW-GTR-435. Portland, OR: U.S. Department of Agriculture, Forest Service, Pacific Northwest Research Station. 123 p.

Donoghue, E. 2003. Social values and compatible forest management. Chapter 15. In: R.A. Monserud, R.W. Haynes, and A.C. Johnson (Eds.), Compatible forest management. Dordrecht, The Netherlands: Kluwer Academic Publishers.

Duncan, S. 2000. Closer to the truth: 75 years of discovery in forest & range research. Miscellaneous publication. Portland, OR: U.S. Department of Agriculture, Forest Service, Pacific Northwest Research Station. 76 p.

Everest, F.H., Swanston, D.N., Shaw, C.G., III, Smith, W.P., Julin, K.R., and Allen, S.D. 1997. Evaluation of the use of scientific information in developing the 1997 forest plan for the Tongass National Forest. General Technical Report PNW-GTR-415. Portland, OR: U.S. Department of Agriculture, Forest Service, Pacific Northwest Research Station. 69 p.

Forest Ecosystem Management Assessment Team (FEMAT). 1993. Forest ecosystem management: an ecological, economic, and social assessment. Washington, DC: U.S. Department of the Interior [and others]. [Irregular pagination].

Forest Practices Code of British Columbia Act. 1994. Statutes of B.C., Bill 40. Victoria, BC: Queen's Printer. Retrieved April 4, 2003 from http://www.legis.gov.bc.ca/1994/3rd_read/gov40-3.htm.

Franklin, J.F. 1988. Pacific Northwest forests. In: M.G. Barbour, and W.D. Billings (Eds.), North American terrestrial vegetation. New York: Cambridge University Press. 434 p.

Franklin, J.F., and Dyrness, C.T. 1973. Natural vegetation of Oregon and Washington. General Technical Report PNW-GTR-8. Portland, OR: U.S. Department of Agriculture, Forest Service, Pacific Northwest Research Station. 417 p.

Franklin, J.F., Norris, L.A., Berg, D.R., and Smith, G.R. 1999. The history of DEMO: an experiment in regeneration harvest of northwestern forest ecosystems. *Northwest Science*, 73(special issue): 3-11.

Franklin, J.F., Spies, T.A., Van Pelt, R., Carey, A., Thornburgh, D.A., Berg, R., Lindenmayer, D.B., Harmon, M.E., Keeton, W.S., Shaw, D.C., Bible, K., and Chen, J. 2002. Disturbances and structural development of natural forest ecosystems with silvicultural implications, using Douglas-fir as an example. *Forest Ecology and Management*, 155: 399-423.

Franklin, J.F., and Waring, R.H. 1981. Distinctive features of the northwestern coniferous forest: development, structure, and function. In: R.H. Waring (Ed.), Forests: fresh perspectives from ecosystem research (pp. 59-86). Corvallis, OR: Oregon State University Press.

Fujimori, T., Kawanabe, S., Saito, H., Grier, C.C., and Shidei, T. 1976. Biomass and primary production in forests of three major vegetation zones of the northwestern United States. *Journal of Japanese Forestry Society*, 58: 360-373.

Gregory, R.G. 1972. Forest resource economics. Chapter 13. New York: Forest Resource Economics. 254 p.

Hall, P. 2000. The issue of scale in the aggregation of data on indicators of sustainable forest management from subnational to national levels. *Forestry Chronicle*, 76(3): 419-422.

Halpern, C.B. 1995. Response of forest communities to green-tree retention harvest: a study plan for the vegetation component of the Demonstration of Ecosystem Management Options (DEMO) study. Seattle, WA: College Forestry Resources, University of Washington. 36 p. plus appendix.

Halpern, C.B., and Raphael, M.G. (Eds.). 1999. Retention harvests in Northwestern forest ecosystems: the demonstration of ecosystem management options (DEMO) study. *Northwest Science*, 73(special issue): 125 p.

Haynes, R.W. (Tech. coord.). 2003. An analysis of the timber situation in the United States: 1952 to 2050. A technical document supporting the 2000 USDA Forest Service RPA assessment. General Technical Report PNW-GTR-560. Portland, OR: U.S. Department of Agriculture, Forest Service, Pacific Northwest Research Station. 254 p.

Haynes, R.W., Adams, D.M., and Mills, J.R. 2003. Contemporary management regimes in the Pacific Northwest: balancing biophysical and economic concerns. Chapter 10. In: R.A. Monserud, R.W. Haynes, and A.C. Johnson (Eds.), Compatible forest management. Dordrecht, The Netherlands: Kluwer Academic Publishers.

Haynes, R.W., and Monserud, R.A. 2002. A basis for understanding compatibility among wood production and other forest values. General Technical Report PNW-GTR-529. Portland, OR: U.S. Department of Agriculture, Forest Service, Pacific Northwest Research Station. 46 p.

Isaac, L.A. 1943. Reproductive habits of Douglas-fir. Washington, DC: Charles Lathrop Pack Forestry Foundation.

Isaac, L.A. 1956. Place of partial cutting in old-growth stands of the Douglas-fir region. Research Paper PNW-RP-16. Portland, OR: U.S. Department of Agriculture, Forest Service, Pacific Northwest Research Station. 48 p.

Johnson, A.C., Haynes, R.W., and Monserud, R.A. (Eds.). 2002. Congruent management of multiple resources: proceedings from the wood compatibility workshop. General Technical Report PNW-GTR-563. Portland, OR: U.S. Department of Agriculture, Forest Service, Pacific Northwest Research Station. 252 p.

Kimmins, H. 1992. Balancing act: environmental issues in forestry. Vancouver, BC: University of British Columbia Press. 245 p.

Loucks, D.M., Knowe, S.A., Shainsky, L.J., and Pancheco, A.A. 1996. Regenerating coastal forests in Oregon: an annotated bibliography of selected ecological literature. Research Contribution 14. Corvallis, OR: Oregon State University, Forest Research Laboratory. 122 p.

McClellan, M.H., Swanston, D.N., Hennon, P.E., Deal, R.L., De Santo, T.L., and Wipfli, M.S. 2000. Alternatives to clearcutting in the old growth forests of southeast Alaska: study plan and establishment report. General Technical Report PNW-GTR-494. Portland, OR: U.S. Department of Agriculture, Forest Service, Pacific Northwest Research Station. 40 p.

Meidinger, D., and Pojar, J. (Eds.). 1991. Ecosystems of British Columbia. Special Report Series 6. Victoria, BC: British Columbia Ministry of Forests. 330 p.

Mendoza, G.A., and Prabhu, R. 2000. Qualitative multi-criteria approaches to assessing indicators of sustainable forest resource management. *Forest Ecology and Management*, 131: 107-126.

Mihajlovich, M. 2001. Does forest certification assure sustainability? *Forestry Chronicle*, 77(6): 994-997.

Mills, J.R., and Kincaid, J.C. 1992. The aggregate timberland assessment system-ATLAS: a comprehensive timber projection model. General Technical Report PNW-GTR-281. Portland, OR: U.S. Department of Agriculture, Forest Service, Pacific Northwest Research Station. 66 p.

Monserud, R.A. 2002. Large-scale management experiments in the moist maritime forests of the Pacific Northwest. *Landscape and Urban Planning*, 59: 159-180.

Monserud, R.A. 2003. Experimental approaches to joint forest production. Chapter 5. In: R.A. Monserud, R.W. Haynes, and A.C. Johnson (Eds.), Compatible forest management. Dordrecht, The Netherlands: Kluwer Academic Publishers.

Monserud, R.A. [In press]. Evaluating forest models in a sustainable forest management context. *Forest Biometry, Modelling and Information Sciences (FBMIS)*. To be available on: http://www.fbmis.info/.

Montgomery, C.A. 2003. The production possibilities approach to understanding and modeling compatibility. Chapter 2. In: R.A. Monserud, R.W. Haynes, and A.C. Johnson (Eds.), Compatible forest management. Dordrecht, The Netherlands: Kluwer Academic Publishers.

Oliver, C.D., and Larson, B.C. 1990. Forest stand dynamics. New York: McGraw-Hill. 467 p.

O'Neil, T.A., Bettinger, K.A., Heyden, M.V., Marcot, B.G., Barrett, C.B., Mellen, K., Vanderhaegen, W.M., Johnson, D.H., Doran, P.J., Wunder, L., and Boula, K.M. 2001. Structural conditions and habitat elements of Oregon and Washington. In: D.H. Johnson, and T.A. O'Neil (Eds.), Wildlife-habitat relationships in Oregon and Washington (pp. 115-139). Corvallis, OR: Oregon State University Press.

Perry, D.A., and Amaranthus, M.P. 1997. Disturbance, recovery, and stability. In: K.A. Kohm, and J.F. Franklin (Eds.), Creating a forestry for the 21st century (pp. 31-56). Washington, DC: Island Press.

Peterson, C.E., and Monserud, R.A. 2002. Compatibility between wood production and other values and uses on forested lands. A problem analysis. General Technical Report PNW-GTR-564. Portland, OR: U.S. Department of Agriculture, Forest Service, Pacific Northwest Research Station. 50 p.

Peterson, E.B., Peterson, N.M., Weetman, G.F., and Martin, P.J. 1997. Ecology and management of Sitka spruce, emphasizing its natural range. Vancouver, BC: University of British Columbia Press. 336 p.

Powers, R.F. 2001. Assessing potential sustainable wood yield. In: J. Evans (Ed.), The forestry handbook. Applying forest science for sustainable management (Vol 2. pp. 105-128). Oxford, UK: Blackwell Science, Ltd.

Rolston, H., and Coufal, J. 1991. A forest ethic and multivalue forest management. *Journal of Forestry*, 89(4): 35-40.

Schlaepfer, R., and Elliott, C. 2000. Ecological and landscape considerations in forest management—The end of forestry? In: K. von Gadow, T. Pukkala, and M. Tomé (Eds.), Sustainable forest management (pp. 1-68). Dordrecht, The Netherlands: Kluwer Academic Publishers.

Schwalm, C., and Ek, A.R. 2001. Climate change and site: relevant mechanisms and modeling techniques. *Forest Ecology and Management*, 150: 241-257.

Smith, D.M. 1986. The practice of silviculture (8th ed.). New York: John Wiley & Sons.

Smith, J.P, Gresswell, R.E., and Hayes, J.P. 1997. A research problem analysis in support of the Cooperative Forest Ecosystem Research (CFER) Program. Corvallis, OR: U.S. Geological Society. 92 p.

Spies, T.A., and Franklin, J.F. 1991. The structure of natural young, mature, and old-growth Douglas-fir forests in Oregon and Washington. In: L.F. Ruggiero, K.B. Aubry, A.B. Carey, and M.H. Huff (Eds.), Wildlife and vegetation of unmanaged Douglas-fir forests (pp. 91-109). General Technical Report PNW-GTR-285. Portland, OR: U.S. Department of Agriculture, Forest Service, Pacific Northwest Research Station.

Spies, T.A., and Franklin, J.F. 1996. The diversity of maintenance of old-growth forests. In: R.C. Szaro, and D.W. Johnson (Eds.), Biodiversity in managed landscapes: theory and practice (pp. 296-314). New York: Oxford.

Spies, T.A., and Johnson, K.N. 2003. The importance of scale in assessing the compatibility of forest commodities and biodiversity. Chapter 8. In: R.A. Monserud, R.W. Haynes, and A.C. Johnson (Eds.), Compatible forest management. Dordrecht, The Netherlands: Kluwer Academic Publishers.

Tappeiner, J.C., Lavender, D., Walstad, J., Curtis, R.O., and DeBell, D.S. 1997. Silvicultural systems and regeneration methods: current practices and new alternatives. In: K.A. Kohm, and J.F. Franklin (Eds.), Creating a forestry for the 21st century. (pp. 151-164). Washington, DC: Island Press.

Tittler, R., Messier, C., and Burton, P.I. 2001. Hierarchical forest management planning and sustainable forest management in the boreal forest. *Forestry Chronicle*, 77(6): 998-1005.

U.S. Department of Agriculture, Forest Service. 1963. Timber trends in western Oregon and western Washington. Research Paper PNW-5. Portland, OR: Pacific Northwest Forest and Range Experiment Station, Division of Forest Economics Research. 154 p.

U.S. Department of Agriculture, Forest Service. 1997a. Land and resource management plan: Tongass National Forest. R10-MB-338dd [Juneau, AK]: Alaska Region.

U.S. Department of Agriculture, Forest Service. 1997b. Tongass land management plan revision: final environmental impact statement [FEIS]. Part 1: Summary, chapters 1 through 3 (physical and biological environment). R10-MB-338b. [Juneau, AK]: Alaska Region.

U.S. Department of Agriculture, Forest Service 1997c. Tongass land management plan revision: record of decision [ROD]. R10-MB-338a. [Juneau, AK]: Alaska Region.

U.S. Department of Agriculture, Forest Service and U.S. Department of the Interior, Bureau of Land Management. 1994a. Final supplemental environmental impact statement on management of habitat for late-successional and old-growth forest related species within the range of the northern spotted owl. Vol. 1 and 2. Washington, DC: U.S. Department of Agriculture, Forest Service and U.S. Department of the Interior, Bureau of Land Management.

U.S. Department of Agriculture, Forest Service and U.S. Department of the Interior, Bureau of Land Management. 1994b. Record of decision for the President's forest plan. Washington, DC: U.S. Department of Agriculture, Forest Service and U.S. Department of the Interior, Bureau of Land Management. 74 p. plus appendices.

von Gadow, K., Pukkala, T., and Tomé, M. (Eds.). 2000. Sustainable forest management. Dordrecht, The Netherlands: Kluwer Academic Publishers. 356 p.

Walter, H. 1985. Vegetation of the Earth and ecological systems of the geobiosphere (3rd ed.). New York: Springer-Verlag. 318 p.

Waring, R.F., and Franklin, J.F. 1979. Evergreen coniferous forests of the Pacific Northwest. *Science*, 204: 1380-1386.

Watson, R.T., Zinyowera, M.C., and Moss, R.H. (Eds.). 1995. Climate change 1995: impacts, adaptations and mitigation of climate change: scientific-technical analyses. Contribution of working group 2 to the second assessment of the intergovernmental panel on climate change. Cambridge, UK: Cambridge University Press. 878 p.

Williamson, R.L. 1973. Results of shelterwood harvesting of Douglas-fir in the Cascades of western Oregon. Research Paper PNW-RP-1161. Portland, OR: U.S. Department of Agriculture, Forest Service, Pacific Northwest Research Station. 13 p.

Wilson, B., and Wang, S. 1999. Sustainable forestry—the policy prescription in British Columbia. In: A. Yoshimoto, and K. Yukutake (Eds.), Global concerns for forest resource utilization—sustainable use and management (pp. 35-45). London: Kluwer Academic Publishers.

World Commission on Environment and Development. 1987. Our common future: report of the World Commission on Environment and Development. New York: Oxford University Press.

Appendix 1. Major classification schemes used to describe forest developmental stages and associated characteristics. Characteristics are illustrated from various ecosystem perspectives by using a Douglas-fir (*Pseudotsuga menziesii*) dominated sere growing in the western hemlock (*Tsuga heterophylla*) zone (Franklin and Dyrness 1988). Characteristics will vary widely based on site location, disturbance history, management, and forest types. This table was developed by B. Kerns (see list of contributors).

Major classification systems and characteristics for forest developmental stages			
Ecosystem perspective → / Forest development stage ↓	Ecological structure and process	Wildlife habitat	Timber production
First	Reorganization (Bormann and Likens 1979) Stand initiation (Oliver and Larson 1990) Establishment (Spies and Franklin 1991) Ecosystem initiation (Carey and Curtis 1996) Disturbance/legacy creation and cohort establishment (Franklin et al. 2002)	Grass/Forb-Open, Grass/Forb-Closed, Shrub/Seedling-Open, Shrub/Seedling-Closed, Sapling/Pole-Open (O'Neil et al. 2001)	Seedling (Haynes 2003) Early seral (FEMAT 1993)
Characteristics	• Pioneer tree cohort established with a range of regeneration densities • Biological legacies present depending on initial disturbance type, intensity and management • Rapid biomass accumulation • Biological legacies retained provide habitat • Above- and below-ground resource availability high • Nutrient transfer from soil to biomass • Possible introduction and spread of exotic/invasive species	• Biodiversity high • Herb and shrub understory may be abundant or persistent • Open canopy conditions important for birds and mammals	Stand age typically 0-15 years Single species tree cohort densely seeded or planted, typically with genetically altered stock Competing vegetation controlled or removed Precommercial Includes first tree age class of seedlings (average age of 5 years)

Appendix 1. Major classification schemes used to describe forest developmental stages and associated characteristics. Characteristics are illustrated from various ecosystem perspectives by using a Douglas-fir (*Pseudotsuga menziesii*) dominated sere growing in the western hemlock (*Tsuga heterophylla*) zone (Franklin and Dyrness 1988). Characteristics will vary widely based on site location, disturbance history, management, and forest types. This table was developed by B. Kerns (see list of contributors). (continued)

	Major classification systems and characteristics for forest developmental stages		
Ecosystem perspective → / Forest development stage ↓	Ecological structure and process	Wildlife habitat	Timber production
Second	Aggradation (Bormann and Likens 1979) Stem (Oliver and Larson 1990) Thinning (Spies and Franklin 1991) Competitive exclusion (Carey and Curtis 1996) Canopy closure (Franklin et al. 2002)	Sapling/Pole-Moderate, Sapling/Pole-Closed (O'Neil et al. 2001)	Poles and saplings (Haynes 2003) Mid-seral (FEMAT 1993)
Characteristics	• Taller vegetation becomes dominant • Leaf area and biomass accumulate • Canopies close on some sites—rate depends on regeneration density and site productivity • Few snags and coarse woody debris (CWD) in managed stands • Rapid understory environment changes • Resource availability decline	• Biodiversity declines • Depending on canopy structure, herb and shrub understory abundance declines • Amphibians associated with closed canopies • Minimize stage through precommercial and variable-density thinning	• Stand age typically 15 to 35 years • Conventional precommercial thinning to maintain evenly spaced trees and promote tree growth • Pole and sapling sized trees usually not merchantable • Commercial thinning can occur depending on market conditions
Third	Aggradation (Bormann and Likens 1979) Stem exclusion (Oliver and Larson 1990) Thinning (Spies and Franklin 1991) Competitive exclusion (Carey and Curtis 1996) Biomass accumulation/competitive exclusion (Franklin et al. 2002)	Small Tree-Single Story-Moderate, Small Tree-Single Story-Closed, Medium Tree-Single Story-Moderate, Medium Tree-Single Story-Closed, Large Tree-Single Story-Moderate, Large Tree-Single Story-Closed (O'Neil et al. 2001)	Young (Haynes 2003) Late seral (FEMAT 1993)

Appendix 1. Major classification schemes used to describe forest developmental stages and associated characteristics. Characteristics are illustrated from various ecosystem perspectives by using a Douglas-fir (*Pseudotsuga menziesii*) dominated sere growing in the western hemlock (*Tsuga heterophylla*) zone (Franklin and Dyrness 1988). Characteristics will vary widely based on site location, disturbance history, management, and forest types. This table was developed by B. Kerns (see list of contributors). (continued)

	Major classification systems and characteristics for forest developmental stages		
Ecosystem perspective → / Forest development stage ↓	Ecological structure and process	Wildlife habitat	Timber production
Characteristics	• Woody biomass development • Tree crown differentiation and lower branch pruning • Low resource availability early, increases later • Density dependent tree mortality with high stand density • Few snags and CWD • Competitive exclusion of many organisms	• Low biodiversity • Depending on canopy structure, herb and shrub abundance may be low • Amphibians associated with closed canopies • Minimize stage through precommercial and variable density thinning	• Stand age typically 45 to 75 years • Pioneer tree cohort dominates site • Sawtimber and nonsawtimber size trees • Conventionally thought of as the culmination of mean annual increment • For many private industrial land-owners, may reflect typical rotation lengths and stand developments ends
Fourth	Transition (Bormann and Likens 1979) Understory reinitiation (Oliver and Larson 1990) Mature (Spies and Franklin 1991) Understory reinitiation, developed understory (Carey and Curtis 1996) Maturation (Franklin et al. 2002)	Small Tree-Single Story-Open, Medium Tree-Single Story-Open, Large Tree-Single Story-Open (O'Neil et al. 2001)	Mature seral (FEMAT 1993) Mature (Haynes 2003)

Appendix 1. Major classification schemes used to describe forest developmental stages and associated characteristics. Characteristics are illustrated from various ecosystem perspectives by using a Douglas-fir (*Pseudotsuga menziesii*) dominated sere growing in the western hemlock (*Tsuga heterophylla*) zone (Franklin and Dyrness 1988). Characteristics will vary widely based on site location, disturbance history, management, and forest types. This table was developed by B. Kerns (see list of contributors). (continued)

Major classification systems and characteristics for forest developmental stages			
Ecosystem perspective → / Forest development stage ↓	Ecological structure and process	Wildlife habitat	Timber production
Characteristics	• Maximum height and crown spread of pioneer tree cohort • Minimal coarse woody debris • Heterogeneous resource availability • Shift to density independent mortality • Sub-lethal tree damage produces greater individual tree conditions and niche diversification	• Extended rotations (> 80 years) to provide habitat • Re-establishment of understory species, including shade-tolerant conifers • Increase in diversity of fauna, especially with multi-stored canopies • Increase habitat through commercial thinning and CWD management	• Stand age typically 85 to 135 years • Less common stage on private industrial lands • Composed mostly of sawtimber size trees • Conventionally thought of as over-culmination of mean annual increment
Fifth	Steady-state (Bormann and Likens 1979) Old-growth (Oliver and Larson 1990) Transitionand shifting-gap (Spies and Franklin 1996) Botanically diverse, niche diversification fully functional (managed) and old-growth (Carey and Curtis 1996) Vertical diversification, horizontal diversification and pioneer cohort loss (Franklin et al. 2002)	Small Tree-Multistory-Open, Small Tree-Multistory-Moderate, Small Tree-Multistory-Closed, Medium Tree-Multistory-Open, Medium Tree-Multistory-Moderate, Medium Tree-Multistory-Closed, Large Tree-Multistory-Open, Large Tree-Multistory-Moderate, Large Tree-Multistory-Closed, Giant Tree-Multistory (O'Neil et al. 2001)	Mature (FEMAT 1993) Old mature stage (Haynes 2003)

Appendix 1. Major classification schemes used to describe forest developmental stages and associated characteristics. Characteristics are illustrated from various ecosystem perspectives by using a Douglas-fir (*Pseudotsuga menziesii*) dominated sere growing in the western hemlock (*Tsuga heterophylla*) zone (Franklin and Dyrness 1988). Characteristics will vary widely based on site location, disturbance history, management, and forest types. This table was developed by B. Kerns (see list of contributors). (continued)

	Major classification systems and characteristics for forest developmental stages		
Ecosystem perspective → / Forest development stage ↓	Ecological structure and process	Wildlife habitat	Timber production
Characteristics	• Slow decline in aboveground biomass • Many substages for long-lived species • Development of late successional and old-growth attributes (Spies and Franklin 1996) • Density-independent mortality increases, large, persistent gaps may form • Accelerated generation of cwd • Highly heterogeneous resource availability • Sub-lethal tree damage continues • Loss of dominants (800 to 1300 yrs.)	• Extended rotations to provide habitat • Large trees, multiple stories, snags, CWD, and closed canopies create habitats for numerous species • Faunal diversity, especially birds and mammals is high	• Stand age typically more than 145 years • Uncommon stage on private industrial lands • Conventionally thought of as past the point where net annual growth has peaked

Section 1—Research Examples

These research examples bring the essential features of compatible forest management into focus. In Chapter 2, Montgomery uses an elegant model to get to the heart of the issues to deliver some concrete examples. Her production possibilities approach is a useful tool for moving from general discussion of the issues into analysis. With the model, people can summarize large amounts of data about competing forest uses and relations among them, identify cases where change could have significant benefits, and describe options that would achieve those benefits. Tradeoffs between competing uses such as wildlife habitats and timber production are delineated clearly.

Two case studies illustrate the main points in Chapter 2. First, in the western Cascade Range of Oregon, is a study of a three-way production possibilities frontier for timber production and two wildlife species, the common porcupine and great horned owl, with competing habitat needs. The production possibilities approach can be used to define the productive capacity of the study area for the modeled uses, evaluate whether management alternatives are efficient, and explore the nature of possible tradeoffs among uses. The second case study is set on the west side of the Oregon Coast Range. It analyzes the tradeoffs between biodiversity, defined as the expected species richness for 166 native terrestrial vertebrate species, and the net present value of timber production, over a 100-year time horizon. All management possibilities start with the realities of the current forest conditions. The production possibilities frontier may reveal new, more efficient options for the future, where nobody is worse off and at least somebody is better off.

The production possibilities approach should be considered in context with other sections in this book. Silvicultural treatments discussed in Section 2 can be used to develop forests toward the options discovered under the production possibilities approach. Scale issues must be considered, however, and these are discussed in Section 3. Some uses are not compatible at the stand level, for example timber production and habitat for certain species. However, both uses may be compatible at the forest or bioregional levels. Because the production possibilities approach makes tradeoffs between competing uses explicit, it can be useful when considering the aspects of management discussed in Section 4, such as wood quality, nontimber forest products, and wildlife. The case studies in Chapter 2 highlight the importance of considering carefully which forest uses are studied. Value judgments are involved in the choice of uses to model and the ways in which those uses are evaluated.

In Chapter 3, the authors show how noncommercial tree species—in this case, red alder in the forests of southeastern Alaska—can have ecological functions that mitigate some effects of timber harvest. Mixed-species young forests and new silvicultural systems may provide sustainable wood production and protection of other forest values. The study examines whether red alder influences understory vegetation, birds, fish, invertebrates, and conifer growth, and ultimately, ecosystem structure and function. These findings likely have relevance in other forests as well. Encouraging mixed-species composition in young-growth forests may help improve compatibility, such as wildlife habitat, food resources for wildlife, and long-term development of more complex structure in mature forests.

The full context for Chapter 3 is rounded out in later sections of this book. Silvicultural options for managing mixed-species young forests are described in Section 2. The mixed red alder-conifer forests offer both benefits and trade-offs. Some of these other management aspects are examined in Section 4. From an ecosystem perspective, however, the benefits seem to outweigh the negative aspects of red alder occurrence in young forests in southeastern Alaska. These benefits may include increased social acceptability, as discussed in Section 5.

Scale, which is examined thoroughly in Section 3, is discussed in terms of both watersheds and time in Chapter 3. The authors report their findings on the functions of red alder in young forests at both the stand and watershed levels. Red alder plays ecological roles from the headwaters and uplands of watersheds downslope to fish-bearing streams lower in the watersheds. In the scale of time, red alders have a very different life history than conifers, and this difference becomes part of their ecological contribution to forests. Red alders grow rapidly as very young trees and die when they are 60 to 100 years old. Conifers, on the other hand, grow slower but eventually overtop alders and continue to live for hundreds of years. Red alders leave legacies in forests long after the trees have died and decayed. The shorter-lived red alders may accelerate the transition of a young stand to mature forest structure.

Chapter 2

The Production Possibilities Approach to Understanding and Modeling Compatibility

Claire A. Montgomery
Department of Forest Resources, College of Forestry,
Oregon State University, Corvallis, OR 97331, USA

1. The Production Possibilities Concept

As demand grows for the multiple goods and services that flow from the forest, it becomes increasingly important that forests be managed carefully for all competing uses and that waste be avoided. This is the motivation for wood compatibility research. The word "compatibility" has a particular meaning to economists. It refers to the attributes of the set of all feasible combinations of outputs when resources are limited. This is known as the production possibilities set. For forestland, that set defines the productive capacity of a site with respect to a set of potential forest uses. If the set is empty, there are no combinations of the forest uses that are feasible; the forest uses are completely incompatible, or mutually exclusive. Most forest uses exhibit some degree of compatibility in the sense that joint production is feasible. The greater the set of possible combinations of forest uses, the more compatible the uses.

Modeling the production possibilities set can be a useful way to frame important questions faced by forestland managers and policy makers. First, can we do better? Are we fully exploiting compatibility? Is it possible to increase at least one of the forest uses that people value without decreasing any others? If so, current management is inefficient and resources are being wasted. Figure 4, Chapter 1 is repeated here as Figure 1(a) for illustration. It shows a hypothetical production possibilities set for ecological and socioeconomic forest conditions. All of the combinations of forest conditions inside and along the frontier of the set are possible. But those inside the frontier (e.g., point X) are inefficient; it is possible to improve socioeconomic conditions at no cost in terms of forgone ecological conditions and, vice versa; it is possible to improve ecological conditions at no cost in terms of forgone socioeconomic

R.A. Monserud, R.W. Haynes and A.C. Johnson (eds.)
Compatible Forest Management, 35-53

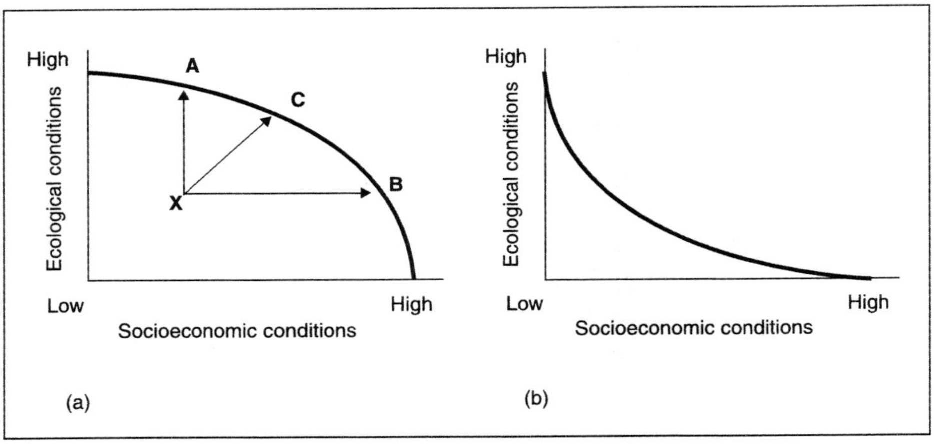

Figure 1. Two hypothetical production possibilities sets illustrate cost-effective management on the frontier of the production possibilities set and inefficient management inside the set.

conditions. To the extent that socioeconomic and ecological conditions can be accurately modeled and the production possibilities set for a site identified, a finding that we are managing inside the frontier suggests that there are opportunities to do better. For example, all of the points between A and B in Figure 1(a) are improvements over point X because they all involve increasing at least one forest use without decreasing the other.

Second, if current management is efficient, increasing one forest use imposes tradeoffs. The production possibilities approach can address the magnitude of the required tradeoffs. In Figure 1(a), the subset of points along the frontier of the production possibilities set is considered efficient in production because resources are used fully and there are ambiguous opportunities for improvement. Any increase in one use requires a decrease in another.

The slope of the production possibilities frontier shows the cost of increasing one use in terms of the necessary reduction in the other use. That is, in moving from point A to point C (thereby improving socioeconomic conditions) ecological conditions are reduced; the slope indicates how much. In Figure 1(a), when ecological conditions are high and socioeconomic conditions are low, it costs little to improve socioeconomic conditions; the reduction in ecological conditions is relatively small. But, as socioeconomic conditions improve, the cost of further improvement increases as well. For example, it costs less to move from A to C than it costs to move from C to B. The frontier of the production possibilities set becomes steeper. Conversely, when socioeconomic conditions are high and ecological conditions are low, it costs little to improve ecological conditions; the reduction in socioeconomic conditions is relatively small. Again, as the level of ecological conditions improve, the cost of further improvement increases as well. With the production possibilities set depicted in Figure 1(a), optimal management will likely involve both forest uses, rather than one or the other. Hence, these uses can be said to be compatible.

In contrast, Figure 1(b) depicts a production possibilities set for a site where socioeconomic and ecological objectives conflict. The set of feasible combinations of the two uses is small relative to Figure 1(a). When socioeconomic conditions are low and ecological conditions are high, improving socioeconomic conditions requires a large reduction in ecological conditions. The cost is relatively high. And, conversely, when ecological conditions are low and socioeconomic conditions are high, a small improvement in ecological conditions is costly; the required reduction in socioeconomic conditions is relatively large. The two uses are less compatible on this site than on the site depicted in Figure 1(a). It is less likely that optimal management will involve both uses.

The different degrees of compatibility depicted in Figure 1 could be due to the nature of the forest uses, the attributes of the sites, or the scale of the analysis. Two uses may not be very compatible at the stand level, so it is optimal to allocate a stand to one use or the other. At the same time, the uses may be compatible at the forest or regional level, with compatibility achieved by allocating individual stands to their highest valued use.

2. Limitations of the Production Possibilities Approach

The production possibilities approach can be used to (1) define the productive capacity of a site with respect to the modeled uses, (2) evaluate whether current management and other management alternatives are inefficient, and (3) explore the nature of the tradeoffs between them. It cannot define optimal management. In Figure 1(a), points A, B, and C are all efficient in production; there is no waste. But there is insufficient information to determine which is preferable, which generates the highest value, or whether the socioeconomic benefit obtained in moving from point A to point B is worth the reduction in ecological conditions. Indeed, it is not even possible to determine whether the benefit obtained by moving from point X, inside the frontier, to a point on the frontier but outside the range from A to B is worth the necessary reduction in one use. In Figure 1(b), although it is likely to be optimal to manage for either socioeconomic conditions or ecological conditions, it is not possible to determine which is preferable.

Valuation is required to identify the preferred mix of uses. If it can be objectively known how society values one use relative to another, different outcomes can be compared and the highest valued alternative identified. But valuation is outside the realm of compatibility research. Compatibility research can provide important information to decisionmakers, but it is insufficient, in itself, to identify the best choice. That choice ultimately rests on social values as perceived by decisionmakers.

When the production possibilities approach was proposed as a way to analyze difficult multiple use problems in forestry, it was thought to be value-free. After all, the approach is to model technical relations between a given set of outputs by using science-based models of timber growth and yield, wildlife population dynamics, ecological processes and so on. But, in fact, it is becoming increasingly apparent that social values are implicitly a part of a production possibilities analysis in its earliest phases. The process of identifying the set of forest uses to model is, itself, value-laden. To ask which forest uses should be included in the analysis is to ask which forest uses matter to people.

3. Applications of the Production Possibilities Approach

The concept of the production possibilities set was developed as part of the microeconomic theory of joint production and welfare economics. As part of the wood compatibility initiative research program, Stevens and Montgomery (2002) reviewed the forest economics literature, searching for studies that applied the production possibilities approach specifically to forest resources.

In one of the earliest references in the forest economics literature, Gregory (1955) illustrates how the concept of the production possibilities from microeconomic theory might be applied to the stand-level multiple-use problem using timber and forage as examples. He describes a hypothetical set of iso-cost curves that are essentially production possibilities frontiers for limited budgets. If relative prices for the two goods are known, an expansion path can be constructed from the iso-cost curves that identifies optimal combinations of the two forest uses for increasing levels of the budget constraint. Gregory's example is interesting because it shows how the shape of the production possibilities frontier (and, hence, the extent of compatibility of two forest uses) can depend on the level of the fixed input. For forest uses, for example, the expanding budget constraint is analogous to increasing the geographic scale of the analysis. More recently, Connaughton and Fight (1984) argue for the usefulness of the production possibilities approach for national forest planning, but note that its practical usefulness is limited by the need to quantify complex production relations between forest uses and to develop optimization methods that can be used to address complex problems.

Most applications of the production possibilities approach to multiple use on forestland have involved evaluating tradeoffs between timber production and an individual wildlife species. And most have demonstrated compatibility between the modeled forest uses, in the sense that joint production is likely to be optimal. That is, the production possibilities frontiers estimated in these studies are convex to the origin as depicted in Figure 1(a).

Muhlenberg (1964) provided one of the first examples of the practical applicability of the approach by using empirical data on pulpwood production and carrying capacity for deer. Jones and Schuster (1985) extended Muhlenberg's approach by applying mixed integer programming to identify highest present value combinations of elk carrying capacity and timber harvest for a hypothetical tract of forestland.

In more recent studies, simulation models for wildlife and timber production have allowed more detailed modeling of production relations between forest uses under a wider range of conditions than the studies that relied on empirical observations of forest use outcomes. Montgomery et al. (1994) used a stochastic population simulation model with spatial juvenile dispersal behavior for the northern spotted owl (*Strix occidentalis caurina*) (Lamberson et al. 1992) to estimate tradeoffs between probability of northern spotted owl survival over a 100-year time horizon and federal timber harvest levels in Oregon and Washington. This was a steady-state-analysis, and timber harvest schedules were nonspatial. Also, outcomes were not optimized so the estimated production relation between federal timber harvest and probability of owl survival most likely lies inside the production possibilities frontier for the two modeled forest uses. In fact, individual stands were allocated either to timber production or owl habitat reserves and the concave (to the origin) shape of the estimated production possibilities frontier arose from the large geographic scale of the analysis. Rohweder et al. (2000) used an individual tree simulation model to model timber stand conditions over 100 years under a range of timber management alternatives. Tradeoffs between timber production, a set of individual wildlife species—elk (*Cervus elaphus*), pileated woodpeckers (*Dryocopus pileatus*), and various songbirds—and insect and fire resistance were evaluated by constructing indices of the nontimber uses that were dependent on the attributes of the trees in the stands and the spatial configuration of the stands. Compatibility of uses was evaluated by estimating the curvature properties of the resulting production possibilities frontiers. The study extended Montgomery et al. (1994) by modeling actual managed stands and their development over time. But, like Montgomery et al. (1994), the estimated production relations were likely inside the production possibilities frontier because no optimization was employed. As a consequence, these studies serve as useful illustrations of marginal analysis of multiple use in forestry, but they do not exploit the power of the production possibilities approach by tracing out the limits on the productive capacity of a site.

Part of the challenge in applying the production possibilities approach has been the lack of optimization methods with the potential to solve real world problems. Arthaud and Rose (1996) developed an algorithm for efficiently searching for points on the production possibilities frontier between aspen pulpwood production and ruffed grouse habitat (*Bonasa umbellus*) in the Lake

States. The attributes of the ruffed grouse habitat (which included spatial configuration) were summarized in a habitat suitability index. But their analysis allowed one of only six possible 45-year timber harvest schedules to be assigned to each of 20 actual stands. Their method performed well in comparison to complete enumeration of the solution space of 6^{20} possible combinations of stands and management regimes. But the authors noted that solution times would increase substantially by increasing the geographic scale, time horizon, complexity of the wildlife model, or flexibility of timber harvest management regimes; they suggested some ideas for keeping the problem tractable. The problem of size, known as the curse of dimensionality, plagues the combinatorial models by which spatial forest-use problems are most conveniently and realistically specified.

Interest in the production possibilities approach to compatibility research appears to be ongoing, partly because of heightened interest in the policy arena in cost-effective management for nonmarket forest uses. At the same time, there continue to be advances in the knowledge, methods, and technology required for realistic modeling of production relations between multiple forest uses, particularly for wildlife populations and timber stand structure. Spatially explicit modeling of actual landscapes has been facilitated by the development of technologies for obtaining spatial data (e.g., remote sensing) and technologies for using that data (e.g., geographic information systems (GIS)). Spatial and dynamic simulation models of relations among different forest uses are growing increasingly sophisticated and reliable. The development of flexible optimization methods, such as stochastic heuristic search algorithms, is making it increasingly possible to solve complex problems, at least approximately.

4. Two Recent Case Studies

Recent studies have attempted to exploit these developments and extend the earlier studies in several ways. In one set of studies, integration of detailed spatial and dynamic simulation models with optimization methods capable of handling large complex problems is allowing more realistic modeling of wildlife and wood production tradeoffs. In another set of studies, wildlife objectives are formulated in terms of multiple species by using biodiversity measures and individual species habitat requirements.

This chapter describes one example of each set of studies. Nalle et al. (2001) represents studies involving individual wildlife species. In that study, a 3-way production possibilities frontier for timber production and two wildlife species with competing habitat needs was developed for the western Cascade Range of Oregon. Lichtenstein and Montgomery (2003) represent studies involving large numbers of species. They developed a production possibilities frontier for timber production and biodiversity in the Oregon Coast Range. Both studies expanded the geographic scale of the analysis beyond earlier studies and increased the flexibility with which management activities are applied by making the timing of treatments endogenous. They also extended earlier studies by using the production possibilities approach to evaluate current management in the study areas. The approach might be further extended to evaluate other alternative management scenarios.

The two studies took the same general approach:

- They integrated biological and economic models in an optimizing framework to address economic issues related to efficiency.
- They used the integrated model to estimate the frontier of the production possibilities set for the modeled forest uses on a particular site to assess its productive capacity—what can be produced?
- They simulated current management under the existing pattern of land ownership, and compared that to the production possibilities frontier to search for opportunities for improvement—what is being produced and is it possible to do better?

4.1. Single-Species Models

Nalle et al. (2001) modeled the tradeoffs between two species—the common porcupine (*Erethizon dorsatum*) and the great horned owl (*Bubo virginianus*)—and timber production on a 1.7 million hectares (ha) landscape on the west side of the central Cascade Range of Oregon over a 100-year time horizon. Roughly half of this area is private land managed primarily for timber production, the other half is public land managed primarily for natural resource protection. Timber production was measured as the discounted sum of consumer and producer surpluses in the regional log market over a 100-year time horizon.

The great horned owl and the common porcupine were selected because they are native to the study area, have negligible predator-prey interactions, and have relatively reliable behavioral and life history data (Bigger and Vesely 2000). Because these two species have different habitat preferences, life spans, reproductive capability, and dispersal behavior (Verts and Carraway 1998), they represent competing wildlife objectives. The great horned owl prefers old mature conifer forest for nesting, but will inhabit younger stands as long as some larger diameter trees are available for nesting. It is typically not found in intermediate-aged dense stands because navigation is difficult for the large predator. It can travel relatively long distances (up to approximately 75 kilometers (km)) in search of suitable habitat. The common porcupine has a preference for younger stands and travels only short distances in search of suitable habitat (up to approximately 2.6 km). These differences were expected to result in different configurations of vegetative cover depending on which species was emphasized in the optimization.

Nalle et al.'s study extended an earlier study by Calkin et al. (published in 2002) that modeled tradeoffs between the northern flying squirrel (*Glaucomys sabrinus*) and the net present value[1] of timber production on a 10 000-ha landscape on the west side of the central Cascade Range of Oregon over a 100-year time horizon. Both studies used a spatial and dynamic biological wildlife population simulation model, PATCH or Program to Assist in the Tracking of Critical Habitat (Schumaker 1998). The PATCH model tracks individual females on a landscape as they are born, move to inhabit a territory, produce offspring, and die. Dispersal and breeding success depends on the quality and configuration of habitat, which changes over time due to timber management and vegetation growth. The studies used neighborhood ring structures of habitat in regression models to quickly predict PATCH simulation outcomes in the optimization algorithm. Endogenous spatial interactions entered this model through the neighborhood habitat structure in the regression model of PATCH outcomes. The optimization procedure identified a set of preliminary solutions that are likely to be close to the production possibility frontier. These preliminary solutions were simulated using PATCH, which is too slow to use directly in the optimization, to finally estimate the production possibilities frontier.

Nalle et al.'s model was solved by searching for the timing and location of timber harvest in each of ten 10-year decision periods to maximize timber value subject to meeting targets for the wildlife species. The wildlife targets were varied over a grid of values for each species and the optimization was solved repeatedly to estimate the production possibilities frontier without regard to landowner boundaries. Timber value was the present value of benefits to consumers (consumer surplus) and benefits to landowners (rent) resulting from timber production during the time horizon plus the present value of the land and timber at the end of the time horizon, assuming it will be managed forever on the management regime assigned during the time horizon. Stumpage prices depended on down-sloping demand for logs and distance-dependent harvest and haul costs. Endogenous intertemporal interactions entered this model in two ways. First, the wildlife population thresholds for the time horizon were specified as geometric means of estimated population size in each 10-year decision period. This form heavily penalizes extinction and favors stable populations over time. Second, downward-sloping log demand also favors stable timber harvest levels because log price in each period depends on the timber harvest level in each period.

The geographic scale of Nalle et al.'s study was large (1.7 million ha divided into 28,251 management units of 0.70 km^2 per unit). This unit size was chosen because it closely approximates the average territory size of the two species. Although management activities in each unit during each period were

[1] All of the studies described in this paper used a real discount rate of 4% in conformance with U.S. Forest Service guidelines. The many data sources are described in the individual papers.

restricted to either clearcut timber harvest or no harvest, there were 127 possible combinations of harvest schedules for each management unit. Because of the spatial component of habitat quality, the management units were not independent. Timber harvest on one unit affected habitat quality on neighboring units.

Heuristic optimization algorithms are becoming popular for this kind of large, complex, spatial forest management problem. Heuristic algorithms are approximate rather than exact optimization methods that use stochastic criteria to "intelligently" search the solution space for good, but not necessarily globally optimal solutions. Examples include Lockwood and Moore (1992), Bettinger et al. (1997), Bettinger et al. (1998), Boston and Bettinger (1999), and Nalle et al. (2002). Although the heuristic method is not exact and, hence, global optimality is not guaranteed, it has been shown capable of producing solutions that are close to optimal (Bettinger et al. 2002) while allowing far more realism and detail than exact methods. In fact, for this study, a hybrid heuristic optimization algorithm was developed that built on the strengths of two commonly used heuristic algorithms–simulated annealing (Metropolis et al. 1953, Kirkpatrick et al. 1983) and tabu search (Glover and Laguna 1997). The hybrid algorithm reduced solution times from days to hours. This was important because many solutions were required to trace out the production possibilities frontier.

In reality, wildlife and timber are not management objectives across the entire landscape. Private landowners, having little incentive to manage for wildlife, emphasize timber production. Policy on federal land evolved, in part, to mitigate the market failure on private land and, hence, emphasizes non-timber forest uses—particularly those related to endangered wildlife species. Current management under this pattern of ownership does not allow the potential compatibility of forest uses to be fully exploited. To evaluate the potential for more efficient management (e.g., in Figure 1(a), moving from a point X inside the production possibilities frontier to points A, B, or C on the production possibilities frontier), we simulated the configuration of land ownership types in the study area with their corresponding management objectives. Private landowners were assumed to manage for maximum timber value. Timber harvest was not allowed on federal land.

Figure 2(a) shows the estimated relations between the three modeled outputs. Porcupines and timber are complements (positively related) over much of the range—up to the maximum timber value solution (point A on all three graphs)—and owls and timber are complements as porcupine population size is increased beyond that point. Hence, timber and porcupines can be considered a composite good up to point A, with the relation between them depicted in the middle graph. Owl population size is mapped on timber in the leftmost graph and it is mapped on porcupine population size in the rightmost graph.

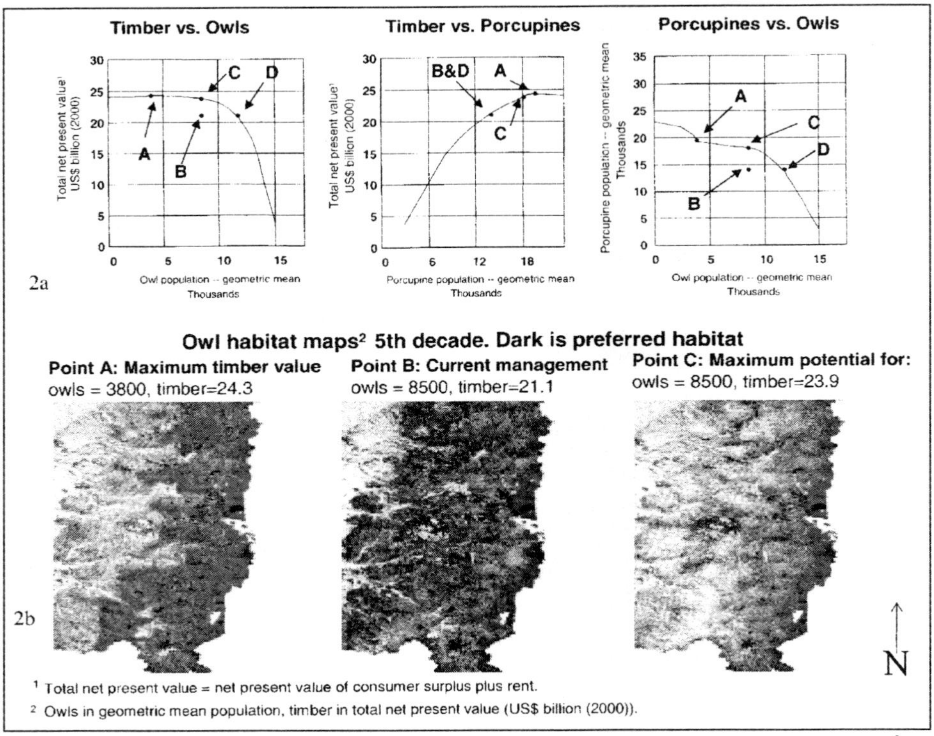

Figure 2. Model solutions for cost-effective production of net present value of timber production, geometric mean owl population size, and geometric mean porcupine population size over 100 years in a 1.7 million-ha study area on the west side of the central Cascade Range of Oregon. Figure 2a illustrates estimated relations between the three modeled outputs. The maps (2b) illustrate the forest conditions that produce points A, B, and C in each of these 2-dimensional slices. Point A = maximum timber value; B = current management; C = the production possibilities frontier solution that corresponds to (B) in geometric mean owl population size (Nalle et al. 2001). Point D = the timber value of the current management solution.

Owl habitat maps in the fifth decade corresponding to three solutions are shown in Figure 2(b): (A) the maximum timber value solution, (B) current management simulated by maximizing the value of timber production on private land and allowing no timber harvest on public land, and (C) the solution that matches the owl population of the current management simulation. One additional point (D) that matches the timber value of the current management solution is labeled on Figure 2(a), but no map is shown.

The result of the current management simulation, point B in Figure 2(a), suggests that it might be possible to do better. As shown, point B lies inside the production possibilities frontier. Owls may increase 38% without reducing timber value or porcupine populations by moving horizontally from point B to point D. Alternatively, without reducing the number of owls, the timber value may increase 14% by moving vertically from the point B to point C, with a corresponding increase in porcupines of 28%. Whereas a move from point B to point C or D clearly improves efficiency, that cannot be said for moves from point B to any point outside that range on the production possibilities frontier. For example, a move from point B to point A, the maximum timber value solution, would increase timber value by 15%, but reduce the owl population by 55%. Without knowledge of the relative values of timber and wildlife to society, one cannot determine which outcome is preferable.

The habitat maps in Figure 2(b) contain information that can help managers understand how efficiency might be improved. The owl habitat map for point C shows owl habitat dispersed more evenly on the landscape than either point A, the maximum timber value solution, or point B, current management. In fact, it can be seen by comparing the habitat maps for points A and B that the habitat configuration produced under current management, point B, resembles A, which emphasizes timber production, more than it resembles C, which matches the owl population of B but increases timber value. That is because the current landowner pattern evolved historically in response to relative timber values across the landscape. Consequently, timber values are lower on federal land, which is concentrated on the eastern edge of the maps than on private land. As a result, owl habitat is concentrated on public land even in the maximum timber value solution. This analysis suggests that current management is only cost-effective if timber production is the primary land use objective.

4.2. Multiple-Species Models

To model biodiversity, it is necessary to summarize data about numerous species, many of which are not well understood. The most common approach is to use a probability-based biodiversity index (e.g., Solow et al. 1993, Bevers et al. 1995, Montgomery et al. 1999).

Montgomery et al. (1999) decomposed a general biodiversity index into four components to be modeled: the effect of management activities on habitat, the link between habitat and each species population, the link between species populations and probability of persistence over some time period, and the contribution of each species to the diversity of the set of species. Although data are insufficient to allow detailed modeling of these four components for most species, simple modeling is possible for many species.

In a case study set in Monroe County, Pennsylvania, Montgomery et al. (1999) evaluated tradeoffs between biodiversity (as indicated by 147 bird species that are native to that county) and market-valued land uses. Management involved allocating land to one of three land uses: commercial/residential, agriculture, and forest/open space. Habitat quality was based on habitat preference rankings over the vegetative cover types represented by each land use class. These were obtained by using a survey of wildlife biologists and a literature review. Species indices, X_s, proportional to the aggregate amount of preference-weighted habitat were constructed for each species. Therefore, changes in habitat resulted in proportional changes in the population index. Simple logistic probability of persistence functions, $V(X_s)$, were constructed from International Union for the Conservation of Nature (IUCN) imperilment ranking guidelines and species initial population size index values were based on IUCN imperilment ranking. Finally, each individual species' contribution to the diversity of the set was represented by a diversity weight, w_s for each species, s, that was based on the taxonomic tree as proposed by Vane-Wright et al. (1991) and modified by May (1990). Species with many close relatives were assigned smaller weights and, hence, contributed less to the value of the diversity index, than species with few close relatives. The resulting biodiversity index, $B(y)$, was weighted species richness:

$$B(y) = \sum_{s=1}^{S} w_s V(X_s(y)) \tag{1}$$

A marginal cost curve for the index in Monroe County showing the cost of increments in the biodiversity index as the value of forgone market-valued land uses was estimated by maximizing market land value plus biodiversity value. Because the value of biodiversity is unknown, the model was solved repeatedly for incrementally increasing "biodiversity prices." The production possibilities frontier underlying the marginal cost curve is shown in Figure 3. In a case study set in the Muddy Creek Basin of Oregon, Shunk (2000) extended Montgomery et al.'s study by adding timber management alternatives as land uses and comparing some scenarios proposed by a land use planning study (Hulse et al. 2000) to the outcomes on the production possibilities frontier.

The usefulness of these two case studies was limited for two reasons. First, neither involved spatial analysis. The wildlife persistence probabilities depended only on the amount and quality of the habitat, not on its location. Hence, important wildlife habitat issues, such as fragmentation, edge, interior area requirements, and proximity to breeding and foraging habitat were ignored. Similarly, land use values depended only on aggregate area allocated to a particular use and not its location. As a result, the outcomes on the production possibilities frontier may not be feasible on the ground. Second, the case studies used static, steady-state analyses. The target land use allocation was assumed to occur at a point in time and be constant thereafter. The conversion period was not considered in the analysis except in the computation of forest land values for different rotation lengths. The biodiversity index, $B(y)$, was a function of an initial land use allocation only.

In an effort to address these limitations, Lichtenstein and Montgomery (2003) modeled tradeoffs between biodiversity, defined as expected species richness for 166 native terrestrial vertebrate species, and the net present value of timber production on a 41 200-ha study area on the west side of the Coast

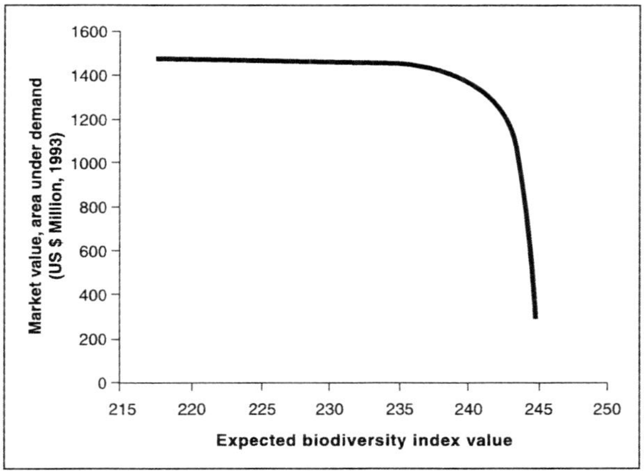

Figure 3. Production possibilities frontier for a biodiversity index of 147 native bird species and market-valued land uses in Monroe County, Pennsylvania (Montgomery et al. 1999).

Range in Oregon. Approximately 40% of the area is public land, managed by the U.S. Forest Service. Almost all the rest is privately owned; approximately 40% of the area is classified as industrial forestland and the remaining 20% is classified as nonindustrial private forestland.

This case study extended earlier studies in two ways. First, the landscape was dynamic: it evolved over the 100-year time horizon as vegetation aged and management activities were applied. Data for the initial landscape, as it exists currently on the ground were obtained from maps of vegetative cover types that had been developed from remote-sensed images (Spies et al. 2002b). The analysis selected the location and timing of management activities for the 100-year analysis period for each of 4,500 conifer forest management units. The resulting time series of vegetative cover maps, X_{st}, were evaluated as habitat for each species, s, in each of the 20, 5-year decision periods. Persistence probabilities, $V(y)$, resulting from management activities, y, were evaluated as the product of short-term probabilities of persistence for each 5-year decision period, $V_5(X_{st}(y))$, and an ending long-term probability of persistence for 100 years beyond the time horizon based on the vegetative cover at the end of the 20th decision period, $V_{100}(X_{s,20}(y))$:

$$V(y) = \left(\prod_{t=1}^{20} V_5(X_{st}(y)) \right) V_{100}(X_{s,20}(y)) \tag{2}$$

The logistic persistence probabilities were the same as those used by Montgomery et al. (1999), calibrated for the different time periods.

Second, wildlife habitat was modeled spatially by requiring minimum contiguous areas of habitat before it could be counted. The minimum area requirements were based on estimates of home range size for each species. With this spatial modeling, management in one unit affects habitat in adjacent units and, hence, the management units are not independent. Other spatial attributes of habitat, such as proximity to breeding and foraging habitat were not modeled in this study, but as more complete data for large numbers of species become available, the approach can be extended to do so in the future.

Timber management activities involved assignment of each management unit to commercial management, mixed management, or reserve status with the timing of precommercial and commercial thinning, fertilization, and clearcut harvest determined endogenously. Mixed management allowed two commercial thins to encourage conifer diameter growth and accelerated movement to "old mature conifer" status as suggested in silvicultural studies of thinning (Hayes et al. 1997, Tappeiner et al. 1997, Bailey and Tappeiner 1998). The study used habitat preference rankings that had been developed for terrestrial vertebrates in the Oregon Coast Range (Hulse et al. 2000). The vegetative cover classes included three conifer forest age classes. Habitat was affected by timber management activities that influenced the age class structure of the forest. Other structural attributes of the stand, such as snags and multiple canopy layers, were not modeled.

The production possibilities frontier was identified by searching for the set of timber management activities that maximizes the net present value of timber production subject to meeting targets for $B(y)$. The targets were increased incrementally from zero and the model was solved repeatedly. Again, due to the spatial dependence of one management unit upon activities in adjacent units, and because of the size of the problem, a heuristic optimization algorithm, simulated annealing, was used. As in Nalle et al. (2001), current management was simulated by maximizing the net present value of timber production on private land and not allowing timber harvest on federal land. Nonindustrial private forest landowners were further assumed to apply management prescriptions in the proportions indicated in an Oregon Department of Forestry survey of those owners.

The resulting estimate of the production possibilities frontier is shown in Figure 4. Point B represents the current management simulation; the production possibilities outcome that achieves the same biodiversity index value as current management is labeled point A. Vegetation maps corresponding to the end of the time horizon for both labeled points are also shown in Figure 4. The landownership pattern is clearly highlighted in the current management simulation by the presence of mature conifer forest on federal land. Along the production possibilities frontier, blocks of forest age classes are smaller and distributed more evenly on the landscape than in the current management scenario. In particular, there are far fewer and smaller blocks of mature conifer forest.

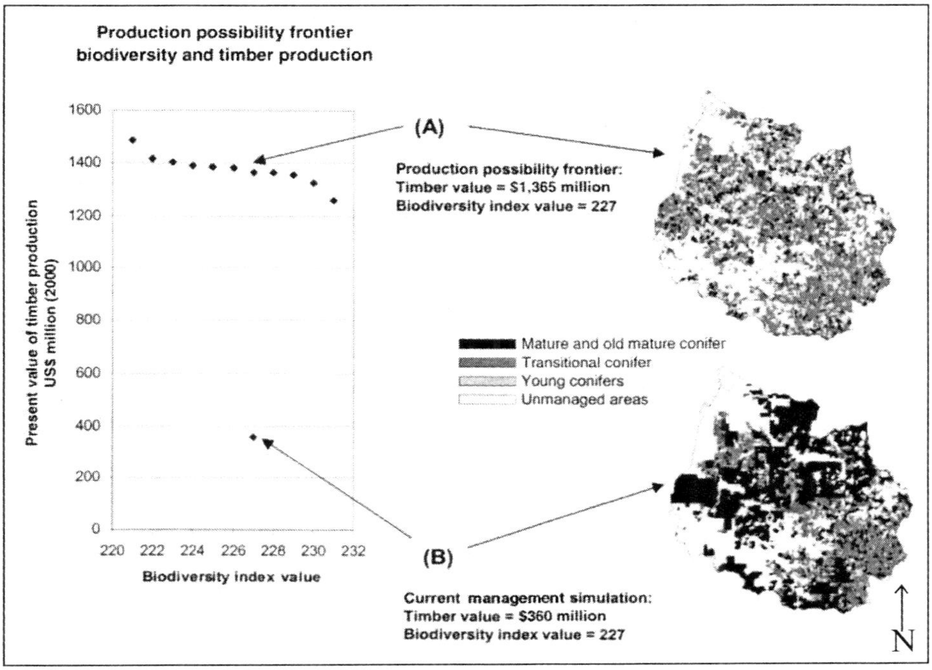

Figure 4. Production possibilities frontier for a biodiversity index of 166 terrestrial vertebrate species and the present value of timber production over 100 years in a 41 200-ha study area on the west side of the Coast Range in Oregon. Vegetation maps correspond to (A) the point on the production possibilities frontier with index value=227 and (B) the current landowner simulation with index value=227 (Lichtenstein and Montgomery 2003).

The biodiversity index value achieved in the current management simulation, $B(y) = 227$, was in the midrange of the production possibilities frontier index values. The timber value, however, was only a quarter of the production possibilities frontier values, suggesting that biodiversity, as measured in this study, could be maintained while quadrupling the net present value of timber production. Much of that gain is due to mature forest that is financially overmature on federal land. This is harvested in the first periods of the analysis. But even after it is harvested, the average annual timber harvest on the production possibilities frontier is nearly double that attained in the current management simulation. Harvesting the large blocks of mature conifer forest that occur under current management imposes relatively little cost in terms of forgone biodiversity because most of the species in the set are either generalists (they do not display strong habitat preferences) or they can inhabit other vegetative cover types besides mature forest. Those that do require large patches of mature forest habitat, such as the pileated woodpecker, the marbled murrelet (*Brachyramphus marmoratus*), and the spotted owl, appear better off under current management than in any of the production possibilities outcomes (e.g., the habitat index, X_{st}, was 30% higher for the pileated woodpecker, more than 100% higher for the marbled murrelet, and 14% higher for the northern spotted owl under current management than in the maximum biodiversity production possibilities solution).

5. Discussion

The production possibilities approach is a potentially powerful policy tool. It makes tradeoffs between competing forest uses explicit so that policymakers can make informed judgments. It allows feasible management alternatives to be compared with unconstrained solutions on the production possibilities frontier. The solutions themselves can yield general information about efficient land management that can help land managers develop feasible management alternatives that exploit compatibility more fully.

There are reasons managers may choose to continue current management even when analysis indicates it may be inefficient or inside the production possibilities frontier. First, as noted in the introductory chapter to this book, some argue that although the two-dimensional representation shown in Figure 1 may be a useful way to frame policy and science issues, it is far too simple to be useful in practice. But this practical limitation does not arise from the concept of a production possibilities set, which can be multidimensional. It arises because of difficulties in identifying the appropriate forest uses to model and modeling them with sufficient detail and realism.

The Lichtenstein and Montgomery (2003) study highlights the importance of considering carefully which forest uses are to be modeled. In that study, biodiversity was defined by an index that places no special value on endangered species. The optimization algorithm searches for species for which persistence probability can be improved at a relatively small cost. With the logistic persistence function, these are likely to be species that are classed as "vulnerable to extinction" where the logistic function is steepest and the marginal return to conservation effort will be the largest, rather than those classed as "imperiled" or "critically imperiled" where the logistic function is flat (the so-called "lost causes"). These are also likely to be species with habitat requirements that are compatible with timber production. But these are not the species that are driving current federal forest policy in Oregon and Washington. Instead, the Pacific Northwest Forest Plan establishes late-successional reserves to increase the area of mature conifer forest and provide protection for the endangered species that rely on old mature forest for survival. This suggests extensions to studies of cost-effective biodiversity protection, such as Lichtenstein and Montgomery (2003), or the inclusion of additional indicators of biodiversity (such as focal species) that emphasize different aspects of biodiversity as people perceive it (Spies et al. 2002a), and diversity of vegetative cover (Haight et al. 2000). As noted earlier, value judgments are implicit in both the selection of forest uses to model and the construction of metrics for evaluating those uses.

The irreversibility of some events (e.g., extinction) may also lead land managers and policy analysts to be cautious in implementing management suggested by modeling when there are uncertainties about model parameters, future events, and disturbances that might occur on the site. Furthermore, there are institutional, as well as operational, costs of change. Current management is the result of the long-term historical development of the institutions of private property and public land management.

Nonetheless, forest and conservation policy will most likely succeed when it is perceived to be cost-effective. When resources are fully employed, there is more available for all uses. The production possibilities approach can be a useful tool for summarizing large amounts of data about competing forest uses and the relations between them, identifying cases where the potential benefit from change is great, and describing management that can achieve that benefit.

6. Acknowledgments

I am grateful to Roger Fight and Tom Spies for their helpful comments. The research described in this paper was supported by funding from U.S. Department of Agriculture, Forest Service, Pacific Northwest Research Station and a joint initiative between the National Science Foundation and U.S. Environmental Protection Agency Star Grant R82-6619. No official endorsement by the funding agencies should be inferred.

7. References

Arthaud, G.J., and Rose, D.W. 1996. A methodology for estimating production possibility frontiers for wildlife habitat and timber value at the landscape level. *Canadian Journal of Forest Research*, 26: 2191-2200.

Bailey, J.D., and Tappeiner, J.C. 1998. Effects of thinning on structural development in 40-to-100-year-old Douglas-fir stands in western Oregon. *Forest Ecology and Management*, 108: 99-113.

Bettinger, P., Graetz, D., Boston, K., Sessions, J., and Chung, W. 2002. Eight heuristic planning techniques applied to three increasingly difficult wildlife planning problems. *Silva Fennica*, 36(2): 561-584.

Bettinger, P., Sessions, J., and Boston, K. 1997. Using tabu search to schedule timber harvests subject to spatial wildlife goals for big game. *Ecological Modelling*, 94: 111-123.

Bettinger, P., Sessions, J., and Johnson, K.N. 1998. Ensuring the compatibility of aquatic habitat and commodity production goals in eastern Oregon with a tabu search procedure. *Forest Science*, 44: 96-112.

Bevers, M., Hof, J., Kent, B., and Raphael, M.G. 1995. Sustainable forest management for optimizing multispecies wildlife habitat: a coastal Douglas-fir example. *Natural Resource Modeling*, 9(1): 1-24.

Bigger, D., and Vesely, D. 2000. Life history database for vertebrates of the Willamette River basin. Corvallis, OR: Pacific Wildlife Research, Inc.

Boston, K., and Bettinger, P. 1999. An analysis of Monte Carlo integer programming, simulated annealing, and tabu search heuristics for solving spatial harvest scheduling problems. *Forest Science*, 45: 292-301.

Calkin, D., Montgomery, C.A., Schumaker, N.H., Polasky, S., Arthur, J.L., and Nalle, D.J. 2002. Developing a production possibility set of wildlife species persistence and timber harvest value. *Canadian Journal of Forest Research*, 32(8): 1329-1342.

Connaughton, K.P., and Fight, R.D. 1984. Applying tradeoff analysis to national forest planning. *Journal of Forestry*, 82(11): 680-683.

Glover, F., and Laguna, M. 1997. Tabu search. Boston, MA: Kluwer Academic Publishers.

Gregory, G.R. 1955. An economic approach to multiple use. *Forest Science*, 1(1): 6-13.

Haight, R.G., ReVelle, C.S., and Snyder, S.A. 2000. An integer optimization approach to the probabilistic reserve site selection problem. *Operations Research*, 48(5): 697-708.

Hayes, J.P., Chan, S.S., Emmingham, W.H., Tappeiner, J.C., Kellogg, L.D., and Bailey, J.D. 1997. Wildlife response to thinning young forests in the Pacific Northwest. *Journal of Forestry*, 95(8): 28-33.

Hulse D., Eilers, J.E., Freemark, K.E., White, D., and Hummon, C. 2000. Planning alternative future landscapes in Oregon: evaluating effects on water quality and biodiversity. *Landscape Journal*, 19(2): 1-19.

Jones, J.G., and Schuster, E.G. 1985. An application of discrete optimization for developing economically efficient multiple-use projects. General Technical Report INT-GTR-178. Ogden, UT: U.S. Department of Agriculture, Forest Service, Intermountain Research Station. 16 p.

Kirkpatrick, S., Gelatt, C., and Vecchi, M. 1983. Optimization by simulated annealing. *Science*, 220: 671-680.

Lamberson, R.H., McKelvey, R., Noon, B.R., and Voss, C. 1992. A dynamic analysis of spotted owl viability in a fragmented forest. *Conservation Biology*, 6: 505-512.

Lichtenstein, M.E., and Montgomery, C.A. 2003. Biodiversity and timber in the Coast Range of Oregon: inside the production possibility frontier. *Land Economics*, 79(1): 56-73.

Lockwood, C., and Moore, T. 1992. Harvest scheduling with spatial constraints: a simulated annealing approach. *Canadian Journal of Forest Research*, 23: 468-478.

May, R.M. 1990. Taxonomy as destiny. *Nature*, 347: 129-130.

Metropolis, N., Rosenbluth, A., Rosenbluth, M., Teller, A., and Teller, E. 1953. Equation of state calculations by fast computing machines. *Journal of Chemical Physics*, 21: 1087-1092.

Montgomery, C., Brown, G.M., Jr., and Adams, D.M. 1994. The marginal cost of species preservation: the northern spotted owl. *Journal of Environmental Economics and Management*, 26: 111-128.

Montgomery, C.A., Pollak, R.A., Freemark, K., and White, D. 1999. Pricing biodiversity. *Journal of Environmental Economics and Management*, 38: 1-19.

Muhlenberg, N. 1964. A method for approximating multiple-use optima. *Forest Science*, 10(2): 209-212.

Nalle, D.J., Arthur, J.L., and Sessions, J. 2002. Designing compact and contiguous reserve networks with a hybrid heuristic approach. *Forest Science*, 48(1): 59-68.

Nalle, D.J., Montgomery, C.A., Arthur, J.L., Schumaker, N.H., and Polasky, S. 2001. Locating production possibilities frontiers of economic and ecological outputs over long horizons. Chapter 5. In: D.J. Nalle, Optimizing spatial and temporal aspects of nature reserve design under economic and ecological objectives. Corvallis, OR: Oregon State University. 227 p. Ph.D. dissertation.

Rohweder, M.R., McKetta, C.W., and Riggs, R. 2000. Economic and biological compatibility of timber and wildlife production: an illustrative use of production possibilities frontier. *Wildlife Society Bulletin*, 28(2): 435-447.

Schumaker, N.H. 1998. A users guide to the PATCH model. EPA/600/R-98/135. Corvallis, OR: U.S. Environmental Protection Agency, Environmental Research Laboratory. 120 p.

Shunk, N.J. 2000. Choosing efficient land allocations and forest management regimes for biodiversity. Corvallis, OR: Oregon State University. 92 p. M.S. thesis.

Solow, A., Polasky, S., and Broadus, J. 1993. On the measurement of biological diversity. *Journal of Environmental Economics and Management*, 24: 60-68.

Spies, T.A., Johnson, K.N., Reeves, G.H., Bettinger, P., McGrath, M., and Olsen, K. 2002a. An evaluation of the compatibility of wood production and ecological integrity in the Oregon Coast Range. In: A.C. Johnson, R.W. Haynes, and R.A. Monserud. (Eds.), Congruent management of multiple resources: proceedings from the Wood Compatibility Initiative workshop (pp. 13-25). General Technical Report PNW-GTR-563. Portland, OR: U.S. Department of Agriculture, Forest Service, Pacific Northwest Research Station.

Spies, T.A., Reeves, G.H., and Burnett, K.M. 2002b. Assessing the ecological consequences of forest policies in a multi-ownership province in Oregon. In: J. Liu, and W.W. Taylor (Eds.), Integrating landscape ecology into natural resource management. Cambridge: Cambridge University Press.

Stevens, J.A., and Montgomery, C.A. 2002. A synthesis of multiresource research with application to the Pacific Northwest: multiple use, tradeoffs, and joint production. General Technical Report PNW-GTR-539. Portland, OR: U.S. Department of Agriculture, Forest Service, Pacific Northwest Research Station. 44 p.

Tappeiner, J.C., Huffman, D., Marshall, D., Spies, T.A., and Bailey, J.D. 1997. Density, ages, and growth rates in old-growth and young-growth forests in coastal Oregon. *Canadian Journal of Forest Research*, 27: 638-648.

Vane-Wright, R.K., Humphries, C.J., and Williams, P.H. 1991. What to protect? Systematics and the agony of choice. *Biological Conservation*, 55: 235-254.

Verts, B.J., and Carraway, L.N. 1998. Land mammals of Oregon. Berkeley, CA: University of California Berkeley Press.

Chapter 3

Compatible Management of Red Alder-Conifer Ecosystems in Southeastern Alaska

Mark S. Wipfli
USDA Forest Service, Pacific Northwest Research Station,
1133 N. Western Ave., Wenatchee, WA 98801, USA
Robert L. Deal
USDA Forest Service, Pacific Northwest Research Station,
620 SW Main St., Suite 400, Portland, OR 97205, USA
Paul E. Hennon, Adelaide C. Johnson, Richard T. Edwards, and Toni L. De Santo
USDA Forest Service, Pacific Northwest Research Station,
2770 Sherwood Lane, Juneau, AK 99801, USA
Takashi Gomi
Department of Geography, University of British Columbia,
1984 West Mall, Vancouver, BC VV6T 1Z2, Canada
Ewa H. Orlikowska, Mason D. Bryant, and Mark E. Schultz
USDA Forest Service, Pacific Northwest Research Station,
2770 Sherwood Lane, Juneau, AK 99801, USA
Christian LeSage and Ryan Kimbirauskus
Department of Entomology, Michigan State University, E. Lansing, MI 48824, USA
David V. D'Amore
USDA Forest Service, Pacific Northwest Research Station,
2770 Sherwood Lane, Juneau, AK 99801, USA

1. Research Proposition and Hypotheses

Forest clearcutting has been the primary timber management practice in forests of southeastern Alaska since commercial timber harvesting began in the 1950s, and the dense, even-aged conifer stands that subsequently developed have broad and undesirable consequences for some nontimber

R.A. Monserud, R.W. Haynes and A.C. Johnson (eds.)
Compatible Forest Management, 55-81

resources–most notably, fish and wildlife. Because a few earlier reports suggested that red alder (*Alnus rubra* Bong.) helps mitigate some negative effects of timber harvesting (Deal 1997, Wipfli 1997, Hanley and Barnard 1998), we studied the influence of red alder on a broad set of nontimber resources in young conifer forests (40-year-old; equivalent to early third stage of Appendix 1, Chapter 1) in southeastern Alaska (Figure 1). It is unclear what the ecological functions of red alder are in young forested ecosystems in southeastern Alaska. Key questions include: Does red alder affect forest understory development, tree growth, and timber production? How does red alder influence

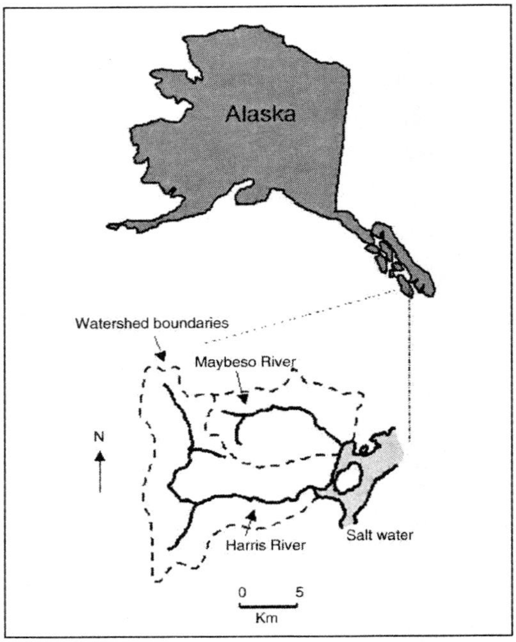

Figure 1. Study area in southeastern Alaska, USA.

food and habitat for fish and wildlife? How does red alder function in stream and riparian habitats? Does red alder influence forest ecosystem diversity and productivity?

This paper is a synthesis of overall findings from a series of studies within an integrated research program. Details of individual studies (methods, data analysis, results) from each research component either have been or are being published in stand-alone papers in various journals, and those papers are identified at various places within this chapter.

The three main themes of the project were (1) influence of red alder on specific ecosystem components, i.e., on vegetation, birds, invertebrates, wood, and fish (Figure 2); (2) flow and use of wood among habitats (Figure 3); and

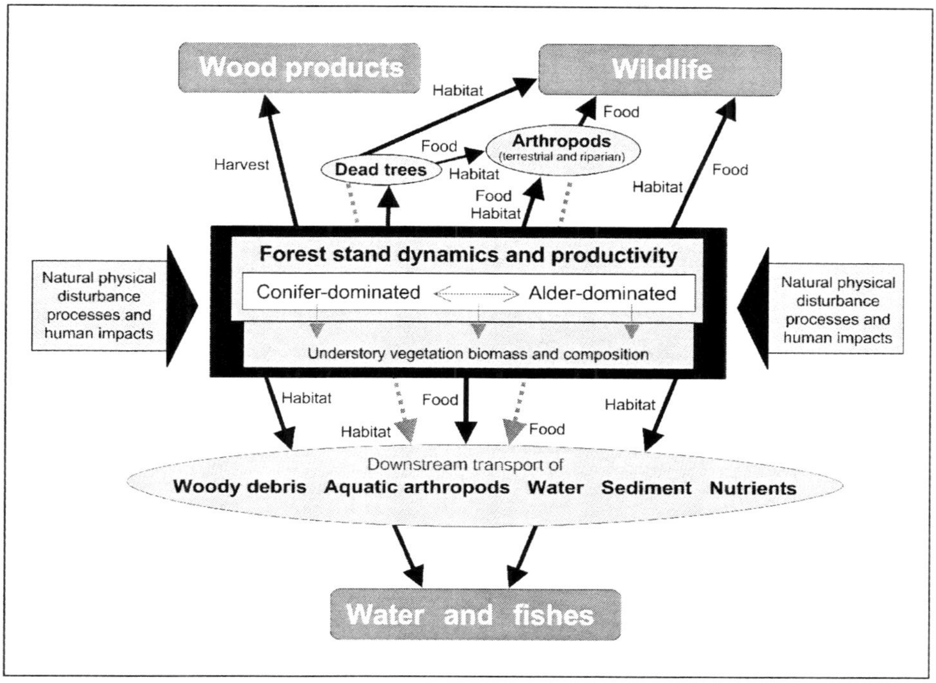

Figure 2. Hypothesized tropic and habitat linkages among prevalent ecosystem components in young-growth stands across a conifer-red alder canopy gradient.

(3) influence of red alder on ecosystem linkages and processes (Figure 4). We hypothesized that red alder increases the abundance and diversity of understory plants, which in turn influence aquatic and terrestrial invertebrate communities (abundance and species richness) and the bird and fish communities that feed on these prey. We also predicted that dead wood, especially red alder, serves important biological functions (e.g., food) affecting invertebrate communities in riparian and aquatic habitats. We investigated interactions in seven major resource areas–geomorphic processes; wood production (live and dead trees); understory vegetation; avian ecology; terrestrial invertebrates; aquatic and riparian ecology; and fish ecology–across non-riparian, riparian, and associated headwater stream habitats.

Despite the unique effects–both positive and negative–of red alder noted in past research, to our knowledge no studies have investigated the ecological functions of red alder concurrently on multiple resources of stream, riparian, and non-riparian habitats in southeastern Alaska. Basic information on silvics, stand growth and yield, tree species mixtures, succession, tree mortality and decay, and understory vegetation is lacking, as well as the effects of red alder on invertebrates, fish, and wildlife. Most information about red alder in Alaska is speculative or based on research from other regions. Further, linkages

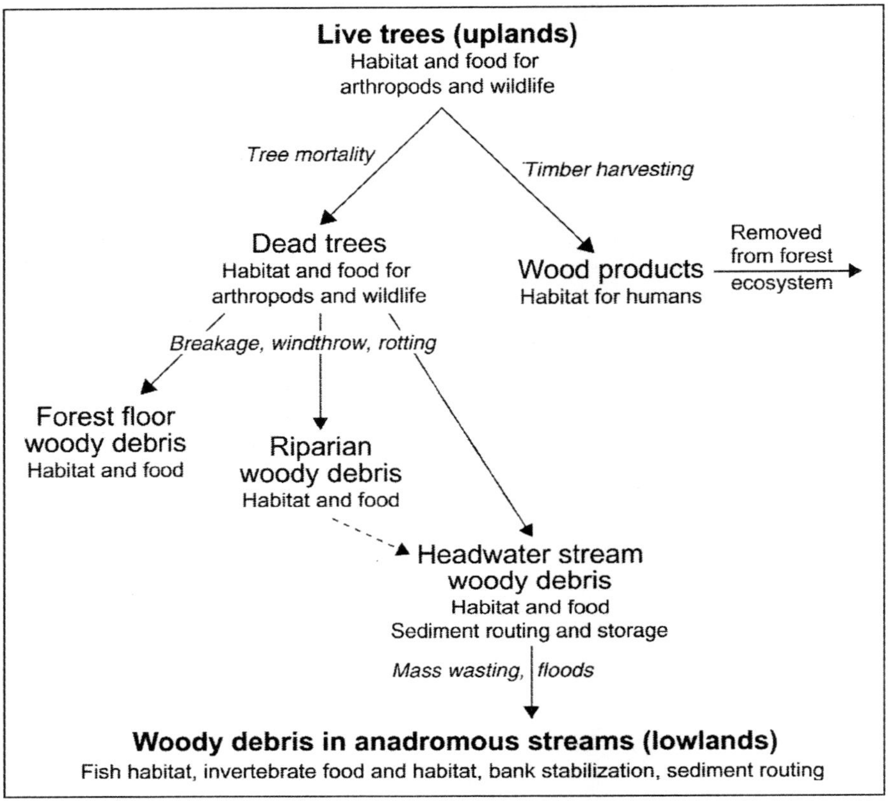

Figure 3. Flow and function of wood among habitats. Processes of delivery in italics.

among resources are important for understanding ecological function and management implications of red alder. Consequently, this interdisciplinary project targeted understanding the function of red alder in young forested ecosystems, and the influence of red alder on the linkages among biologic, hydrologic, and geomorphic processes, and between terrestrial and aquatic environments.

2. Trends in Forest Management in Alaska, Oregon, and Washington

Forest ownership in Alaska, Oregon, and Washington is dominated by Federal and other public ownership where it comprises 72%, 64%, and 55% of total forestland, respectively (Smith et al. 2001). Recent management of these public forests has emphasized policy and guidelines that accelerate the development of late-succession stand characteristics (Franklin et al. 1981, Franklin et al. 1997). Forest management has increasingly emphasized multiple

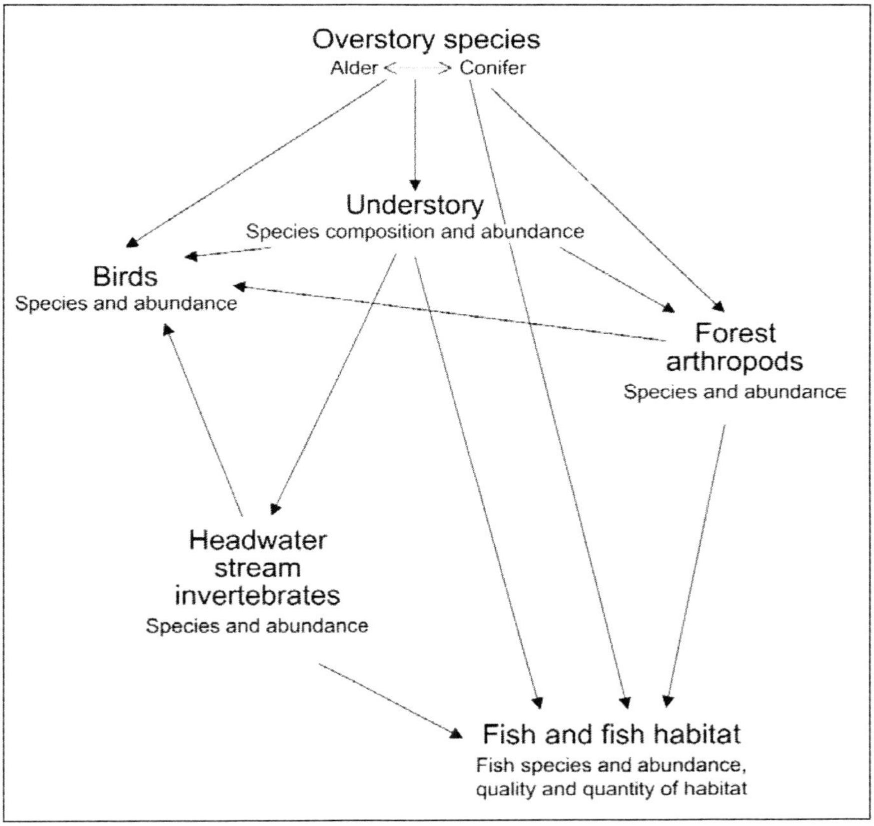

Figure 4. Hypothesized influence of red alder on ecosystem linkages and processes.

resource management, and recent region-wide forest management plans have prescribed policy and silvicultural guidelines to manage forests based on ecosystem management (Salwasser 1994, USDA Forest Service 1997).

New silvicultural systems and mixed-species forests that provide sustainable wood production and protection of other forest values are also being developed in southeastern Alaska (McClellan et al. 2000, Deal et al. 2002, Wipfli et al. 2002). However, even-aged silvicultural systems that utilize clearcutting have been implemented almost exclusively in the western hemlock (*Tsuga heterophylla* (Raf.) Sarg.)–Sitka spruce (*Picea sitchensis* (Bong.) Carr.) forests of southeastern Alaska since the establishment of large-scale timber operations in the early 1950s (Harris and Farr 1974). Little is known about the consequences of forest management practices in this region other than even-aged silvicultural systems for hemlock-spruce forests. Basic information on even-aged forests with red alder is lacking. Most information on red alder in Alaska is speculative or based on research from other regions. The information that does exist for southeastern Alaska indicates that red alder may enhance some ecosystem components (Deal 1997, Wipfli 1997, Hanley and Barnard 1998, Piccolo and Wipfli 2002). It is important for resource managers to understand the ecological function of red alder in even-aged forest ecosystems so they can potentially mitigate some of the negative impacts of clearcutting on forest, wildlife, and aquatic resources.

3. Ecological Consequences of Forest Clearcutting

Forest clearcutting in southeastern Alaska began in the 1950s. Logging in the first two decades was particularly intense; large clearcuts, log dragging, and cutting on steep slopes caused heavy soil disturbance. Nearly 200 000 hectares (ha) in southeastern Alaska were harvested this way (USDA Forest Service 2001) and are now dominated by dense, even-aged conifer stands (Figure 5A). This has had undesirable consequences for some nontimber resources such as fish and wildlife (Wallmo and Schoen 1980, Thedinga et al. 1989, Hanley 1993, Dellasala et al. 1996). Canopy closure in these stands generally occurs 25 to 35 years after cutting, and is followed by a nearly complete elimination of understory vegetation for 100 years or longer (Alaback 1982, 1984; Tappeiner and Alaback 1989). These even-aged conifer forests have simple, uniform stand structures, notably lack the diverse structures of mature forests, and are poorly suited for many wildlife species (Franklin et al. 1981).

The effects of clearcutting on streams include increased short-term primary production, reduced allochthonous (leaf litter) inputs, and loss of woody debris for sediment retention and fish habitat (Bilby and Bisson 1998, Bisson and Bilby 1998). Short-term increases in primary production result from increased light and temporary increases in nutrient inputs from decomposing logging debris (Fisher and Likens 1973). Longer term consequences include reduced invertebrate abundance as canopies overgrow the stream, leading to less food available for fish and other consumers (Piccolo and Wipfli 2002, Wipfli and Musslewhite, in review). Removal of streamside timber also reduces the amount and size of large wood in streams, which leads to a subsequent loss of fish habitat (Bryant 1985, Bisson et al. 1987). Changes in forest structure alter the supply, storage, and transport of woody debris and sediment through mass wasting, blowdown and other forms of tree mortality, and stream channel erosion (Swanston and Marion 1991, Smith et al. 1993, Bilby and Bisson 1998, Johnson et al. 2000). Consequently, there is increasing interest in forest management practices that maintain or enhance productivity and biodiversity, and assure long-term sustainability of forest products, wildlife, and aquatic resources.

Figure 5. Biologically simplified, 40-year-old, even-aged conifer ecosystem (A) and a mixed red alder-conifer ecosystem of the same age (B) showing the more diverse stand structure and abundant understory.

4. Red Alder Distribution and Biology

Red alder is the most common hardwood tree in the Pacific Northwest (Harrington 1990), extending from southern California to southeastern Alaska (Hibbs et al. 1994) (Figure 1, Chapter 1). It is a relatively short-lived, shade-intolerant pioneer with rapid juvenile height growth. Alder, in association with symbiotic *Frankia* spp., fixes atmospheric nitrogen that often improves soil fertility. Red alder can be an aggressive pioneer on avalanche paths, roadsides, log landings, log skid trails, and other areas where mineral soil has been freshly exposed to seedfall, forming both pure and mixed stands (Harris and Farr 1974, Harrington et al. 1994). In Alaska, alder often grows with Sitka spruce, black cottonwood (*Populus balsamifera* L. ssp. *trichocarpa* (Torr. and Gray ex

Figure 6. Distribution and occurrence of red alder in a section of the Maybeso drainage, Prince of Wales Island, southeastern Alaska.

Hook.) Brayshaw) and willows (*Salix* spp.) (Harris and Farr 1974, Ruth and Harris 1979). Extensive soil disturbance caused by mass wasting, road building, and tractor logging since the 1950s in southeastern Alaska has increased the abundance of red alder (Ruth and Harris 1979) (Figure 6).

5. Red Alder as a Tool to Mitigate Negative Effects of Timber Harvest

Red alder appears to benefit forested ecosystems throughout the Pacific Northwest in several ways, and may hold promise for mitigating some negative impacts of timber harvesting. Conifers often grow faster in the presence of alder (Miller and Murray 1978) because of improved soil fertility (Bormann et al. 1994). In southeastern Alaska, understory plant coverage in even-aged, red alder-conifer stands was more than six times that of a nearby even-aged, spruce-hemlock stand (Deal 1997). Hanley and Hoel (1996) found no significant differences in total understory biomass between 40-year-old red alder riparian forests and mature riparian and upland forests, which are considered benchmark standards for suitable vegetation composition and biomass for deer.

The importance of red alder to birds breeding in coniferous forests of the Pacific Northwest (western Oregon and Washington, Coastal British Columbia, and southeastern Alaska) is not clear. Dense conifer stands appear to have low bird diversity (Kessler and Kogut 1985) and abundance (Schwab 1979); higher densities of live deciduous trees have been shown to support greater diversity and abundance of birds in young-growth forests (Gilbert and Allwine 1991, Huff and Raley 1991, McComb 1994). The simple stand structure and sparse forest understory of dense young-growth conifer stands appear to limit the number of nest sites available to birds, and reduce the concealment of nests, a feature associated with successful nesting for some open-cup nesting species (reviewed by Kelly 1993, Martin 1993). Deciduous vegetation in coniferous stands is important for successful nesting for some songbirds in Douglas-fir plantations in British Columbia (Easton and Martin 2002). Red alder may be an important foraging substrate (Gilbert and Allwine 1991, McComb 1994), but this relationship has not been studied. Because insects from foliage are a primary food source for insectivorous birds (Airola 1979, Holmes 1990) and insect biomass is greater in deciduous vegetation (Stiles 1980, Willson and Comet 1996), it follows that birds may be more successful when foraging for insects in red alder trees. Therefore, mixed red alder-conifer stands potentially provide more food, more and safer nest sites, and reduced susceptibility to nest predation for certain bird species.

Riparian management that alters plant communities may change food resources within a stream and its surroundings (Cummins et al. 1989). Little is known about the influence of these developing riparian communities on headwater stream productivity and downstream fish. Red alder could have a variety of impacts on streams, such as changes in light penetration and allochthonous inputs. These effects could lead to changes in aquatic productivity and terrestrial invertebrate inputs to streams (Wipfli 1997). Some riparian tree species contribute more terrestrial invertebrate mass to streams than others (Mason and Macdonald 1982), and red alder appears to support relatively high levels of prey for fish (Wipfli 1997, Piccolo and Wipfli 2002, Allan et al., in press). This is significant because over half of the prey biomass ingested by juvenile salmonids in southeastern Alaska is terrestrial and originates from adjacent riparian vegetation (Wipfli 1997).

Riparian red alder and other vegetation types along streams can also have major influences on stream riparian soil nutrient levels. For example, red alder stands have increased soil nitrogen levels from a few years after harvest until the stand senesces and is replaced by conifers (Luken and Fonda 1983). Groundwater and hyporheic flows move red alder-derived nitrogen into adjacent streams and can increase nutrient stocks and algal standing stocks in streams with sufficient light levels (Fevold 1998).

Habitats immediately adjacent to headwater streams may have a significant influence on foodwebs in fish-bearing reaches downstream. Over 70% of streams within a drainage network are headwater streams (Benda and Dunne 1997), and are sources of sediment, water, nutrients, and organic matter for downstream habitats (Gomi et al. 2002). Transport of organic and inorganic material in headwater streams draining upland forests may affect physical and biological processes in downstream reaches. For instance, sediment produced in headwater systems moves through channel networks and alters channel morphology (Benda and Dunne 1997), creating various channel environments and modifying patterns of riparian structure and hyporheic exchange in downstream reaches (Gregory et al. 1991, Swanson et al. 1998). Movement of detritus and invertebrates from headwaters to downstream food webs may alter productivity, population density, and community structure of stream biota in downstream reaches (Wipfli and Gregovich 2002). Red alder in headwater stream riparian zones may influence physical and biological processes in these systems. Because red alder decays faster than conifers and is a desirable food source for invertebrates (McComb 1994), it likely affects the detritus and invertebrates exported from headwaters to downstream habitats.

An increase in red alder in riparian forest canopies may have both positive and negative impacts on aquatic communities. Higher aquatic productivity and more food for fish and wildlife may be outweighed by loss of fish habitat (Hicks 1989, Bilby and Ward 1991, Ralph et al. 1994) resulting from decreased large woody debris inputs (Grette 1985, Andrus et al. 1988, McDade et al. 1990), wood longevity (Anderson et al. 1978, Benda and Cundy 1990, Keim et al. 2000), and increased sediment loading from fewer large conifers (Megahan 1982). Understanding where red alder occurs both naturally and as a result of management may explain the morphological, functional, and community-level changes that occur in streams as succession from red alder to conifer occurs.

6. Research Approach

We investigated geomorphic processes, wood production (live and dead trees), understory vegetation, avian ecology, terrestrial invertebrate communities, aquatic and riparian ecology, and fish ecology, across non-riparian, riparian, and associated headwater stream habitats in young (40-year-old), regenerating forests. We sampled selected response variables (see Wipfli et al. 2002) within these components in sub-basins encompassing a range of red alder abundance (measured as a proportion of live stand basal area and percentage of canopy coverage) in two adjoining watersheds during 2000-2001 on Prince of Wales Island, southeastern Alaska (Figure 1). We selected two distinct types of study sites: non-riparian where the focus was to evaluate the influence of red alder on vegetation, birds, and invertebrates; and stream-riparian where the focus was the effect of red alder on vegetation, stream nutrients, organic detritus, invertebrates, woody debris, and fish (Figures 5, 6, 7). In

Figure 7. Headwater stream draining an upland forest where it has already transitioned into a lower gradient, fish-bearing reach.

some cases, these were common sites (a given study plot included both site types), and in other cases they were neighboring (but not necessarily contiguous). Wipfli et al. (2002) provide detailed objectives, hypotheses, rationales, methods, and site descriptions.

One independent variable was common to all aspects of sampling: the proportion of red alder (i.e., as a percentage of basal area of live trees, or percentage of canopy cover for stream studies). One of our goals was to construct an empirical model to show how resources might vary along the continuum of red alder dominance. We intended for managers to be able to use this model as a predictive tool for selecting the proportion of red alder to be managed in association with single or multiple resource objectives, at the stand, stream reach, or watershed scale.

To minimize among-site variability, we selected study sites in neighboring Maybeso and Harris watersheds by using composition of tree species (percentage of red alder in young-growth conifer stands) as the primary site selection criterion. These watersheds have large areas of relatively uniform stand age and productivity, a wide range of red alder-conifer mixtures, and are representative of many forests in southeastern Alaska. They were among the first areas harvested in southeastern Alaska when commercial timber harvesting began in the 1950s (USDA Forest Service 1997).

7. Research Findings

Most of the red alder in these drainages occurred in areas with slope gradients >15° (Johnson and Edwards 2002). Approximately 90% of this red alder occurred as a consequence of soil mass wasting (debris avalanches and debris flows) and 10% was a result of cable logging or road building. Streams influenced by mass wasting had significantly more red alder than streams with no apparent soil mass wasting history. About one-third of the red alder occurring in areas with slope gradients <15° was associated with logging roads, most of which are now abandoned. The remaining red alder on the lower gradient slopes typically grew on alluvial fans, landslide deposits, flood-reworked landslide deposits, and areas disturbed by logging tractors (Figure 6).

Soil particle size and moisture, influenced by landslides and slope position, also affected red alder distribution. Red alder was commonly found along zones of shallow subsurface flow or discharge. Larger particle-sized soils, commonly found in the alluvial fans, supported more red alder than sites with finer soils. Mineral soils associated with the various soil disturbances (e.g., abandoned logging roads, landslides, stream corridors) supported more red alder than sites on organic soil material. Lower water retention in coarse-textured soils also favored alder regeneration.

Responses to the presence of red alder varied considerably (Figure 8). Although most of the resources increased with the proportion of red alder (e.g., understory vegetation abundance and browse for wildlife, songbird and aquatic invertebrate abundance, terrestrial arthropod abundance, red alder woody debris in streams), some decreased at higher red alder densities (e.g., total wood production), while still others declined or remained unchanged (e.g., total large woody debris, fish densities). Red alders in non-riparian stands had different tree diameter and height distributions than conifers, and mixed red alder-conifer stands provided greater complexity than stands dominated by conifer or red alder (Deal et al., in review). The diameter distribution of live red alder trees approached a normal distribution, but the distribution was skewed towards smaller trees for conifers. Height growth of red alder was initially much greater than that of conifers, but conifers emerged from red alder overstory 20 to 30 years after logging and averaged 4 to 9 m above the red alder canopy. The more evenly mixed red alder-conifer stands had a wider variety of tree sizes than either the conifer-dominated sites or the red alder-dominated sites. Overall, these mixed red alder-conifer stands provided more heterogeneous structures than even-aged conifer stands that are common after disturbances such as clearcutting. Total understory plant cover and herbaceous

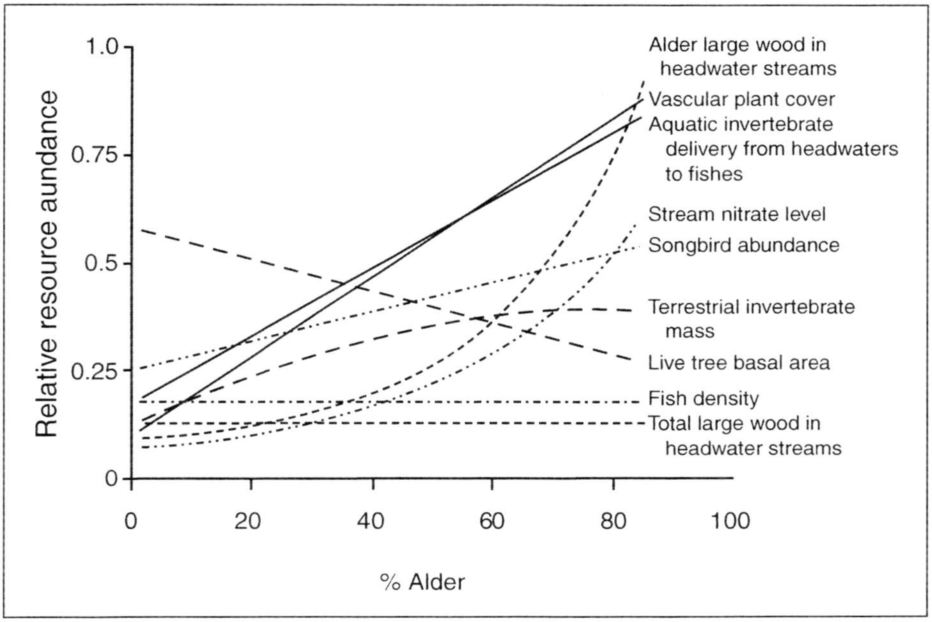

Figure 8. Responses of forest ecosystem components to the proportion of red alder in mixed red alder-conifer forests in southeastern Alaska. Responses start out from the y-axis at an approximate position that would be expected in unharvested systems in southeastern Alaska and progress along the x-axis (red alder gradient) that reflects their relative response across the red alder gradient.

cover increased with increasing proportions of red alder (Figure 8). Understory plant communities were more diverse and contained more biomass in mixed red alder-conifer stands than those in pure conifer stands (Deal et al., in review).

Total basal area (wood production) decreased significantly with increasing proportion of red alder basal area (Deal et al., in review) (Figure 8). In contrast, tree density did not vary with red alder composition. Average diameter of all live conifers significantly decreased with increasing proportion of red alder basal area but there was no significant difference in mean diameter of the largest conifer trees in these stands. These large-diameter trees have important future ecological contributions, particularly for providing large woody debris in streams and riparian habitats.

Some vegetation responses differed between riparian corridors and non-riparian stands. In riparian stands, red alder was highly clumped and was concentrated along stream margins. Tree density, basal area of live and dead trees, and mean diameter of live conifers were not significantly related to the proportion of red alder within riparian forests (Orlikowska et al., in review). The number of large-diameter conifers was similar in riparian stands across the full range of red alder-conifer mixtures. Mean diameter of dead conifers increased slightly with increasing proportion of red alder. Similar to non-riparian sites, most dead trees within riparian stands were small and had died standing.

Few large trees were found dead in any of our sampled forests; most dead trees were small standing stems (Deal et al., in review). This pattern of dead tree occurrence was similar between riparian and non-riparian sites. There was no correlation between increasing proportion of red alder basal area and basal area of dead trees. There was a significant decrease in dead tree density with increasing proportion of red alder basal area. The number of dead conifer stems generally exceeded that of dead red alder even in stands of relatively balanced composition. More than 60% of all dead trees were found standing, with small percentages each of broken boles, down trees with broken roots, and uprooted trees. This pattern of tree death was similar for both conifers and red alders, and for small and large trees. Small dead conifers were by far the most abundant on a stem density basis (Deal et al., in review). The dominant agent of decay in dead red alder and conifer was white rot, for both small and large trees.

The greater diversity and abundance of vascular plants found with more alder improved habitat for small mammals (Figure 8). Deer forage also significantly increased with increasing proportion of red alder basal area (Deal et al., in review), thereby increasing summer deer-carrying capacity (Hanley and Hoel 1996). There was no significant relationship between the percentage of red alder and songbird diversity (De Santo, in prep.). Although most songbird species did not vary with the percentage of red alder, the abundance of the Pacific-slope flycatcher (*Empidonax difficilis*), an aerial insectivore, increased with the percentage of alder in forest stands, resulting in higher overall bird abundance (Figure 8). Nesting density of all songbird species combined was also positively associated with the percentage of red alder, and given that nesting success did not differ between stand types, total stand-level nesting success was presumably higher in the red alder stands (De Santo, in prep.).

The biomass of flying insects, primarily Diptera, Lepidoptera, and Plecoptera, significantly increased as the percentage of basal area of red alder increased (Figure 8) (Schultz, in prep.). Malaise trap catch of winged invertebrates in the nearly pure conifer stands was 60% and 35% of that caught in stands with a high proportion (70%) of red alder in 2000 and 2001, respectively. Further, riparian red alder leaf litter samples had higher invertebrate taxonomic richness than samples of conifer leaf litter. Highly decayed conifer wood had higher abundances of invertebrates within the wood than less decayed conifer or red alder in any decay class. And highly decayed red alder wood supported greater taxonomic richness than either conifer decay class (LeSage 2003).

Streams with more red alder in their adjacent riparian zones transported significantly more invertebrates and organic detritus from headwaters to downstream, fish-bearing habitats than streams with predominantly conifers along their banks (Figure 8). Streams at the higher end of the red alder density continuum delivered about four times more invertebrate biomass than streams canopied predominantly with conifers (Wipfli and Musslewhite, in review). Aquatic invertebrate mass composed 3/4 of the fluvial export from headwaters to fish-bearing habitats, while terrestrial species composed 1/4 of the export. Midges were the most abundant invertebrates in the stream transport, followed by baetis mayflies and nemourid stoneflies. The amount of organic detritus transported downstream was also significantly greater in streams with more red alder growing along their banks than streams with little red alder (Wipfli and Musslewhite, in review).

Nitrate concentrations within study streams varied from 4 to 387 μg nitrate-nitrogen per liter (Figure 8) (Johnson and Edwards 2002). Streams with the highest nitrate concentrations had high riparian red alder basal area coverage, but not all streams with extensive red alder had high nitrate concentrations. Streams originating in alpine areas generally had larger catchments and higher discharge, which apparently diluted the nitrogen originating from red alder-influenced soils. Those streams had lower nitrate concentrations than smaller streams with similar alder densities (Johnson and Edwards 2002).

The number of red alder, large woody debris pieces in streams increased with increasing proportion of red alder in the riparian stand (Figure 8) (Gomi et al., in review). However, significant relationships between the volume of large woody debris and sediment stored behind these pieces and the proportion of red alder in riparian zones were not found. Woody debris and sediment supply appeared to be governed by other factors such as recruitment, transport capacity, and decomposition rates. Recruitment of large woody debris from the hillslopes to the stream channels was related to a number of factors including valley topography (e.g., valley incision), past logging practices, and natural physical disturbance history. Much woody debris from the original old-growth forest (Appendix 1, Chapter 1-fifth development stage) still persisted in the stream channels and provided physical function in streams (Gomi et al., in review).

Red alder appeared to have no direct relationship with salmonid densities (Figure 8) (Bryant et al., in review). Dolly Varden char (*Salvelinus malma* Walbaum) and juvenile coho salmon (*Oncorhynchus kisutch* Walbaum) densities were related to stream habitat conditions: including gradient, pool abundance, and the amount of large wood. Dolly Varden char were found in all reaches of the streams and in all streams where fish were present, whereas juvenile coho salmon were found in some but not all streams, and were most abundant in the lower gradient reaches. The density of all species decreased as gradient increased. The relatively low densities of all fish species combined with the limited amount of habitat suggests that the amount and quality of habitat were the primary factors limiting salmonid populations in these streams.

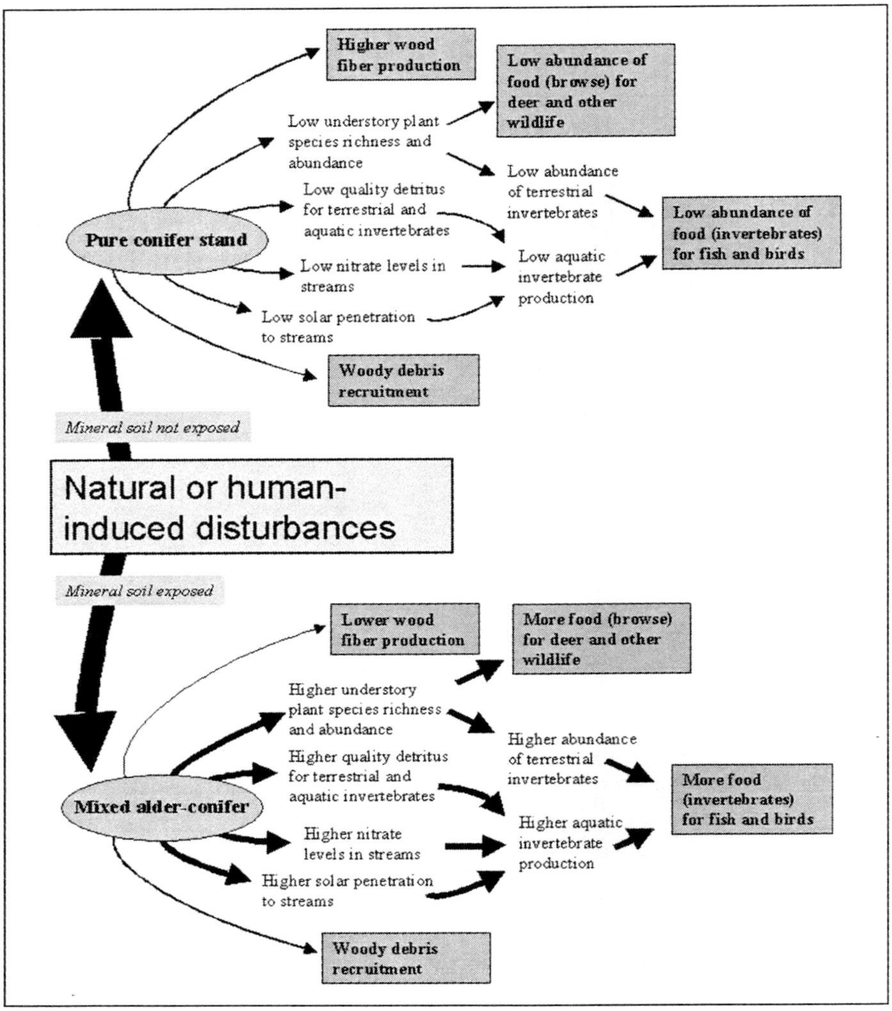

Figure 9. Disturbance-induced trajectories of conifer or mixed alder-conifer forests, and the subsequent influence on the linkages within associated communities. Arrow thickness represents relative magnitude of response.

8. Integration and Synthesis

Young, regenerating forests progress along one of several trajectories following forest clearcutting or other disturbance (Figure 9). If the trajectory is toward a pure or nearly pure conifer forest (resulting from lack of adequate soil disturbance), the subsequent forest community develops a sparse and non-diverse plant understory, which in turn leads to little wildlife browse and sparse foliage for herbivorous invertebrates (Figures 5A, 9). Nearly pure conifer forests also provide less nitrate and light for stream producers and consumers, which can lead to fewer aquatic invertebrates in the associated headwater streams, ultimately providing less food for birds and downstream fish. They also provide fewer nesting sites for songbirds, and support lower songbird density.

Alternatively, if the forest regeneration trajectory includes red alder (that develops as a result of adequate soil disturbance), a markedly different community develops. This regenerating forest is characterized by a more species-rich and abundant understory plant community (habitat and browse for wildlife such as deer), more terrestrial invertebrates, higher nitrate levels and more light reaching streams, which in turn leads to more aquatic invertebrates that are prey for selected bird species and downstream fish (Figures 5B, 9). This second scenario appears to support a greater overall diversity and productivity in the forest ecosystem.

An interesting pattern of red alder regeneration can be observed throughout harvested landscapes of southeastern Alaska: the tree is common in young-growth stands that regenerated before 1970 and is rare or absent on sites regenerated after that date, except along headwater streams and in landslide areas. Prior to 1970, harvesting and yarding practices commonly used tractors to drag logs across forest floors and streams. After 1970, there was greater use of high-lead cable logging systems that partially suspended logs above ground and stream surfaces (J. Russell, USDA, Forest Service, Sitka, Alaska, personal communication, 2002). Much less soil disturbance occurred in the post-1970 logging, reducing the amount of subsequent red alder regrowth. The current trend towards partial harvest or clearcutting with helicopter yarding will also result in less soil disturbance. Harvests that avoid streams and steep slopes will further reduce the chance for red alder regeneration. Although pre-1970 logging practices were later viewed as destructive to components of the forest ecosystem, the promotion of red alder regeneration clearly had some compensatory effects on forest ecosystem processes that are missing from current practices.

We documented that the life history of red alder in southeastern Alaska operates at a faster pace than conifers, which is consistent with what has been found elsewhere (Franklin and Pechanec 1968, Miller and Murray 1978). The rapid, early growth of red alder allows it to initially outgrow conifers in height, although conifers eventually overtopped red alders (Deal and Orlikowska 2002). Because of their relative intolerance to shade, the small red alder trees in our 40-year-old forests died while conifers of the same size persisted, resulting in different diameter and height distributions for red alders and conifers, both live and dead trees. The larger red alders will probably eventually succumb to low light levels and die at stand ages 60 to 100 years, and overstory conifer death will likely begin much later. These different stand development patterns of red alders and conifers appear to provide a wider range of forest structures in mixed red alder-conifer stands than expected for pure conifer or pure red alder stands, and benefit the plant and animal communities living there.

The forest structures of mixed red alder-conifer stands versus pure conifer stands have significant effects on forest ecosystem resources. Mature forests have complex forest structures including a wide range of tree sizes, snags, logs, and multiple canopy layers (Franklin et al. 1981). A major reason for improved understory plant diversity and abundance in partially cut stands is the wider range of tree size in these stands, compared to the simpler even-aged forests that develop after clearcutting (Deal and Tappeiner 2002). Mixed red alder-conifer stands also have a wide range of tree sizes and contain more open forest canopies than even-aged conifer forests, providing the complex and diverse forest structures important for wildlife.

Red alder stands may leave a legacy in these forests long after they have died and decayed. Gaps created where one or more large red alders have died leave conifer stands more open (Newton and Cole 1994). These gaps may also develop a second canopy by invasion of new trees. Longer term stand development is less certain, but without additional soil disturbances and enlargement of openings, western hemlock or other conifers are more likely to colonize the opening than red alder. The resulting complex canopy and open forest structure may occur decades earlier than in pure conifer forests where overstory tree death may not begin until stands are 150 years old. The longer term benefit of red alder may be that its death accelerates the transition to a mature forest structure.

The fate of dead trees, and the pathways from tree to stream woody debris are complex (Bisson et al. 1987, Benda and Cundy 1990, Bilby and Bisson 1998) but we gained more insight on them for both red alder and conifer. Nearly all of the dead alder and conifer wood recruited from these young-growth forests was small. This small wood entered streams and riparian forest floors, but appeared to play a minor role in retaining and routing stream sediment (Gomi et al., in review). Its main function is probably biological, providing substrate and food for invertebrates as in the case with leaf litter (Richardson 1992). The legacy wood from mature trees before the forests were cut provided physical functions (Figure 10). Although the larger live red alder and conifer trees in these stands may become future woody debris, it will take decades before the larger dominant and co-dominant trees begin to die. Because of their different life histories, we anticipate that larger red alders will die and fall to the forest floor and into streams well before larger conifers do, serving as an intermediate source of woody debris.

Red alders apparently serve important biological functions in forest ecosystems by providing nitrogen, high-quality organic matter for decomposers and invertebrates, and better light penetration through the forest canopy than pure conifer stands (Hibbs et al. 1994, McComb 1994). We observed more invertebrates in streams with more riparian red alder; and red alder stems and

Figure 10. Red alder and conifer woody debris provides both physical and biological functions in streams and their riparian zones.

leaf litter supported greater invertebrate species richness than conifers. The smaller wood pieces–typically red alder–provided more habitat for invertebrates than larger pieces because of their higher surface-to-volume ratios. Management that balances the biological benefits of red alder with the physical benefits of larger conifers in young successional forests should offer the most benefits–both biological and physical–to stream and forest ecosystems.

Red alder increased the abundance of food in its immediate vicinity and in downstream reaches. Within red alder stands, vegetation for both invertebrate and vertebrate herbivores was increased by the presence of red alder, an observation consistent with previous work in southeastern Alaska (Hanley 1996, Deal 1997, Wipfli 1997, Hanley and Barnard 1998, Piccolo and Wipfli 2002). Increased prey availability likely affects many more species, such as bats and other small mammals, birds, and other wildlife in these young forested ecosystems. Also noteworthy was the trophic influence that upland red alder had on more distant habitats. Fish habitats downstream of red alder-dominated reaches were supplied with invertebrates (food for fish) and organic detritus (food for invertebrates) from upstream headwaters (Piccolo and Wipfli 2002, Wipfli and Gregovich 2002). And aquatic food webs downstream of red alder received more prey and detritus than those below conifer-dominated reaches (Wipfli and Musslewhite, in review).

Although these results are a snapshot in time approximately 40 years after timber harvest, the influence that these differing forest regeneration trajectories have is dramatic. Divergence between red alder- and conifer-dominated stands probably occurred early in post harvest development, and will likely continue for at least the next several decades before red alder die out (Deal and Orlikowska 2002). The influence of red alder on soil nitrogen supplies and plant associates may linger decades beyond the life span of the red alder itself.

9. Management Implications

These results suggest that red alder has an important ecological function in forested ecosystems in southeastern Alaska and corroborates other findings (Pechanec and Franklin 1968, Tarrant et al. 1983, McComb et al. 1993, Deal 1997, Wipfli 1997, Hanley and Barnard 1998). Several ecosystem responses showed positive relationships with increasing proportions of red alder, while a few did not. There are both benefits and tradeoffs of mixed red alder-conifer stands, but from an ecosystem perspective, benefits of including red alder in young-growth conifer forests seem to outweigh the negative aspects of red alder occurrence. Rather than viewing red alder as an undesirable, early-successional tree and managing it accordingly, there may be instances where managers can manipulate sites or existing vegetation to favor the establishment or maintenance of red alder. Altering the composition of young-growth forests to encourage a component of red alder may help improve habitat and food resources for deer, songbirds, and fish thereby offsetting some of the negative consequences and main criticisms of timber harvesting, particularly clearcutting.

Based on these findings, how would a resource manager manipulate the composition of a stand? Opportunities for increasing red alder in forests involve encouraging its reproduction (e.g., seeding, planting, regeneration through soil disturbance) in newly harvested or disturbed areas, and thinning to favor alder where the species already occurs. If deliberate red alder regeneration could be incorporated into present harvest plans while minimizing the mass wasting and sedimentation that resulted from the large-scale disturbances that favored red alder growth in the past, managers might be able to exploit the positive benefits of red alder and improve structure and function of young forests. Deliberately disturbing the soil surface enough to promote red alder regeneration at an appropriate scale and pattern, depending upon local site conditions could be beneficial in established even-aged forests that contain little or no red alder.

Although these results are from research conducted in the temperate rainforests of southeastern Alaska, they have much broader geographic and ecologic implications. Red alder is widely distributed throughout the west coast of North America (Harrington 1990), and other species of alder occur throughout the world (Murai 1964). These alder species undoubtedly have biological properties similar to red alder in southeastern Alaska, and presumably have similar ecological effects in other ecoregions and ecosystems where they persist. Developing or maintaining alder beyond the young regenerating forested ecosystems as demonstrated here will likely affect associated plant and animal species, populations and communities, and ecosystem structure and function.

10. Acknowledgments

We thank Ellen Anderson, Dave Bassett, Becca Cameron, Alison Cooney, Amy Cunkleman, John Frisch, Jake Musslewhite, Sarah Newton, Jeff Nichols, Kim Obermeyer, Yuho Okada, Aaron Rinker, Troy Seiler, Tanya Skurski, Dave Sperry, Renee Vincent, Robert Walton, Brenda Wright, and Nik Zymonas for field, laboratory, editorial, and technical assistance. We acknowledge the Craig Ranger District of the USDA Forest Service for logistical support. We also thank Sandy Boyce, Bob Bugert, Richard Haynes, Winifred Kessler, Bob Monserud, and Richard Woodsmith for constructive reviews of earlier drafts. Finally, we acknowledge funding from the Wood Compatibility Initiative (WCI), Pacific Northwest Research Station, Portland, Oregon.

11. References

Airola, D.A. 1979. Foraging and habitat relations of insectivorous birds in a managed Sierra Nevada mixed conifer forest. Berkeley, CA: University of California. 59 p. M.S. thesis.

Alaback, P.B. 1982. Dynamics of understory biomass in Sitka spruce-western hemlock forests of southeast Alaska. *Ecology*, 63: 1932-1948.

Alaback, P.B. 1984. Secondary succession following clearcut logging in the Sitka spruce-western hemlock forests of southeast Alaska: implications for management. General Technical Report PNW-GTR-173. Portland, OR: U.S. Department of Agriculture, Forest Service. Pacific Northwest Research Station.

Allan, J.D., Wipfli, M.S., Caouette, J.P., Prussian, A., and Rodgers, J. [In press]. Influence of streamside vegetation on terrestrial invertebrate subsidies to salmonid food webs. *Canadian Journal of Fisheries and Aquatic Sciences*.

Anderson, N.H., Sedell, J.R., Roberts, L.M., and Triska, F.J. 1978. The role of aquatic invertebrates in processing of wood debris in coniferous forest streams. *American Midland Naturalist*, 100: 64-82.

Andrus, C.W., Long, B.A., and Froelich, H.A. 1988. Woody debris and its contribution to pool formation in a coastal stream 50 years after logging. *Canadian Journal of Fisheries and Aquatic Sciences*, 45: 2080-2086.

Benda, L.E., and Cundy, T.W. 1990. Predicting deposition of debris flows in mountain channels. *Canadian Geotechnical Journal*, 27: 409-417.

Benda, L.E., and Dunne, T. 1997. Stochastic forcing of sediment routing and storage in channel networks. *Water Resource Research*, 33: 2865-2880.

Bilby, R.E., and Bisson, P.A. 1998. Function and distribution of large woody debris. In: R.J. Naiman, and R.E. Bilby (Eds.), River ecology and management (pp. 324-346). New York: Springer-Verlag.

Bilby, R.E., and Ward, J.W. 1991. Characteristics and function of large woody debris in streams draining old-growth, clearcut, and second-growth forests in southwestern Washington. *Canadian Journal of Fisheries and Aquatic Sciences*, 48: 2499-2508.

Bisson, P.A., and Bilby, R.E. 1998. Organic matter and trophic dynamics. In: R.J. Naiman, and R.E. Bilby (Eds.), River ecology and management (pp. 373-398). New York: Springer-Verlag.

Bisson, P.A., Bilby, R.E., Bryant, M.D., Dollof, C.A., Grette, G.B., House, R.A., Murphy, M.L., Koski, K.V., and Sedell, J.R. 1987. Large woody debris in the Pacific Northwest: past, present, and future. In: E.O. Salo, and T.W. Cundy (Eds.), Streamside management: forestry and fishery interactions (pp. 143-190). Contribution 57. Seattle, WA: Institute of Forest Resources, University of Washington.

Bormann, B.T., Cromack, K., and Russell, W.O. 1994. Influences of red alder on soils and long-term ecosystem productivity. In: D.E. Hibbs, D.S. DeBell, and R.F. Tarrant (Eds.), The biology and management of red alder (pp. 47-56). Corvallis, OR: Oregon State University Press.

Bryant, M.D. 1985. Changes 30 years after logging in large woody debris, and its use by salmonids. In: R.R. Johnson, C.D. Ziebell, D.R. Patton, P.F. Ffolliott, and R.H. Hamre (Eds.), Riparian ecosystems and their management: reconciling conflicting uses. General Technical Report RM-120. U.S. Department of Agriculture, Forest Service, Rocky Mountain Research Station.

Bryant, M.D., Zymonas, N., and Wright, B. [In review]. Salmonids on the fringe: distribution, habitat use, and response of salmonids to upslope riparian forest composition in high gradient headwater streams, southeast Alaska.

Cummins, K.W., Wilzbach, M.A., Gates, D.M., Taliaferri, J.B., and Perry, W.B. 1989. Shredders and riparian vegetation: leaf litter that falls into streams influences communities of stream invertebrates. *BioScience*, 39: 24-30.

De Santo, T.L. [In prep.]. Occurrence and abundance of songbirds in young-growth forests on Prince of Wales Island, Alaska.

Deal, R.L. 1997. Understory plant diversity in riparian alder-conifer stands after logging in southeast Alaska. Research Note PNW-RN-523. Portland, OR: U.S. Department of Agriculture, Forest Service, Pacific Northwest Research Station.

Deal, R.L., Hennon, P.H., Orlikowska, E.H., and D'Amore, D.V. [In review]. Structure and dynamics of mixed red alder-conifer young-growth forests of southeast Alaska. *Canadian Journal of Forest Research*.

Deal, R.L., and Orlikowska, E.H. 2002. Development of mixed red alder-conifer stands in southeast Alaska. In: A.C. Johnson, R.W. Haynes, and R.A. Monserud (Eds.), Congruent management of multiple resources: proceedings from the wood compatibility workshop. General Technical Report PNW-GTR-563. Portland, OR: U.S. Department of Agriculture, Forest Service, Pacific Northwest Research Station.

Deal, R.L., and Tappeiner, J.C. 2002. The effects of partial cutting on stand structure and growth of western hemlock-Sitka spruce stands in southeast Alaska. *Forest Ecology and Management*, 159: 173-186.

Deal R.L., Tappeiner J.C., and Hennon, P.E. 2002. Developing silvicultural systems based on partial cutting in western hemlock—Sitka spruce stands in southeast Alaska. *Forestry*, 75(4): 425-431.

Dellasala, D.A., Hagar, J.C., Engel, K.A., McComb, W.C., Fairbanks, R.L., and Campbell, E.G. 1996. Effects of silvicultural modifications of temperate rainforest on breeding and wintering bird communities, Prince of Wales Island, Southeast Alaska. *Condor*, 98: 706-721.

Easton, W.E., and Martin, K. 2002. Effects of thinning and herbicide treatments on nest-site selection by songbirds in young managed forests. *Auk*, 119: 685-694.

Fevold, K. 1998. Sub-surface controls on the distribution of benthic algae in floodplain back channel habitats of the Queets River. Seattle, WA: University of Washington. 72 p. M.S. thesis.

Fisher, S.G., and Likens, G.E. 1973. Energy flow in Bear Brook, New Hampshire: an integrative approach to stream ecosystem metabolism. *Ecological Monographs*, 43: 421-439.

Franklin, J.F., Berg, D.R., Thornburgh, D.A, and Tappeiner, J.C. 1997. Alternative silvicultural approaches to timber harvesting: variable retention harvest systems. In: K.A. Kohm, and J.F. Franklin (Eds.), Creating a forestry for the 21st century (pp. 111-139). Washington, DC: Island Press.

Franklin, J.F., Cromack, K.J., Denison, W., McKee, A., Maser, C., Swanson, F., and Juday, G. 1981. Ecological characteristics of old-growth Douglas-fir forests. General Technical Report. PNW-GTR-118. Portland, OR: U.S. Department of Agriculture, Forest Service, Pacific Northwest Research Station.

Franklin, J.F., and Pechanec, A.A. 1968. Comparison of vegetation in adjacent alder, conifer, and mixed alder-conifer communities. I. Understory vegetation and stand structure. In: J.M. Trappe, J.F. Franklin, R.F. Tarrant, and G.M. Hansen (Eds.), Biology of alder (pp. 37-43). Portland, OR: U.S. Department of Agriculture, Forest Service, Pacific Northwest Forest and Range Experiment Station.

Gilbert, F.F., and Allwine, R. 1991. Spring bird communities in the Oregon Cascade Range. In: L.F. Ruggiero, K.B. Aubry, A.B. Carey, and M.H. Huff (Tech. coords.), Wildlife and vegetation of unmanaged Douglas-fir forests. General Technical Report PNW-GTR-285. Portland, OR: U.S. Department of Agriculture, Forest Service, Pacific Northwest Research Station.

Gomi, T., Johnson, A.C., Deal, R.L., Hennon, P.E., Orlikowska, E.H., and Wipfli, M.S. [In review]. Mixed red alder-conifer riparian forests of southeastern Alaska: Implications for the accumulations of woody debris, organic matter, and sediment in headwater streams.

Gomi, T., Sidle, R.C., and Richardson, J.S. 2002. Understanding processes and downstream linkages of headwater systems. *BioScience*, 52: 905-916.

Gregory, S.V., Swanson, F.J., McKee, W.A., and Cummins, K.W. 1991. An ecosystem perspective of riparian zones. *BioScience*, 41: 540-551.

Grette, G.B. 1985. The role of large organic debris in juvenile salmonid rearing habitat in small streams. Seattle, WA: University of Washington. 105 p. M.S. thesis.

Hanley, T.A. 1993. Balancing economic development, biological conservation, and human culture: the Sitka black-tailed deer (*Odocoileus hemionus sitkensis*) as an ecological indicator. *Biological Conservation*, 66: 61-67.

Hanley, T.A. 1996. Small mammals of even-aged, red alder-conifer forests in southeastern Alaska. *Canadian Field-Naturalist*, 10: 626-629.

Hanley, T.A., and Barnard, J.C. 1998. Red alder, *Alnus rubra*, as a potential mitigating factor for wildlife habitat following clearcut logging in southeastern Alaska. *Canadian Field-Naturalist*, 112: 647-652.

Hanley, T.A., and Hoel, T. 1996. Species composition of old-growth and riparian Sitka spruce-western hemlock forests in southeastern Alaska. *Canadian Journal of Forest Research*, 26: 1703-1708.

Harrington, C.A. 1990. *Alnus rubra* Bong. (red alder). In: R.M. Burns, and B.H. Honkala (Eds.), Silvics of North America. (Vol. 2 pp. 116-123). Washington, DC: U.S. Department of Agriculture, Forest Service.

Harrington, C.A., Zasada, J.C., and Allen, E.A. 1994. Biology of red alder (*Alnus rubra* Bong.). In: D.E. Hibbs, D.S. DeBell, and R.F. Tarrant (Eds.), The biology and management of red alder (pp. 3-22). Corvallis, OR: Oregon State University Press.

Harris, A.S., and Farr, W.A. 1974. The forest ecosystem of southeast Alaska. General Technical Report PNW-GTR-25. Portland, OR: U.S. Department of Agriculture, Forest Service, Pacific Northwest Research Station.

Hibbs, D.E., DeBell, D.S., and Tarrant, R.F. 1994. The biology and management of red alder. Corvallis, OR: Oregon State University Press. 256 p.

Hicks, B.J. 1989. The influence of geology and timber harvest on channel morphology and salmonid populations in Oregon Coast Range streams. Corvallis, OR: Oregon State University. 199 p. Ph.D. dissertation.

Holmes, R.T. 1990. The structure of a temperate deciduous forest bird community: variability in time and space. In: A. Keast (Ed.), Biogeography and ecology of forest bird communities (pp. 121-139). Hague, Netherlands: SPB Publishing.

Huff, M.H., and Raley, C.M. 1991. Regional patterns of diurnal breeding bird communities in Oregon and Washington. In: Wildlife and vegetation of unmanaged Douglas-Fir forests. General Technical Report PNW-GTR-285. Portland, OR: U.S. Department of Agriculture, Forest Service, Pacific Northwest Research Station.

Johnson, A.C., and Edwards, R.T. 2002. Physical and chemical processes in headwater channels with red alder. In: A.C. Johnson, R.W. Haynes, and R.A. Monserud (Eds.), Congruent management of multiple resources: proceedings from the wood compatibility workshop. (pp. 101-108). General Technical Report PNW-GTR-563. Portland, OR: U.S. Department of Agriculture, Forest Service, Pacific Northwest Research Station.

Johnson, A.C., Swanston, D., and McGee, K.E. 2000. Landslide initiation, runout, and deposition within clearcuts and old-growth forests of Alaska. *Journal of American Water Resources Association*, 6: 17-30.

Keim, R.F., Skaugset, A.E., and Bateman, D.S. 2000. Dynamics of coarse woody debris placed in three Oregon streams. *Forest Science*, 46: 13-22.

Kelly, J.P. 1993. The effect of nest predation on habitat selection by dusky flycatchers in limber pine-juniper woodland. *Condor*, 95: 83-93.

Kessler, W.B., and Kogut, T.E. 1985. Habitat orientations of forest birds in southeastern Alaska. *Northwest Science*, 59: 58-65.

LeSage, C.M. 2003. Headwater riparian arthropod community changes in response to red alder stand composition in southeast Alaska. E. Lansing, MI: Michigan State University, M.S. thesis.

Luken, J.O., and Fonda, R.W. 1983. Nitrogen accumulation in a chronosequence of red alder communities along the Hoh River, Olympic National Park, Washington. *Canadian Journal of Forest Research*, 13: 1228-1237.

Martin, T.E. 1993. Nest predation among vegetation layers and habitat types: revising the dogmas. *American Naturalist*, 141: 898-913.

Mason, C.F., and Macdonald, S.M. 1982. The input of terrestrial invertebrates from tree canopies to a stream. *Freshwater Biology*, 12: 305-311.

McClellan, M.H., Swanston, D.N., Hennon, P.E., Deal, R.L., De Santo, T.L., and Wipfli, M.S. 2000. Alternatives to clearcutting in the old-growth forests of southeast Alaska: study plan and establishment report. General Technical Report PNW-GTR-494. Portland, OR: U.S. Department of Agriculture, Forest Service, Pacific Northwest Research Station.

McComb, W.C. 1994. Red alder: interactions with wildlife. In: D.E. Hibbs, D.S. DeBell, and R.F. Tarrant (Eds.), The biology and management of red alder (pp. 131-138). Corvallis, OR: Oregon State University Press.

McComb, W.C., Chambers, C.L., and Newton, M. 1993. Small mammal and amphibian communities and habitat associations in red alder stands, Central Oregon Coast Range. *Northwest Science*, 67: 181-188.

McDade, M.H., Swanson, F.J., McKee, W.A., Franklin, J.F., and Van Sickle, J. 1990. Source distances for coarse woody debris entering small streams in western Oregon and Washington. *Canadian Journal of Forest Research*, 20: 326-330.

Megahan, W.F. 1982. Channel sediment storage behind obstructions in forested drainage basins draining the granitic bedrock of the Idaho batholith. In: F.J. Swanson, R.J. Janda, T. Dunne, and D.N. Swanston (Eds.), Sediment budgets and routing in forested drainage basins. General Technical Report PNW-GTR-141. Portland, OR: U.S. Department of Agriculture, Forest Service, Pacific Northwest Research Station.

Miller, R.E., and Murray, M.D. 1978. The effect of red alder on growth of Douglas-fir. In: D.G. Briggs, D.S. DeBell, and W.A. Atkinson (Eds.), Utilization and management of alder. General Technical Report PNW-GTR-70. Portland, OR: U.S. Department of Agriculture, Forest Service, Pacific Northwest Research Station.

Murai, S. 1964. Phytotaxonomical and geobotanical studies on genus *Alnus* in Japan. (III) Taxonomy of whole world species and distribution of each sect. Government Forest Experiment Station Bulletin (Japan), 171: 1-107.

Newton, M., and Cole, E.C. 1994. Stand development and successional implications: pure and mixed stands. In: D.E. Hibbs, D.S. DeBell, and R.F. Tarrant (Eds.), The biology and management of red alder (pp. 106-115). Corvallis, OR: Oregon State University Press.

Orlikowska, E.H., Deal, R.L., Hennon, P.E., and Wipfli, M.S. [In review]. Stand structure of red alder-conifer riparian forests in southeastern Alaska.

Pechanec, A.A., and Franklin, J.F. 1968. Comparison of vegetation in adjacent alder, conifer, and mixed alder-conifer communities. Epiphytic, epixylic, and epilithic cryptogam. In: M. Trappe, J.F. Franklin, R.F. Tarrant, and G.M. Hansen (Eds.), Biology of alder (Vol. 2 pp. 85-98). Portland, OR: U.S. Department of Agriculture, Forest Service, Pacific Northwest Range and Experiment Station.

Piccolo, J.J., and Wipfli, M.S. 2002. Does red alder (*Alnus rubra*) in upland riparian forests elevate macroinvertebrate and detritus export from headwater streams to downstream habitats in southeastern Alaska? *Canadian Journal of Fisheries and Aquatic Sciences*, 59: 503-513.

Ralph, S.C., Poole, G.C., Conquest, L.L., and Naiman, R.J. 1994. Stream channel morphology and woody debris in logged and unlogged basins of western Washington. *Canadian Journal of Fisheries and Aquatic Sciences*, 51: 37-51.

Richardson, J.S. 1992. Food, microhabitat, or both? Macroinvertebrate use of leaf accumulations in a montane stream. *Freshwater Biology*, 27: 169-176.

Ruth, R.H., and Harris, A.S. 1979. Management of western hemlock-Sitka spruce forests for timber production. General Technical Report PNW-GTR-88. Portland, OR: U.S. Department of Agriculture, Forest Service, Pacific Northwest Range and Experiment Station.

Salwasser, H. 1994. Ecosystem management: can it sustain diversity and productivity? *Journal of Forestry*, 92: 6-10.

Schultz, M.E. [In prep.]. Terrestrial arthropods in young-growth forests on Prince of Wales Island, Alaska.

Schwab, F.E. 1979. Effect of vegetation structure on breeding bird communities in the dry zone Douglas-fir forests of southeastern British Columbia. Vancouver, BC: University of British Columbia. 135 p. M.S. thesis.

Smith, R.D., Sidle, R.C., and Porter, P.E. 1993. Effects on bedload transport of experimental removal of woody debris from a forest gravel-bed stream. *Earth Surface Processes and Landforms*, 18: 455-468.

Smith, W.B., Vissage, J.S., Darr, D.R., and Sheffield, R.M. 2001. Forest resources of the United States, 1997. General Technical Report NC-GTR-219. St. Paul, MN: U.S. Department of Agriculture, Forest Service, North Central Research Station.

Stiles, E.W. 1980. Bird community structure in alder forests in Washington. *Condor*, 82: 20-30.

Swanson, F.J., Johnson, S.L., Gregory, S.V., and Acker, S.A. 1998. Flood disturbance in a forested mountain landscape. *BioScience*, 48: 681-689.

Swanston, D.N., and Marion, D.A. 1991. Landslide response to timber harvest in southeast Alaska. In: Proceedings of the Fifth Interagency Sedimentation Conference. Las Vegas, NV: Federal Energy Regulatory Commission, 10: 49-58.

Tappeiner, J.C., II, and Alaback, P.B. 1989. Early establishment and vegetative growth of understory species in the western hemlock-Sitka spruce forests of southeast Alaska. *Canadian Journal of Botany*, 67: 318-326.

Tarrant, R.F., Bormann, B.T., DeBell, D.S., and Atkinson, W.A. 1983. Managing red alder in the Douglas-fir region: some possibilities. *Journal of Forestry*, 81: 787-792.

Thedinga, J.F., Murphy, M.L., Heifetz, J., Koski, K.V., and Johnson, S.W. 1989. Effects of logging on size and age composition of juvenile coho salmon (*Oncorhynchus kisutch*) and density of pre-smolts in southeast Alaska streams. *Canadian Journal of Fisheries and Aquatic Sciences*, 46: 1383-1391.

U.S. Department of Agriculture, Forest Service. 1997. Record of Decision for Tongass National Forest Land and Resource Management Plan Revision, Alaska. R10-MB-338a, Juneau, AK: U.S. Department of Agriculture, Forest Service. 44 p.

U.S. Department of Agriculture, Forest Service. 2001. Unpublished data. Data query conducted by Gary Fisher, Geographic Information Analyst, information was derived from the Managed Stands Layer, May 2001.

Wallmo, O.C., and Schoen, J.W. 1980. Response of deer to secondary forest succession in southeast Alaska. *Ecology*, 26: 448-462.

Willson, M.F., and Comet, T.A. 1996. Bird communities of northern forests: patterns of diversity and abundance. *Condor*, 98: 337-349.

Wipfli, M.S. 1997. Terrestrial invertebrates as salmonid prey and nitrogen sources in streams: contrasting old-growth and young-growth riparian forests in southeastern Alaska, USA. *Canadian Journal of Fisheries and Aquatic Science*, 54: 1259-1269.

Wipfli, M.S., Deal, R.L., Hennon, P.E., Johnson, A.C., De Santo, T.L., Hanley, T.A., Schultz, M.E., Bryant, M.D., Edwards, R.T., Orlikowska, E.H., and Gomi, T. 2002. Managing young upland forests in southeast Alaska for wood products, wildlife, aquatic resources, and fishes: problem analysis and study plan. General Technical Report PNW-GTR-558. Portland, OR: U.S. Department of Agriculture, Forest Service, Pacific Northwest Research Station.

Wipfli, M.S., and Gregovich, D.P. 2002. Export of invertebrates and detritus from fishless headwater streams in southeastern Alaska: implications for downstream salmonid production. *Freshwater Biology*, 47: 957-969.

Wipfli, M.S. and Musslewhite, J. [In review]. Trophic subsidy from fishless headwaters to downstream fish habitats: riparian red alder (*Alnus rubra*) augments invertebrate and detritus delivery to salmonid food webs.

Section 2—Silviculture and Modeling

Silvicultural science and practices can be used to increase compatibility among forest resources. In Chapter 4, Hummel discusses how the broadening of management objectives is leading to a broadening of silvicultural practices. During much of the 20th century, the silvicultural objective was to convert structurally and compositionally diverse forests to regulated, uniform stands of commercially valuable Douglas-fir trees. These practices simplified forest structure and composition. Now, at the beginning of the 21st century, structural and compositional diversity are part of most forest management objectives. Hummel reviews silvicultural treatments and the responses of other forest resources, such as species richness of birds, small mammals, invertebrates, and understory plants to changes in forest structures.

In Chapter 5, Monserud describes large-scale management experiments in joint production of forest resources. After World War II, even-aged silviculture and plantation management were widely adopted in the Douglas-fir region of the Pacific Northwest without systematic scientific studies. During the 1990s, a number of large-scale management experiments were started in the region to examine silvicultural alternatives to traditional even-aged management. These operational-size studies emphasize joint production and alternatives to clearcutting, concepts central to compatible forest management. They involve various types of partial cuts and variable retention, and most treatments emphasize accelerating the development of old-forest structural characteristics and retaining biological legacies.

In both chapters, the authors discuss links between silvicultural treatments and other resources. Many of these other resources are discussed further in Section 4. For example, Chapter 12, discusses studies on plant species richness associated with silvicultural treatments, focusing on herbaceous understory species. Small mammal responses to silvicultural treatments are examined in Chapter 14. The responses of aquatic resources are integrated into the two silviculture chapters and throughout Section 4.

In Chapters 6 and 7, Monserud discusses computer models and their uses in compatible forest management. Modeling began with early yield tables developed more than a century ago. In Chapter 6, he reviews models of forest stand dynamics for their utility in addressing forest management questions regarding compatibility. None of the existing models is fully adaptable for alternative silvicultural treatments, although the newest models are promising in that managers can more accurately evaluate management alternatives based on mixed-species stands and heterogeneous structures.

In Chapter 7, Monserud examines landscape-level modeling by considering several different approaches at different scales. These models are designed to analyze the effects of forest management across large and complex ecosystems. Stand-based landscape models, visualization tools, successional pathways models, and landscape optimization models all offer different ways to look at long-term changes and forest patterns across large landscapes. Regional and national models are typically used as planning and policy tools, and most global vegetation models have been developed to predict how vegetation zones or biomes would evolve or shift under climate change.

Visualization models are particularly helpful for showing the public how the landscape would look decades into the future if it were managed in different ways. The newest visualization tools translate technical stand data into credible, technically accurate pictures. The models can be valuable tools for working with the public to design compatible management options, and thus the models are likely to help with social acceptability issues discussed in Section 5.

Scale is considered in Chapters 4 and 5. Silvicultural actions are used at the stand level, but their consequences on landscape patterns are major issues. In Chapter 4, Hummel begins at the scale of the forest stand, the standard unit of forest management, and then moves up to the watershed scale. Section 3 further develops the idea that a diversity of stand structures and ages are needed across the landscape to conserve biodiversity or achieve compatibility. An advantage of the operational-size studies discussed in Chapter 5 is that the results can be interpreted directly at the scale of the study, with no risk of bias by scaling up from very small plots to an operational level. These studies are essentially well-executed case studies designed to demonstrate feasibility of approaches that will produce a desired mix of conditions.

The temporal scale is also considered. Silviculture starts with the existing forest condition that is the result of historical land ownership patterns and past management. Forests grow at a slow pace compared to human desires for results and many of the studies described in these two chapters have not been completed.

Chapter 4

Managing Structural and Compositional Diversity with Silviculture

Susan Stevens Hummel
USDA Forest Service, Pacific Northwest Research Station,
620 SW Main St., Suite 400, Portland, OR 97205, USA

1. Introduction

Ecology, economy, and demography interact to affect forest management objectives. In the temperate rainforests of northwestern North America (Franklin and Halpern 1988), the outcome of this interaction for most of the 20th century was a management emphasis on wood production (Curtis et al. 1998, Haynes et al. 2003). Because of production efficiencies, even-aged, clearcut systems favoring harvest and regeneration of Douglas-fir (*Pseudotsuga menziesii* (Mirb.) Franco) dominated regional silvicultural practices for decades (Tesch 1995, Curtis et al. 1998) (see Appendix 1 for all common and Latin names). By century end, however, economic and social conditions had changed (Haynes et al. 2003), and so had knowledge about forest ecology (Perry 1994, Franklin et al. 2002). Not surprisingly, different management objectives emerged. Indeed, although forest management in the 21st century still includes wood production, it is now just one of many objectives for regional forests (Kohm and Franklin 1997). Because management objectives and silvicultural practices are intertwined (Baker 1934, Daniel et al. 1979), this broadening of objectives is accompanied by a broadening of practices. Whereas earlier silvicultural practices simplified forest structure (arrangement and variety of elements) and composition (variety and amount of species present), the emphasis now is on managing diversity (O'Hara 2001). It is the concept of managing structural and compositional diversity that links silviculture to compatible forest management, which explicitly includes biological diversity (biodiversity) as an objective (Haynes et al 2002; Figure 5, Chapter 1).

R.A. Monserud, R.W. Haynes and A.C. Johnson (eds.)
Compatible Forest Management, 85-119

Silvicultural practices patterned on natural disturbance dynamics may conserve biodiversity in managed landscapes (Franklin et al. 2002, Lindenmayer and McCarthy 2002). This is a proposition being tested in the Pacific Northwest (PNW) (western Oregon and Washington, coastal British Columbia and southeastern Alaska; see Figure 1, Chapter 1), where silvicultural treatments are being developed to create conditions reminiscent of the biological legacies associated with natural disturbances (Cissel et al. 1998, Beese et al. 2001). Contemporary treatments, which are being applied both experimentally (Monserud 2002) and operationally (Beese et al. 2001), do not always coincide with treatments used in traditional even- and uneven-aged silvicultural systems (Mitchell and Beese 2002). Some of the treatments, for example, do not emphasize timber harvest or stand regeneration (Franklin et al. 1997), but instead create dead trees or other structures to benefit specific animal populations (Bull et al. 1995, Chambers et al. 1997).

Silvicultural treatments are often applied at the stand scale, but many can be used at other spatial scales as changing knowledge and objectives warrant (Curtis et al. 1998). This is important because the focus of forest management in the PNW is shifting from stands to watersheds and landscapes as knowledge about structural legacies and regional disturbance dynamics evolves (Swanson et al. 2003). Indeed, Spies and Johnson (2003) suggest that compatibility among forest resources may be scale dependent. Silvicultural treatments to enhance resource compatibility at the stand level thus may not do so at a different spatial scale, and vice versa.

This chapter discusses the importance of silvicultural science and practice for managing PNW forests to increase compatibility among resources. Compatibility occurs when forest management simultaneously produces wood and another resource without contributing to a decline in any other resource (Haynes et al. 2003). Section 2 reviews traditional silvicultural terms and systems and discusses their use in the PNW. Section 3 introduces contemporary silvicultural terms and treatments used when forest management objectives include biodiversity conservation and wood production. Section 4 considers evidence for potential compatibility among resources by reviewing the literature from regional studies of silvicultural treatments that include response variables related to biodiversity, aquatic resources, and wood—the key objectives of compatible forest management (Haynes et al. 2002; Figure 5, Chapter 1). The review identifies studies that include multiple resources and topics on which such studies are lacking (Appendix 2). The last section discusses information currently available, and that which is still needed, to manage structural and compositional diversity in PNW forests in support of enhanced compatibility among resources. Of particular importance is information about patterns and ranges of desired conditions that, in combination with site-specific information, will help silviculturists translate the concept of diversity into a system of treatments over space and time.

2. Traditional Silvicultural Terms and Their Use in the Pacific Northwest

Many silvicultural terms used in the PNW have remained essentially unchanged since their introduction to the English language (Curtis et al. 1998, O'Hara 2002) from Latin and German (USDA 1893, Troup 1928, James 1996). This chapter refers to such terms as traditional. A few are defined in this section (in **bold**), to give background for subsequent discussion on how traditional terms differ from, and contribute to, some contemporary terms now used in the PNW.

Silvics focuses on principles underlying the growth and development of single trees and of the forest as a biological unit (Hawley 1937, Smith 1986); it is synonymous with forest ecology (Toumey and Korstian 1937). By the 1920s, local research on the silvics of PNW trees began (Duncan and Miner 2000). As a result, extensive information is now available on the silvics of the conifers that dominate regional forests (Minore 1979, Burns and Honkala 1990a, Tesch 1995), particularly those with commercial value like Douglas-fir. In recent decades, knowledge on the silvics of other tree species (e.g., Sitka spruce, red alder, Oregon white oak) and associated vegetation has increased (e.g., Burns and Honkala 1990b, Tappeiner and Zasada 1993, Harrington and Kallas 2002, Wipfli et al. 2003).

Silviculture (*silva* = forest (Latin) + culture (English)) is the practice of altering forest ecosystems by using silvical knowledge and economic and social considerations (Daniel et al. 1979, Smith 1986). Most definitions of silviculture include three concepts: manipulating woody vegetation; influencing the establishment, growth, and composition of forests or woodlands; and achieving the goals of landowners and of society (e.g., Daniel et al. 1979, Smith 1986, Dawkins and Philip 1998, Helms 1998). Silvicultural science and practice provide information and methods for managing forested ecosystems over time and space in support of the objectives that people want.

A forest **stand** is a spatially continuous group of trees and associated vegetation growing on a site of similar soil and climate and having similar structure (Oliver and Larson 1990). **Structure** includes the horizontal and vertical distribution of forest components, including the heights and diameters of live and dead trees and the arrangement of foliage, crown layers, herbaceous plants, and down wood (Helms 1998, Spies 1998). A shift in community ecology in the 1980s (Perry and Amaranthus 1997) influences current views on forest stand development (Fujimori 2001). Patterns of stand development vary by species and site, yet are grouped into several stages, which describe key structures and processes typical of each stage. Stand development classifications used in the PNW include those by Oliver and Larsen (1990) and Franklin and Spies (1991); several others are displayed and compared in Appendix 1, Chapter 1. Although such classification methods can help link ecosystem processes and stand development (Fujimori 2001), they also have limitations. For example, structure-based classifications group continuous variables like diameter and foliage distribution, light conditions, and inter- and intra-specific competition into categories. Conceptual limitations resulting from categorizing forest ecosystems may now be impeding, rather than facilitating communication (Bunnell and Huggard 1999, Franklin et al. 2002) because of the various ways that different scientific disciplines regard structure (Appendix 1, Chapter 1). Further, such classification categories are usually qualitative rather than quantitative, and thus lack sufficient detail to be related empirically to biodiversity measures, such as species richness.

In this chapter, the first stage of stand development refers to the germination and establishment of trees and associated herbaceous vegetation. In the second stage, trees dominate a site, compete for growing space and nutrients, and crowns close to form a continuous canopy. In the third stage, tree crowns separate into different strata as density-dependent mortality occurs. Stage four shifts to density-independent mortality, and new trees and understory vegetation establish in available growing space. The fifth stage includes gap formation, continued mortality, and accretion of fallen dead trees. In the PNW, stages four and five are sometimes referred to collectively as "old-growth" forest. Because of the age-class distribution of forests available for studies of stand development after the 1980s, most existing data come from planted early—(first and second) stage or unmanaged late—(fourth and fifth) stage stands. Knowledge of the establishment and early development of existing late-stage forests in the PNW is derived from observational and reconstructive studies (Tappeiner et al. 1997a, Poage and Tappeiner 2002).

A **silvicultural system** is a planned series of treatments for a forest stand (Smith 1986), and implies a process for creating target conditions over time. The systems are classified either according to the regeneration methods used or by the number of tree age classes that result (Matthews 1989). **Even-aged** silvicultural systems secure a single cohort, or age class, of regeneration. **Age class** is variously defined, but generally implies that the difference in age between the oldest and youngest trees is no more than 20 years (Baker 1934), or not more than 20% of the rotation length (Smith 1986). **Rotation** is a term usually associated with even-aged systems and refers to the time between establishment and final harvest (Helms 1998). **Clearcutting** is an even-aged system that removes almost all trees, creating a fully exposed microclimate for a new age class to develop. In the PNW, clearcutting has predominated for a century (Tesch 1995, Curtis et al. 1998), and over this time rotation lengths have progressively decreased to 40 or 50 years (Curtis 1997, Wilhere 2003). After harvest, methods of site preparation and seedling planting ensure stand establishment is consistent with various state and provincial forest practices laws (Cleary et al. 1978, Arnott 1986, Cafferata 1986). Dependable methods for regenerating harvest units have removed the size restrictions on clearcuts that were previously imposed by the need for adjacent seed blocks for natural regeneration (Curtis and Carey 1996). Herbicides and fertilizers are sometimes applied, depending on landowner objectives, and various intermediate vegetation management treatments are typically done after stand establishment and before harvest, including weeding and release (Cafferata 1986). The latter are generally done in the first 5 years after planting, when sprouting shrubs or hardwoods (e.g., salal, vine maple) grow in conifer plantations (Tappeiner and Zasada 1993, Cafferata 1986). **Seed tree** systems also result in a single age class of regeneration. In this system, almost all trees are removed at harvest save a few kept for regenerating a new age class. Seed tree systems are not common in the PNW.

Two-aged silvicultural systems regenerate a stand with two age classes. **Shelterwood** systems, for example, cut most trees but leave some to shade the new age class establishing underneath. A shelterwood is distinguished from a seed tree system by the intentional use of shade, which gives desired species a growth advantage over competing vegetation during establishment. In contrast to trees retained in a shelterwood, which are generally harvested after a new age class is established, **reserve trees** are not typically removed. In the PNW, the main reason for experimenting with shelterwood systems was to observe regeneration responses in different forest types, including Douglas-fir (Williamson 1973), and western hemlock (Williamson and Ruth 1976). Considerable research on regeneration using shelterwood has been done (Tappeiner et al. 1997b), but use of this system has been limited to sites where ameliorating harsh conditions is important for stand establishment, such as in southwestern Oregon (Hobbs et al. 1992).

Other systems investigated have included strip thinning (McCreary and Perry 1983) and seed tree with reserves (Franklin 1963). The response variables of interest in these alternatives to clearcutting were often the quantity and distribution of natural reproduction (Seidel and Cooley 1974; Seidel 1979a,b, 1983).

Uneven-aged silvicultural systems regenerate a stand with three or more age classes. The objective is a forest with trees of different ages or sizes intermingled, typically accomplished with some form of selection system. In these systems, mature and immature trees are felled to create or maintain uneven-aged stands. **Single tree selection** fells individual trees and generally tends to increase the proportion of shade-tolerant species in mixed-species stands. **Group selection** cuts trees in units and therefore maintains a higher proportion of shade-intolerant species in mixed species stands than does individual tree selection. **Partial cutting** is a general term denoting something other than a clearcut; it can include selection systems. Experience with uneven-age silvicultural systems is limited in the PNW. In the decade before World War II, the regional forester instructed the supervisors of national forests in Oregon and

Washington to implement individual tree and small group selection with infrequent clearcut areas of "as much as 5 or 10 acres" (Buck 1934). This mandate to replace clearcutting in old-growth Douglas-fir forests with selection systems was controversial (Steen 1998, O'Hara 2002). By the 1940s, uneven-aged systems were abandoned in the PNW (Tappeiner et al. 1997b, Curtis et al. 1998), perhaps because of similarities with all-age systems used in Germany at the time (e.g., Dauerwald, Plenterwald) (Vogel 1990, Schabel and Palmer 1999). The current lack of information on the use of uneven-age silvicultural systems in the PNW is largely due to this earlier decision, which was based on limited data (Tappeiner et al. 1997b).

Plant ecology (Harper 1977, Koslowski et al. 1991) provides a basis for silvicultural theories on size-density relations (Hummel 2002). According to silvicultural theory, individual tree height and height growth are relatively unaffected by stand density, but tree diameter growth decreases with increasing density for both intraspecific and interspecific competition (e.g., see Knowe and Hibbs 1995 for review). Size-density relations imply that in plant populations—such as forest trees—growth reaches a point where some trees must die for others to get bigger (e.g., Weller 1987, Lonsdale 1990, Jack and Long 1996). Tree diameter is commonly used as a proxy for size in the PNW because it is easily measured and because early research in regional forests established relations between diameter and density (Reineke 1933).

Quantitative silviculture applies principles and models from plant ecology, size-density relations, and biometrics to make predictions about stand development (Jack and Long 1996). **Density management diagrams**, a quantitative tool used in the PNW (e.g., Drew and Flewelling 1979, Jack and Long 1996, Newton 1997), illustrate volumes and diameters associated with different numbers of trees per area and indicate how a stand is stocked relative to a maximum density (e.g., McCarter and Long 1986, Long et al. 1988). The diagrams are species-specific, and derived primarily from research in even-aged, single-species stands. They are used to guide spacing and stocking decisions and to design thinning schedules for various objectives, such as timber production or habitat structure (Smith and Long 1987, McTague and Patton 1989, Sturdevant et al. 1996, Ferguson and Archibald 2001).

Thinning is a treatment to reduce tree density (Helms 1998). **Precommercial thinning** implies that trees cut are not merchantable, whereas in **commercial thinning** the trees are valuable enough to recover harvest costs (Helms 1998). When trees are cut from the lower crown classes to favor those in the upper classes, the term used in the PNW is **thinning from below**. In contrast, **thinning from above** involves cutting dominant and co-dominant trees to favor the best trees of these classes (Helms 1998). By the 1950s, precommercial thinning was adopted throughout the PNW as a way to regulate spacing in plantations and to select future crop trees. Commercial thinning was not widely adopted, however, (Curtis et al. 1998) because selling old-growth timber and establishing short-rotation (40 to 60 year) plantations was more profitable than selling logs from commercial thinning operations (Curtis and Carey 1996). Now, however, this situation has changed and there is interest in combining commercial thinning with rotations of 100 years or more (Curtis 1997).

3. Contemporary Silvicultural Terms and Their Use in the Pacific Northwest

Ecological knowledge about the dynamics of PNW forests gained in recent decades is influencing silvicultural science and practice. As a result, scientific and forest management literature from the PNW contains a mix of traditional and contemporary silvicultural terms. The latter are emerging as treatments develop in response to forest management objectives that include biodiversity conservation and wood production. This section introduces some contemporary silvicultural terms (in *italics*) and associated treatments now being used in the PNW. The treatments share the objective of creating structural and compositional diversity but have different names, owing to the place or date of their origin. Further, use of treatments with the same name varies between agencies such as the U.S. Departments of Agriculture and the Interior (USDA/USDI 1994). The lack of a common lexicon and usage is confusing; it is also evidence of the transition occurring in PNW forest management. Ongoing discussion about traditional versus contemporary silvicultural terms (e.g., Franklin et al. 1997, Buermeyer and Harrington 2002) stems in part from debate about contemporary treatments and whether these treatments describe a target condition or a system by which a target condition is perpetuated (Mitchell and Beese 2002).

3.1. Biodiversity Conservation

Conservation of biodiversity as a goal of forest management is widely (e.g., Fenger 1996, Lähde et al. 1999) but not completely (Zavala and Oria 1995) accepted, in part because relations among biodiversity, ecosystem resilience and stability, and disturbance remain unclear (Perry and Amaranthus 1997, Symstad et al. 2003). For this reason, Simberloff (1999) argues that conservation of biodiversity in managed forests must rely on a commitment to biodiversity as an intrinsic value.

Several contemporary silvicultural treatments used in the PNW are designed to support a management objective of conserving biodiversity. *Green-tree retention*, for example, is a stand management treatment that leaves live trees, snags, and down wood within harvest units to provide structure over the next management cycle (USDA/USDI 1994). Green trees are not removed after a new age class has established but instead are left on site (Beese and

Bryant 1999) to provide structures thought to be vital for birds and other animals (Franklin et al. 1997). Green-tree retention may be considered synonymous with reserve trees (Helms 1998), but these two terms are not interchangeable in PNW literature (Franklin et al. 1997, Buermeyer and Harrington 2002). Green trees are sometimes deliberately killed to create snags (Chambers et al. 1997) and, depending on how they are arrayed spatially, the terms dispersed retention or aggregated retention are used (Franklin et al. 1997). *Dispersed retention* refers to retaining live and dead trees distributed evenly over a harvest unit, whereas in *aggregated retention* the trees are clumped (USDA/USDI 1994, Franklin et al. 1997). The two methods carry different implications for the costs, safety, and effects of harvest operations (Moore et al. 2002, Wilhere 2003), and may provide different habitat values for various species (McComb and Lindenmayer 1999).

On forests in the United States within the range of the northern spotted owl (*Strix occidentalis caurina*), a federal plan mandates different land allocations or management zones (USDA/USDI 1994). The zone where scheduled timber harvest can occur, the matrix land, offers an example of green-tree retention (USDA/USDI 1994). On matrix land administered by the U.S. Department of Agriculture, the general prescription requires that 15% of the area of each cutting unit be left unharvested. At least 70% of this leave area must be covered by patches bigger than 1 hectare (ha) (aggregated retention), and the rest must be spread across the cutting unit as single trees or smaller patches (dispersed retention). In addition to trees, down wood must also be retained: 240 linear feet of logs per acre (181 meters (m)/ha) greater than 20 feet (6.1 m) long and 20 inches (50.1 centimeters (cm)) in diameter. In contrast, on land administered by the U.S. Department of the Interior in the southern portion of the PNW, the general prescription calls for retaining 16 to 25 large green trees per acre (40 to 60 trees per ha) in harvest units, without mention of retention pattern (USDA/USDI 1994).

Variable retention is a term used when trees are left distributed throughout harvest areas to provide structural diversity. Variable retention is a form of partial cutting (Beese et al. 2001) and is based on the concept that natural disturbances leave residual structures thought to be important for ecosystem function and biological diversity (Franklin et al. 1997). The structures emphasized are live and dead trees of varying sizes, multiple canopy layers and coarse woody debris (Franklin et al. 1997, Beese and Bryant 1999). Retention of these structural features during logging operations is hypothesized to sustain biological diversity by providing a place for species that used the features prior to harvest; indeed a key assumption is that these retained features will be the basis for reestablishing species once the harvested area has grown into a suitable condition (Franklin et al. 1997).

The term variable retention was created because traditional silvicultural terms were not legally adequate to describe management objectives in coastal

British Columbia (Mitchell and Beese 2002). The definition, now codified in British Columbia legislation, includes two features that characterize variable retention: (1) long-term structural diversity and (2) sufficient structure remaining to maintain forest influence on the harvested area (Mitchell and Beese 2002). The former is provided by trees retained in either groups (aggregated retention) or as individuals (dispersed retention) for at least one rotation, whereas the latter results from the requirement that no more than one tree length exist between retained trees (Mitchell and Beese 2002). Variable retention is being combined with traditional silvicultural systems such as shelterwood and group selection (Beese et al. 2001). Variable retention is, therefore, not synonymous with uneven-aged systems. Rather, variable retention describes a target condition that avoids structural simplification regardless of the silvicultural system used.

In 1998, the forest products company McMillan Bloedel adopted variable retention to guide management on private and tenured Crown forests in British Columbia. The decision was made in response to social concerns (Beese et al. 2001), and was continued by Weyerhaeuser when it acquired McMillan Bloedel a year later. Although monitored at the stand scale, variable retention on public land managed by Weyerhaeuser is being applied in landscape zones that range from 5000 to 50 000 ha. Three zones exist, each with a different emphasis: old growth, habitat, and timber; Beese et al. (2001) describe the retention and harvest intensity gradient that occurs among them. Ten percent of land managed by Weyerhaeuser is in the old-growth zone. Here, 33% of the landscape is available for group selection and irregular shelterwood harvest, but 20% retention is required at the unit level. In the habitat zone, which covers 25% of land managed by Weyerhaeuser, 70% is available for harvest using a range of silvicultural systems; within units, 15% retention is required. Finally, the most intensive zone is timber, which covers 65% of land managed by Weyerhaeuser. In the timber zone, 80% of the total area is available for harvest through group retention (10%), dispersed retention (5%), and shelterwood systems.

Another contemporary term encountered in PNW literature is *variable density* (USDA FS 2003), which describes thinning treatments that avoid regular spacing among residual trees. It has arisen from concern over the extensive area of structurally uniform, second stage Douglas-fir plantations in the region that are a result of post-World War II clearcutting practices (Hunter 2001). Variable density thinning is a treatment being used in early-stage forests to accelerate the development of structural and compositional diversity observed in late-stage forests (Muir et al. 2002), and to investigate associated effects on small mammal populations (Carey 2001).

These contemporary treatments share the premise that structural and compositional diversity in managed forest ecosystems is better than uniformity. Diversity is the target condition. Translating this target into a process by which diversity is created and maintained over time and space requires site-specific information. It also requires using, observing, and adjusting treatments over time; a practice sometimes referred to as adaptive management (e.g., Nyberg and Taylor 1995) or monitoring (Mitchell and Beese 2002). Actual outcomes will likely differ from expectations as, for example, when converting even-aged uniform stands to uneven-aged species mixtures (Nyland 2001, O'Hara 2001). As experience is gained, and as data are available that relate specific silvicultural treatments to biodiversity over space and time, modifications to the treatments currently being applied will occur and silvicultural systems for managing structural and compositional diversity will be developed. These could include extending rotations over several hundred years in combination with commercial thinning (Curtis 1995, Curtis 1997, Busing and Garman 2002), relying on natural regeneration from group selection (Worthington 1953) or retention (Buermeyer and Harrington 2002), and favoring species mixtures that include hardwoods.

3.2. Wood Production

Management primarily for wood production is concentrated on private industrial land in the PNW (Curtis et al. 1998, Haynes et al. 2003), and associated silvicultural practices often include intensive site preparation and vegetation management. State and provincial forest management is less intensive but still includes wood production; federal forest management has a decreasing emphasis on wood production. Wilhere (2003) gives an example of practices on industrial forestland in the PNW: (1) 2-year-old Douglas-fir seedlings from a nursery are planted at 1075 trees per hectare (tph); (2) at age 15, stand is precommercially thinned from below to a residual density of approximately 740 tph; (3) at age 30, stand is commercially thinned from below to a residual density of approximately 340 tph; and (4) at age 50, stand is clearcut.

Quantitative silvicultural tools developed from single-species, even-aged stands make it possible to estimate yield associated with various species, site conditions, and stand ages by using density management diagrams and growth and yield models (Monserud 2003). If, however, the objective is wood production from mixed-species or uneven-aged stands, then comparable tools are not yet available. Their development requires better knowledge about the dynamics of mixed species stands and the response of such stands to specific treatments. For example, data on the effects of residual trees on the growth and yield of regeneration are limited (Zenner et al. 1998, Mitchell 2001), as is information about maximum density in mixed species stands (Puettmann et al. 1992, Shaw 2000).

Forest management in the PNW is creating divergence in forest age classes (Harris 1984, Figure 3, Chapter 1). Older age classes (fourth and fifth stages) tend to be on public land whereas the younger age classes (first and second stages) tend to be on private land. Recent studies indicate this trend will continue (Spies and Johnson 2003). This predicted age class distribution, and potential age class gap (third stage), is prompting discussion about management practices that might support a full array of forest age classes, structures, and processes in the region.

4. Evidence for Compatibility Among Different Objectives

The previous section described silvicultural treatments for achieving either biodiversity conservation or wood production in the PNW. Compatibility implies not having to choose between these two objectives but managing for both simultaneously without a decline in other forest resources (Stevens and Montgomery 2002, Haynes et al. 2003). This section reviews the literature from regional studies of silvicultural treatments that include multiple response variables related to biodiversity, aquatic resources, and wood because these are the key objectives of compatible forest management (Haynes et al. 2002; Figure 5, Chapter 1). The objective is to identify results from PNW studies that directly address potential compatibility among forest resources.

4.1. Biodiversity

Biodiversity includes genetic, species, and ecosystem diversity (Wilson 1988). Compositional diversity refers to the number of elements in a system; structural diversity refers to the physical organization; and functional diversity refers to the different processes (Zavala and Oria 1995). In this chapter, species richness is adopted as a proxy for compositional diversity; therefore, studies that focus on species abundance or dominance are not emphasized. Silvicultural treatments directly affect stand composition and structure, so most studies investigating effects of silvicultural treatments on biodiversity measure compositional or structural alpha diversity; a few investigate treatment effects on processes such as decomposition (Appendix 2).

Birds, small mammals (Carey 2003), and understory plants (Kerns et al. 2003) are the focus of most studies investigating compositional diversity associated with silvicultural treatments (Appendix 2). Studies of structural diversity, including wildlife habitat elements, generally investigate the vertical and horizontal distribution of living or dead trees and shrubs. Most designed experiments and observational studies investigating compositional and structural diversity associated with silvicultural treatments in the PNW are less than 100 ha (Appendix 2). The studies cover a range of forest types (key tree species are noted parenthetically).

Most studies that apply silvicultural treatments along a gradient of tree removal intensity and compare species richness values to an untreated control focus on birds (Appendix 2). None of these studies report the wood volume, quality, or potential product values associated with the different treatments. Treatment effects on bird species richness are mixed; some studies report no difference whereas others report increases or decreases. For example, in a replicated study on Vancouver Island, British Columbia, shelterwood, patch clearcut, and clearcut with reserves (green-tree) treatments were applied to coastal montane forest (*Abies amabilis/Tsuga heterophylla/Thuja plicata/Chamaecyparis nootkatensis*). After 3 years, the species richness of breeding birds was reduced in all treatments (Beese and Bryant 1999). The uncut control had significantly more breeding bird species than any treated unit; the shelterwood and green-tree retention were similar in terms of richness. Likewise, the species richness of winter resident birds was higher in the uncut control than in treated units, which were ranked in descending order (patch cut, shelterwood, green-tree). In another replicated study, this one from Douglas-fir dominated forests of the Oregon Coast Range (*Pseudotsuga menziesii/Abies grandis/Acer macrophyllum/Quercus garryana*), wintering (Chambers and McComb 1997) and breeding (Chambers et al. 1999) bird responses to three silvicultural treatments (small-patch group selection, green-tree, and modified clearcut) were measured. No treatment effects on species richness of breeding birds were detected after 2 years, although more species were observed in the green-tree than in the other treatments, and the small-patch group selection was most similar in composition to control stands (Chambers et al. 1999). Like Beese and Bryant (1999), Chambers et al. (1997) report treatment effects on the species richness of winter resident birds, but the latter report more species in the small-patch group selection than in control and modified clearcut stands. In a retrospective study comparing commercially thinned and unthinned Douglas-fir stands in the Oregon Coast Range, Hagar et al. (1996) found that breeding and winter bird species richness did not appear affected by thinning that occurred 4 to 15 years prior to their observations. However, the density of hardwoods 31 to 43 cm diameter at breast height (dbh) was an important predictor of bird species richness, and this density did not differ between thinned and unthinned stands (Hagar et al. 1996).

Despite a lack of long-term experiments on patterns of regional bird species richness, Olson et al. (2001) provide some general observations:

- Species richness tends to be similar in early (first stage) and late (fourth and fifth) stage forests and lowest in mid stages (second and third)
- Mean species richness of diurnal breeding birds is higher in the Coast Range than in the Cascade Range (owing to a generally negative correlation with latitude and positive correlation with longitude in western Oregon and Washington)
- Species richness increases with vegetation height

Experimentally investigating these observations in association with different types and intensities of treatment will contribute information that can be used to evaluate potential compatibility among forest resources. Overall, results to date indicate that the response of breeding and wintering birds to forest structures created via silvicultural treatments differs by species. There are too few data from long-term, replicated studies in the PNW, however, to identify if specific response patterns exist and, if so, what they might be. This suggests that a mix of stand developmental phases and structures are important for bird species diversity, but does not identify the specific mixture or spatial pattern. Information on demographic performance of species (e.g., reproduction and survival) is needed (DeStefano 2002) to assess potential treatment effects that are not evident in species richness measures, such as nest predation (Tittler and Hannon 2000) and parasitism (Chambers et al. 1999). For example, an increase in brown-headed cowbird (*Molothrus ater*) abundance was noted by Chambers et al. (1999) in green-tree retention treatments of Douglas-fir; a similar observation was made by Bull et al. (1995) in a case study of a modified selection treatment in mixed-conifer forests of northeastern Oregon (*Abies grandis/Pseudotsuga menziesii/Pinus contorta/Larix occidentalis/Pinus ponderosa*). Brown-headed cowbirds are an obligate brood parasite that can contribute to declines in bird populations, but parasitic effects associated with different treatments are not clear (Bull et al. 1995, Chambers et al. 1999).

Information on species richness associated with different types and intensities of silvicultural treatments is sparse for birds but it is even sparser for other vertebrate and invertebrate species. Studies of mammals generally report species abundance (e.g., Wolff and Zasada 1975, Cole et al. 1998, Carey 2001, Gitzen and West 2002) rather than richness, and are often observational (Appendix 2). None of the published studies on mammals report the wood volume, quality, or potential product values associated with silvicultural treatments. Arthropod diversity is described by Schowalter (1995) for partially-harvested, late-stage and early-stage Douglas-fir forests in the central Oregon Cascade Range.

Research on plant species richness associated with silvicultural treatments focuses on herbaceous understory species (Appendix 2) (Kerns et al. 2003). Treatment effects on understory species richness are mixed, but greater numbers are often associated with shelterwood systems. Species identification may be more important than mere enumeration if management objectives include minimizing the establishment or spread of non-native (exotic) species. For example, Bailey et al. (1998) observed that understory plant species richness was greater in thinned than in unthinned Douglas-fir forests in western Oregon; but they noted that exotic species contributed to observations of increased richness. In a replicated study on Vancouver Island, British Columbia, 3 years after treatments were applied to coastal montane forest (*Abies amabilis/Tsuga heterophylla/Thuja plicata/Chamaecyparis nootkatensis*), Beese and Bryant (1999) found that understory plant species richness in the shelterwood was significantly higher after treatment than before, but no significant difference existed among the clearcut, patch clearcut, or green-tree treatments. South of the PNW region, in the mixed-species conifer forests of the northern California Sierra Mountains, Battles et al. (2001) compared understory vascular plant diversity in stands treated with shelterwood, single tree selection, group selection, and no active management (reserve units). They found that even-age methods were associated with the highest species richness, whereas reserve units and single tree selection methods were the least rich. Uneven-age methods had greater proportion of late-seral species and fewer exotic species than shelterwoods.

4.2. Structural Diversity

Information about the effects of silvicultural treatments on forest structures is needed to evaluate the proposition that treatments designed to create structural diversity, such as variable retention, conserve biodiversity. Structural elements hypothesized to be important for biodiversity conservation include live and dead trees of different sizes, multiple canopy layers, and coarse woody debris; hence they are emphasized in current PNW literature. Most published studies focus on the effects different treatments and intensities have on these structural elements; none relate their presence or abundance to species richness of multiple taxa (Appendix 2), although Chambers et al. (1997) investigated the use of artificially created Douglas-fir snags by cavity-nesting birds. Evidence is currently insufficient, therefore, to evaluate whether silvicultural treatments to create diverse forest structures conserve biodiversity better than those that result in structural simplification. Data from several regional long-term, large-scale experiments installed in the 1990s (Monserud 2002) are becoming available, however, and continued investment in these studies will help evaluate relations between forest structure and biodiversity.

Reports of structural elements associated with silvicultural treatments focus on standing or down dead trees and on woody debris (Halpern and McKenzie 2001), conifer regeneration in montane British Columbia (Mitchell 2001) and the Cascade Range in southwestern Washington (Buermeyer and Harrington 2002), live tree growth and density in Alaska (Deal and Tappeiner 2002), and understory plant strata (Bailey et al. 1998) (Appendix 2). Halpern and McKenzie (2001) considered effects on the structural attributes of ground cover, slash depth, and woody debris associated with the retention level (15% and 40%) and pattern (aggregated versus dispersed) on four treatment units at six locations in the Cascade Range of Washington. At the lowest retention level there was greater depth and cover of slash, and less woody debris; in general, aggregated retention had higher covers of slash and dispersed retention had greater variation in slash cover and depth (Halpern and McKenzie 2001). Down wood is created indirectly after treatment via windthrow, which Beese and Bryant (1999) discuss in relation to shelterwood, patch clearcut, and clearcut with reserves (green-tree) in three replicates of coastal montane forest in British Columbia. Three years after treatment, 25% of the reserve trees in the green-tree retention unit were down, compared to 5% in the shelterwood; more trees in the intermediate crown class (22%) blew down in the green-tree retention unit than did trees in dominant (7%) or codominant (10%) crown classes (Beese and Bryant 1999). Although these windthrown trees add to woody debris levels, these measurements were not reported.

In a retrospective study in southeastern Alaska, Deal and Tappeiner (2002) investigated species composition, understory plant diversity, and tree size diversity in Sitka spruce-dominated forests. Sixty years after harvest, Deal and Tappeiner (2002) detected no significant changes between uncut and partially cut stands in tree species composition and stand structure. Further, the species richness and community structure of understory plants were similar among uncut and partially cut plots, although stands in which more than 50% of the basal area had been cut had significantly different plant community structure. Other elements of forest structure that have received attention are gap size and dynamics (Coates and Burton 1997, Ott and Juday 2002), the relation between gaps and tree species diversity (Coates and Burton 1997), structural and compositional diversity (Wardman and Schmidt 1998), and small mammal communities (Gitzen and West 2002).

4.3. Functional Diversity

Few published studies have specifically investigated functional diversity associated with different silvicultural systems (Appendix 2). In British Columbia, Marshall (2000) focused on nutrient cycling, which involves most of the soil biota, and nitrogen fixation because nitrogen (N) is often a limiting factor in PNW forests. Marshall (2000) reviewed the literature to report the response of various taxa (rotifers, protozoans, arthropods, bacteria, annelids, etc.) to clearcut, shelterwood, and a modified group selection. The only soil organisms studied in association with uneven-age regeneration methods are nematodes. Prescott (1997) investigated decomposition and N mineralization in an unreplicated study on Vancouver Island, British Columbia, and reported that the decomposition of needle litter was faster in the uncut, old-growth stand than in the units that received patch, shelterwood, and green-tree treatments. No significant differences in N concentrations were measured among the treatments (Prescott 1997).

4.4. Aquatic Resources

Most published research on aquatic resources focuses on physical elements like water, wood, and sediment (Harr 1980, Beschta et al. 2000). In recent years, studies have also included biological elements such as aquatic vertebrates and macro-invertebrates (Gregory 1997, Kauffmann et al. 2001, Olson et al. 2002) as research on riparian ecology has increased (Cunningham 2002). The link between terrestrial forest ecosystems and aquatic ecosystems is the riparian zone, which exists on a continuum from headwaters to estuaries (Kauffman et al. 2001). It appears that differences in headwater assemblages of both vertebrates and invertebrates exist longitudinally with stream flow gradients and latitudinally from streams (Olson et al. 2002). Despite the importance of riparian zones for fish and wildlife, data on the dynamics of riparian forests, their role in larger landscapes, and their responses to silvicultural treatments are limited (Bisson et al. 2002, Wipfli et al. 2003).

Research on silvicultural treatments and riparian ecosystems has focused on clearcutting in forested areas outside the riparian zone, rather than on direct treatment of riparian vegetation (Cunningham 2002). Controversy exists about the influence of clearcutting on peak flows at both the watershed and headwater scale. Beschta et al. (2000) focused on large peak flow events at watershed (60 to 101 ha) and at basin (62 to 640 km^2) scales because these flows are important, both ecologically for fish habitat, water quality, and riparian vegetation and physically for sediment transport and channel morphology. They did not find strong evidence for peak flow increases on large basins associated with clearcutting. Evidence for treatments other than clearcutting is limited (Appendix 2) despite decreasing emphasis on this silvicultural system in the PNW. One study, with 13 sites in western Oregon, varies riparian buffer widths and density management treatments (Olson et al. 2002). Initial reports describe predicted wood volume associated with the treatments and effects on aquatic vertebrates and invertebrates (Olson et al. 2002). Data on the effects of silvicultural treatments on aquatic resources are limited, but information is available on stream dynamics, fish communities, and species richness in the PNW (Reeves et al. 1998).

4.5. Wood

Estimates of wood quality or volume (Barbour et al. 2002, Busing and Garman 2002) and net revenues (Hummel et al. 2001, Marzluff et al. 2002,) associated with silvicultural treatments designed to enhance wildlife habitat (Bull et al. 1995) via retention of structural diversity (Weigand and Burditt 1992) are derived either from simulation (Barbour et al. 1997, Busing and Garman 2002, Cissel et al. 2002) or from retrospective studies (Barbour et al. 2002), but not from designed experiments (Appendix 2). Instead, computer models are being used to predict the potential effects of different silvicultural treatments on stand structure and wood quality (Busing and Garman 2002), but this opportunity is limited by the applicability of available models (Monserud 2003). Models can provide insight into relative differences associated with the spatial and temporal application of silvicultural treatments over landscapes comprising many stands (Hummel et al. 2002, Spies and Johnson 2003, Swanson et al. 2003); however, model development requires data from well-designed field studies measured on spatial and temporal scales relevant to the variable of interest. Estimates of variation in response variables are also important because they will differ by forest type, season, location, and other factors.

5. Discussion

Silvicultural treatments can be used to manipulate the structure and composition of forests to achieve specific objectives. As such, silviculture has an important role in forest management, whether the objective is increased compatibility among resources, site restoration, or wood production. Effective forest management—accomplishing objectives with minimum cost—requires information about the responses of various biotic and abiotic ecosystem elements to silvicultural treatments. This information is commonly a mixture of operational results, observational studies, and designed experiments. The weight given to these various sources of information depends on forest management objectives, on the legal and financial obligations of different forest owners and users, and on social values (Donoghue 2003). For example, a family managing a 30-ha forest will obtain and use information differently than an industrial or public forest owner with many thousands of hectares under management. In the PNW, most forest is publicly owned (Haynes et al. 2003), which means that setting objectives and evaluating information is subject to public review and comment. In this context, empirical data are often given greater weight in decisionmaking than knowledge obtained from other sources.

In the PNW, empirical data are currently insufficient to evaluate whether silvicultural treatments that create diversity in forest structure and composition conserve biodiversity. Data from experiments that examine the relation of forest structures and plant composition to biodiversity, aquatic resources, and wood production in association with silvicultural treatments are lacking for spatial scales larger than 100 ha (Appendix 2). Further, it is difficult to compare data from smaller scale studies because of differences in the (1) silvicultural treatments applied, (2) forest types studied, and (3) response variables measured. Few studies report on wood removals associated with silvicultural treatments, and even fewer discuss the costs of their implementation. Pre- and post-treatment basal area, species composition, and density of trees and snags are sometimes given, but information on the volume and value of wood removed is not. Data on the species richness of small mammals, understory vegetation, and birds provide a mixed picture of treatment effects for each stage of forest development. It may be that more studies are needed or perhaps species richness needs to be augmented with other diversity indices to measure biodiversity in meaningful ways (Reeves et al. 1998) and at meaningful spatial scales (He et al. 2002). Potential latitudinal and longitudinal effects on species richness (Olson et al. 2001) suggest that differences in biodiversity may exist independent of treatment effect among forests in the northern (Alaska) and southern (SW Oregon) extremes of the PNW and forests of coastal (Coast Range) and inland mountains (Cascade Range). Studies that characterize outcomes associated with different silvicultural treatments over areas comprising multiple stand types are needed to measure and interpret effects over time.

Despite a lack of empirical data from long-term studies relating specific silvicultural treatments to compatible forest management objectives in the PNW, both observation and simulation suggest that a mixture of forest development stages is important for conserving biodiversity (Chambers et al. 1999, Spies and Johnson 2003). Insight into patterns associated with the spatial and temporal application of silvicultural treatments over landscapes (Hummel et al. 2002, Spies and Johnson 2003, Swanson et al. 2003) may come from hybrid models (Monserud 2003), which together with growth and yield models can help generate testable hypotheses and guide management decisions. Additional data are necessary to identify the ranges in forest structure and composition desired at various geographic scales to accomplish specific management objectives. This is important because managing structure and composition at fine scales can be expensive, and managing them at coarse scales can create multijurisdiction ownership issues (Spies and Johnson 2003). In the 19th century, rectilinear surveys were adopted in the United States and Canada (Kain and Baigent 1992) instead of watersheds (Powell 1879) to demarcate ownership; this geometric legacy affects 21st century forest management. Square parcels governed by different laws and managed for different objectives often fragment watersheds in the PNW.

The methods used to manage structural and compositional diversity depend on site-specific conditions as well as on management objectives. For example, thinning is one silvicultural technique applied in various ways and for various objectives. It is not a new technique, yet it is being done in new ways because forest management objectives in the PNW are changing, particularly on public land. By targeting specific diameter classes or strata, thinning can be used to alter wildfire behavior (Agee 1996, Graham et al. 1999), to recruit specific understory plants (Kerns et al. 2003), to provide thermal cover (Smith and Long 1987, McTague and Patton 1989), and to grow trees with specific characteristics for wildlife habitat (Ferguson and Archibald 2001). Although contemporary practices like variable-density thinning do not have the benefit of long-term data supporting their intensity or design, they do rely on established theories of plant ecology and size-density relations. Further, extensive silvical information on the shade tolerance and growth form of various PNW species can be used to predict responses of mixed-species stands to various types and intensities of thinning. Current knowledge may be adequate, depending on the particular objective and the scale at which it is manifested. At the tree level, for example, if large branches and deep crowns are desired, then tree density should be lower than if small knot size is the goal. In contrast, wildfire (Agee 1993) and large carnivore (Ruggiero et al. 1994) home ranges generally occur at spatial scales larger than a stand in PNW temperate rainforests, and knowledge about within- and between-stand variation in forest structure and composition relate to organisms or to phenomena occurring across multiple stands is limited (DeStefano 2002).

Diversity rather than simplicity in forest structure and composition is not a new target condition (Gayer 1886, Matthews 1989, James 1996, Lähde et al. 1999, O'Hara 2001), nor is the concept of maintaining continuous forest cover (Schütz 2002, Heitzman 2003). However, these concepts are just beginning to be adapted to PNW forests, as an outcome of regional economic and demographic change. In the 19th and 20th centuries, the PNW was a frontier, and prosperity and a continued timber supply were challenges of the time. It was believed that the way to meet these challenges was to convert structurally and compositionally diverse forests to regulated, uniform stands of commercially valuable Douglas-fir trees. In contrast, at the beginning of the 21st century the PNW is no longer a frontier. Diversity in structure and composition is now part of regional forest management objectives, but both the scale at which this diversity should be determined and the way it should be managed remains unclear. Increased clarity in the range and pattern of conditions desired over time, combined with site-specific information, will help silviculturists translate the concept of diversity into a system of treatments that support increased compatibility among forest resources.

6. References

Agee, J.K. 1993. Fire ecology of Pacific Northwest forests. Washington, DC: Island Press. 493 p.

Agee, J.K. 1996. The influence of forest structure on fire behavior. In: Proceedings, 17th Annual Forest Vegetation Management Conference (pp. 52-68). Redding, CA.

Arnott, J.T. 1986. Douglas-fir stand establishment overview: Coastal British Columbia. In: C.D. Oliver, D.P. Hanley, and J.A. Johnson (Eds.), Douglas-fir: stand management for the future. (pp. 219-229). Seattle, WA: College of Forest Resources, University of Washington Institute of Forest Resources.

Bailey, J.D., Marysohn, C., Doescher, P., St. Pierre, E., and Tappeiner, J.C. 1998. Understory vegetation in old and young forests in western Oregon. *Forest Ecology and Management*, 112: 289-302.

Baker, F.S. 1934. Theory and practice of silviculture. New York: McGraw-Hill Book Company, Inc. 502 p.

Barbour, R.J., Johnston, S., Hayes, J.P., and Tucker, G.F. 1997. Simulated stand characteristics and wood product yields from Douglas-fir plantations managed for ecosystem objectives. *Forest Ecology and Management*, 91: 205-219.

Barbour, R.J., Marshall, D.D., Parry, D.L., and Christensen, G. 2002. Do large trees always have higher wood product value? In: A.C. Johnson, R.W. Haynes, and R.M. Monserud (Eds.), Congruent management of multiple resources: proceedings from the wood compatibility initiative workshop. General Technical Report PNW-GTR-563 (pp. 135-144). U.S. Department of Agriculture, Forest Service, Pacific Northwest Research Station.

Battles, J.J., Shlisky, A.J., Barrett, R.H., Heald, R.C., and Allen-Diaz, B.H. 2001. The effects of forest management on plant species diversity in a Sierran conifer forest. *Forest Ecology and Management*, 146: 211-222.

Beese, W.J., and Bryant, A.A. 1999. Effect of alternative silvicultural systems on vegetation and bird communities in coastal montane forests of British Columbia, Canada. *Forest Ecology and Management*, 115: 231-242.

Beese, W.J., Dunsworth, G., and Perry, J. 2001. The forest project: three-year review and update. *Ecoforestry*, 16(4): 10-17.

Beschta, R.L., Pyles, M.R., Skaugset, A.E., and Surfleet, C.G. 2000. Peakflow responses to forest practices in the western Cascades of Oregon USA. *Forest Ecology and Management*, 233: 102-120.

Bisson, P.A., Raphael, M.G., Foster, A.D., and Jones, L.L.C. 2002. Influence of site and landscape features on vertebrate assemblages in small streams. In: A.C. Johnson, R.W. Haynes, and R.M. Monserud (Eds.). Congruent management of multiple resources: proceedings from the wood compatibility initiative workshop (pp. 61-72). General Technical Report PNW-GTR-563. Portland, OR: U.S. Department of Agriculture, Forest Service, Pacific Northwest Research Station.

Buck, C.J. 1934. Unpublished memo to Region 6 forest supervisors on sales policy and selective logging. October 16, 1934. Portland, OR. On file with: S. Hummel, Portland Forestry Sciences Laboratory, P.O. Box 3890, Portland, OR 97205, USA.

Buermeyer, K.R., and Harrington, C.A. 2002. Fate of overstory trees and patterns of regeneration 12 years after clearcutting with reserve trees in southwest Washington. *Western Journal of Applied Forestry*, 17(2): 78-85.

Bull, E.L., Torgerson, T.R., Blumton, A.K., McKenzie, C.M., and Wyland, D.S. 1995. Treatment of an old-growth stand and its effects on birds, ants, and large woody debris: a case study. General Technical Report PNW-GTR-353. Portland, OR: U.S. Department of Agriculture, Forest Service, Pacific Northwest Research Station. 12 p.

Bunnell, F.L., and Huggard, D.J. 1999. Biodiversity across spatial and temporal scales: problems and opportunities. *Forest Ecology and Management*, 115: 113-126.

Burns, R.M., and Honkala, B.H. 1990a. Silvics of North America: Vol. 1, Conifers. Agriculture Handbook 654. Washington, DC: U.S. Department of Agriculture, Forest Service. 675 p.

Burns, R.M., and Honkala, B.H. 1990b. Silvics of North America: Vol. 2, Hardwoods. Agriculture Handbook 654. Washington, DC: U.S. Department of Agriculture, Forest Service. 877 p.

Busing, R.T., and Garman, S.L. 2002. Promoting old-growth characteristics and long-term wood production in Douglas-fir forests. *Forest Ecology and Management*, 160: 161-175.

Cafferata, S.L. 1986. Douglas-fir stand establishment overview: western Oregon and Washington. In: C.D. Oliver, D.P. Hanley, and J.A. Johnson (Eds.), Douglas-fir: stand management for the future (pp. 211-219). Institute of Forest Resources 55. Seattle, WA: College of Forest Resources, University of Washington.

Carey, A.B. 2001. Experimental manipulation of spatial heterogeneity in Douglas-fir forests: effects on squirrels. *Forest Ecology and Management*, 152: 13-30.

Carey, A.B. 2003. Managing for wildlife: a key component for social acceptance of compatible forest management. Chapter 14. In: R.A. Monserud, R.W. Haynes, and A.C. Johnson (Eds.), Compatible forest management. Dordrecht, The Netherlands: Kluwer Academic Publishers.

Chambers, C.L., Carrigan, T., Sabin, T.E., Tappeiner, J., and McComb, W.C. 1997. Use of artificially created Douglas-fir snags by cavity nesting birds. *Western Journal of Applied Forestry*, 12(3): 93-97.

Chambers, C.L., and McComb, W.C. 1997. Effects of silvicultural treatments on wintering bird communities in the Oregon Coast Range. *Northwest Science*, 71(4): 298-304.

Chambers, C.L., McComb, W.C., and Tappeiner, J.C., II. 1999. Breeding bird responses to three silvicultural treatments in the Oregon Coast Range. *Ecological Applications*, 9(1): 171-185.

Cissel, J.H., Mayo, J.H., Garman, S.L. and Swanson, F.J. 2002. Application of landscape objectives to stand-level silviculture: Blue River, Oregon. In: S. Parker, and S.S. Hummel (Comps.), Beyond 2001: silvicultural odyssey to sustaining terrestrial and aquatic ecosystems—proceedings of the 2001 National Silviculture Workshop (pp. 21-31). General Technical Report PNW-GTR-546. Portland OR: U.S. Department of Agriculture, Forest Service, Pacific Northwest Research Station.

Cissel, J.H., Swanson, F.J., Grant, G.E., Olson, D.H., Gregory, S.V., Garman, S.L., Ashkenas, L.R., Hunter, M.G., Kertis, J.A., Mayo, J.H., McSwain, M.D., Swetland, S.G., Swindle, K.A., and Wallin, D.O. 1998. A landscape plan based on historic fire regimes for a managed forest ecosystem: the Augusta Creek Study. General Technical Report PNW-GTR-422. Portland, OR: U.S. Department of Agriculture, Forest Service, Pacific Northwest Research Station. 82 p.

Cleary, B.D., Greaves, R.D., and Hermann, R.K. 1978. Regenerating Oregon's forests. Corvallis, OR: Oregon State University Extension Service. 287 p.

Coates, K.D., and Burton, P.J. 1997. A gap-based approach for development of silvicultural systems to address ecosystem management objectives. *Forest Ecology and Management*, 99: 337-354.

Cole, E.C., McComb, W.C., Newton, M., Leeming, J.P., and Chambers, C.L. 1998. Response of small mammals to clearcutting, burning and glyphosate application in the Oregon Coast Range. *Journal of Wildlife Management*, 62(4): 1207-1216.

Cunningham, P.G. 2002. A survey of research on riparian responses to silviculture. In: A.C. Johnson, R.W. Haynes, and R.A. Monserud (Eds.), Congruent management of multiple resources: proceedings from the wood compatibility workshop (pp. 73-79). General Technical Report PNW-GTR-563. Portland, OR: U.S. Department of Agriculture, Forest Service, Pacific Northwest Research Station.

Curtis, R.O. 1995. Extended rotation and culmination age of coast Douglas-fir: old studies speak to current issues. Research Paper PNW-RP-485. Portland, OR: U.S. Department of Agriculture, Forest Service, Pacific Northwest Research Station. 49 p.

Curtis, R.O. 1997. The role of extended rotations. In: K.A. Kohm, and J.F. Franklin (Eds.), Creating a forestry for the 21st century (pp. 165- 171). Covelo, CA: Island Press.

Curtis, R.O., and Carey, A.B. 1996. Timber supply in the Pacific Northwest: managing for economic and ecological values in Douglas-fir forests. *Journal of Forestry*, 94(9): 4-7, 35-37.

Curtis, R.O., DeBell, D.S., Harrington, C.A., Lavender, D.P., St. Clair, J.B., Tappeiner, J.C., and Walstad, J.D. 1998. Silviculture for multiple objectives in the Douglas-fir region. General Technical Report PNW-GTR-435. Portland, OR: U.S. Department of Agriculture, Forest Service, Pacific Northwest Research Station. 123 p.

Daniel, T.W., Helms, J.A., and Baker, F.S. 1979. Principles of silviculture. New York: McGraw-Hill, Inc.

Dawkins, H.C., and Philip, M.S. 1998. Tropical moist forest silviculture and management: a history of success and failure. Wallingford, UK: CAB International. 359 p.

Deal, R.L., and Tappeiner, J.C. 2002. The effects of partial cutting on stand structure and growth of western hemlock-Sitka spruce stands in southeast Alaska. *Forest Ecology and Management*, 159: 173-186.

DeStefano, S. 2002. Regional and national issues for forest wildlife research and management. *Forest Science*, 48(2): 181-189.

Donoghue, E.M. 2003. Social values and compatible forest management. Chapter 15. In: R.A. Monserud, R.W. Haynes, and A.C. Johnson (Eds.), Compatible forest management. Dordrecht, The Netherlands: Kluwer Academic Publishers.

Drew, T.J., and Flewelling, J.W. 1979. Stand density management: an alternative approach and its application to Douglas-fir plantations. *Forest Science*, 25(3): 518-532.

Duncan, S., and Miner, C. 2000. Closer to the truth: 75 years of discovery in forest & range research. Portland, OR: U.S. Department of Agriculture, Forest Service, Pacific Northwest Research Station. 76 p.

Fenger, M. 1996. Implementing biodiversity conservation through the British Columbia Forest Practices Code. *Forest Ecology and Management*, 85: 67-77.

Ferguson, S.H., and Archibald, D.J. 2001. The ¾ power law in forest management: how to grow dead trees. *Forest Ecology and Management*, 169: 283-292.

Franklin, J.F. 1963. Natural regeneration of Douglas-fir and associated species using modified clear-cutting systems in the Oregon Cascades. Research Paper PNW-RP-3. Portland, OR: U.S. Department of Agriculture, Forest Service, Pacific Northwest Forest and Range Experiment Station. 14 p.

Franklin, J.F., Berg, D.R., Thornburgh, D.A., and Tappeiner, J.C. 1997. Alternative silvicultural approaches to timber harvesting: variable retention harvest systems. In: K.A. Kohm, and J.F. Franklin (Eds.), Creating a forestry for the 21st century (pp. 107-139). Covelo, CA: Island Press.

Franklin, J.F., and Halpern, C.B. 1988. Pacific Northwest Forests. In: M.G. Barbour and W.D. Billings (Eds.), North American terrestrial vegetation (pp. 123-159). Cambridge: Cambridge University Press.

Franklin, J.F., and Spies, T.A. 1991. Composition, function, and structure of old-growth Douglas-fir forests. In: L.F. Ruggiero, K.B. Aubry, A.B. Carey, and M.H. Huff (Eds.), Wildlife and vegetation of unmanaged Douglas-fir forests (pp. 71-81). General Technical Report PNW-GTR-285. Portland, OR: U.S. Department of Agriculture, Forest Service, Pacific Northwest Research Station.

Franklin, J.F., Spies, T.A., Van Pelt, R., Carey, A.B., Thornburgh, D.A., Berg, D.R., Lindenmayer, D.B., Harmon, M.E., Keeton, W.S., Shaw, D.C., Bible, K., and Chen, J. 2002. Disturbances and structural development of natural forest ecosystems with silvicultural implications, using Douglas-fir forests as an example. *Forest Ecology and Management*, 155: 399-423.

Fujimori, T. 2001. Ecological and silvicultural strategies for sustainable forest management. London: Elsevier.

Gayer, K. 1886. Der gemischte Wald [The mixed forest]. Berlin: Paul Parey Verlag. 168 p.

Gitzen, R.A., and West, S.D. 2002. Small mammal response to experimental canopy gaps in the southern Washington Cascades. *Forest Ecology and Management*, 168: 187-199.

Graham, R.T., Harvey, A.E., Jain, T.B., and Tonn, J.R. 1999. The effects of thinning and similar stand treatments on fire behavior in western forests. General Technical Report PNW-GTR-463. Portland, OR: U.S. Department of Agriculture, Forest Service, Pacific Northwest Research Station. 27 p.

Gregory, S.V. 1997. Riparian management in the 21st century. In: K.A. Kohm, and J.F. Franklin (Eds.). Creating a forestry for the 21st century (pp. 69-85). Covelo, CA: Island Press.

Hagar, J.C., McComb, W.C., and Emmingham, W.H. 1996. Bird communities in commercially thinned and unthinned Douglas-fir stands of western Oregon. *Wildlife Society Bulletin*, 24(2): 353-366.

Halpern, C.B., and McKenzie, D. 2001. Disturbance and post-harvest ground conditions in a structural retention experiment. *Forest Ecology and Management*, 154: 215-225.

Harper, J.L. 1977. Population biology of plants. Oxford: Oxford Academic Press. 892 p.

Harr, R.D. 1980. Streamflow after patch logging within the Bull Run municipal watershed, Oregon. Research Paper PNW-RP-268. Portland, OR: U.S. Department of Agriculture, Forest Service, Pacific Northwest Research Station.

Harrington, C.A., and Kallas, M.A. 2002. A bibliography for *Quercus garryana* and other geographically associated and botanically related oaks. General Technical Report PNW-GTR-554. Portland, OR: U.S. Department of Agriculture, Forest Service, Pacific Northwest Research Station. 115 p.

Harris, L.D. 1984. The fragmented forest: island biogeography theory and the preservation of biotic diversity. Chicago: University of Chicago Press. 211 p.

Hawley, R.C. 1937. The practice of silviculture with particular reference to its application in the United States of America. New York: John Wiley & Sons. 252 p.

Haynes, R.W., Monserud, R.A., and Johnson, A.C. 2002. Emergent results from the wood compatibility initiative. In: A.C. Johnson, R.W. Haynes, and R.M. Monserud (Eds.), Congruent management of multiple resources: proceedings from the wood compatibility initiative workshop (pp. 1-7). General Technical Report PNW-GTR-563. Portland, OR: U.S. Department of Agriculture, Forest Service, Pacific Northwest Research Station.

Haynes, R.W., Monserud, R.A., and Johnson, A.C. 2003. Compatible forest management: background and context. Chapter 1. In: R.A. Monserud, R.W. Haynes, and A.C. Johnson (Eds.), Compatible forest management. Dordrecht, The Netherlands: Kluwer Academic Publishers.

He, F., LaFrankie, J.V., and Song, B. 2002. Scale dependence of tree abundance and richness in a tropical rainforest, Malaysia. *Landscape Ecology*, 17: 559-568.

Heitzman, E. 2003. New forestry in Scotland. *Journal of Forestry*, 101: 36-39.

Helms, J.A. (Ed.). 1998. The dictionary of forestry. Bethesda, MD: Society of American Foresters. 210 p.

Hinkelmann, K., and Kempthorne, O. 1994. Design and analysis of experiments: Vol. 1., Introduction to experimental design. New York: John Wiley & Sons. 495 p.

Hobbs, S. (Ed.). 1992. Reforestation practices in southwestern Oregon and northern California. Corvallis, OR: Oregon State University Press. 465 p.

Hummel, S. 2002. Size-density relations in tropical forests: a role for research. *Journal of Tropical Forest Science*, 14(2) 277-281.

Hummel, S.S., Barbour, R.J., Hessburg, P.F., and Lehmkuhl, J.F. 2001. Ecological and financial assessment of late-successional reserve management. Research Note PNW-RN-531. Portland, OR: U.S. Department of Agriculture, Forest Service, Pacific Northwest Research Station. 25 p.

Hummel, S., Calkin, D., and Barbour, J. 2002. Landscape analysis with FVS and optimization techniques: efficient management planning for the Gotchen late successional reserve. In: N.L. Crookston, and R.N. Havis (Comps.), Second Forest Vegetation Simulator Conference Proceedings (pp. 78-82). RMRS-P-25. Ogden, UT: U.S. Department of Agriculture, Forest Service, Rocky Mountain Research Station.

Hunter, M.G. 2001. Management in young forests. Communique No. 3. Cascade Center for Ecosystem Management. R6-WILL-139. Eugene, OR: U.S. Department of Agriculture, Forest Service, Willamette National Forest. 28 p.

Jack, S.B., and Long, J.N. 1996. Linkages between silviculture and ecology: an analysis of density management diagrams. *Forest Ecology and Management*, 86: 205-220.

James, N.D.G. 1996. A history of forestry and monographic forestry literature in Germany, France, and the United Kingdom. In: P. McDonald, and J. Lassoie (Eds.), The literature of forestry and agroforestry (pp. 15-44). Ithaca, NY: Cornell University Press.

Kain, R.J.P., and Baigent, E. 1992. The cadastral map in the service of the state: a history of property mapping. Chicago: University of Chicago Press. 423 p.

Kauffman, J.B., Mahrt, M., Mahrt, L., and Edge, W.D. 2001. Wildlife of riparian habitats. In: D.H. Johnson, and T.A. O'Neil (Managing directors), Wildlife-habitat relationships in Oregon and Washington (pp. 361-388). Corvallis, OR: Oregon State University Press.

Kerns, B.K., Pilz, D., Ballard, H., and Alexander, S.J. 2003. Compatible management of understory forest resources and timber. Chapter 12. In: R.A. Monserud, R.W. Haynes, and A.C. Johnson (Eds.), Compatible forest management. Dordrecht, The Netherlands: Kluwer Academic Publishers.

Knowe, S.A., and Hibbs, D.E. 1995. Stand structure and dynamics of young red alder as affected by planting density. *Forest Ecology and Management*, 82: 69-85.

Kohm, K.A., and Franklin, J.F. 1997 (Eds.). Creating a forestry for the 21st century: the science of ecosystem management. Covelo, CA: Island Press. 475 p.

Koslowski, T.T., Kramer, P.J., and Pallardy, S.J. 1991. The physiological ecology of woody plants. New York: Academic Press. 657 p.

Lähde, E., Laiho, O., and Norokorpi, Y. 1999. Diversity-oriented silviculture in the boreal zone of Europe. *Forest Ecology and Management*, 118: 223-243.

Lindenmayer, D., and McCarthy, M.A. 2002. Congruence between natural and human forest disturbance: a case study from Australian montane ash forests. *Forest Ecology and Management*, 155: 319-335.

Long, J.N., McCarter, J.B., and Jack, S.B. 1988. A modified density management diagram for coastal Douglas-fir. *Western Journal of Applied Forestry*, 3: 88-89.

Lonsdale, W.M. 1990. The self-thinning rule: Dead or alive? *Ecology*, 71(4): 1373-1388.

Marshall, V.G. 2000. Impacts of forest harvesting on biological processes in northern forest soils. *Forest Ecology and Management*, 133: 43-60.

Marzluff, J.M., Millspaugh, K.R., Ceder, K.R., Oliver, C.D., Withey, J., McCarter, J.B., Mason, C.L., and Comnick, J. 2002. Modeling changes in wildlife habitat and timber revenues in response to forest management. *Forest Science*, 48: 191-202.

Matthews, J.D. 1989. Silvicultural systems. Oxford: Clarendon Press. 284 p.

McCarter, J.B., and Long, J.N. 1986. A lodgepole pine density management diagram. *Western Journal of Applied Forestry*, 1: 6-11.

McComb, W., and Lindenmayer, D. 1999. Dying, dead, and down trees. In: M. Hunter, (Ed.), Maintaining biodiversity in forest ecosystems (pp. 335-372). Cambridge: Cambridge University Press.

McCreary, D.D., and Perry, D.A. 1983. Strip thinning and selective thinning in Douglas-fir. *Journal of Forestry*, 81(6): 375-377.

McTague, J.P., and Patton, D.R. 1989. Stand density index and its application in describing wildlife habitat. *Wildlife Society Bulletin*, 17: 58-62.

Minore, D. 1979. Comparative autecological characteristics of northwestern tree species: a literature review. General Technical Report PNW-GTR-87. Portland, OR: U.S. Department of Agriculture, Forest Service, Pacific Northwest Forest and Range Experiment Station. 72 p.

Mitchell, A.K. 2001. Growth limitations for conifer regeneration under alternative silvicultural systems in a coastal montane forest in British Columbia, Canada. *Forest Ecology and Management*, 145: 129-136.

Mitchell, S.J., and Beese, W.J., 2002. The retention system: reconciling variable retention with the principles of silvicultural systems. *The Forestry Chronicle*, 78(3): 397-403.

Monserud, R.A. 2002. Large-scale management experiments in the moist maritime forests of the Pacific Northwest. *Landscape and Urban Planning*, 59: 159-180.

Monserud, R.A. 2003. Modeling stand growth and management. Chapter 6. In: R.A. Monserud, R.W. Haynes, and A.C. Johnson (Eds.), Compatible forest management. Dordrecht, The Netherlands: Kluwer Academic Publishers.

Moore, J.R., Maguire, D.A., Phillips, D.L., and Halpern C.B. 2002. Effects of varying levels and patterns of green-tree retention on amount of harvesting damage. *Western Journal of Applied Forestry*, 17(4): 202-206.

Muir, P.S., Mattingly, R.L., Tappeiner, J.C., II, Bailey, J.D., Elliott, W.E., Hagar, J.C., Miller, E.B., Peterson, E.B., and Starkey, E.E. 2002. Managing for biodiversity in young Douglas-fir forests of western Oregon. Biological Science Report USGS/BRD/BSR-2002-0006. [Place of publication unknown]: U.S. Geological Survey, Biological Resources Division. 76 p.

Newton, P.F. 1997. Stand density management diagrams: review of their development and utility in stand-level management planning. *Forest Ecology and Management*, 98: 251-265.

Nyberg, J.B. and Taylor, B. 1995. Applying adaptive management in British Columbia's forests. In: Proceedings of the FAO/ECE/ILO International Forestry Seminar (pp. 239-245). British Columbia Ministry of Forestry. Retrieved April 2003 from http: www.for.gov.bc.ca/hfp/amhome/apply/apply.htm.

Nyland, R.D. 2001. Even- to uneven-aged: the challenges of conversion. *Forest Ecology and Management*, 172: 291-300.

O'Hara, K.L. 2001. The silviculture of transformation—a commentary. *Forest Ecology and Management*, 151: 81-86.

O'Hara, K.L. 2002. The historical development of uneven-aged silviculture in North America. *Forestry*, 75(4): 339-346.

Oliver, C.D., and Larson, B.C. 1990. Forest stand dynamics. New York: McGraw-Hill, Inc. 467 p.

Olson, D.H., Chan, S.S., and Thompson, C.R. 2002. Riparian buffers and thinning design in western Oregon headwaters accomplish multiple resource objectives. In: A.C. Johnson, R.W. Haynes, and R.M. Monserud (Eds.), Congruent management of multiple resources: proceedings from the wood compatibility initiative workshop (pp. 81-91). General Technical Report PNW-GTR-563. Portland, OR: U.S. Department of Agriculture, Forest Service, Pacific Northwest Research Station.

Olson, D.H., Hagar, J.C., Carey, A.B., Cissel, J.H., and Swanson, F.J. 2001. Wildlife of west-side and high montane forests. In: D.H. Johnson, and T.A. O'Neil (Managing directors), Wildlife-habitat relationships in Oregon and Washington (pp. 187-212). Corvallis, OR: Oregon State University Press.

Ott, R.A., and Juday, G.P. 2002. Canopy gap characteristics and their implications for management in the temperate rainforests of southeast Alaska. *Forest Ecology and Management*, 159: 271-291.

Perry, D.A. 1994. Forest ecosystems. Baltimore, MA: Johns Hopkins University Press. 649 p.

Perry, D.A., and Amaranthus, M.P. 1997. Disturbance, recovery, and stability. In: K.A. Kohm, and J.F. Franklin (Eds.), Creating a forestry for the 21st century (pp. 31-56). Covelo, CA: Island Press.

Poage, N.J., and Tappeiner, J.C., II. 2002. Long-term patterns of diameter and basal area growth of old-growth Douglas-fir trees in western Oregon. *Canadian Journal of Forest Resources*, 32: 1232-1243.

Powell, J.W. 1879. Report on the lands of the arid region of the United States with a more detailed account of the lands of Utah. Washington, DC: Government Printing Office. 195 p.

Prescott, C.E. 1997. Effects of clearcutting and alternative silvicultural treatments on rates of decomposition and nitrogen mineralization in a coastal montane coniferous forest. *Forest Ecology and Management*, 95: 253-260.

Puettmann, K.J., Hibbs, D.E., and Hann, D.W. 1992. The dynamics of mixed stands of *Alnus rubra* and *Pseudotsuga menziesii*: extension of size-density analysis to species mixture. *Journal of Ecology*, 80: 449-458.

Reeves, G.H., Bisson, P.A., and Dambacher, J.M. 1998. Fish communities. In: R.J. Naiman, and R.E. Bilby (Eds.), River ecology and management (pp. 200-234). New York: Springer.

Reineke, L.H. 1933. Perfecting a stand-density index for even-aged forests. *Journal of Agricultural Resources*, 46: 627-638.

Ruggiero, L.F., Aubry, K.B., Buskirk, S.W., Lyon, L.J., and Zielinski, W.J. 1994. The scientific basis for conserving forest carnivores: American marten, fisher, lynx, and wolverine in the western United States. General Technical Report RM-GTR-254. Fort Collins, CO: U.S. Department of Agriculture, Forest Service, Rocky Mountain Research Station. 184 p.

Schabel, H.G., and Palmer, S.L. 1999. The Dauerwald: its role in the restoration of natural forests. *Journal of Forestry*, 97: 20-25.

Schowalter, T.D. 1995. Canopy arthropod communities in relation to forest age and alternative harvest practices in western Oregon. *Forest Ecology and Management*, 78: 115-125.

Schütz, J.P. 2002. Silvicultural tools to develop irregular and diverse forest structures. *Forestry*, 75(4): 329-337.

Seidel, K.W. 1979a. Natural regeneration after shelterwood cutting in a grand-fir-shasta red fir stand in Central Oregon. Research Paper PNW-RP-259. Portland, OR: U.S. Department of Agriculture, Forest Service, Pacific Northwest Forest and Range Experiment Station. 23 p.

Seidel, K.W. 1979b. Regeneration in mixed conifer shelterwood cuttings in the Cascade Range of eastern Oregon. Research Paper PNW-RP-264. Portland, OR: U.S. Department of Agriculture, Forest Service, Pacific Northwest Forest and Range Experiment Station.

Seidel, K.W. 1983. Regeneration in mixed conifer and Douglas-fir shelterwood cuttings in the Cascade Range of Washington. Research Paper PNW-RP-314. Portland, OR: U.S. Department of Agriculture, Forest Service, Pacific Northwest Forest and Range Experiment Station. 17 p.

Seidel, K.W., and Cooley, R. 1974. Natural reproduction of grand fir and mountain hemlock after shelterwood cutting in central Oregon. Research Paper PNW-RN-229. Portland, OR: U.S. Department of Agriculture, Forest Service, Pacific Northwest Forest and Range Experiment Station. 10 p.

Shaw, J.D. 2000. Application of stand density index to irregularly structured stands. *Western Journal of Applied Forestry*, 15: 40-42.

Simberloff, D. 1999. The role of science in the preservation of forest biodiversity. *Forest Ecology and Management*, 115: 101-111.

Smith, D.M. 1986. The practice of silviculture. (8th ed.). New York: John Wiley & Sons. 527 p.

Smith, F.W., and Long, J.N. 1987. Elk hiding and thermal cover guidelines in the context of lodgepole pine stand density. *Western Journal of Applied Forestry*, 2(1): 6-10.

Spies, T.A. 1998. Forest structure: a key to the ecosystem. *Northwest Science*, 72: 34-39.

Spies, T.A., and Franklin, J.F. 1991. The structure of young, mature, and old-growth Douglas-fir forests in Oregon and Washington. In: L.F. Ruggiero, K.B. Aubry, A.B. Carey, and M.H. Huff (Eds.), Wildlife and vegetation of unmanaged Douglas-fir forests (pp. 91-109). General Technical Report PNW-GTR-285. Portland, OR: U.S. Department of Agriculture, Forest Service, Pacific Northwest Research Station.

Spies, T.A., and Johnson, K.N. 2003. The importance of scale in assessing the compatibility of forest commodities and biodiversity. Chapter 8. In: R.A. Monserud, R.W. Haynes, and A.C. Johnson (Eds.), Compatible forest management. Dordrecht, The Netherlands: Kluwer Academic Publishers.

Steen, H.K. 1998. Forest Service research: finding answers to conservation's questions. Durham, NC: Forest History Society. 102 p.

Stevens, J.A., and Montgomery, C.A. 2002. Understanding the compatibility of multiple uses on forest land: a survey of multiresource research with application to the Pacific Northwest. General Technical Report PNW-GTR-539. Portland, OR: U.S. Department of Agriculture, Forest Service, Pacific Northwest Research Station. 44 p.

Sturdevant, B.R., Bissonette, J.A., and Long, J.N. 1996. Temporal and spatial dynamics of boreal forest structure in western Newfoundland: silvicultural implications for marten habitat management. *Forest Ecology and Management*, 87: 13-25.

Swanson, F.J., Cissel, J.H., and Reger, A. 2003. Landscape management: diversity of approaches and points of comparison. Chapter 9. In: R.A. Monserud, R.W. Haynes, and A.C. Johnson (Eds.), Compatible forest management. Dordrecht, The Netherlands: Kluwer Academic Publishers.

Symstad, A.J., Chapin, F.S., III, Wall, D.H., Gross, K.L., Huenneke, L.F., Mittelbach, G.G., Peters, D.P.C., and Tilman, D. 2003. Long-term and large-scale perspectives on the relationship between biodiversity and ecosystem functioning. *BioScience*, 53 (1): 89-98.

Tappeiner, J.C., II, Huffman, D., Marshall, D., Spies, T.A., and Bailey, J.D. 1997a. Density, ages, and growth rates in old-growth and young-growth forests in coastal Oregon. *Canadian Journal of Forest Resources*, 27: 638-648.

Tappeiner, J.C., Lavender, D.C., Walstad, J., Curtis, R.O., and DeBell, D.S. 1997b. Silvicultural systems and regeneration methods: current practices and new alternatives. In: K.A. Kohm, and J.F. Franklin (Eds.), Creating a forestry for the 21st century (pp.151-164). Covelo, CA: Island Press.

Tappeiner, J.C., and Zasada, J.C. 1993. Establishment of salmonberry, salal, vine maple and bigleaf maple seedlings in the coastal forests of Oregon. *Canadian Journal of Forest Resources*, 23 1775-1780.

Tesch, S.D. 1995. The Pacific Northwest Region. In: J.W. Barrett (Ed.), Regional silviculture of the United States (3rd ed.). (pp. 499-558). New York: Wiley & Sons, Inc.

Tittler, R., and Hannon, S.J. 2000. Nest predation in and adjacent to cutblocks with variable tree retention. *Forest Ecology and Management*, 136: 147-157.

Toumey, J.W., and Korstian, C.F. 1937. Foundations of silviculture upon an ecological basis. New York: John Wiley & Sons. 456 p.

Troup, R.S. 1928. Silvicultural systems. Clarendon Press, Oxford, England. 199 p.

U.S. Department of Agriculture [USDA]. 1893. Report of the Secretary of Agriculture. Washington, DC: Government Printing Office. 608 p.

U.S. Department of Agriculture, Forest Service [USDA]. 2003. Potential pitfalls in applying variable density management. Pacific Northwest Research Station. Retrieved March 20, 2003 from http://www.fs.fed.us/pnw/olympia/silv/selected_studies/variable/pitfalls.htm.

U.S. Department of Agriculture, Forest Service; U.S. Department of the Interior, Bureau of Land Management [USDA/USDI]. 1994. Record of decision for amendments to Forest Service and Bureau of Land Management planning documents within the range of the northern spotted owl. [Place of publication unknown]. 74 p. [plus Attachment A: standards and guidelines].

Vogel, G. 1990. Der einfluss des Nationalsocialismus auf die deutsche Forstwirtschaft als Voraussetzung fuer die Entstehung der "Zweiten" Dauerwaldbewegung [The influence of National Socialism on German forestry as a prerequisite for the genesis of the second Dauerwald movement]. *Forst und Holz*, 7: 171-176.

Wardman, C.W., and Schmidt, M.G. 1998. Growth and form of Douglas-fir adjacent to persistent vine maple gaps in southwestern British Columbia. *Forest Ecology and Management*, 106: 223-233.

Weigand, J.F., and Burditt, A.L. 1992. Economic implications for management of structural retention on harvest units at the Blue River Ranger District, Willamette National Forest, Oregon. Research Note PNW-RN-510. Portland, OR: U.S. Department of Agriculture, Forest Service, Pacific Northwest Research Station. 17 p.

Weller, D.E. 1987. A reevaluation of the –3/2 power rule of plant self-thinning. *Ecological Monographs*, 57(1): 23-43.

Wilhere, G.F. 2003. Simulations of snag dynamics in an industrial Douglas-fir forest. *Forest Ecology and Management*, 174: 521-539.

Williamson, R.L. 1973. Results of shelterwood harvesting of Douglas-fir in the Cascades of western Oregon. Research Paper PNW-RP-161. Portland, OR: U.S. Department of Agriculture, Forest Service, Pacific Northwest Forest and Range Experiment Station. 13 p.

Williamson, R.L., and Ruth, R.H. 1976. Results of shelterwood cutting in western hemlock. Research Paper PNW-RP-201. Portland, OR: U.S. Department of Agriculture, Forest Service, Pacific Northwest Forest and Range Experiment Station. 25 p.

Wilson, E.O. 1988. Biodiversity. Washington, DC: National Academy Press. 521 p.

Wipfli, M.S., Deal, R.L, Hennon, P.E., Johnson, A.C., Edwards, R.T., DeSanto, T.L., Takashi, G., Orlikowska, E.H., Bryant, M.D., Schultz, M.E., LeSage, C., Kimbirauskus, R., and D'Amore, D.V. 2003. Compatible management of red alder-conifer ecosystems in southeastern Alaska. Chapter 3. In: R.A. Monserud, R.W. Haynes, and A.C. Johnson (Eds.), Compatible forest management. Dordrecht, The Netherlands: Kluwer Academic Publishers.

Wolff, J.O., and Zasada, J.C. 1975. Red squirrel response to clearcut and shelterwood systems in interior Alaska. Research Note PNW-RN-255. Portland, OR: U.S. Department of Agriculture, Forest Service, Pacific Northwest Forest and Range Experiment Station. 7 p.

Worthington, N.P. 1953. Reproduction following small group cuttings in virgin Douglas-fir. Research Note PNW-RN-84. Portland, OR: U.S. Department of Agriculture, Forest Service, Pacific Northwest Forest and Range Experiment Station. 5 p.

Zavala, M.A., and Oria, J.A. 1995. Preserving biological diversity in managed forests: a meeting point for ecology and forestry. *Landscape and Urban Planning*, 31: 363-378.

Zenner, E.K., Acker, S.A., and Emmingham, W.H. 1998. Growth reduction in harvest-age, coniferous forests with residual trees in the western central Cascade Range of Oregon. *Forest Ecology and Management*, 102: 75-88.

Appendix 1. List of species used in text.

Common name	**Latin name**
Plants	
Alaska cedar	*Chamaecyparis nootkatensis* (D. Don) Spach
Bigleaf maple	*Acer macrophyllum* Pursh
Douglas-fir	*Pseudotsuga menziesii* (Mirb.) Franco
Grand fir	*Abies grandis* (Dougl. ex D. Don) Lindl.
Lodgepole pine	*Pinus contorta* Dougl. ex Loud.
Oregon white oak	*Quercus garryana* Dougl. ex Hook.
Pacific silver fir	*Abies amabilis* Dougl. ex Forbes
Ponderosa pine	*Pinus ponderosa* Dougl. ex Laws.
Red alder	*Alnus rubra* Bong.
Salal	*Gaultheria shallon* Pursh
Sitka spruce	*Picea sitchensis* (Bong.) Carr.
Vine maple	*Acer circinatum* Pursh
Western hemlock	*Tsuga heterophylla* (Raf.) Sarg
Western larch	*Larix occidentalis* Nutt.
Western redcedar	*Thuja plicata* Donn ex D. Don
Animals	
Brown-headed cowbird	*Molothrus ater*
Northern spotted owl	*Strix occidentalis caurina*
Red squirrel	*Tamiasciurus hudsonicus*

Appendix 2. Silvicultural studies from the temperate coastal rainforest region of North America that include response variables other than (or in addition to) wood.

Citation	**Region**[1]	**Study type**[2]	**Size (and total number of treated units)**	**Silvicultural treatment**[3]	**Response variable(s)**[4]
Bailey et al. 1998	OCA, OCR (c)	OBS	>10 ha (80)	Thinned from below	Cover, abundance, and species richness of understory vascular plants
Battles et al. 2001	SONC (b)	OBS	<0.1-8 ha (16)	Clearcut, shelterwood, group selection, single tree selection	Species richness of understory vascular plants
Beese and Bryant 1999	BC (b)	DE (3)	8.6-69 ha (11)	Shelterwood, patch clearcut, green-tree retention	Cover, frequency and number of understory plant species; breeding bird species richness and abundance
Beschta et al. 2000	OCA (b)	OBS	60 ha-640 km (2)	Clearcut and broadcast burn, with and without road construction	Peakflow

Appendix 2. Silvicultural studies from the temperate coastal rainforest region of North America that include response variables other than (or in addition to) wood (continued).

Citation	**Region**[1]	**Study type**[2]	**Size (and total number of treated units)**	**Silvicultural treatment**[3]	**Response variable(s)**[4]
Buermeyer and Harrington 2002	WCA (a)	OBS	15 ha (1)	Clearcut with reserves	Overstory tree mortality and conifer regeneration
Bull et al. 1995	BM (a)	OBS	15-22 ha (2)	Thinned from below	Abundance of birds, ants, dead wood; timber volume and value
Busing and Garman 2002	OCA (b)	SIM	Not applicable	Clearcut, thinning from below	Wood volume and quality, canopy height diversity
Carey 2001	WCO (b)	DE (4)	13 ha (16)	Variable density thinning	Squirrel abundance
Chambers and McComb 1997	OCR (b)	DE (3)	5-18 ha (12)	Group selection, green-tree, shelterwood	Abundance of wintering birds

Appendix 2. Silvicultural studies from the temperate coastal rainforest region of North America that include response variables other than (or in addition to) wood (continued).

Citation	Region[1]	Study type[2]	Size (and total number of treated units)	Silvicultural treatment[3]	Response variable(s)[4]
Chambers et al. 1997, 1999	OCR (b)	DE (3)	5.5-17.8 ha (29)	Group selection, green-tree, shelterwood	Abundance of breeding birds, cavity-nesting birds, condition and use of artificially created snags
Cole et al. 1998	OCR (c)	DE (3)	8-12 ha (9)	Clearcut; clearcut and bum; clearcut, burn and glyphosate	Small mammal capture rates
Deal and Tappeiner 2002	AK (c)	OBS	10 ha (18)	Partial cutting	Species composition and richness, tree cohorts and growth, and understory vascular plants in Sitka spruce/western hemlock forest

Appendix 2. Silvicultural studies from the temperate coastal rainforest region of North America that include response variables other than (or in addition to) wood (continued).

Citation	Region[1]	Study type[2]	Size (and total number of treated units)	Silvicultural treatment[3]	Response variable(s)[4]
Gitzen and West 2002	WCA (b)	OBS	38-1963 m^2 (33)	Experimental gap	Abundance, persistence, and occurrence of small mammals
Hagar et al. 1996	OCR (c)	OBS	65-510 ha (16)	Thinned from below	Abundance and diversity of breeding and wintering birds
Halpern and McKenzie 2001	WCA, SONC (d)	DE (6)	13 ha (24)	Aggregated retention, dispersed retention	Ground cover, slash depth, woody debris
Harr 1980	OCA (b)	OBS	59-253 ha (3)	Clearcut and road construction	Peakflow
Mitchell 2001	BC (b)	DE (3)	9 ha (12)	Clearcut, patch cut, green tree, shelterwood	Foliage morphology and physiology of planting seedlings

Appendix 2. Silvicultural studies from the temperate coastal rainforest region of North America that include response variables other than (or in addition to) wood (continued).

Citation	Region[1]	Study type[2]	Size (and total number of treated units)	Silvicultural treatment[3]	Response variable(s)[4]
Olson et al. 2002	OCR (c)	OBS	Variable width stream buffers (6-126 m (13)	Variable retention thinning	Aquatic vertebrates and macroinvertebrates, microsite conditions, estimated wood volume
Prescott 1997	BC (b)	DE (3)	>9 ha (9)	Patch cut, green-tree retention, shelterwood	Litter decomposition and nitrogen mineralization
Schowalter 1995	OCA (b)	OBS	10-25 ha (24)	Partial cutting; clearcut	Canopy arthropod population intensities and herbivory
Weigand and Burditt 1992	OCA (b)	OBS	7-33 ha (4)	Structural (green-tree) retention	Net potential merchantable volume, net potential lumber value
Wolff and Zasada 1975	AK (a)	OBS	1-2.4 ha (8)	Clearcut, shelterwood	Red squirrel (*Tamiasciurus hudsonicus*) middens

Appendix 2. Silvicultural studies from the temperate coastal rainforest region of North America that include response variables other than (or in addition to) wood (continued).

[1] Region: AK= Alaska; BC = British Columbia: BM = Blue Mountains (eastern Oregon); OCA = Oregon Cascade Range; OCR = Oregon Coast Range; SONC = southern Oregon/northern California; WCA= Washington Cascade Range; WCO = Washington Coast/Olympic Peninsula. a = one location, b = more than one location within a geographically restricted area (e.g., an experimental forest), c = multiple locations within part of a region, and d = several locations distributed across one or more regions.
[2] Study type: DE = designed experiment (uses standard statistical principles (e.g., randomization) for scientific experimentation (see Hinkelmann and Kempthome 1994 for discussion)). Number of replicates given in parentheses. OBS = observational study (not a designed experiment); SIM = treatments "applied" using computer simulation.
[3] Silvicultural treatment: refer to the accompanying text for descriptions and definitions of treatments. Simulation (SIM) studies "apply" treatments using computer simulation.
[4] Response variables: Studies that include only wood quality or volume as response variables are not included.

Chapter 5

Experimental Approaches to Joint Forest Production

Robert A. Monserud
USDA Forest Service, Pacific Northwest Research Station,
620 SW Main St., Suite 400, Portland, OR 97205, USA

1. Introduction

This chapter examines large-scale management experiments in joint production of forest resources. This is a somewhat limited body of knowledge, mostly because the emphasis on joint production of multiple products and values from the forest is so recent (Stevens and Montgomery 2002). Large-scale management experiments emphasize joint production and alternatives to clearcutting, therefore they lie at the heart of compatible forest management. Scale becomes an issue because compatibility is easier to reach at larger scales, such as the watershed or the landscape (Haynes and Monserud 2002). Classical experimentation (using randomization, replication, and controls) in ecological management, at scales as large as a landscape is nearly impossible because of logistical and administrative problems (Carpenter 1998, Monserud 2002). Therefore, I begin at the scale of the forest stand, the unit of forest management, and then move on to the watershed scale.

Experimentation is the heart of science. In all experiments the goal is to reduce the number of competing hypotheses and make correct population inferences by controlling important factors through the classical methods of randomization, replication, and controls developed by Fisher (1925). Manipulative experiments, if properly designed and executed, can identify a causal link between a manipulated variable and some measured response (Tilman 1989). Observational, correlative studies cannot establish a causal link, nor can chronosequences. Although the theoretical advantage of large-scale manipulative experiments is great, a plethora of problems awaits the experimenter at the large scale.

R.A. Monserud, R.W. Haynes and A.C. Johnson (eds.)
Compatible Forest Management. 121-143

Large-scale field experiments are unusual in science, but are critically needed to improve enviromental management and policies (Carpenter 1998). In such experiments, the treatment unit is large enough to include the relevant physical, chemical, and biotic context of the processes being studied. Because the large-scale context is designed into the experiment, results can be directly interpreted at the scale of management that produced the manipulation (Carpenter 1998). In contrast, comparative or retrospective studies examine context without direct manipulation or control of key factors, and small-scale experiments study mechanisms out of context (Carpenter 1998). Small-scale experiments introduce a change-of-scale bias when they are applied at the larger operational scale of management (Monserud 2002).

Hurlbert (1984) rattled a hornet's nest by pointing out the prevalence of pseudoreplication in field ecology, and that it invalidated statistical inference. Replicates that are not independent are pseudoreplicates. They artificially increase sample size, bias hypothesis tests, and give a false sense of the power of the test of the experiment (Hurlbert 1984). In short, the strength of the experimental evidence is overstated (Hargrove and Pickering 1992). Although the existence of a spatial environmental gradient is a common cause of pseudoreplication, a more serious cause rests with the scientist: poor experimental design or execution, and improper statistical analysis (Hargrove and Pickering 1992). Repeated measurements over time also are pseudoreplicates because they are not independent, although sound statistical methods are available for accounting for the time dependency (e.g., Proc MIXED in SAS version 8.1; SAS 2000).

Randomized block designs (Fisher 1925) are an excellent tool for countering spatial pseudoreplication. Blocking ensures that treatment units are interspersed rather than segregated, replication increases the precision of the estimate, randomization eliminates possible bias and improves the accuracy of the estimate, and controls provide the proper treatment difference (Monserud 2002).

With these general comments in mind, I now move on to specific applications, namely, a series of joint production forestry experiments in the Pacific Northwest (western Oregon and Washington, coastal British Columbia, and southeastern Alaska).

2. Large-Scale Management Experiments in the Pacific Northwest

Even-age plantation management has been the dominant silvicultural system in the Douglas-fir (*Pseudotsuga menziesii* (Mirb.) Franco) region (western Oregon and Washington) for the past 50 years (Curtis 1998). The management history of the coastal forests of British Columbia and southeastern Alaska is quite similar to that of the Douglas-fir region; it is characterized by a strong reliance on clearcut logging (especially old growth) followed by even-age plantation management (Beese and Bryant 1999, McClellan et al. 2000). Collectively, these forests dominate the Pacific Northwest essentially as a continuous temperate rainforest. Although several experiments with shelterwood cutting in mature and old-growth stands (e.g., Williamson 1973) have been done, well-documented comparative trials of other possible silvicultural systems are lacking in the region (Curtis 1996). The consequent lack of research into alternatives to clearcutting has severely handicapped current efforts to meet changing objectives and public concerns (Curtis 1998).

A frequent misperception of traditional forest management is that it is an outgrowth of scientific experimentation (Franklin et al. 1999). Rather, it is rare that silvicultural systems are formulated from a systematic scientific approach tested prior to widespread application (Franklin et al. 1999). Dispersed patch clearcutting was widely adopted in the Douglas-fir region after World War II without experimental testing, largely with the support of Isaac (1943) and other forestry experts of the day. Alternative silvicultural approaches were rejected. Since then, foresters have accumulated an enormous amount of information and experience on even-age silviculture and plantation management (Curtis et al. 1998). Nevertheless, this scientific basis was lacking when widespread implementation of modern plantation management began decades earlier in the Douglas-fir region.

Today, forest managers are faced with enormous social pressure to find alternatives to clearcutting (Peterson and Monserud 2002). Again, there are few experimental guidelines to draw upon. One stark example is provided by the Weyerhaeuser Company's purchase of MacMillan Bloedel Limited in 1999. Months earlier, MacMillian had made a public commitment to end clearcutting on their lands in British Columbia. In effect, Weyerhaeuser purchased this social contract along with the forestry assets. The lack of a well-developed scientific basis for alternatives to clearcutting did not allow this implicit contract to be suspended or broken. As a result, conversion to a variable-retention management system that addresses multiple forest values is being implemented in British Columbia, and pilot research programs are in place (Weyerhaeuser 2000).

Monserud (2002) examines several large-scale, multidisciplinary silvicultural experiments from the Pacific Northwest. These are stand-level experiments with various types of partial cuts and variable retention. They involve various thinnings, from patch cuts to variable-density regimes. All of these studies are recent (begun in the 1990s) and have not been completed (Figure 1).

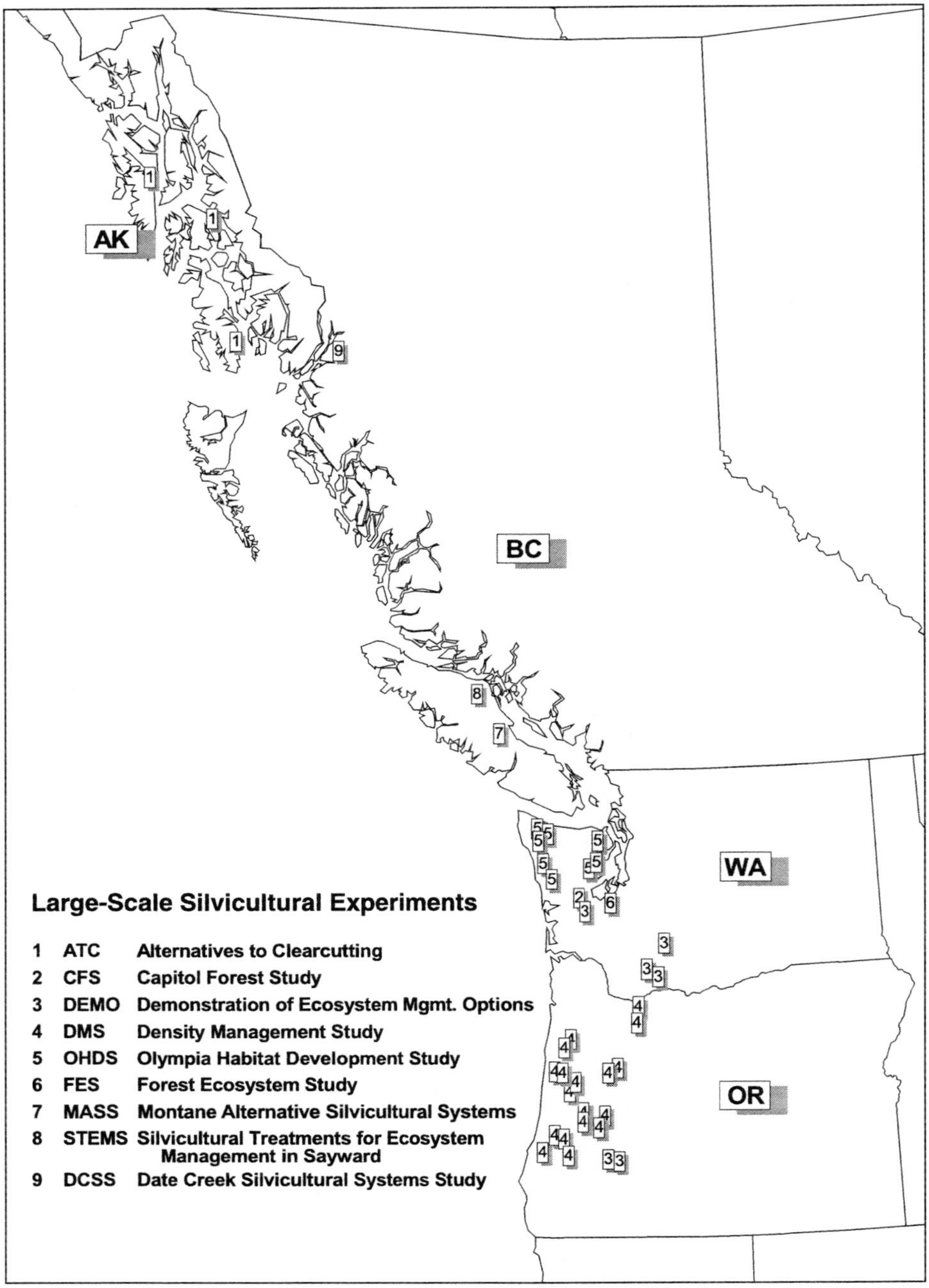

Figure 1. Location of the blocks in each of the large-scale silvicultural experiments listed in Table 1.

Large-scale silvicultural experiments in the Pacific Northwest:

Acronym	**Description**
ATC	Alternatives to Clearcutting (Alaska)
CFS	Washington Department of Natural Resources Capitol Forest Study (Washington)
DEMO	Demonstration of Ecosystem Management (Washington, Oregon)
DMS	Density Management Study (Oregon)
OHDS	Olympic Habitat Development Study (Washington)
FES	Forest Ecosystem Study (Washington)
MASS	Montane Alternative Silvicultural Systems (British Columbia)
STEMS	Silvicultural Treatments for Ecosystem Management in Sayward (British Columbia, Canada)
DCSS	Date Creek Silvicultural Systems Study (British Columbia, Canada)

Preliminary reports are available for most of these studies (e.g., Coates and Burton 1997, Beese and Bryant 1999, Carey et al. 1999, Halpern and Raphael 1999, McClellan et al. 2000). All are implementing silvicultural alternatives to the widely used plantation management of the previous 50 years. Some experiments are attempting to hasten the approach to old-growth structure and composition. Several are trying to increase within-stand heterogeneity. Others are examining the social acceptability of wood production (CFS, DEMO, STEMS). All the experiments are multidisciplinary and examine particular forest values such as biodiversity or wildlife habitat in addition to wood production (Table 1). All use replicated randomized block designs with controls (Fisher 1925) and all are on public land of some sort (state, federal, or provincial).

One of the most unusual and important aspects of this collection of silvicultural experiments is that the treatment units are large enough to be commercially operational (6 to 32 hectares (ha), with most between 13 and 20 ha). Using large, operational units as treatment areas has several important advantages over small research plots:

(1) It is easier to generalize management results to the watershed and landscape scale because the spatial variation is accurately represented by the experimental units (Carpenter 1998),

(2) Visual acceptance can be determined by direct observation of the treatments on the landscape,

(3) Larger units can cover the home range of focus animals (e.g., northern flying squirrel (*Glaucomys sabrinus*)) better than small research plots, and

Table 1. Summary of stand-level experiments linking wood production with at least one additional resource outcome (wildlife, aquatics, biodiversity, or social values). An UPPER CASE convention indicates studies with a given outcome as a major factor, and lower case indicates that the factor is minor or merely a survey (e.g., ATC vs. atc). The *All* level yields no wood production but is the necessary experimental control, and *None* indicates a clearcut.

	Large-scale management experiments[1]				
	Residual density levels				
Outcome	**All**	**High**	**Medium**	**Low**	**None**
Wood * *Wildlife*	atc	atc	atc	atc	atc
	mass		mass	mass	mass
	DCSS	DCSS		DCSS	DCSS
	OHDS		OHDS		
	DEMO	DEMO	DEMO	DEMO	
	FES		FES		
	cfs	cfs		cfs	cfs
	dms	dms	dms	dms	
Wood * *Aquatics*	ATC	ATC	ATC	ATC	ATC
	mass		mass	mass	mass
	dcss	dcss		dcss	dcss
	DMS	DMS	DMS		
Wood * *Biodiversity*	MASS		MASS	MASS	MASS
	STEMS	STEMS	STEMS	STEMS	STEMS
	DCSS	DCSS		DCSS	DCSS
	OHDS		OHDS		
	DEMO	DEMO	DEMO	DEMO	
	FES		FES		
	cfs	cfs		cfs	cfs
	DMS	DMS	DMS	DMS	
Wood * *Social*	STEMS	STEMS	STEMS	STEMS	STEMS
	CFS	CFS		CFS	CFS
	DEMO	DEMO	DEMO	DEMO	
	dms	dms	dms	dms	

[1] Codes:
ATC: Alternatives to Clearcutting (AK)
CFS: Washington DNR Capitol Forest Study (WA)
DEMO: Demonstration of Ecosystem Management (WA, OR)
DMS: Density Management Study (OR)
OHDS: Olympic Habitat Development Study (WA)
FES: Forest Ecosystem Study (WA)
MASS: Montane Alternative Silvicultural Systems (BC)
STEMS: Silvicultural Treatments for Ecosystem Management in Sayward (BC)
DCSS: Date Creek Silvicultural Systems Study (BC)

(4) They demonstrate that the management treatments are both economical and feasible to implement because—by definition—they are operational.

If treatment units are to be commercially operational, the scientist must design treatments that are easily understood and applied. This operational requirement generally precludes uniform research plots, and is likely to introduce considerable variation among study sites and treatment replications. Combining the operational size requirement within the context of a well-designed experiment should ensure that sound statistical inferences can be drawn from the results (Carpenter 1998), and that the results can provide a scientific basis for new forest management systems.

In the past, silvicultural experiments were rather small and rarely operational. For example, the 1925 Wind River Douglas-fir plantings were made in 1.1-ha blocks (Reukema 1979), and the levels-of-growing-stock plots were 0.08 ha (Curtis and Marshall 1986). Because small plots are more uniform, experiments that use them should have a smaller variance than the large operational units discussed here. A small variance is useful for detecting treatment differences, but the lack of uniform conditions, especially in mountainous terrain, means that results from small experimental plots may not generalize to the forest landscape with the same precision (Hobbs 1999). Basically, this is a bias due to a change of scale. Experiments that use large, operational plots accept the tradeoff of a larger variance (both within-block and among-block) to gain greater assurance that the experimental results will generalize better to the population as a whole. The crucial advantage is that results can be directly interpreted at the scale of management that produced the manipulation because the large-scale context is designed into the experiment.

All of these studies are trials looking for viable alternative silvicultures that could replace clearcutting. Whether called green-tree retention (DEMO), variable-density thinning (OHDS) and (FES), or variable retention (MASS), the types of silviculture under examination are alternatives to traditional even-age management. They explicitly consider structural and spatial diversity unlike the spatially uniform stand treatments common with silvicultural systems favored in the Pacific Northwest during the past 50 years. Examination of the silvicultural treatments in the experiments reveals a broad search for structural diversity. A few examples of experimental designs illustrate this:

- ATC (McClellan et al. 2000): Nine 18-ha randomized treatment units per block consisting of an uncut control (100% retention), four treatments with uniform spacing of residual trees (0%, 5%, 25%, 75% basal area retention), and four treatments with a variety of patches of trees either retained (25%, 75%) or removed (25%, 75%) (see Figures 2 and 3). Thus, the factorial design is simultaneously testing for residual density, spatial pattern (uniform vs. patchy), and the type and size of patches (gaps or uncut clumps).

- DEMO (Aubry et al. 1999): Six square 13-ha randomized treatment units per block consisting of a control (100% retention), 75% aggregated retention (three circular 1-ha patches removed), 40% aggregated retention (five undisturbed circular 1-ha patches are retained), 15% aggregated retention (two undisturbed circular 1-ha patches are retained), 40% and 15%

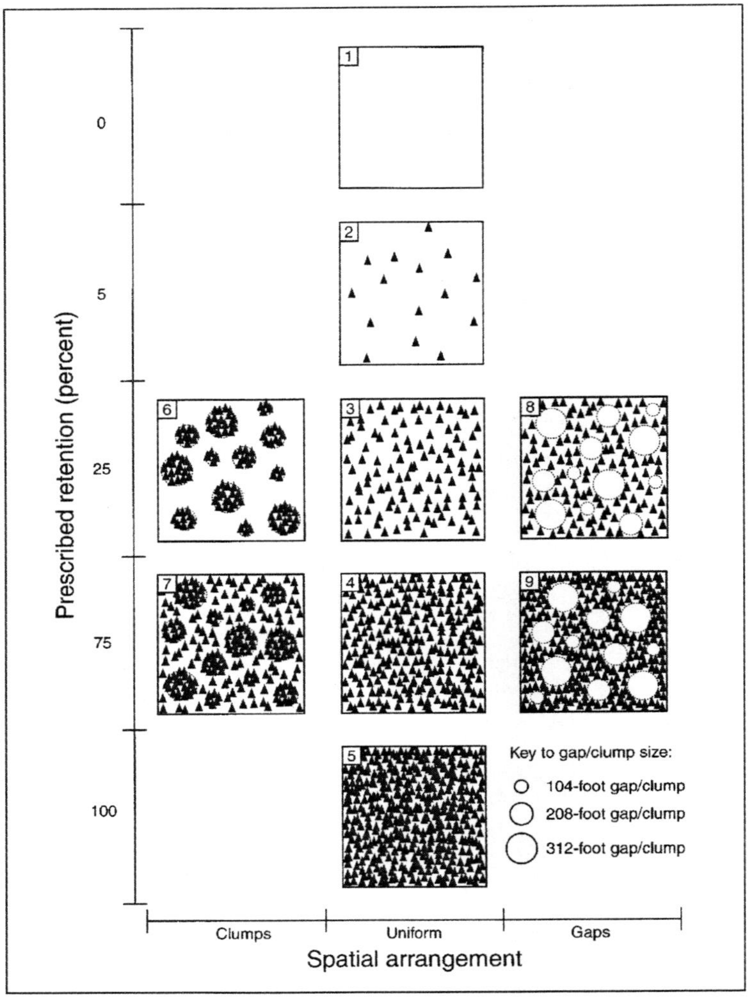

Figure 2. Schematic layout of the ATC replicates (McClellan et al. 2000).

dispersed retention (residual trees evenly spaced). This experiment is testing the effects of alternative harvesting treatments (especially green-tree retention) on wildlife, vegetation, and hydrology (Halpern and Raphael 1999). From the air, the circular patches on a square field make the aggregated retention treatments look like dice (Figure 4).

- FES (Carey et al. 1999): Four 13-ha randomized treatment units per block consisting of an uncut control (100% retention), variable-density thinning with three thinning intensities and underplanting, den-augmentation-only (cavities in live trees and nest boxes), and den-augmentation plus variable-density thinning and underplanting. The variable-density thinnings are

Figure 3. A unit at Hanus Bay, Alaska after treatment (25% retention) in the Alternatives to Clearcutting study of McClellan et al. (2000).

Figure 4. Two contrasting 13-ha replicates in the DEMO study (Halpern and Raphael 1999). The left side shows the 75% retention treatment: three small circular harvest areas (1 ha each) were cut, leaving the remaining 75% of the area uncut. Future regeneration cuttings may occur in 20- to 30-year cycles, and some thinnings may be done at the same time. The right side shows the 40% retention (aggregated) treatment: five circular patches of forest, each 1 ha in size, are left uncut and the rest of the 13-ha stand harvested. The uncut patches, as an undisturbed reservoir of biological material, may lead to rapid recolonization and recovery of species where trees were cut.

applied by randomly assigning one of four different relative density levels (Curtis 1982) to each 40-m by 40-m (meter) square cell across the entire 13-ha treatment unit. The FES study tests factors for both residual density and wildlife habitat simultaneously. The STEMS project on Vancouver Island is using essentially the same design.

The goal in these experiments is to find alternative silviculture treatments that enhance wildlife habitat, biodiversity, or the conservation of aquatic resources in a manner that is socially acceptable (Table 1). All have well-designed replicated experiments to evaluate various intermediate thinning levels and alternatives to clearcutting.

All of these studies examine the effect of silvicultural treatments on both wildlife habitat and biodiversity. In most cases, the silvicultural treatments place considerable emphasis on accelerating development of old-forest structural characteristics and retaining biological legacies. The interaction of social acceptance and wood production is a major component in three of these studies (CFS, DEMO, STEMS). A test of the visual quality and social acceptance of silvicultural treatments is probably the easiest component to add to an existing study. Only two studies (ATC, DMS) have a major focus on the interaction of aquatics and wood production. Even though the aquatic conservation strategy (USDA and USDI 1994b) is an important driving component of the Northwest Forest Plan, the DMS is the only large-scale experiment examining the interaction of density management and aquatics in western Washington and Oregon.

Economic analyses of joint forest production are very limited. Only the ATC, CFS, and MASS studies were designed to collect economic information to evaluate the wood production component of the study, and those were not joint production economic analyses. The MASS study has comparative harvesting costs for alternative silviculture vs. even-age management using clearcutting. Only the FES study is examining the direct tradeoffs between wood and any other resource (viz., the influence on wildlife by managing second-growth forest for timber with multiple commercial thinnings vs. managing with legacy retention followed by protection with no active management).

The power of a test is the probability of correctly rejecting the null hypothesis (Cohen 1988). Sample size is a crucial determinant of power in large-scale field experiments. Although the size of the treatment blocks is large in all the experiments, the geographic scope of the studies varies greatly. Four studies are quite dispersed geographically (ATC, DEMO, DMS, OHDS), and can support broader inferences than the studies with only one location (CFS, FES, MASS, DCSS). Only the DMS has a sample size (18 blocks) large enough to have strong power for generalizing across a geographic range of site conditions.

However, some of these studies are designed to have strong power to test specific hypotheses. The DEMO study, for example, has 36 randomized replicates across 6 blocks. Because each replicate (13 ha) is larger than the home range of the vertebrates under investigation (Halpern and Raphael 1999), power should be high for testing the effect of silvicultural treatments on the target wildlife populations. Because the replicates are spread over 6 widely dispersed blocks, pseudoreplication (Hurlbert 1984) should not be a serious problem. The STEMS project took a different approach. This study intentionally used the same design as the CFS to increase power in a joint analysis of common hypotheses (a meta-analysis).

When blocks are located in a relatively small geographic area, potential problems with pseudoreplication become a concern (Hurlbert 1984). Consider the MASS, CFS, and FES studies, which use randomized block designs in one forest each. Block independence should be carefully addressed when stating probability levels for statistical tests.

The ability to detect significant differences depends not only on the number of replicates but also on the variability of the population of interest. Harrington and Carey (1997) expected that 8 to 10 replications in their OHDS study would allow them to detect treatment-induced differences of 20% or more in small animal populations. Use of fewer replications would require larger differences between treatments to achieve statistical significance and thus confidence in the results.

The level we accept as significant also affects our ability to detect differences. In the face of few replicates and high variability (both within and among treatments), the field scientist must consider the possibility that differences may only be detectable at untraditionally high significance levels, such as $\alpha=0.2$ or more.

It is usually not possible to randomly choose block locations (sites) in forestry. It is more desirable and realistic to use stratification to locate blocks across a range of some important geographic, physiographic, or other site characteristic. The key to strong statistical inference is to randomly assign all treatments, including controls, and have sufficient replication (blocks) to allow for detecting significant differences among treatments (Fisher 1925). The randomization of treatments is a fundamental attribute of all designs reviewed here. All studies have at least three full replications of each treatment, and all but MASS have randomized controls for measuring treatment differences.

Cooper et al. (2000) found that the power of most large-scale manipulative experiments is low, primarily because of the expense and difficulty of establishing an experiment with a large number of replicates. They recommend standardizing methods and designs among studies so that a meta-analysis (Fernandez-Duque and Valeggia 1994) can be performed on common research hypotheses. The meta-analysis effectively increases sample size—and therefore power—while accounting for the different degrees of variation across studies. All of these studies are examining the effect of silvicultural treatments on wildlife habitat and biodiversity, with an emphasis on hastening the approach to old-growth characteristics and retaining biological legacies. This is the most promising area for a meta-analysis.

The long-term nature of experiments such as these (slow processes with strong transient dynamics) is a long-standing problem (Walters and Holling 1990, Carpenter 1998). Investigators are faced with institutional and academic demands for short-term results that not only are publishable but also justify the large investments (Franklin 1989). All of these studies are recent (begun in the 1990s) and have not been completed.

When the experimental blocks are large enough to be commercially operational (e.g., 50 to 200 ha), the scientist is faced with the realities of the timber-sale process. The possibility of lawsuits and appeals is perhaps one of the most serious institutional challenges, especially on federal land. An experimental block that is held up in appeal prevents the full experimental design from being implemented, reducing the power of key hypothesis tests. The OHDS and DEMO studies both have two blocks that have not been implemented for this reason.

The concept of randomization has serious consequences in large experimental units. The very nature of randomization means that a clearcut or treatment unit with heavy removal could fall anywhere in the block, and some locations might be far more sensitive or visible than others to the public. Concerns over randomization of treatments were real issues for several of these large-scale experiments (ATC, DEMO, STEMS), and contributed to the clearcut in MASS not being randomized. Franklin et al. (1999) describe the decision to exclude a clearcut (0% retention) as a treatment in DEMO, "The discomfort of many National Forest personnel was accentuated by the knowledge that they had no control over where the 0% retention treatment would be placed since it was to be randomly assigned." A related issue with randomization is the risk of wind damage, especially on coastal or exposed stands. The solution is to find suitable watersheds for experimentation so that no treatment units are at high risk, regardless of the outcome of randomization. Admittedly, this is a difficult goal to reach.

As scientists, it is tempting to believe that the world will wait patiently for our well-designed experiments to be completed before conclusions can be reached. Land managers have a more practical view. The demonstration that a new type of management is operationally feasible is sometimes sufficient for managers to seriously consider it (Monserud 2002). They are not waiting to see if treatment differences are significant.

3. Additional Problems with Large-Scale Field Experiments

Control over individual variables becomes increasingly difficult as we move to broader scales with experimentation. True replication becomes almost impossible, covarying factors confound the treatment effects, and the results of an experiment become difficult to interpret (Hobbs 1999, Wiens 1999). As the size of experimental units increases beyond small ecosystems to include entire landscapes, the logistics of experimental manipulation requires some separation over time, creating temporal pseudoreplication (Hurlbert 1984). The alternative is to closely space the treatments, which creates spatial pseudoreplication (Hurlbert 1984). Because of this dilemma, Hargrove and Pickering (1992) conclude that pseudoreplication is almost inevitable in landscape ecology and in experimentation at broad scales. Because of the difficulty of controlling factors over a large scale, confounding factors threaten to obscure treatment response even in well-designed experiments.

In the opinion of Hobbs (1999), dogmatic adherence to classical experimentation as the primary scientific method leads almost invariably to fine-scale reductionism, with results that are difficult to generalize to a large-scale context. Carpenter (1998) finds that measuring the magnitude of response to manipulation and understanding our capacity to manage are the two most common reasons for ecosystem experimentation, rather than classical statistical inference. This was clearly borne out by managers in southeastern Alaska, who are already implementing prescriptions from the alternatives to clearcutting experiment (McClellan et al. 2000) decades before even early results are available (Monserud 2002). Both Hobbs (1999) and Pickett et al. (1994) contend that experimentation and falsification are only part of the scientific process, and that methodologies such as descriptive studies, correlation, spatial statistics, and modeling are also essential at the landscape scale. Walters (1997) points out two situations when modeling rather than experimentation is almost the only option for scientific insight: when spatial replication is impossible because the area is unique (there is only one Lake Baikal or Grand Canyon), and when experimentation is quite undesirable because management treatments are effectively irreversible. Hurlbert (1984) and Wiens (1999) both emphasize that sound scientific conclusions can be reached even when replication becomes prohibitive or impossible for large-scale experiments. Hurlbert (1984) cites the famous Hubbard Brook deforestation experiment (Likens et al. 1970) as a example of an unreplicated study that convincingly demonstrated the effects of the experimental variables without resorting to inferential statistical tests that would have been inappropriate.

Because of Hurlbert's (1984) seminal paper, field ecologists have widely recognized the problems of pseudoreplication, especially in large-scale experiments (Tilman 1989, Walters et al. 1989, Hargrove and Pickering 1992,

Walters 1997, Carpenter 1998, Hobbs 1999, Wiens 1999). Although some have seen this pseudoreplication issue as the cause of a harmful backlash (Hawkins 1986), most ecologists have responded to Hurlbert's paper with a positive reexamination of both the limitations of manipulative experimentation and the conduct of science in large-scale studies. The Experimental Lakes Area (Schindler 1991) in northwest Ontario, Canada, is a case in point. Lakes are one of the few ecosystems with well-defined borders and are, therefore, amenable to experimental manipulation (Hargrove and Pickering 1992). Even with a large number of similar lakes to work with (58 lakes, from 1 to 84 ha), Schindler (1991) concedes that differences among lakes are great enough that pseudoreplication is always a problem, and that the requirements of a classical analysis of variance are difficult to meet.

Four main conceptual problems must be surmounted when designing forest experiments (Tilman 1989): transient dynamics, indirect or feedback effects, environmental variability, and multiple stable equilibria.

(1) Transient dynamics occur after the ecosystem is perturbed by the experimental manipulation and before it reaches the new, experimentally imposed state (Tilman 1989). Transient dynamics can also be induced by some uncontrolled environmental disturbance such as an extreme weather event or gradual climate change. Thus, almost all field experiments are time dependent because the experimental manipulations are perturbations that lead to a period of transient dynamics (Walters and Holling 1990). Partly because the transient dynamics of a manipulative field experiment are conditioned both by site-specific initial conditions and a unique chain of weather, Taylor (1989) contends that the experiment cannot be replicated; it can only be repeated.

(2) Indirect or feedback effects are difficult to guard against, especially in multispecies communities. The temptation is to infer a direct cause-and-effect response when there may be several indirect or feedback-loop explanations that are equally plausible (Tilman 1989, Lindenmayer 1999). Statistical significance cannot ensure that the investigator's logic and conclusions are correct.

(3) Environmental variability is a third problem confounding field experiments. Weather is especially variable and unpredictable from year to year. Because all field experiments are dependent on climate over time, we do not know if the conclusions drawn from an experiment started in one year will apply to the same experiment started in a different year (Tilman 1989). One solution is to start these replicate experiments over time in a staircase design (Walters et al. 1989) so that results reflect a variety of climatic starting conditions. If the experiment is long-term, this multiple-start solution will eventually be dominated by the common climatic signal, which actually is an ever-changing transient response.

(4) A fourth problem is multiple stable equilibria, which could be dependent on the history of the site. Forest ecologists are aware that several successional pathways can exist, depending on site conditions and availability of germinants (Franklin 1989). Old-field forest succession, for example, can be markedly different than succession on nearby forested land. Recent results from fragmentation experiments in the Amazon indicate that strikingly different successional pathways can result from forest clearing in the tropics (Laurance et al. 1997).

Long-term study is unusual in science (Taylor 1989). Tilman (1989) discusses the appropriate length of time for a field experiment to run before testing the hypothesis of interest. If the hypothesis involves the new state of the system, then the experimenter must wait until the transient dynamics die out and the system stabilizes. He concludes that most manipulative field experiments in ecology are much too short, with less than 2% lasting at least 5 field seasons. In systems with slow dynamics such as forests, silvicultural manipulations can trigger a long-term chain of successional dynamics, the transient response to the manipulation. Results of the experiment are strongly dependent on time, and can change or even contradict earlier results as the successional dynamics progress. Thus, a well-replicated forestry experiment could register a statistically significant short-term response to a manipulation that contradicts the eventually obtained, and equally significant, long-term response (Tilman 1989).

Perhaps the oldest and most famous field experiments are the Park Grass Experiments at Rothamsted, England, which have been in operation since 1856 (Lawes and Gilbert 1880). They illustrate the importance of long-term experimentation and long-lasting transient dynamics. These plots have had nutrient applications for nearly 150 years. They show (among other things) a long period of transient dominance by species that are later displaced (Tilman 1989). The experimental manipulation triggered a long period of succession (the transient dynamics) that easily could be misinterpreted as a climax, with conclusions that do not generalize to the long-term. A second famous example is the long-term study of wolf–moose dynamics on Isle Royale in Lake Superior (Peterson et al. 1984). A conclusion that wolf and moose populations were unrelated was one of several erroneous conclusions reached at earlier points in the study (Franklin 1989).

Other features of ecological systems that call for long-term studies are slow dynamics, rare events, episodic phenomena, high variability, subtle processes, and complex phenomena (Franklin 1989). Many or all of these features can confound treatment effects. Time is the answer that solves many of these problems (Franklin 1989). Institutional constraints work against the luxury of time, especially when ecological dynamics take longer than the careers of investigators. The academic reward system favors studies that are sharply focused, short-term, and reductionist, which is counter to the requirements for broad-scale and long-term multidisciplinary experiments in highly variable ecosystems (Walters and Holling 1990).

Field experiments require land for an experiment, land that might be tied up for decades. This calls for strong cooperation between scientists and land managers, who have different professional and institutional goals. Successful large-scale experiments are the result of effective partnerships and communication between science and management, with both parties profiting from the collaboration.

And finally, we have the problem of cost. Large-scale field experiments are often exceedingly expensive to establish, implement, and maintain (Tilman 1989, Monserud 2002). Several of the experiments reviewed here cost well over $1 million (M)/block. Expenditures since plot establishment are approximately US$6M for ATC, $12M for DEMO, $1M for OHDS, and $4.5M for FES. One consequence is that scientists are forced to consider a smaller number of promising contrasts, potential treatments, and number of blocks. Jeffers (1988) emphasizes that the more costly the research, the more important it is that the experimental design be statistically sound and efficient, and that the results are capable of valid interpretation.

4. Demonstration Projects

Given the considerable impediments to large-scale forest management experimentation, many managers and scientists are pursuing demonstration projects instead. These often have some, but not all of the necessary features of a classical experiment: randomization, replication, and controls. Their goal is not to test a hypothesis but rather to implement one or more management alternatives so that feasibility can be determined or demonstrated. These are case studies. If executed well, they can be extremely valuable for demonstrating the potential of new management methods. As mentioned earlier, the Hubbard Brook deforestation experiment (Likens et al. 1970) is a fine example of an unreplicated study that convincingly demonstrated the effects of the experimental variables.

Compatibility among forest resources is more easily achieved at scales larger than the individual stand (Haynes et al. 2003, Monserud et al. 2003). Experimentation becomes infeasible after the watershed or landscape scale is reached (Carpenter 1998, Monserud 2002). Large-scale demonstration areas become valuable tools for considering management strategies that focus on the interaction and development of several forest resources on a large number of contiguous stands across a landscape. A few examples follow.

4.1. Pathways to Biodiversity

The past decade has seen the ascendance of broad biodiversity elements as an indicator of forest health in Pacific Northwest forests (Haynes and Monserud 2002). Often, biodiversity and wood production are portrayed as stark tradeoffs. Carey and Curtis (1996) contend that conflicts between conserving biodiversity and maintaining wood production disappear if it is recognized that the conservation of biodiversity is the foundation for sustainable forestry.

To examine this premise, Carey et al. (1996) implemented a Pathways to Biodiversity program on Washington Department of Natural Resources (DNR) trust land. Their research goal is to find forest management strategies that will meet the public desire for reduced visual effects from forestry operations and for maintenance of species diversity, wildlife, and other environmental values, while providing continued timber production at a relatively high and sustainable level. Simply put, they are trying to develop biodiversity pathways that lead to old-growth characteristics more quickly than even-age plantation management.

Their most significant result indicates that forest habitat with old-growth characteristics is achieved more quickly by maximizing biodiversity through forest management than by other management strategies, and thereby producing significant economic benefit (Carey et al. 1996). They found that protecting an area by excluding management after a timber harvest appears to delay development of old-growth characteristics significantly longer than if thinning or other management techniques are allowed. This no-management strategy also appears to fragment the remaining forest.

Implications for compatible forest management are clear: active management designed to produce a desired mix of conditions can be more effective and less costly than blanket attempts at "preservation" that eliminate human intervention. Various management techniques across the land base are recommended, such as a shift to extended rotations combined with increased use of commercial thinning, or a shift to regeneration systems other than large clearcuts. By adopting biodiversity management, Curtis and Carey (1996) note that long-term commodity output is maintained, and the health and function of the forest ecosystem is protected and enhanced.

4.2. The Blue River Watershed Management Plan

In the 1990s, landscape management in federally managed forests in the Douglas-fir region reached a crossroad (Cissel et al. 1999). Listing of the northern spotted owl (*Strix occidentalis caurina*) as a threatened species and public dissatisfaction with clearcutting culminated in the Northwest Forest Plan (USDA and USDI 1994a), the overriding plan for millions of hectares of federally managed forestland. This plan, with its roots in the old-growth and spotted owl issues, emphasizes static forest reserves, wildlife corridors, and standardized matrix harvest prescriptions.

At the same time, concepts emerged concerning use of information on historical disturbance regimes and recognition of the dynamic and variable character of many forest landscapes (Cissel et al. 1998, 1999). These approaches use information on historical and current landscape conditions, disturbance history, and social goals to set objectives for future landscape structures that provide desired plant and wildlife habitat, watershed protection, timber, and other functions. The intent was not to mimic historical conditions, but rather to use them as a reference in developing and evaluating management alternatives to meet these objectives (Cissel et al. 1999).

Cissel et al. (1999) describe a landscape management plan based on interpretations of historical fire disturbance regimes. The study area is the 23 900-ha Blue River subwatershed, located within the McKenzie River watershed in the Cascade Range of western Oregon. The plan contains a reserve system and other landscape areas where three distinct types of timber harvest are prescribed. These timber harvest prescriptions approximate the frequency, severity, and spatial extent of past fires. Future harvest blocks are mapped and used to project forest patterns 200 years forward. The plan is currently being implemented.

Cissel et al. (1999) interpret historical fire frequency by using fire events from tree origin and fire scar dates, statistically modeling point estimates of fire frequency as a function of environmental variables, and then using the resulting predictive algorithms and other observations to map predicted fire frequency over the study area. They then derive a generalized map of fire frequency. Fire regimes are further defined by assigning fire severity classes to areas of different fire frequency. Fire regime descriptions are completed by associating mortality patch size with fire frequency. Spatial pattern objectives at the landscape level are developed from analysis of individual fire events and mortality patch sizes resulting from historical fires in each landscape area. The landscape management strategy calls for a range of patch sizes (10 to 160 ha), roughly corresponding with the size of many individual mortality patches from past fires and excluding the infrequent large fires that historically created patches that were thousands of hectares in size.

In terms of disturbance frequency, management disturbance does not completely substitute for fire, but it does serve as a guide for the structural size distribution of forest stands across the landscape. Additional components of the plan include an analysis of selected sensitive-species habitat, an evaluation of the aquatic ecosystem objectives in the Northwest Forest Plan, and watershed restoration.

Cissel et al. (1999) compare this plan with an alternative plan for the same area based on the extensive reserves and prescriptions for matrix lands in the

Northwest Forest Plan. Their management approach based on historical disturbance patterns produced more late-successional habitat, more overstory structure in young stands, larger patches, and less edge between young and old forest. Although landscape structures resulting from both plans are historically unprecedented, Cissel et al. (1999) suggest that landscape management plans incorporating key aspects of ecosystem disturbance history and variability may pose less risk to native species and ecological processes than those called for in the Northwest Forest Plan.

A similar disturbance-based approach was used to develop a management plan for the nearby Augusta Creek Watershed (Cissel et al. 1998). Primary objectives included the maintenance of native species, ecosystem processes and structures, and long-term ecosystem productivity in a federally managed landscape where substantial acreage was allocated for timber harvests. Management objectives and prescriptions were evaluated in light of the long-term range of natural variability of landscape conditions and disturbance processes, including a 500-year fire history record reconstructed by using dendrochronology. The Augusta Creek project is intended to provide a test of alternative management approaches through comparison of the reserve-matrix system emphasized by the Northwest Forest Plan (USDA and USDI 1994a) with the disturbance-based approach developed for the Augusta area.

Rather than fragmenting management (and the landscape) by imposing the default stream buffer-width recommendations throughout the entire riparian system, Cissel et al. (1998) instead formed landscape blocks (20 to 150 ha, based on disturbance risk and landform) as the basic management units. When contrasted with the prescriptions from the Northwest Forest Plan (200-year simulations), these two approaches result in strikingly different potential landscapes, especially with regard to aquatic reserves and timber harvest. All prescriptions are spatially linked to specific blocks of land to provide an efficient transition to site-level planning and project implementation.

The landscape management implications are clear: aquatic conservation strategy goals (USDA and USDI 1994b) can be met by minimizing long-term risk to the watershed as a whole. This is accomplished by managing large landscape blocks rather than fragmenting management by imposing the default stream buffer-width recommendations (Cissel et al. 1998). This study provides an example of how ecosystem management could be applied in a particular landscape by using the results of watershed analysis.

These examples from Blue River and Augusta Creek offer a promising alternative to the static reserves and standard matrix prescriptions in the Northwest Forest Plan. They are based on historical landscape patterns and disturbance regimes (Cissel et al. 1998, 1999). By using historical information to guide management, the dynamic and variable character of the landscape is addressed, and our ability to meet ecosystem management objectives may be improved. It remains to be seen if the public will find this approach any more socially acceptable than the Northwest Forest Plan or earlier management systems geared to maximize timber production.

This chapter began with an examination of operational forestry experiments in joint production and the myriad problems associated with classic experimentation at scales as large as the watershed. Experimentation is essentially impossible at scales larger than the watershed. This leaves us with demonstration projects as the most promising method for testing the feasibility of new management approaches. As scientists, we are reassured by significant differences that can be detected in a well-designed experiment. However, managers are often far more interested in a straightforward demonstration of feasibility when it comes to new management approaches.

5. References

Aubry, K.B., Amaranthus, M.P., Halpern, C.B., While, J.D., Woodard, B.L., Peterson, C.E., Lagoudakis, C.A., and Horton, A.J. 1999. Evaluating the effects of varying levels and patterns of green-tree retention: experimental design of the DEMO study. *Northwest Science*, 73(special issue): 12-26.

Beese, W.J., and Bryant, A.A. 1999. Effect of alternative silvicultural systems on vegetation and bird communities in coastal montane forests of British Columbia, Canada. *Forest Ecology and Management*, 115: 231-242.

Carey, A.B., and Curtis, R.O. 1996. Conservation of biodiversity: a useful paradigm for forest ecosystem management. *Wildlife Society Bulletin*, 24(4): 610-620.

Carey, A.B.; Elliott, C.; Lippke, B.R., Chambers, C.J., Oliver, C.D., Franklin, J.F., and Raphael, M.G. 1996. A pragmatic ecological approach to small-landscape management: Washington forest landscape management project—final report. Project Report No. 2. Washington Forest Landscape Management. Olympia, WA: Department of Natural Resources. 99 p.

Carey, A.B., Thysell, D.R., and Brodie, A.W. 1999. The forest ecosystem study: background, rationale, implementation, baseline conditions, and silvicultural assessment. General Technical Report PNW-GTR-457. Portland, OR: U.S. Department of Agriculture, Forest Service, Pacific Northwest Research Station. 129 p.

Carpenter, S.R. 1998. The need for large-scale experiments to assess and predict the response of ecosystems to perturbation. In: M.L. Pace, and P.M. Groffman (Eds.), Successes, limitations, and frontiers in ecosystem science (pp. 287-312). New York: Springer-Verlag.

Cissel, J.H., Swanson, F.J., Grant, G.E. [and others]. 1998. A landscape plan based on historical fire regimes for a managed forest ecosystem: the Augusta Creek study. General Technical Report PNW-GTR-422. Portland, OR: U.S. Department of Agriculture, Forest Service, Pacific Northwest Research Station. 82 p.

Cissel, J.H., Swanson, F.J., and Weisberg, P.J. 1999. Landscape management using historical fire regimes: Blue River, Oregon. *Ecological Applications*, 9(4): 1217-1231.

Coates, K.D., and Burton, P.J. 1997. A gap-based approach for development of silvicultural systems to address ecosystem management objectives. *Forest Ecology and Management*, 99: 337-354.

Cohen, J. 1988. Statistical power analysis for the behavioral sciences (2nd ed.) Hillsdale, NJ: L. Erlbaum Assoc.

Cooper, R.J., Gale, G.A., and Brennan, L.A. 2000. Answering questions in management and research using large-scale manipulative experiments. In: Bonney et al. (Eds.), Strategies for bird conservation: the partners in flight planning process (pp. 220-224). RMRS-P-16. Fort Collins, CO: U.S. Department of Agriculture, Forest Service, Rocky Mountain Research Station.

Curtis, R.O. 1982. A simple index of stand density for Douglas-fir. *Forest Science*, 28: 92-94.

Curtis, R.O. 1996. Silvicultural options for harvesting young-growth production forests. Prospectus for DNR/PNW Cooperative Project. 5 p. Unpublished document. On file with: U.S. Department of Agriculture, Forest Service, Pacific Northwest Research Station, Forestry Sciences Laboratory, 3625 93rd Avenue SW, Olympia, WA 98512-9193.

Curtis, R.O. 1998. "Selective cutting" in Douglas-fir: history revisited. *Journal of Forestry*, 96(7): 40-46.

Curtis, R.O., and Carey, A.B. 1996. Timber supply in the Pacific Northwest: managing for economic and ecological values. *Journal of Forestry*, 94(9): 4-7, 35-37.

Curtis, R.O., and Marshall, D.D. 1986. Levels-of-growing-stock cooperative study in Douglas-fir: Report 8—The LOGS study: twenty-year results. Research Paper PNW-356. Portland, OR: U.S. Department of Agriculture, Forest Service, Pacific Northwest Research Station. 113 p.

Curtis, R.O., DeBell, D.S., Harrington, C.A., Lavender, D.P., St. Clair, J.B., Tappeiner, J.C., and Walstad, J.D. 1998. Silviculture for multiple objectives in the Douglas-fir region. General Technical Report PNW-GTR-435. Portland, OR: U.S. Department of Agriculture, Forest Service, Pacific Northwest Research Station. 123 p.

Fernandez-Duque, E., and Valeggia, C. 1994. Meta-analysis: a valuable tool in conservation research. *Conservation Biology*, 8: 555-561.

Fisher, R.A. 1925. Statistical methods for research workers. London: Oliver and Boyd. 318 p.

Franklin, J.F. 1989. Importance and justification of long-term studies in ecology. In: G.E. Likens (Ed.), Long-term studies in ecology: approaches and alternatives (pp. 3-19). New York: Springer-Verlag.

Franklin, J.F., Norris, L.A., Berg, D.R., and Smith, G.R. 1999. The history of DEMO: an experiment in regeneration harvest of northwestern forest ecosystems. *Northwest Science*, 73(special issue): 3-11.

Halpern, C.B., and Raphael, M.G. (Eds.). 1999. Retention harvests in northwestern forest ecosystems: the demonstration of ecosystem management options (DEMO) study. *Northwest Science*, 73(special issue). 125 p.

Hargrove, W.W., and Pickering, J. 1992. Pseudoreplication: a *sine qua non* for regional ecology. *Landscape Ecology*, 6(4): 251-258.

Harrington, C.A., and Carey, A.B. 1997. The Olympic habitat development study: conceptual study plan. Unpublished manuscript. On file with: U.S. Department of Agriculture, Forest Service, Pacific Northwest Research Station, Forestry Sciences Laboratory, 3625 93rd Avenue, Olympia, WA 98512-9193. 38 p.

Hawkins, C.P. 1986. Pseudo-understanding of pseudoreplication: a cautionary note. *Bulletin of the Ecological Society of America*, 67(2): 184-185.

Haynes, R.W., and Monserud, R.A. 2002. A basis for understanding compatibility among wood production and other forest values. General Technical Report PNW-GTR-529. Portland, OR: U.S. Department of Agriculture, Forest Service, Pacific Northwest Research Station. 46 p.

Haynes, R.W., Monserud, R.A., and Johnson, A.C. 2003. Compatible forest management: background and context. Chapter 1. In: R.A. Monserud, R.W. Haynes, and A.C. Johnson (Eds.), Compatible forest management. Dordrecht, The Netherlands: Kluwer Academic Publishers.

Hobbs, R.J. 1999. Clark Kent or Superman: Where is the phone booth for landscape ecology? In: J.M. Klopatek, and R.H. Gardner (Eds.), Landscape ecological analysis: issues and applications (pp. 11-23). New York: Springer-Verlag.

Hurlbert, S.H. 1984. Pseudoreplication and the design of ecological field experiments. *Ecological Monographs*, 54(2): 187-211.

Isaac, L.A. 1943. Reproductive habits of Douglas-fir. Washington, DC: Charles Lathrop Pack Forestry Foundation.

Jeffers, J.N.R. 1988. Statistical and mathematical approaches to issues of scales in ecology. In: T. Rosswall, R.G. Woodmansee, and P.G. Risser (Eds.), SCOPE 35: Scales and global change: spatial and temporal variability in biospheric and geospheric processes (Chapter 4). Chichester, UK: Wiley.

Laurance, W.F., Laurance, S.G., Ferreira, L.V., Rankin de Merona, J., Gascon, C., and Lovejoy, T.E. 1997. Biomass collapse in Amazonian forest fragments. *Science*, 278: 1117-1118.

Lawes, J., and Gilbert, J. 1880. Agricultural, botanical and chemical results of experiments on the mixed herbage of permanent grassland, conducted for many years in succession on the same land. *Philosophical Transactions of the Royal Society*, 171: 189-416.

Likens, G.E., Bormann, F.H., Johnson, N.M., Fisher, D.W., and Pierce, R.S. 1970. Effects of forest cutting and herbicide treatment on nutrient budgets in the Hubbard Brook watershed ecosystem. *Ecological Monographs*, 40: 23-47.

Lindenmayer, D.B. 1999. Future directions for biodiversity conservation in managed forests: indicator species, impact studies and monitoring programs. *Forest Ecology and Management*, 115(2-3): 277-287.

McClellan, M.H., Swanston, D.N., Hennon, P.E., Deal, R.L., De Santo, T.L., and Wipfli, M.S. 2000. Alternatives to clearcutting in the old growth-forests of southeast Alaska: study plan and establishment report. General Technical Report PNW-GTR-494. Portland, OR: U.S. Department of Agriculture, Forest Service, Pacific Northwest Research Station. 40 p.

Monserud, R.A. 2002. Large-scale management experiments in the moist maritime forests of the Pacific Northwest. *Landscape and Urban Planning*, 59(3): 159-180.

Monserud, R.A., Haynes, R.W., and Johnson, A.C. 2003. The search for compatibility: What have we learned? Chapter 17. In: R.A. Monserud, R.W. Haynes, and A.C. Johnson (Eds.), Compatible forest management. Dordrecht, The Netherlands: Kluwer Academic Publishers.

Peterson, C.E., and Monserud, R.A. 2002. Compatibility between wood production and other values and uses on forested lands: a problem analysis. General Technical Report PNW-GTR-564. Portland, OR: U.S. Department of Agriculture, Forest Service, Pacific Northwest Research Station. 50 p.

Peterson, R.O., Page, R.E., and Dodge, K.M. 1984. Wolves, moose, and the allometry of population cycles. *Science*, 224: 1350-1352.

Pickett, S.T.A., Kolasa, J., and Jones, C.G. 1994. Ecological understanding: the nature of theory and the theory of nature. New York: Academic Press.

Reukema, D.I. 1979. Fifty-year development of Douglas-fir stands planted at various spacings. Research Paper PNW-253. Portland, OR: U.S. Department of Agriculture, Forest Service, Pacific Northwest Forest and Range Experiment Station. 21 p.

SAS Institute. 2000. The SAS system. SAS Online Doc.® Version 8.1. HTML Format [CD-ROM]. Cary, NC.

Schindler, D.W. 1991. Whole-lake experiments at the Experimental Lakes Area. In: H.A. Mooney, E. Medina, D. Schindler, E.D. Schulze, and B.H. Walker (Eds.), SCOPE 45: ecosystem experiments (Chapter 7). Chichester, UK: Wiley.

Stevens, J.A., and Montgomery, C.A. 2002. Understanding the compatibility of multiple uses on forest land: a survey of multiresource research with application to the Pacific Northwest. General Technical Report PNW-GTR-539. Portland, OR: U.S. Department of Agriculture, Forest Service, Pacific Northwest Research Station. 44 p.

Taylor, L.R. 1989. Objective and experiment in long-term research. In: G.E. Likens (Ed.), Long-term studies in ecology: approaches and alternatives (pp. 20-70). New York: Springer-Verlag.

Tilman, D. 1989. Ecological experimentation: strengths and conceptual problems. In: G.E. Likens (Ed.), Long-term studies in ecology: approaches and alternatives (pp. 136-157). New York: Springer-Verlag.

U.S. Department of Agriculture and U.S. Department of the Interior. 1994a. Final supplemental environmental impact statement on management of habitat for late-successional and old-growth forest related species within the range of the northern spotted owl. Washington, DC: U.S. Department of Agriculture, Forest Service; U.S. Department of the Interior, Bureau of Land Management. 520 p. [plus appendices]. 2 vol.

U.S. Department of Agriculture and U.S. Department of the Interior. 1994b. Record of decision for the President's forest plan. Washington, DC: U.S. Department of Agriculture, Forest Service; U.S. Department of the Interior, Bureau of Land Management. 74 p. [plus appendices].

Walters, C. 1997. Challenges in adaptive management of riparian and coastal ecosystems. *Conservation Ecology*, 1(2): 1-19. Retrieved March 2003 from http://www.consecol.org/vol1/iss2/art1.

Walters, C.J., Collie, J.S., and Webb, T.S. 1989. Experimental designs for estimating transient responses to management disturbances. *Canadian Journal of Fisheries and Aquatic Science*, 45: 530-538.

Walters, C.J., and Holling, C.S. 1990. Large-scale management experiments and learning by doing. *Ecology*, 71(6): 2060-2068.

Weyerhaeuser. 2000. TFL 39 Enhanced forest management pilot project. EFMPP Report No. 1. 4 p. Retrieved July 2000 from http://www.weyerhaeuser.com/ourproducts/timberland sp/bc/new_efmpp/pdf/efm_brochure.pdf.

Wiens, J.A. 1999. The science and practice of landscape ecology. In: J.M. Klopatek, and R.H. Gardner (Eds.), Landscape ecological analysis: issues and applications (pp. 371-383). New York: Springer-Verlag.

Williamson, R.L. 1973. Results of shelterwood harvesting of Douglas-fir in the Cascades of western Oregon. Research Paper PNW-RP-1161. Portland, OR: U.S. Department of Agriculture, Forest Service, Pacific Northwest Research Station. 13 p.

Chapter 6

Modeling Stand Growth and Management

Robert A. Monserud
USDA Forest Service, Pacific Northwest Research Station,
620 SW Main St., Suite 400, Portland, OR 97205, USA

1. Introduction

Compatible forest management involves the simultaneous production of wood and other forest resources in a socially acceptable manner (Haynes and Monserud 2002). The goal is to expand information on the options that may increase the compatibility between wood production and other societal values derived from forestlands. In the opening chapter, we asked: *How can the various forest management practices be related to an array of associated goods and services such as aquatic conservation, biodiversity, wildlife habitat, or social acceptance?* This involves the quantitative modeling of the effects of different approaches to forest management. Recall that the silvicultural treatments that comprise a forest management regime are applied stand by stand, even when the management plan is for an entire watershed or landscape (e.g., Cissel et al. 1999, Greenough et al. 1999). Because management is usually implemented at the stand level, modeling of stand dynamics is necessary. This chapter will review stand simulation models and analysis tools for evaluating stand-level management options.

I begin with an overview of the major approaches to modeling forest stand dynamics. These methodologies are examined for their utility in addressing forest management questions regarding compatibility. Because several, if not most of the key values considered in compatible forest management are dependent on the size structure and species composition determined by the constituent trees in the stand, strong emphasis is placed on the individual tree as a fundamental modeling unit. Many of these models are capable of simulating detailed silvicultural prescriptions over the course of the management period, which provides the connection between modeling and forest management.

R.A. Monserud, R.W. Haynes and A.C. Johnson (eds.)
Compatible Forest Management. 145-175

As discussed in Chapter 1, emphasis on timber-stand management in the Pacific Northwest (western Oregon and Washington, coastal British Columbia, and southestern Alaska) declined during the 1990s, especially on public forest-land. The traditional goal of efficient wood production through even-aged plantations is shifting toward multi-resource ecosystem management, with related goals favoring old-growth characteristics (see Appendix 1, Chapter 1), protecting endangered species, fish habitat, and promoting biodiversity (FEMAT 1993, Clayoquot Scientific Panel 1995). Science-based silvicultural practices and management regimes are needed to reduce conflicts among user groups while providing concurrent production of the many values associated with forest lands on a biologically and economically sustainable basis (Curtis et al. 1998, Committee of Scientists 1999).

Because the goals of land managers have changed, silviculturists are examining new management methods, including alternatives to clearcutting such as variable-density thinning, green-tree retention, and variable retention (Arnott and Beese 1997, Coates et al. 1997, Franklin et al. 1997, Curtis et al. 1998, Monserud 2002, Hummel 2003). Often, the goal is to increase rather than decrease structural heterogeneity within a stand and a watershed (Monserud 2002). In a strong break from the uniform plantation management of the past, these silvicultural alternatives often include mixed-species compositions and attempt to obtain uneven-aged structures (e.g., by retaining legacy trees). The ability to simulate stand growth under increasingly heterogeneous structures and mixed-species compositions is an important aspect of evaluating the compatibility of management alternatives.

2. Classes of Forest Models

For stand management purposes, five broad classes of forest stand models exist (Monserud, in press):

1. Forest Yield Models
2. Ecological Gap Models (population succession)
3. Ecological Compartment Models (resources fluxes)
4. Process and Mechanistic Models
5. Hybrid Models

In the next section, I review the major features of each class in turn, usually providing one example of each. This overview is followed with an in-depth examination of specific models available in the Pacific Northwest.

2.1. Forest Yield Models

These include some of the oldest forest management models, dating from the first yield tables in the 18th and 19th centuries (Monserud, in press). They were intended to predict an expected yield over the management regime.

Foresters assumed that the management associated with such yields was sustainable. Now, modern computers have made individual tree modeling possible. Many stand simulation models have been developed in the last 30 years (see compilations by Dudek and Ek 1980, Vanclay 1995, Ritchie 1999). Yield models are available for almost all ecosystems in the world where sustainable forest management is practiced.

Almost all yield models share a common philosophy: the prediction of volume yield over time, usually as a function of site productivity and density (Monserud, in press). Site index (dominant height at an index age) is the standard measure of productivity, a measure used only by foresters. Most forest yield modelers take great care to model the complex relation between stand density and tree growth (both in diameter (D) and in height (H)). These models are almost always species-specific. Growth is also a function of a simple crown measure such as crown ratio, if it is available (e.g., Wykoff et al. 1982). Considerable work on modeling mortality in relation to stand dynamics has also brought progress to accurate stand forecasting. Monserud and Sterba (1999), for example, predict tree mortality as a species-specific function of tree size, tree crown, and stand density. Because of the attention devoted to stand density and growth relations, the range of potential stand dynamics that can be handled by forest yield models over time is quite rich, and is capable of providing accurate prediction of stem sizes over time (Yaussey 2000).

One of the most widely used forest management yield models is the Forest Vegetation Simulator (FVS), a.k.a. the Stand Prognosis model (Wykoff et al. 1982, Teck et al. 1997, Crookston and Havis 2002). FVS is an individual-tree, nonspatial, stand growth model built around a set of empirically derived equations of diameter growth, height growth, crown ratio, regeneration, and mortality. The key function is the individual tree basal area increment model (Wykoff 1990); growth is a function of tree size, relative size, crown size, stand density, and site-specific characteristics. All models are species-specific, commonly with 12 or more species in a given variant. The original northern Idaho variant (Wykoff et al. 1982) has a rich regeneration module (both natural and artificial) for all available species in the northern Rocky Mountains (Ferguson and Crookston 1991). The northern Idaho variant is also one of the few yield models to eschew the foresters' traditional stand productivity measures: site index and stand age. Because most stands in the northern Rockies are irregular, stand age can be meaningless, and site index unreliable. Site descriptors (vegetation habitat type, elevation, slope, aspect) are used as surrogates for site productivity, with no loss of generality (Wykoff and Monserud 1988). FVS currently has 20 variants, representing most forested regions of the United States, as well as variants in Canada and Austria. FVS simulates a wide array of different forest processes other than simple volume accretion (Teck et al. 1997). Natural disturbance processes linked to FVS include several insect and disease modules, and a fire and fuels extension.

Because they share a common philosophy, yield models share several management-oriented features (Robinson and Monserud 2003). Almost all have site calibration, can simulate basic silviculture, can predict yield in terms of wood products, and often accommodate a variety of sample designs for different inventories (Robinson and Monserud 2003). Few model natural regeneration, although all plantation models allow for the specification of initial density and various site preparations and treatments. Uncertainty is occasionally modeled (e.g., Wykoff et al. 1982), but not often. The real advantage of forest yield models is their ability to make detailed predictions of tree and stand dynamics, particularly stem size distribution predictions that aggregate accurately to the stand level. A second advantage is that tree mortality is often modeled in detail, incorporating both competitive effects at the individual tree level and stand level constraints such as the -3/2 power law (see Yoda et al. 1963). This detail on stand structure, species composition, and silvicultural treatment response make this class of models well suited to analyzing alternative methods for compatible forest management.

However, forest yield models are not linked to the underlying causes of productivity: the carbon and nutrient cycles, the moisture regime, and climate (Monserud, in press). This is their fundamental disadvantage. Because they are environmentally static, they cannot be extrapolated to a new set of boundary conditions that climate change might bring (Schwalm and Ek 2001). They are limited to the range of environmental conditions represented in the underlying database used to build the model. For this reason, such models are often referred to as empirical, occasionally in a derogatory sense. The advantage of yield models is the detailed representation of both tree and stand dynamics, including mortality. This has long provided forest managers with useful predictions of future stand development for specific locations.

2.2. Ecological Gap Models

Ecologists have a rich 30-year history of modeling population succession by the device of a gap in the forest that is formed when a large tree dies (Shugart 1984). All of these models are genealogically related to JABOWA,[1] and the first-born, FORET (see Botkin et al. 1972, Shugart 1984, Botkin and Schenk 1996, Bugmann 2001). Currently, more than 50 variants exist. Their common modeling philosophy is to test ecological population theory. This goal is very different from the site-specific prediction goal of the foresters'

[1] Complete model names are listed in Appendix 1.

yield models. Basically, the development of gap models has taken place independently of traditional forest yield model research, with almost no cross breeding (Sievänen et al. 2000).

The key variable in JABOWA and its variants is tree diameter growth. It is modeled as the product of potential diameter growth and three modifiers: a light response function, a degree-day parabola, and a stand density factor (Hinckley et al. 1996). Potential diameter growth is a function of the current diameter and the maximum diameter observed for the species. The conceptual foundation assumes that tree growth is directly proportional to the abundance of leaves and inversely proportional to the amount of respiring, nonphotosynthetic tissue (Hinckley et al. 1996). Although tree leaf area, height, and species' maximum height are listed in the basic diameter growth equation, they are all simply a function of current diameter. One consequence is that the relation between height and diameter is density independent (Botkin et al. 1972, Lindner et al. 1997). This H/D invariance severely limits the range of possible tree size development within a stand. Species-specific regeneration in the gap is a function of light reaching the forest floor and not dependent on available seed sources. All leaves are assumed to be in a disk at the top of each tree. Temperature and soil moisture also affect regeneration success. Mortality is completely stochastic; the rate is set so that individuals have a 2% chance of reaching the oldest observed age for that species. Mortality rate increases for trees growing less than 0.01 centimeter (cm), an arbitrary constant that serves to speed the death of suppressed trees.

Gap models have no typical forestry inputs or outputs. Uncertainty is not modeled, although the ZELIG variants (Urban 1993) simulate multiple gap replicates. Perhaps the most unusual attribute of gap models is that plot size (gap size) strongly influences the predicted population dynamics, and is therefore a growth parameter (Shugart 1984). In contrast, forest yield models are generally unaffected by a change in plot size (but see Stage and Wykoff 1998).

Gap models typically make predictions for several centuries, well beyond the realm of the historical record. This makes validation of gap models a difficult proposition, one that is usually not attempted. Reasonable model behavior is often determined by expert opinion on predicted species succession patterns rather than an on objective comparison to observed data.

As reviews by Hinckley et al. (1996), Lindner et al. (1997), and Yaussey (2000) reveal, gap models have several structural problems. First, the invariant H/D relation produces unnatural combinations of tree and stand development. Second, the stochastic 2% mortality function is too simplistic; background tree mortality is actually density dependent and size dependent (Sievänen et al. 2000). The unrealistic representation of density independence in H/D and tree mortality severely limits the range of possible stand dynamics and stand structures. Hinckley et al. (1996) questioned the validity of temperature use in the model, both with regard to regeneration vs. mature tree development, and the temperature parabola. Lindner et al. (1997) provide concrete suggestions for improving several of these problems areas, especially when using the second-generation FORSKA model. An additional problem is a lack of connection with forest management needs, either with regard to model input or output.

2.3. Ecological Compartment Models

Ecologists have a second major paradigm: a connected series of boxes. In this view, the world is a series of fluxes across connected resource compartments. Examples of this modeling view are: CENTURY (Parton et al. 1988); FOREST-BGC (Running and Coughlan 1988); and PNET (Aber and Federer 1992). In some sense these are process models but their basic unit is an ecosystem compartment, and their focus is on predicting fluxes from one compartment to another (Godfrey 1983). Rarely are such models tied directly to forest management, although Korol et al. (1996) extended the big-leaf model of FOREST-BGC to predict tree and stand growth with TREE-BGC.

The FORCYTE/FORECAST system of Kimmins et al. (1999) is a well-known exception. Forest biomass compartments are modeled over time, with a strong focus on nutrient availability and cycling. Productivity is modeled as net primary productivity. Although the model is designed to start from a clearcut or stand-replacing fire, natural regeneration is not modeled. Substantially more data are required to run the model (Kimmins et al. 1999) than in comparable gap models such as LINKAGES (Pastor and Post 1985, 1986).

Because biomass compartments rather than trees are the primary modeling unit in the Kimmins et al. system, tree dimensions are inferred from the status of the forest biomass compartments. In spite of this, a broad array of silvicultural systems, management activities, and natural disturbance events can be simulated by FORECAST. From a library of stand table data tree sizes are imputed (stem size class distributions) for a series of stand ages and site qualities. The model uses this information to determine the rate at which individual stems of a single cohort diverge in size, resulting in a predicted distribution of stems. The total stand biomass is distributed among the individual stems in proportion to the relative size of each.

Ecological compartment models are often specialized for one process rather than the whole forest system. As such, they are not amenable to stand management. FORCYTE/FORECAST (Kimmins et al. 1999) is an exception, because it was developed primarily to address forest management problems. The main disadvantages are the voluminous data input required to run the model, and the location-specific nature of that input.

2.4. Process and Mechanistic Models

The common philosophy in mechanistic modeling is to build a tool for scientific explanation rather than prediction. Modelers attempt to explicitly represent causality between variables, thereby reaching toward generality. They set out to model one or more of the key growth processes and underlying causes of productivity: photosynthesis and respiration (carbon allocation), nutrient cycles, climate effects, and the moisture regime and water stress. From a systems analysis point of view, these are the functional components of the biological system of interest. They are usually chosen at a hierarchical level that is one level below the level of the system (Mäkelä et al. 2000a). Isebrands et al. (1990) point out that, paradoxically, process modelers can increase their understanding of the overall system only if it is possible to describe it as connected subsystems that are well understood. Because process models usually have a strong basis in physiological theory, Isebrands et al. (1990) contend that process models should have reasonable prediction accuracy outside the range of the original data used to calibrate the model. Process or mechanistic models have been used for many decades in plant physiology, biochemistry, and agronomy (Isebrands et al. 1990). They have reached critical mass in forestry only in the past 15 years (see Dixon et al. 1990).

The pipe model theory of Shinozaki et al. (1964a,b) is one of the oldest successful paradigms in mechanistic modeling. Valentine (1990) organized this theory into a system of differential equations in PIPESTEM. He used the theory for the structural framework and dry-matter allocation rules needed for carbon-balance models of tree growth. The tree comprises leaves, feeder roots, active pipes, and disused pipes; the pipes represent all woody components of the tree (branches, bole, and support roots) (Valentine 1990). The model tree maintains a constant amount of leaf dry matter (carbon) per unit of cross-sectional area of active pipes. To maintain balance between leaves and roots, a constant amount of feeder-root dry matter per unit of active pipe cross-sectional area is assumed. The challenge is incorporating competitive effects from neighboring pipe-model trees. Although possible, this problem becomes analogous to the famous n-body problem in physics and celestial mechanics. As n increases, the difficulty of simultaneously solving for all interacting bodies (trees) increases exponentially.

It has been difficult for process modelers to demonstrate accurate stand development, primarily because the physiological observations that could provide critical confirmation are so difficult to measure. Speaking of process models in general, Hinckley et al. (1996) comment that process models typically involve complex calculations on massive amounts of detailed environmental data. Consequently, the lack of available data and computing requirements limit the models' ability to scale up to the stand, regional, or global level. They also found that several important physiological processes are so poorly understood that an accurate physiological model is not yet possible.

Process and mechanistic models are quite promising. The possibility of incorporating theory and causality into growth predictions is attractive. Mäkelä et al. (2000a) point out that some very simple models with a mechanistic photosynthesis component have been able to capture some of the essential qualitative features of the differences in productivity between sites (e.g., Landsberg and Waring 1997) and trees. One impediment has been substantial data requirements. Another has been the lack of connection between model outputs and the needs of managers. The difficulty of model validation is a third problem. Recently, several process model developers tested their models with independent data and in regions outside the original area of development (Host and Isebrands 1994, Mäkelä et al. 2000b, Sievänen et al. 2000). Success demonstrates the generality of the model formulation. Mäkelä et al. (2000a) maintain that the carbon balance approach is well developed, and should provide a strong theoretical framework for predicting tree growth.

2.5. Hybrid Models

Although most process models are research tools, some process modelers saw the need to make their models amenable to typical forest management situations and problems (Mäkelä et al. 2000a). One strategy is to merge the best features of a process-based physiological model with an empirically based yield model. The result is a hybrid model, which is an attempt to build a forest management model able to address the effect of a changing environment at a fundamental level. A hybrid model is thus a mixture of both causal and empirical elements. Mäkelä et al. (2000a) finds that modeling the carbon balance of trees and forests based on photosynthetic production provides a workable framework for building management-oriented hybrid models. Furthermore, work will be accelerated if it can be demonstrated clearly that empirical (yield) models can be improved through incorporation of mechanistic (causal) functions, and that process models can be improved by incorporating system-level empirical elements and constraints (Mäkelä et al. 2000a). Hybrid models hold great promise because they are predicated on the goal of developing an operational model with useful products for the manager.

Mäkelä and her team of Finnish scientists are developing PipeQual, a hybrid model to predict internal wood structure and timber quality in stands at harvest (Mäkelä et al. 2000a). The goal is to assess the quality distribution of stems from a particular stand, and to plan forest management adjustments to long-term changes in the forestry market. Tree growth, including stem form and crown structure, is based on CROBAS, the individual-tree carbon balance model of Mäkelä (1997). Pipe model theory is used (Shinozaki et al. 1964a,b) with several of the same assumptions and constraints used by Valentine (1990) in PIPESTEM. An allometric relation is employed between crown surface and foliage, as well as an assumption of constant crown shape for Scots pine (*Pinus sylvestris* L.). Trees compete through shading, as mediated through photosynthetic rate and a self-pruning coefficient. The hybrid system uses empirical submodels for branch numbers, locations, and inclinations. An unusual strategy is the parallel development of empirical models for the same quality characteristics using MELA, the existing forest-yield planning simulator for Finland (Mäkelä et al. 2000a).

A second well-known model is 3-PG (Physiological Principles in Predicting Growth), developed by Landsberg and Waring (1997). It is a stand growth model based on both physiological processes and empirical constants and relationships, and is intended to provide utility as a forest management model. 3-PG calculates total carbon fixed (gross primary production (GPP)) from photosynthetically active radiation absorbed by the forest canopy. It corrects for the effects of soil drought, atmospheric vapor pressure deficits, and stand age. GPP is also estimated from an empirical constant: canopy quantum efficiency. They assume that the ratio of net to gross primary production is constant for trees, eliminating the need to calculate respiration and allowing for a direct estimation of net primary production (NPP). 3-PG uses a simple relationship to estimate below-ground carbon, and empirical allometric ratios to allocate carbon to foliage and stems and constrain tree growth patterns. The effects of nutrition are incorporated through the carbon allocation procedure; the amount of carbon allocated below ground will increase with decreasing soil fertility. Changes in stem populations (self-thinning) are based on the -3/2 power law, combined with stem growth rates. The model requires weather data as input and works on monthly time steps. 3-PG can be run from remotely sensed estimates of leaf area index coupled with weather data and readily available information about soils and stand characteristics. Recently, Coops et al. (1998) used 3-PG with satellite data and monthly weather data to estimate forest productivity (NPP) in Australia and New Zealand. Landsberg et al. (in press) report on a validation test from diverse sites in Sweden, Africa, and Australia. The model did not compare favorably in a recent review of how climate change effects are incorporated into 12 process or hybrid models (Schwalm and Ek 2001); most of the important climate change effects were uncoupled from the 3-PG system.

The great appeal of process models is that they try to simulate some physiological theory, without reliance on empirical results. Unfortunately, there will always be a parameter that cannot be determined from the definitions of the process (e.g., canopy quantum efficiency in 3-PG). To compensate, the model must include some empiricism to be consistent. The need for empiricism always appears when scaling up to higher levels of organization in the system being modeled (leaf to tree, or tree to stand). The hybrid modelers seem to accept this, and openly embrace relevant empirical results from yield models. They use them to supply missing parameters unaccounted for by their theories. For example, both PipeQual and 3-PG use empirical allometric relations to supply estimates of key crown attributes or tree dimensions. This is especially important for carbon allocation. At a higher level, empirical stand density constraints from yield models can also be used to bound the stand-level predictions of hybrid models. The parallel modeling of the yield model MELA and the process model CROBAS in the hybrid system PipeQual is a clever strategy for ensuring that process predictions are not drifting out of bounds. Milner and Coble (1995) use the same strategy to bound the hybrid Stand-BGC with stand dynamics from FVS. Conditioning mechanistic stand predictions with parallel results from forest yield models under current climatic conditions should help ensure that the mechanistic parameterization will still hold when boundary conditions change significantly from the calibration data set.

3. Stand Models Calibrated for the Pacific Northwest

The ability to simulate stand growth and development under increasingly heterogeneous stand structures and mixed-species compositions is an important aspect of evaluating the compatibility of management alternatives. Partially in response to contentious resource management debates, managers are examining alternative silvicultural treatments to traditional clearcut plantation management (Hummel 2003). Often, the goal is to jointly produce wood while enhancing ecological integrity (e.g., wildlife habitat, biodiversity, or the conservation of aquatic resources), all in a manner that is socially acceptable. Whether they are called green-tree retention, variable-density thinning, or variable retention (Franklin et al. 1997), the silvicultural treatments under examination are intermediate between the traditional extremes of even-age plantation management, uneven-age selection management, and no management (see Hummel 2003). They explicitly consider structural and spatial diversity as values, rather than the spatially uniform stand treatments common with traditional silvicultural systems.

Stand simulation modeling has a rich history in the Pacific Northwest (Monserud 2001). The first individual-tree stand simulator was developed here by Newnham (1964) for Douglas-fir (*Pseudotsuga menziesii* (Mirb.) Franco). Since then, stand growth models have been developed for most forests of the world. Today, a forest manager has a wide choice of simulation models for assisting with evaluating management prescriptions and comparing alternatives (Ritchie 1999). I will briefly examine a few of the major models in the region. Because of the need for evaluating a rich set of silvicultural options and stand structures, I only examine stand growth models that use the individual tree as the fundamental modeling unit. In this type of model the simulation begins with a tree list, proceeds by updating the size characteristics of each of these trees by using tree-level equations, and produces an updated tree list and summary statistics.

Of the stand models I review, the first six are forest yield models, and the final two are ecological gap (succession) models (see also Robinson and Monserud 2003). Ritchie (1999) is a good source for further information about many of these stand models and includes extensive literature citations.

CACTOS. The California Conifer Timber Output Simulator was developed by a research cooperative and parameterized with private data from northern California timber companies (Ritchie 1999). It is an individual-tree nonspatial forest growth simulator for the mixed conifers of northern California (Wensel et al. 1986). The model was designed for heavily managed industrial forests, and model exclusions accurately reflect the needs of the designers; for example, there is no ingrowth component. CACTOS will simulate the effects of harvesting and thinning as silvicultural treatments. CACTOS supports various sampling designs, including fixed- and variable-radius plots, as well as mixed designs. Uncertainty propagation is put into effect using record quintupling, a variant of Stage's (1973) record-tripling swindle. CRYPTOS (Cooperative Redwood Yield Project's Timber Output Simulator) is a closely related product used to predict the development of young-growth redwood (*Sequoia sempervirens* (D. Don) Endl.)/Douglas-fir stands (Krumland and Wensel 1982).

FPS. The Forest Projection System is a forest planning tool integrated with a suite of forest management tools (Arney and Milner 2000). This is a proprietary model of Forest Biometrics, a forest consulting company in Montana. FPS uses a unique response-surface library for simulation; it requires Microsoft Access®[2] for database capabilities. It also supports links to the ArcInfo® and MapInfo® geographic information systems, as well as a rich economics module. Instead of using parameters or models, the simulator uses a library that summarizes stand development response surfaces, which are used as lookup tables. Therefore, there are no equations built into FPS, only interpolation routines for the response surfaces. Uncertainty is not propagated through the model. For silviculture, the effects of fertilization, pruning, thinning, and harvesting can be simulated. Simulations can begin with planted or natural stands. FPS uses an index of clumpiness, a function of the standard deviation of the diameter distribution, to simulate local competition (see Walters and Arney 2002).

[2] The use of trade or firm names in this publication is for reader information and does not imply endorsement by the U.S. Department of Agriculture of any product or service.

FVS. Originally known as Prognosis (Stage 1973, Wykoff et al. 1982), the Forest Vegetation Simulator is an individual-tree nonspatial forest growth model. Its basic operation is described earlier in the *Forest Yield Model* section. Two of the 20 variants cover the Douglas-fir region (Pacific Northwest Coast (PN) and West Cascades (WC)). There are also variants for the Siskiyou and Klamath Mountains in southwestern Oregon and northern California, as well as for southeastern Alaska. The utility of FVS depends to some extent on the variant. For example, the southeastern Alaska variant simulates natural regeneration, whereas the PN and WC variants do not. FVS is unique for offering a rich assortment of extensions for natural disturbances: insect outbreaks (defoliators, bark beetles), diseases (dwarf mistletoe (*Arceuthobium douglasii*), root rot spread), and fire effects (Teck et al. 1997, Crookston and Havis 2002). It also has powerful extensions such as the Event Monitor, which permits simulation of management decisions that respond to stand development (Crookston 1990), and the Parallel Processor, which allows simultaneous simulation of numerous stands that may interact (Crookston and Stage 1991).

ORGANON. The ORegon Growth ANalysis and projectiON model is a nonspatial, statistically based model of individual tree growth and mortality. There are three variants: one for young mixed-conifer stands in southwestern Oregon (SWO), one for young Douglas-fir stands in northwestern Oregon (NWO), and one for young Douglas-fir and western hemlock stands in southwestern British Columbia, western Washington, and northwestern Oregon (SMC) (Hann et al. 1997). The model was originally developed in southwestern Oregon to simulate the growth of both even-aged and uneven-aged mixed-species stands (Ritchie 1999). The data used to estimate parameters for two of the variants were taken principally (NWO) or exclusively (SMC) from even-aged stands. A wide variety of sample designs can be forecast, including variable-radius plots and nested plots of different sizes with different diameter limits. The model uses the tree-tripling swindle (see Stage 1973) to provide stochastic predictions. If more than one point or plot is entered for a stand, confidence intervals for estimates of population parameters can be determined. Management options include thinning, pruning, and fertilization. Ingrowth is not modeled, although tree lists from regeneration models may be added to an execution. Natural disturbances are not modeled. The model uses considerable information on crown dynamics such as branch diameter and whorl location (e.g., Maguire and Hann 1990), which is one reason it is the only model that predicts wood quality (Hann et al. 1997). Although the code is proprietary for most practical purposes, the model is freely available on the Internet as a dynamic load library (DLL). This DLL deployment strategy treats the simulator as a black box, with the source code unavailable.

SPS. The Stand Projection System is a proprietary forest growth model complemented by forest management tools called the Stand Inventory System (SIS) (Mason, Bruce and Girard, Inc. 1988). SIS requires Microsoft Access® and relies on its database handling capacities. SPS is unusual in that it projects the growth of top height first, with diameter growth following as a function of relative tree size. SPS will impute any missing measurements on height, age, crown, and taper by using regressions. Likewise, a stand table can be imputed from simple stand characteristics by using a Weibull distribution. For silvicultural purposes, the effects of fertilization, thinning, and harvesting can be simulated. Planted and natural regeneration are not simulated, nor are natural disturbances or uncertainty. The economics module is very rich, with the costs of silvicultural treatments and harvesting included in an analysis.

TASS and TIPSY. The Tree and Stand Simulator and the Table InterPolation for Stand Yields models generate growth and yield information for even-aged stands of pure coniferous species with commercial importance in coastal and interior forests of British Columbia (Mitchell 1975, BC Ministry of Forestry 2001). TASS is a spatially explicit individual-tree model. The crowns of individual trees add a shell of foliage each year and either expand or contract asymmetrically in response to height growth, physical restrictions imposed by the crowns of competitors, environmental factors (site quality, defoliation, pest damage), and cultural practices (thinning, pruning, fertilization). Basically, the tree crowns fight for space in each of the three-dimensional grid cells that constitute the above-ground stand environment. TASS does not predict the yield of mixed-species, irregular, or uneven-aged stands. TASS remains proprietary, with the code unavailable. To accommodate various purposes, it was used to generate a yield table database for TIPSY (Mitchell et al. 1992). TIPSY retrieves and interpolates yield tables from this database for a specific site, species and management regime. In the same vein, TADAM-df and TADAM-p are TASS-derived growth models for British Columbia coastal Douglas-fir and interior lodgepole pine (*Pinus contorta* (Dougl. ex Loud.) plantations, respectively (UNBC 2002). They are based on a stand-level dynamic (differential equation system) approximation to output from TASS, and can be run on a Palm Pilot® or a Microsoft Excel® spreadsheet (Garcia 2001).

SORTIE/BC. This gap-phase model was originally written and parameterized for the deciduous forests of the northeastern United States (Pacala et al. 1993, 1996). Recently, the model has been extensively modified for the mixed-conifer forests (nine tree species) of northwestern British Columbia. SORTIE is an individual-tree spatial model that owes much to the gap model tradition of JABOWA and FORET (Botkin et al. 1972, Shugart 1984); it differs from this tradition in that plot size is not a growth parameter. Thus, plot size can be quite large (e.g., 9 hectare (ha)). The British Columbia version of the model simulates four processes: competition for light (Canham et al. 1999), tree growth as a function of light availability (Wright et al. 1998) and tree history (Wright et al. 2000), mortality (Kobe and Coates 1997), and seedling recruitment as a function of parental abundance and seedbed substrate (LePage et al. 2000). SORTIE/BC allows a wide range of silvicultural treatments. For example the model allows incorporation of a wide range of partial cutting strategies (e.g., understory protection, diameter limit, shelterwood, variable retention) and clearcutting. SORTIE/BC has a natural regeneration module and the ability to plant seedlings by tree species on a grid or random distribution at any step during a simulation.

ZELIG.PNW. This succession model is in the JABOWA and FORET gap-phase tradition, is based on ZELIG (Urban 1993), and is parameterized for the Oregon Coast Range (Busing and Garman 2002). A gap-phase model is intended to simulate succession following the death of a large tree, which creates a gap (0.4 ha) in the forest (Botkin et al. 1972, Shugart 1984). The specific locations of trees are not known within the gap; spatial location of all the gaps is known. Like all such gap models, it was designed to test ecological theory rather than be a forest management tool. However, ZELIG.PNW has been used to examine large-scale silvicultural policy (Hansen et al. 1995) and it is being used by Coastal Landscape Analysis and Modeling Study (Spies et al., in press), which is a province-level forest management study for the entire Oregon Coast Range. The model does not simulate any sample designs, and does not report sample-based or model-based uncertainty. Silvicultural options (thinning, harvesting) can be differentially applied to various portions of the stand, which enables the simulation of a complex mosaic of treatments (Hansen et al. 1995). The model simulates natural and planted regeneration. Although the source code for ZELIG.PNW is freely available, it is not considered highly portable because a wide range of parameters is required. For example, version 2 requires 15 growth parameters per tree species modeled, 10 site parameters, 8 climate parameters per month, and a complete array of soil information, including 5 parameters per soil type/layer (Urban 1993).

Three additional well-known models are available in the Northwest: the Douglas-fir stand yield model DFSIM (Curtis et al. 1981), the resource-flux model FORECAST (Kimmins et al. 1999), and the process/hybrid model 3-PG (Landsberg and Waring 1997). However, these models were not considered further because they are not based on individual-tree dynamics and are single-species models. Because of the importance of mixed-species compositions when evaluating compatible management alternatives, single-species simulators were excluded from consideration. Because of the importance of

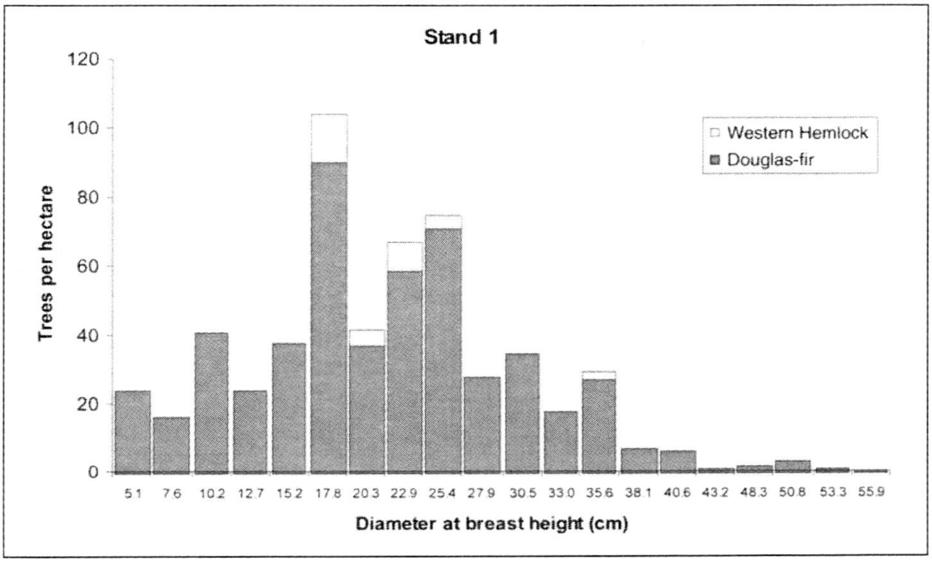

Figure 1. Diameter at breast height distribution of stand 1 (from Johnson 2002).

heterogeneous silvicultural treatments and irregular stand structures in the evaluation of alternatives, models not driven by individual tree dynamics were not considered sensitive to the proposed silvicultural treatments.

4. Variation Across Growth Models

In the winter of 2001-02, the Growth Model User Group conducted an experiment comparing growth models in the Pacific Northwest (Johnson 2002). G. Johnson (personal communication, Weyerhaeuser Co., Albany, Oregon, 2003) provided stand and tree inventory data for three typical Douglas-fir stands under management. User Group members were invited to make projections by using the stand growth simulation model of their choice (ergo, the Growth Model Runoff experiment). The goal was to evaluate the repeatability of projections by the same model among expert users, and to examine the variability and consistency of projecting the same stands across models. Each participant was provided with the same tree list for a given stand (see Figure 1), including heights, crown ratios, and expansion factors. Site index (King 1966) of Douglas-fir and total age were provided. Members were asked to project stand development in 10-year increments to age 60. Sixteen members responded, and eight simulation models were used: CRYPTOS, DFSIM, FPS, FVS-PN, FVS-WC, ORGANON, SPS, and TADAM (a TASS-derived model). The exercise was complicated by an intentional lack of certain information: geographic location, clumpiness (Walters and Arney 2002), plot-level information, stand treatment history, and maximum stand density index

(SDI) were not specified. Furthermore, site index was not completely consistent with some stands' exhibited height. A final complication for the Douglas-fir models (DFSIM, TADAM, CRYPTOS) was that two of the Douglas-fir stands contained an admixture of western hemlock.

Descriptions of the three stands at the start of the projections follow. Except for the individual tree list data, this was the only information provided for the experiment:

Characteristic	**Stand 1**	**Stand 2**	**Stand 3**
Age (years)	30 (from planting)	40 (natural stand)	30 (from planting)
Site index: dominant height (m) at age 50	35	30	38
Trees/ha	568	581	591
Basal area (m^2/ha)	24.3	53.9	21.8
Quadratic mean dbh (cm)	23.4	23.9	21.6
Douglas-fir composition (%)	95	100	89
Western hemlock (%)	5	0	11

Results: The initial diameter distribution (Figure 1) and predictions (Figures 2 through 5) for the first stand are typical of the other stands. Regional (location) effects can be large. FVS, in particular can more than double the predicted volume between the West Cascades (WC) variant and the Pacific Northwest Coast (PN) variant because of much higher growth rates in the Coast Range. Clearly, these two variants are different models. This variation arose because geographic location was not specified. On the other hand, CRYPTOS (a northern California model) projections were very similar to the growth models developed for forests to the North.

There was variation among users running the same model and the same version (Figures 2 through 5). This was quite surprising. We like to think of models as fixed boxes that consistently give the same results. We forget that different users can interpret the same modeling task differently. This can include choosing different growth models for the same problem, or different initial conditions when there is ambiguity in the problem information (e.g., geographic location). Regardless of the variation across models and users, all models maintain a remarkably consistent relation between stand height and volume (Figure 6). Johnson (2002) observed that all simulation model predictions appear to conform to Eichorn's rule (growth trends between dominant height and volume productivity are about the same for all sites) across all three stands.

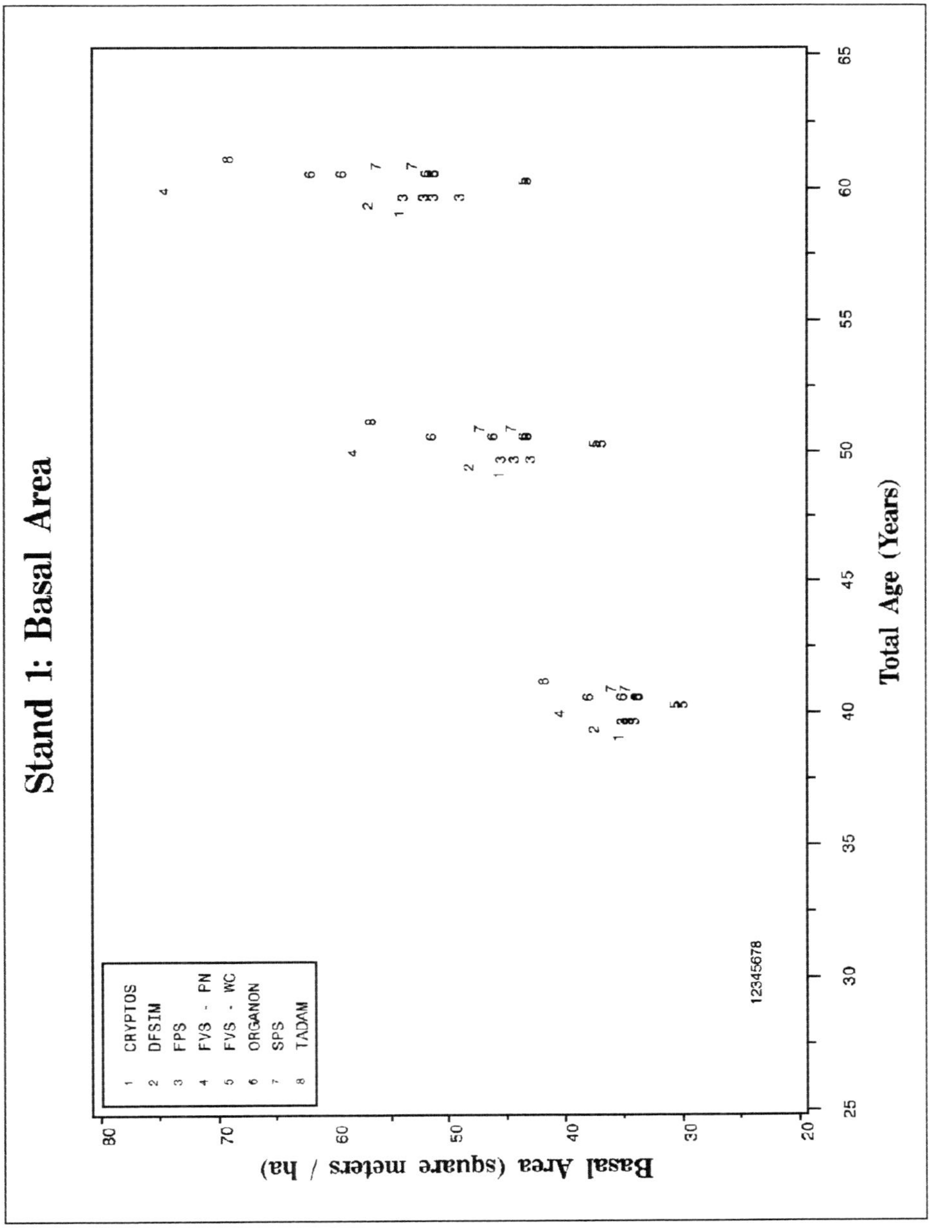

Figure 2. Basal area projections (m^2/ha) for 30 years, using several different stand growth models (from Johnson 2002). For readability, the 10-year projections are slightly offset left and right, in the order of the listed simulation model.

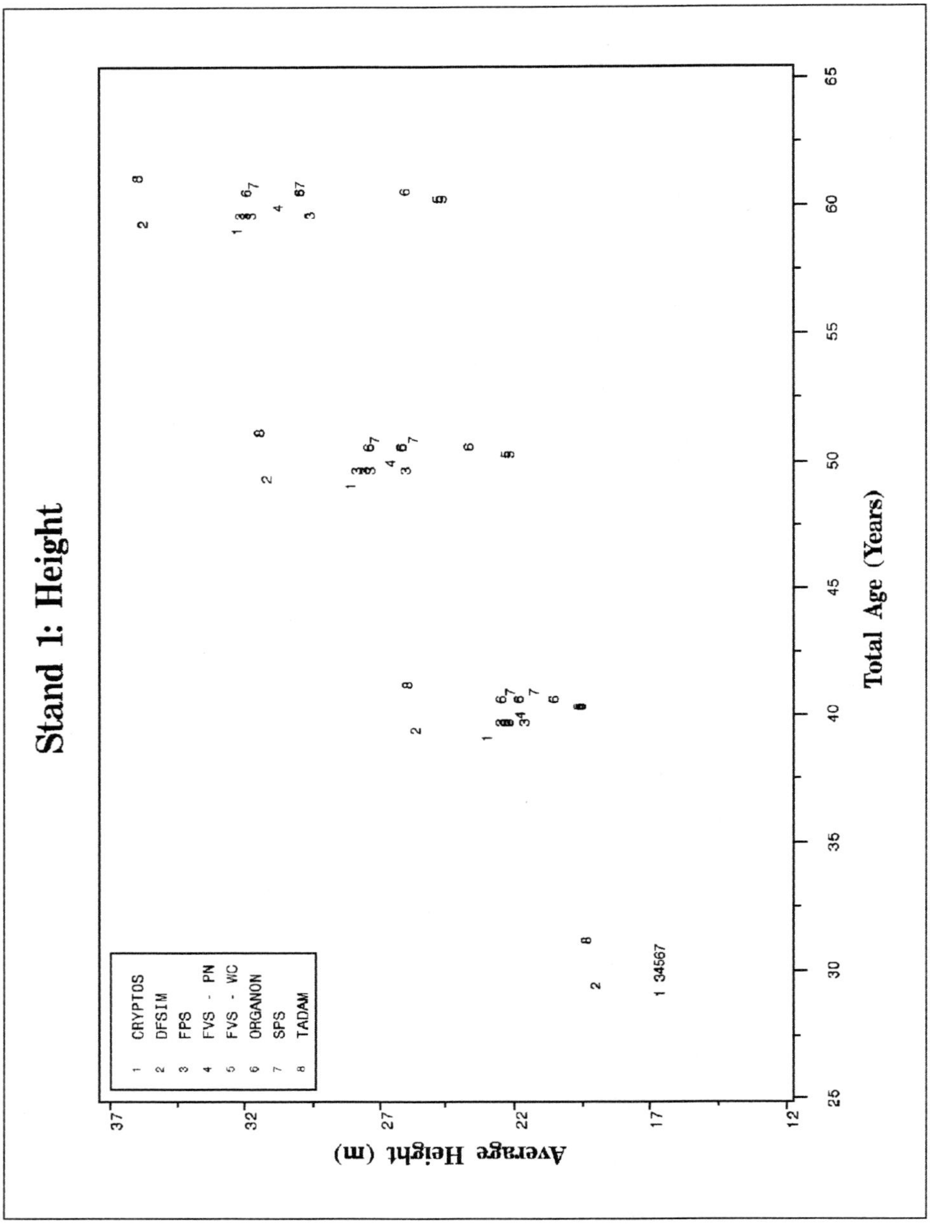

Figure 3. Average height projections (m) for 30 years, using several different stand growth models (from Johnson 2002). For readability, the 10-year projections are slightly offset left and right, in the order of the listed simulation model.

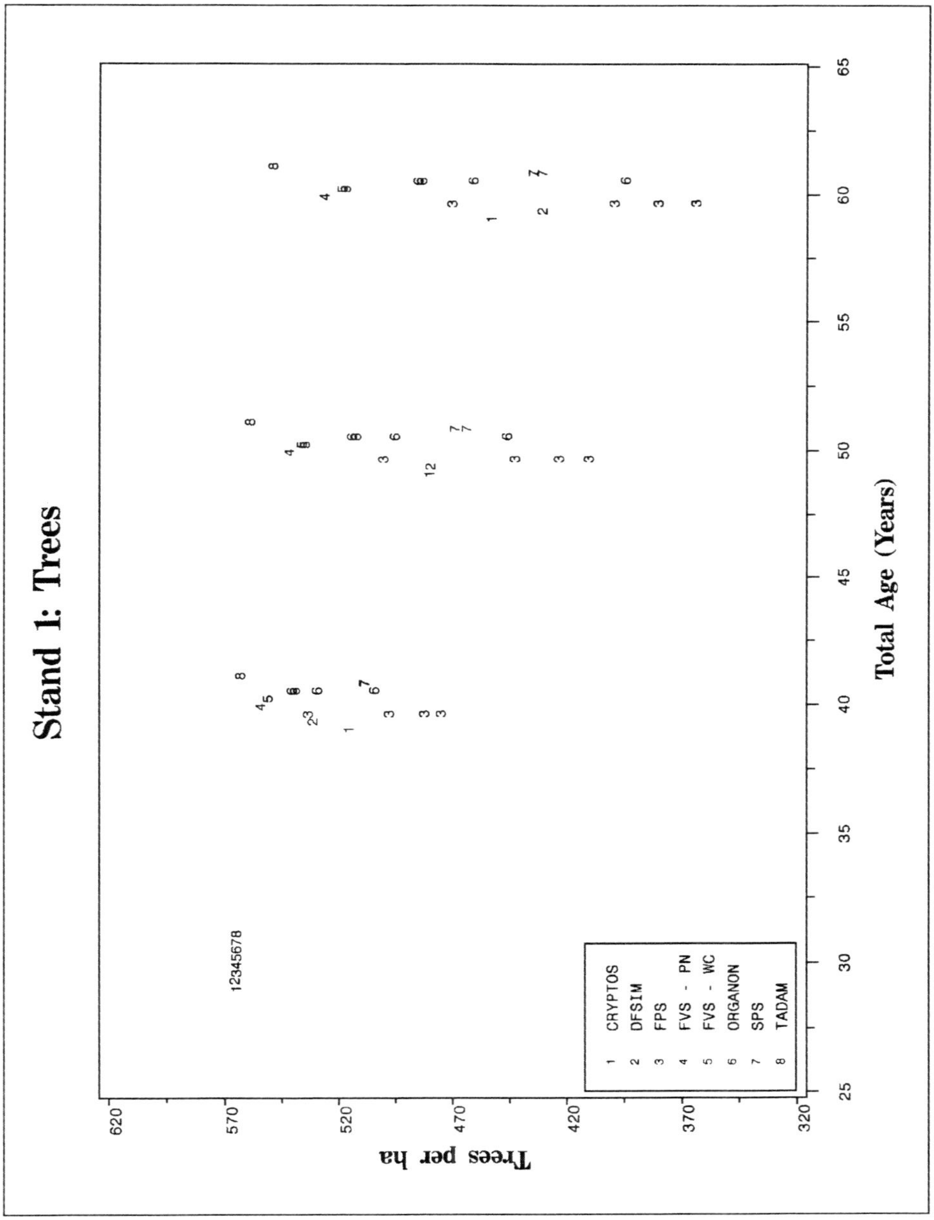

Figure 4. Trees/ha projections for 30 years, using several different stand growth models (from Johnson 2002). For readability, the 10-year projections are slightly offset left and right, in the order of the listed simulation model.

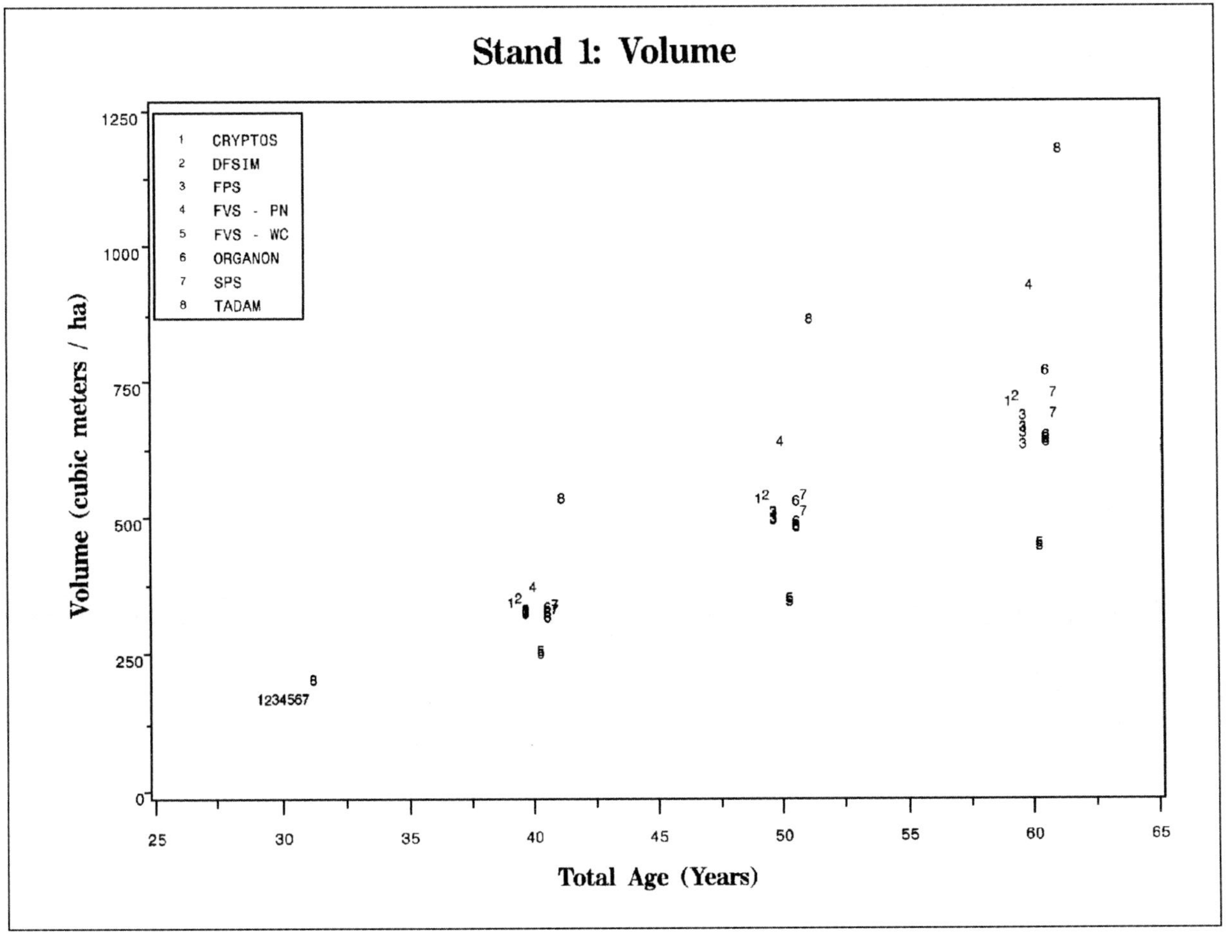

Figure 5. Volume projections (m^3/ha) for 30 years, using several different stand growth models (from Johnson 2002). For readability, the 10-year projections are slightly offset left and right, in the order of the listed simulation model.

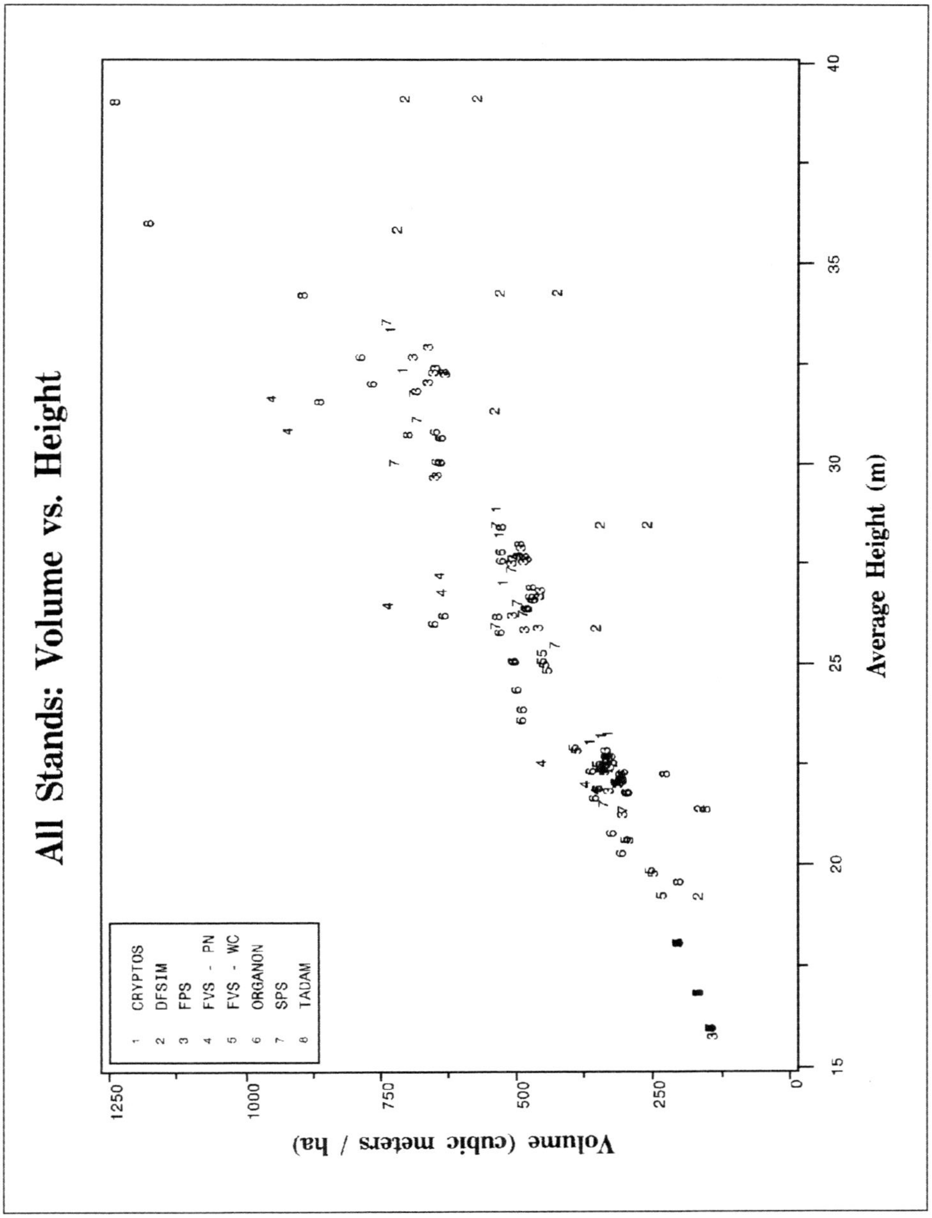

Figure 6. Volume (m^3/ha) vs. height (m) growth projections at age 60 for each of three stands and a variety of stand growth models (from Johnson 2002).

5. Discussion

Robinson and Monserud (2003) evaluated most of these stand growth models for their adaptability and utility in modeling alternative silvicultural systems under consideration in the Pacific Northwest. They examined only models that could simulate a full range of stand structures, species compositions, and age distributions, including both mixed-species uneven-aged stands, as well as the even-aged pure stands typically created in the Douglas-fir region in the previous 50 years. Because these models were mostly developed by using even-aged or plantation growth data, it is not clear how robust the model projections would be for the complex stand structures and compositions under consideration by silviculturists in the Pacific Northwest today.

Based on the criteria developed by Robinson and Monserud (2003), none of the models is fully adaptable or entirely appropriate for the alternative silvicultures being proposed for the Pacific Northwest. Of the eight candidate models, FVS was the most adaptable. It checks favorably on almost all criteria: open source code, a broad geographic coverage, and a well-documented parameter estimation process. It also has a wide range of powerful extensions for simulating both stand disturbances and dynamic management (triggered by stand structure rather than at specific times). The original northern Idaho variant (NI) has undergone extensive critical examination and field-testing for over 25 years. The other geographic variants of FVS remain less scrutinized. The NI variant was designed to simulate the mixed-species compositions and heterogeneous, uneven-aged structures that are increasingly prescribed by silviculturists in the Pacific Northwest. However, other models have facilities that would strengthen FVS, such as ORGANON's wood quality module. FVS also would be improved if the effects of fertilization and pruning could be simulated, and if a natural regeneration module were available for all FVS variants.

The eight simulation models I discussed earlier fall into two general categories: forest succession models (SORTIE, ZELIG.PNW) and forest yield models (the others). Although SORTIE and ZELIG.PNW were not designed as forest management tools, there is increasing interest in applying them and hybrid models like 3-PG (Landsberg and Waring 1997) to forest management problems (e.g., Spies et al., in press). Robinson and Monserud (2003) evaluated these models relative to the expectations and needs of forest managers. They found that forest yield models accommodate a broader array of inputs and outputs important to forest planning and operations: the models are site-specific and allow a variety of sample designs, range of silviculture, wood products, and economic calculations.

Spatial heterogeneity within stands is one of the goals in several alternative silvicultural experiments currently underway in the Pacific Northwest (Monserud 2002). The Alternatives to Clearcutting (ATC) study in southeastern Alaska is examining different spatial patterns of retained trees at a given density: uniform dispersal, clumps within a uniform matrix, and gaps within a uniform matrix (McClellan et al. 2000). Both the Olympic Habitat Development Study (OHDS; see Harrington and Carey 1997) and the Forest Ecosystem Study (FES; see Carey et al. 1999) examine several spatial patterns: variable-density thinning with 10% of area in uncut patches, 15% of area in openings (small patch cuts), and 25 to 30% basal area removal in the remaining stand. These alternatives eschew homogeneous stand structures, and intentionally seek to create a non-uniform spatial distribution within the stand (Monserud 2003). Because most stand simulation models used for forest management are non-spatial models, the tenuous assumption of homogeneous spatial distribution calls for close examination of these management models.

In the 1960s and 1970s, several spatial simulation models were developed (see Biging and Dobbertin 1995), but only a few survived the expensive requirement of supplying mapped tree locations before growth could be predicted (e.g., the TASS model (Mitchell 1975) is still in use). Most ended as academic exercises, but some new approaches are quite promising (e.g., SORTIE/BC). Biging and Dobbertin (1995) found no advantage to using spatial information to predict growth, provided that crown information was incorporated into the prediction. Clearly, a careful reevaluation of the importance of spatial information for stand prediction is needed, especially in relation to alternative silvicultural treatments that aim for heterogeneous stand structures. The several large-scale experiments testing alternative silvicultural methods (e.g., Harrington and Carey 1997, Carey et al. 1999, McClellan et al. 2000) will eventually provide crucial observations for testing the value of spatial information for predicting stand growth under alternative silviculture; the experiments are currently too new for a clear test.

As a spatial model, SORTIE/BC presents a unique combination of assets for simulating the complex silvicultural treatments being examined as alternatives to clearcutting. These include large plot sizes; mixed-species combinations; a wide range of partial cutting strategies, including variable retention; both natural and artificial regeneration at any time; and spatially sensitive treatments. The model does not appear to suffer from extreme site-specificity, which can require large numbers of difficult-to-observe input variables, as both ZELIG.PNW and FORECAST do. SORTIE/BC is a simulation approach that clearly holds promise for examining the silvicultural methods of the future.

Wilhere (2003) provides one of the most striking examples of the utility of certain stand models for addressing aspects of compatible management. He used the Event Monitor (Crookston 1990) to program an ad hoc snag dynamics simulator into FVS. The resulting product allowed him to use FVS to simulate and examine snag dynamics in Douglas-fir forests in the Pacific Northwest (Wilhere 2003). Snags provide important habitat for wildlife, but only recently have forest managers explicitly considered them in planning (Franklin et al. 1997). Wilhere's (2003) results allowed him to make management recommendations for snag management, an area noted for its lack of information. The ability to simulate wildlife habitat in conjunction with forest management options goes to the heart of addressing certain compatible forest management questions.

6. References

Aber, J.D., and Federer, C.A. 1992. A generalized, lumped-parameter model of photosynthesis, evapotranspiration and net primary production in temperate and boreal forests. *Oecologia*, 92: 463-474.

Arney, J.D., and Milner, K.S. 2000. Biometrics of forest inventory, forest growth, and forest planning. Technical Report 10. Forest Biometrics Library. 279 p.

Arnott, J.T., and Beese, W.J. 1997. Alternatives to clearcutting in BC coastal montane forests. *Forestry Chronicle*, 73: 670-678.

Biging, G.S., and Dobbertin, M. 1995. Evaluation of competition indices in individual tree growth models. *Forest Science*, 41: 360-377.

Botkin, D.B., Janak, J.F., and Wallis, J.R. 1972. Rationale, limitations, and assumptions of a Northeastern forest growth simulator. *IBM Journal of Research and Development*, 16(2): 101-116.

Botkin, D.B., and Schenk, H.J. 1996. History and rationales of the JABOWA model of forest growth. NCASI Technical Bulletin 717. I: 1-62 [plus appendix].

British Columbia Ministry of Forestry. 2001. Tree and stand simulator. Retrieved April 10, 2003 from http://www.for.gov.bc.ca/research/gymodels/tass/.

Bugmann, H. 2001. A review of forest gap models. *Climatic Change*, 51: 259-305.

Busing, R.T., and Garman, S.L. 2002. Promoting old-growth characteristics and long-term wood production in Douglas-fir forests. *Forest Ecology and Management*, 160(1-3): 161-175.

Canham, C.D., Coates, K.D., Bartemucci, P., and Quaglia, S. 1999. Measurement and modeling of spatially-explicit variation in light transmission through interior cedar-hemlock forests of British Columbia. *Canadian Journal of Forest Research*, 29: 1775-1783.

Carey, A.B., Thysell, D.R., and Brodie, A.W. 1999. The forest ecosystem study: background, rationale, implementation, baseline conditions, and silvicultural assessment. General Technical Report PNW-GTR-457. Portland, OR: U.S. Department of Agriculture, Forest Service, Pacific Northwest Research Station. 129 p.

Cissel, J.H., Swanson, F.J., and Weisberg, P.J. 1999. Landscape management using historical fire regimes: Blue River, Oregon. *Ecological Applications*, 9(4): 1217-1231.

Clayoquot Scientific Panel. 1995. Scientific panel for sustainable forest practices in Clayoquot Sound, Report 5, sustainable ecosystem management in Clayoquot Sound: planning and practices, Victoria, BC. 296 p.

Coates, K.D., Banner, A., Steventon, J.D., LePage, P., and Bartemucci, P. 1997. The Date Creek silvicultural systems study in the interior cedar-hemlock forests of northwestern British Columbia: overview and treatment summaries. Land Management Handbook. Victoria, BC: British Columbia Ministry of Forests. 38 p.

Committee of Scientists. 1999. Sustaining the people's lands: recommendations for stewardship of the national forests and grasslands into the next century. Washington, DC: U.S. Department of Agriculture. Retrieved May 31, 2002 from http://www.fs.fed.us/forum/nepa/rule/cosreport.shtml.

Coops, N.C., Waring, R.H., and Landsberg, J.J. 1998. Assessing forest productivity in Australia and New Zealand using a physiologically-based model driven with averaged monthly weather data and satellite-derived estimates of canopy photosynthetic capacity. *Forest Ecology and Management*, 104(1-3): 113-127.

Crookston, N.L. 1990. User's guide to the event monitor: part of Prognosis Model, version 6. General Technical Report INT-275. Ogden, UT: U.S. Department of Agriculture, Forest Service, Intermountain Research Station. 21 p.

Crookston, N.L., and Havis, R.N. (Comps.). 2002. Second forest vegetation simulator (FVS) conference; Proceedings RMRS-P-25. Ogden, UT: U.S. Department of Agriculture, Forest Service, Rocky Mountain Research Station. 208 p.

Crookston, N.L., and Stage, A.R. 1991. User's guide to the Parallel Processing extension of the Prognosis model. General Technical Report INT-281. Moscow, ID: U.S. Department of Agriculture, Forest Service, Intermountain Forest and Range Experiment Station. 88 p.

Curtis, R.O., Clendenen, G.W., and DeMars, D.J. 1981. A new stand simulator for coast Douglas-fir: DFSIM user's guide. General Technical Report PNW-GTR-128. Portland, OR: U.S. Department of Agriculture, Forest Service, Pacific Northwest Research Station. 79 p.

Curtis, R.O., DeBell, D.S., Harrington, C.A., Lavender, D.P., St. Clair, J.B., Tappeiner, J.C., and Walstad, J.D. 1998. Silviculture for multiple objectives in the Douglas-fir region. General Technical Report PNW-GTR-435. Portland, OR: U.S. Department of Agriculture, Forest Service, Pacific Northwest Research Station. 123 p.

Dixon, R., Meldahl, R., Ruark, G., and Warren, W. 1990. Forest growth: process modeling of responses to environmental stress. Portland, OR: Timber Press. 441 p.

Dudek, A., and Ek, A.R. 1980. A bibliography of worldwide literature on individual tree based forest stand growth models. Staff Paper Series 12. St. Paul, MN: Department of Forest Resources, University of Minnesota. 33 p.

Ferguson, D.F., and Crookston, N.L. 1991. User's guide to version 2 of the regeneration establishment model: part of the prognosis model. General Technical Report INT-279. Ogden, UT: U.S. Department of Agriculture, Forest Service, Intermountain Research Station. 34 p.

Forest Ecosystem Management Assessment Team [FEMAT]. 1993. Forest ecosystem management: an ecological, economic, and social assessment. Portland, OR: U.S. Department of Agriculture; U.S. Department of the Interior [and others]. [Irregular pagination].

Franklin, J.F., Berg, D.R., Thornburg, D.A., and Tappeiner, J.C. 1997. Alternative silvicultural approaches to timber harvesting: variable retention harvest systems. In: K.A. Kohm, and J.F. Franklin (Eds.), Creating a forestry for the 21st century (pp. 111–140). Washington, DC: Island Press.

Garcia, O. 2001. On bridging the gap between tree-level and stand-level models. In: K. Rennolls (Ed.), Proceedings of IUFRO 4.11 Conference: Forest biometry, modelling and information science, University of Greenwich. Retrieved January 2003 from http://cms1.gre.ac.uk/conferences/iufro/proceedings/.

Godfrey, K. 1983. Compartmental models and their applications. New York: Academic Press. 291 p.

Greenough, J.A., Robinson, D.C.E., Kurz, W.A., Beukema, S.J., Densmore, N., Winter, R., and Snowdon, B. 1999. Use of the prognosis EI model in balancing timber and environmental values at the watershed-level. Vancouver, BC: ESSA Technologies Ltd. 30 p.

Hann, D.W., Hester, A.S., and Olsen, C.L. 1997. ORGANON User's Manual. Corvallis, OR: Oregon State University.

Hansen, A.J., Garman, S.L., Weigand, J.F., Urban, D.L., McComb, W.C., and Raphael, M.G. 1995. Alternative silvicultural regimes in the Pacific Northwest: simulations of ecological and economic effects. *Ecological Applications*, 5(3): 535-554.

Harrington, C.A., and Carey, A.B. 1997. The Olympic habitat development study: conceptual study plan. Unpublished manuscript. 38 p. On file with: U.S. Department of Agriculture, Forest Service, Pacific Northwest Research Station, Olympia Forestry Sciences Laboratory, 3625 93rd Avenue, Olympia, WA 98512-9193.

Haynes, R.W., and Monserud, R.A. 2002. A basis for understanding compatibility among wood production and other forest values. General Technical Report PNW-GTR-529. Portland, OR: U.S. Department of Agriculture, Forest Service, Pacific Northwest Research Station. 46 p.

Hinckley, T.M., Sprugel, D.G., Batista, J.L.F., Brooks, J.R., Brubaker, L.B., Compton, J., Erickson, H.E., Little, R.L., Maguire, D.A., McCarter, J.B., McKay, S.J., Pass, D., Peterson, D.W., Reed, J.P., Tacey, W., Wilkinson, L.E., and Whytemare, A. 1996. Use of the JABOWA family of individual-tree based models for exploration of forest responses to global climate change. NCASI Technical Bulletin 717. II: 1-54.

Host, G.E., and Isebrands, J.G. 1994. An interregional validation of ECOPHYS, a growth process model of juvenile poplar clones. *Tree Physiology*, 14: 933-945.

Hummel, S.S. 2003. Managing structural and compositional diversity with silviculture. Chapter 4. In: R.A. Monserud, R.W. Haynes, and A.C. Johnson (Eds.), Compatible forest management. Dordrecht, The Netherlands: Kluwer Academic Publishers.

Isebrands, J.G., Rauscher, H.M., Crow, T.R., and Dickmann, D.J. 1990. Whole-tree growth process models based on structure—functional relationships. In: R. Dixon, R. Meldahl, G. Ruark, and W. Warren (Eds.), Forest growth: process modeling of responses to environmental stress (pp. 96-112). Portland, OR: Timber Press.

Johnson, G. 2002. Growth model run-off. Growth model user group (GMUG). Retrieved January 22, 2002 from http://www.growthmodel.org/.

Kimmins, J.P., Mailly, D., and Seely, B. 1999. Modelling forest ecosystem net primary production: the hybrid simulation approach used in FORECAST. *Ecological Modelling*, 122: 195-224.

King, J.E. 1966. Site index curves for Douglas-fir in the Pacific Northwest. Weyerhaeuser Forest Paper 8. Federal Way, WA: Weyerhaeuser Co.

Kobe, R.K., and Coates, K.D. 1997. Models of sapling mortality as a function of growth to characterize interspecific variation in shade tolerance of eight tree species of northwestern British Columbia. *Canadian Journal of Forest Research*, 27: 227-236.

Korol, R.L., Milner, K.S., and Running, S.W. 1996. Testing a mechanistic model for predicting stand and tree growth. *Forest Science*, 42: 139-153.

Krumland, B.E., and Wensel, L.C. 1982. CRYPTOS—User's guide, cooperative redwood yield project timber output simulator, version 4.0. Co-op. Research Note 20. Berkeley, CA: Redwood Yield Research Project, University of California.

Landsberg, J.J., and Waring, R.H. 1997. A generalized model of forest productivity using simplified concepts of radiation-use efficiency, carbon balance and partitioning. *Forest Ecology and Management*, 95(3): 209-228.

Landsberg, J.J., Waring, R.H., and Coops, N.C. [In press]. Performance of the forest productivity model 3-PG applied to a wide range of forest types. *Forest Ecology and Management*.

LePage, P., Canham, C.D., Coates, K.D., and Bartemucci, P. 2000. Seed source versus substrate limitation of seedling recruitment in interior cedar-hemlock forests of British Columbia. *Canadian Journal Forest Research*, 30: 415-427.

Lindner, M., Sievänen, R., and Pretzsch, H. 1997. Improving the simulation of stand structure in a forest gap model. *Forest Ecology and Management*, 95: 183-195.

Maguire, D.A., and Hann, D.W. 1990. Constructing models for direct prediction of 5-year crown recession in southwestern Oregon Douglas-fir. *Canadian Journal of Forest Research*, 20: 1044-1052.

Mäkelä, A. 1997. A carbon balance model of growth and self-pruning in trees based on structural relationships. *Forest Science*, 43: 7-24.

Mäkelä, A., Landsberg, J.J., Ek, A.R., Burk, T.E., Ter-Mikaelian, M., Ågren, G.I., Oliver, C.D., and Puttonen, P. 2000a. Process-based models for forest ecosystem management: current state of the art and challenges for practical implementation. *Tree Physiology*, 20: 289-298.

Mäkelä, A., Sievänen, R., Lindner, M., and Lasch, P. 2000b. Application of volume growth and survival graphs in the evaluation of four process-based forest growth models. *Tree Physiology*, 20: 347-355.

Mason, Bruce and Girard, Inc. 1988. SPS user's guide, version 2.4. Portland, OR: Mason, Bruce and Girard, Inc.

McClellan, M.H., Swanston, D.N., Hennon, P.E., Deal, R.L., De Santo, T.L., and Wipfli, M.S. 2000. Alternatives to clearcutting in the old-growth forests of southeast Alaska: study plan and establishment report. General Technical Report PNW-GTR-494. Portland, OR: U.S. Department of Agriculture, Forest Service, Pacific Northwest Research Station. 40 p.

Milner, K.S., and Coble, D.W. 1995. A mechanistic approach to predicting the growth and yield of stands with complex structures. In: K.L. O'Hara (Ed.), Uneven-aged management: opportunities, constraints and methodologies. MFCES Miscellaneous Publication 56. Missoula, MT: School of Forestry, University of Montana.

Mitchell, K.J. 1975. Dynamics and simulated yield of Douglas-fir. *Forest Science Monograph*, 17: 39.

Mitchell, K.J., Grout, S.E., Macdonald, R.N., and Watmough, C.A. 1992. User's guide for TIPSY: a table interpolation for stand yields. Victoria, BC: British Columbia Ministry of Forestry. 78 p.

Monserud, R.A. 2001. A brief history of stand growth modeling in the Northwest. *Western Forester*, 45(4): 6-7.

Monserud, R.A. 2002. Large-scale management experiments in the moist maritime forests of the Pacific Northwest. *Landscape and Urban Planning*, 59(3): 159-180.

Monserud, R.A. 2003. Experimental approaches to joint forest production. Chapter 5. In: R.A. Monserud, R.W. Haynes, and A.C. Johnson (Eds.), Compatible forest management. Dordrecht, The Netherlands: Kluwer Academic Publishers.

Monserud, R.A. [In press]. Evaluating forest models in a sustainable forest management context. *Forest Biometry, Modelling and Information Sciences. Sciences (FBMIS)*. To be available on: http://www.fbmis.info/

Monserud, R.A., and Sterba, H. 1999. Modeling individual tree mortality for Austrian forest species. *Forest Ecology and Management*, 113(2-3): 109-123.

Newnham, R.M. 1964. The development of a stand model for Douglas-fir. Vancouver, BC: University of British Columbia. 201 p. Ph.D. dissertation (University Microfilms 64-4535).

Pacala, S.W., Canham, C.D., Saponara, J., Silander, J.A., Jr., Kobe, R.K., and Ribbens, E. 1996. Forest models defined by field measurements: II. Estimation, error analysis, and dynamics. *Ecological Monographs*, 66: 1-43.

Pacala, S.W., Canham, C.D., and Silander, J.A., Jr. 1993. Forest models defined by field measurements: I. The design of a northeastern forest simulator. *Canadian Journal of Forest Research*, 23(10): 1980-1988.

Parton, W.J., Stewart, J.W.B., and Cole, C.V. 1988. Dynamics of C, N, P and S in grassland soils—a model. *Biogeochemistry*, 5: 109-132.

Pastor, J., and Post, W.M. 1985. Development of a linked forest productivity-soil process model. ORNL/TM-9519. Oak Ridge, TN: Oak Ridge National Laboratory.

Pastor, J., and Post, W.M. 1986. Influence of climate, soil moisture, and succession on forest carbon and nitrogen cycles. *Biogeochemistry*, 2: 3-27.

Prentice, I.C., and Leemans, R. 1990. Pattern and process and the dynamics of forest structure: a simulation approach. *Journal of Ecology*, 78: 340-355.

Prentice, I.C., Sykes, M.T., and Cramer, W. 1993. A simulation model for the transient effects of climate change on forest landscapes. *Ecological Modelling*, 65: 51-70.

Ritchie, M.W. 1999. A compendium of forest growth and yield simulators for the Pacific Coast states. General Technical Report PSW-GTR-174. Albany, CA: U.S. Department of Agriculture, Forest Service, Pacific Southwest Research Station. 59 p.

Robinson, A.P., and Monserud, R.A. 2003. Criteria for comparing the adaptability of forest growth models. *Forest Ecology and Management*, 172(1): 53-67.

Running, S.W., and Coughlan, J.C. 1988. A general model of forest ecosystem processes for regional applications. I. Hydrologic balance, canopy gas exchange and primary production processes. *Ecological Modelling*, 42: 125-154.

Schwalm, C., and Ek, A.R. 2001. Climate change and site: relevant mechanisms and modeling techniques. *Forest Ecology and Management*, 150: 241-257.

Shinozaki, K., Yoda, K., Hozumi, K., and Kira, I. 1964a. A quantitative analysis of plant form: the pipe model theory. I. Basic Analyses. *Japanese Journal of Ecology*, 14(3): 97-105.

Shinozaki, K., Yoda, K., Hozumi, K., and Kira, I. 1964b. A quantitative analysis of plant form: the pipe model theory. II. Further evidence of the theory and its application in forest ecology. *Japanese Journal of Ecology*, 14(4): 133-139.

Shugart, H.H. 1984. A theory of forest dynamics: the ecological implications of forest succession models. New York: Springer. 278 p.

Sievänen, R., Lindner, M., Mäkelä, A., and Lasch, P. 2000. Volume growth and survival graphs: a method for evaluating process-based forest growth models. *Tree Physiology*, 20: 357-365.

Siitonen, M. 1993. Experiences in the use of forest management planning models. *Silva Fennica*, 27: 167-178.

Spies, T.A., Reeves, G.H., Burnett, K.M., McComb, W.C., Johnson, K.N., Grant, G., Ohmann, J.L., Garman, S.L., and Bettinger, P. [In press]. Assessing the ecological consequences of forest policies in a multi-ownership province in Oregon. In: J. Liu, and W.W. Taylor (Eds.), Integrating landscape ecology into natural resource management. Cambridge, UK: Cambridge University Press.

Stage, A.R. 1973. Prognosis model for stand development. Research Paper INT-137. Ogden, UT: U.S. Department of Agriculture, Forest Service, Intermountain Forest and Range Experiment Station. 32 p.

Stage, A.R., and Wykoff, W.R. 1998. Adapting distance-independent forest growth models to represent spatial variability: effects of sampling design on model coefficients. *Forest Science*, 44(2): 224-238.

Teck, R., Moeur, M., and Adams, J. (Eds.) 1997. Proceedings: forest vegetation simulator conference. General Technical Report INT-GTR-373. U.S. Department of Agriculture, Forest Service, Intermountain Research Station. 222 p.

University of Northern British Columbia [UNBC]. 2002. TADAM. Retrieved April 10, 2003 from http://www.unbc.ca/forestry/forestgrowth/tadam/.

Urban, D.L. 1993. A user's guide to ZELIG version 2 with notes on upgrades from version 1. Fort Collins, CO: Colorado State University.

Valentine, H. 1990. A carbon-balance model of tree growth with a pipe-model framework. In: R. Dixon, R. Meldahl, G. Ruark, and W. Warren (Eds.), Forest growth: process modeling of responses to environmental stress (pp. 33-40). Portland, OR: Timber Press.

Vanclay, J.K. 1995. Growth models for tropical forests: a synthesis of models and methods. *Forest Science*, 41: 7-42.

Walters, D., and Arney, J. 2002. Exploring the clumpiness factor in FPS: a dialog. Retrieved April 10, 2002 from http://www.growthmodel.org/.

Wensel, L.C., Daugherty, P.J., and Meerschaert, W.J. 1986. CACTOS user's guide: the California Conifer Timber Output Simulator. Bulletin 1920. Berkeley, CA: Agricultural Experiment Station, University of California.

Wilhere, G.F. 2003. Simulations of snag dynamics in an industrial Douglas-fir forest. *Forest Ecology and Management*, 174: 521-539.

Wright, E.F., Canham, C.D., and Coates, K.D. 2000. Effects of suppression and release on sapling growth for eleven tree species of northern, interior British Columbia. *Canadian Journal of Forest Research*, 30: 1571-1580.

Wright, E.F., Coates, K.D., Canham, C.D. and Bartemucci, P. 1998. Species variability in growth response to light across a climatic gradient in northwestern British Columbia. *Canadian Journal of Forest Research*, 28: 871-886.

Wykoff, W.R. 1990. A basal area increment model for individual conifers in the northern Rocky Mountains. *Forest Science*, 36: 1077-1104.

Wykoff, W.R., Crookston, N.L., and Stage, A.R. 1982. User's guide to the stand prognosis model. General Technical Report GTR-INT-133. Ogden, UT: U.S. Department of Agriculture, Forest Service, Intermountain Forest and Range Experiment Station. 112 p.

Wykoff, W.R., and Monserud, R.A. 1988. Representing site quality in increment models: a comparison of methods. In: A.R. Ek, S.R. Shifley, and T.E. Burk (Eds.), Forest growth and modelling and prediction (Vol. 1, pp. 184-191). General Technical Report NC-120. St. Paul, MN: U.S. Department of Agriculture, Forest Service, North Central Forest Experiment Station.

Yaussy, D.A. 2000. Comparison of an empirical forest growth and yield simulator and a forest gap simulator using actual 30-year growth from two even-aged forests in Kentucky. *Forest Ecology and Management*, 126: 385-398.

Yoda, K., Kira, T., Ogawa, H., and Hozumi, K. 1963. Self-thinning in overcrowded pure stands under cultivated and natural conditions. *Journal of Biology*, Osaka City University, 14: 107-129.

Appendix 1. Key to Model Acronyms and Names, with References

3-PG. . . . . Physiological Principles In Predicting Growth (Landsberg and Waring 1997)
CACTOS . . . . . CAlifornia Conifer Timber Output Simulator (Wensel et al. 1986)
CENTURY. . . . Soil organic matter dynamics (Parton et al. 1988)
CROBAS . . . . . CROwn BASe model (Mäkelä 1997)
CRYPTOS. . . . Cooperative Redwood Yield Project's Timber Output Simulator (Krumland and Wensel 1982)
DFSIM. . . . . . . Douglas-Fir SIMulator. (Curtis et al. 1981)
FORCYTE. . . . Stand-level forest ecosystem model (Kimmins et al. 1999)
FORECAST. . . Stand-level forest ecosystem model (Kimmins et al. 1999)
FOREST-BGC. .Forest-BioGeoChemical cycles (Running and Coughlan 1988)
FORET. . . . . . . Forests of Eastern Tennessee (Botkin et al. 1972)
FORSKA . . . . . Foret for Scandinavia (Prentice and Leemans 1990; Prentice et al. 1993)
FPS. . . . . Forest Projection System (Arney and Milner 2000)
FVS. . . . . .Forest Vegetation Simulator (PN = Pacific Northwest Coast, WC = West Cascades variants). (Teck et al. 1997, Crookston and Havis 2002)
JABOWA. . . . . Janek-Botkin-Wallis (Botkin et al. 1972)
LINKAGES. . . Linked forest productivity-soil process model (Pastor and Post 1985, 1986)
MELA. . . . . . . .MEtsä LAskelma ("forest calculation") (Siitonen 1993)
ORGANON. . . .ORegon Growth ANalysis and projectiON (Hann et al. 1997)
PipeQual. . . . . .PIPESTEM and timber quality (Mäkelä et al. 2000a)
PIPESTEM. . . .Pipe theory model (Valentine 1990)
PNET. . . . . Photosynthesis (Pn) and EvapoTranspiration (Aber and Federer 1992)
Prognosis. . . . . .Stand Prognosis model (Stage 1973, Wykoff et al. 1982)
SPS. . . . . Stand Projection System (Mason, Bruce and Girard, Inc. 1988)
Stand-BGC. . . . Stand-BioGeoChemical cycles. (Milner and Coble 1995)
SORTIE. . . . . . Forest dynamics model for the NE (sortie is synonymous with foray, a pun on FORET; Pacala et al. 1993, 1996)
SORTIE/BC. . . SORTIE for British Columbia. (Wright et al. 1998)
TADAM . . . . . . TASS approximation by A Dynamical Aggregated Model (Garcia 2001)
TASS. . . . . .Tree And Stand Simulator (Mitchell 1975)
TIPSY. . . . . . . .Table Interpolation For Stand Yields from TASS (Mitchell et al. 1992)

TREE-BGC. . . .Tree-BioGeoChemical cycles. (Korol et al. 1996)
ZELIG. . . . . . . .The ZELIG Tree Simulator Model (Urban 1993)
ZELIG.PNW. . .ZELIG for the Pacific Northwest (Busing and Garman 2002)

Chapter 7

Modeling Landscape Management

Robert A. Monserud
USDA Forest Service, Pacific Northwest Research Station,
620 SW Main St., Suite 400, Portland, OR 97205, USA

1. Introduction

The second question we asked in the opening chapter was: *How do different approaches to forest management affect relatively large and complex ecosystems?* This calls for a landscape analysis of the constituent stands. The goal is to expand information on the options that may increase the compatibility between wood production and other societal values derived from forestlands (Haynes and Monserud 2002). The nature of the problem focuses our attention on multidisciplinary solutions across a broad area (large scale). Planning based on single-resource issues can prove costly, or, in some cases, infeasible (Barrett et al. 2000). Forest planning models typically predict forest inventory, harvest levels, and economic outputs over time. Some landscape forest planning models also predict the spatial configuration of stand vegetation and other resources over time. Such models can then make spatially dependent predictions such as insect contagion effects, habitat availability for wildlife species, potential fire risk, aesthetics, or reduction in timber yields because of management constraints on adjacent land (Barrett 1997).

Because of the geographic scope of this book, most models discussed in this chapter are from North America, with particular emphasis placed on models from the Pacific Northwest (western Oregon and Washington, coastal British Columbia, and southeastern Alaska). Similar stand-based landscape models, visualization tools, and forest management models can be found for forests in other countries. The western United States and Canada may have a disproportionate share of these models because the need to integrate wood production and other societal values has been particularly intense in these regions recently. Although I focus on a relatively small number of available models, they reflect a variety of approaches at different scales.

R.A. Monserud, R.W. Haynes and A.C. Johnson (eds.)
Compatible Forest Management. 177-207

Fortunately, a large number of decision support systems are available to aid in forest management planning. They include such models as ArcForest, ARGIS, EnVision, FETM, IMPLAN, INFORMS, KLEMS, LANDIS, LANDSUM, LMS, MAGIS, NED, PPE, Prognosis-EI, RELMdss, SIMPPLLE, SNAP, SPECTRUM, SVS, TEAMS, TELSA, UTOOLS/UVIEW, Vantage Point, and VDDT (see Mowrer 1997, Rauscher 1999, Barrett et al. 2000, and Appendix 1 for a full list of references). Although all of these decision support systems are planning tools, they require different specifics, depending on the interests and intentions of the developers. In general, most share the following traits (Barrett et al. 2000):

- Vegetation change can be projected for a landscape over a planning horizon
- The spatial resolution of this change is set by polygons, which may be based on patches, stands, or management units
- The temporal resolution of this change is set by discrete time periods to provide snapshots of future landscapes
- The systems have direct or indirect links to geographic information systems (GIS) and visualization systems
- The systems are used to analyze management options, with heavy emphasis on silvicultural treatments
- Along with projecting vegetation change, the systems often are used to analyze economic effects and commodity outputs

2. Stand-Based Landscape Models

I begin by summarizing a small set of landscape modeling tools. They all employ individual tree stand models (FVS or ORGANON) for forecasting change across the landscape. FVS is the Forest Vegetation Simulator (Wykoff et al. 1982, Teck et al. 1997, Crookston and Havis 2002); ORGANON is the Oregon Growth Analysis and Projection Model (Hann et al. 1997). A rich and essentially limitless choice of management alternatives is available with either model because silvicultural prescriptions can be tailored to the specific list of trees in a given stand. These landscape models populate the network of stands in the landscape with relevant projections through time. Spatial information (e.g., stand neighbors) is inherently built into the exercise, but whether or not it will be used depends on the objective of a given analysis. For example, an economic analysis such as Lippke and Oliver (1993a,b) need not use all the spatial information available in the landscape system.

PPE. The Parallel Processor Extension of FVS is a multistand simulation model that provides the following capabilities (Crookston and Stage 1991):

- Simulation of the simultaneous development of a thousand or more forest stands for up to 400 years into the future, including dynamic interactions

between stands. The system can simulate effects of contagious pests and management prescriptions that depend on conditions in surrounding stands and at the landscape level.
- Comparison of the effects of applying alternative management policies to a collection of stands. Alternatives may differ in the conditions under which specific treatments will be applied, in the way stands are classified for prescription, and in the timing of application. Numerous policies may be evaluated in the same simulation.
- Processing and displaying alternatives by using decision trees. If some of the alternatives represent outcomes of uncertain events, then expected values can be calculated by using the probabilities associated with the uncertainty as weights in the expectation.

Thus, the PPE can simulate multistand treatment scheduling at the landscape level, accomplish resource supply analyses, simulate contagion in pest-dynamic models, and create and analyze decision trees (Crookston and Stage 1991). The PPE was designed to evaluate management policies. A stand management policy is defined as a set of rules for stand management. A multistand treatment policy is a rule that applies to groups of stands.

An example application is Beukema et al. (1997). They developed a Westwide Pine Beetle Model (WPBM) employing spatially explicit contagion mortality. The WPBM uses FVS to project stand development and represent timber management activities. The PPE is used to manage the simultaneous projection of hundreds of stands in a landscape. The WPBM itself simulates the impact of pine beetles and other disturbance agents on each stand, and modifies the FVS tree list to account for these effects. In addition, the WPBM incorporates within itself a model of all the forestland outside the simulated landscape that is near enough to affect beetle dispersal—the so-called "outside world." The model assumes that the stockable portion of the outside world consists of homogenous forests in a specified condition. It then calculates the quantity of beetles that will disperse between the forests of the outside world and each stand in the landscape. Using the Nez Perce National Forest in Idaho as a test case, the model effectively simulates the movement of beetles (contagion) between the stands in the landscape, as well as to and from the forests of the outside world beyond this landscape. Incorporating the model's projections of beetle impacts on each stand results in a more rapid loss of pine basal area from the landscape than is projected without the landscape contagion effects. This successional change in species composition over time better matches the conditions observed in the real world (Beukema et al. 1997).

Prognosis EI. The Prognosis Environmental Indicators (EI) Model is designed to assist forest management planners by providing environmental impact projections that are comparable to the available timber projections (Greenough et al. 1999b). The goal was to create a single tool for simultaneously projecting both timber and nontimber attributes at the landscape level (Greenough et al. 2002). The approach was designed to integrate appropriate existing FVS tools, and to create additional quantitative indicators where necessary, based primarily on information already available in other FVS tools. The model can provide detailed, quantitative environmental impact projections and expected timber flows under user-defined scenarios for entire watersheds (up to 1,500 stands). Prognosis EI is actually a linked set of models—including Prognosis (now called FVS), its root disease and fire model extensions, and the new environmental indicators model—all operating within the PPE that coordinates their operations on hundreds of stands simultaneously. As part of the EI model, two ArcView GIS post-processing tools were also created (Greenough et al. 2002). One tool uses the projected information and performs various spatial analyses to produce output on landscape characteristics. The other tool allows users to easily display maps and graphs of various indicators over time, or at selected points in time. As a result, Prognosis EI can provide both timber and environmental forecasts for a number of alternatives at the watershed level (Greenough et al. 1999b). Although the model is calibrated for the southern interior of British Columbia, the system is designed to be readily transportable to other regions.

Thorough analysis depends on the development of appropriate and relevant environmental indicators. The EI model combines FVS's forecast of stand conditions with other ecological data to project future environmental values under the same scenarios. Prognosis EI supports the decisionmaking process by quantifying the expected impacts of the available options on each of the available environmental indicators (Greenough et al. 1999b). Prognosis EI is able to develop species-specific wildlife habitat interpretations based on the stand and watershed data available in the model.

LMS. The Landscape Management System (LMS) is a framework that integrates forest inventory information, spatial information, growth models, computer visualization software, and analysis software for a landscape-level analysis (McCarter 1997). The primary objective of LMS is to provide a tool for users to easily investigate landscape-level management options and compare the consequences of alternative management scenarios (McCarter 1997). LMS assists in stand and landscape ecosystem analyses by coordinating the processes of forest growth and management simulations, tabular data summarization, and stand and landscape visualization (McCarter et al. 1998). Over 40 separate programs integrate these tasks. These programs include forest stand growth models (both FVS and ORGANON), harvest simulation programs, data summary programs, and stand- and landscape-level visualization software (SVS and EnVision). Underlying data for LMS are organized into a landscape.

These include forest inventory data (stand-by-stand tree lists), stand-level data (such as site index and age), topographic data (slope, aspect, and elevation), as well as GIS data in digital terrain model format and GIS features such as streams and roads (Ceder and Marzluff 2002).

LMS uses two established stand growth simulators, ORGANON and FVS. Rather than embedding these stand models as executable programs within the LMS framework, specific growth functions are extracted, reprogrammed, and run separately by LMS. Each growth period is independent; tree lists are read in and predictions are written out every cycle, as in an inventory-updating program. This makes the system input-output intensive. LMS also removes important stand modeling features such as site-specific increment calibration (see Wykoff et al. 1982). As a consequence, LMS will not produce the same predictions as FVS with the same input data (McCarter 1997). The keyword system in FVS (Wykoff et al. 1982) is also lost when LMS reprograms FVS functions, as is the Event Monitor for controlling silvicultural specifications and run flow (Crookston 1990).

LMS has been used in several landscape analyses (see McCarter et al. 1998). Useful outputs include graphs over time showing the proportion of the landscape in different structural stages of development or different risk classes, depending on the analysis (McCarter et al. 1998). Ceder and Marzluff (2002) conducted a habitat evaluation procedure by using LMS to evaluate habitat suitability for four indicator wildlife species in western Washington. Because their habitat suitability index models used tree-based measures, they were able to link wildlife habitat models with forest growth models within a forest simulation system. Their joint analysis concluded that LMS was a tool that allows forest managers to assess changes in wildlife habitat caused by potential forest management at the stand and ownership levels.

2.1. Visualization of Stand-Based Landscape Models

Approaches to spatial simulation of forest landscape dynamics are increasing (Mladenoff and Baker 1999) as we face the need to understand complex ecological dynamics over large spatial scales and longer temporal domains. At the same time, forest management is under increasing pressure to incorporate new ecological knowledge while protecting a variety of values and sustaining forest productivity (Aplet et al. 1993). Spatial simulation models enable us to assess management scenarios and environmental change hypotheses at spatial and temporal scales that are otherwise difficult or impossible to evaluate (Gustafson et al. 2000). A visualization system is now a crucial component enhancing forest projections by supplying the visual context.

The previous three landscape models all rely on visualization software to display results at either the tree/stand level (McGaughey 1997) or the full landscape scale (McGaughey 2001). Because individual tree models drive them, they have the rich structural detail of the predicted tree list available at each cycle. This allows for greater detail than available in a conventional GIS mapping system.

EnVision. The appearance of landscapes and individual stands after harvest operations is critical to public acceptance of timber harvest practices (McGaughey 1998). Foresters, biologists, and others charged with selecting stands for treatment and designing silvicultural prescriptions often find it difficult to imagine the complex interactions that occur across landscapes (McGaughey 2001). Few people—even forestry professionals—can look at a data table describing the trees in a timber stand and visualize what it tells them (McGaughey 2001). It is even harder to visualize how that timber stand or landscape would look in the future if it were managed in different ways (Figure 1).

The challenge is not only for forest managers to translate data into visuals for themselves, but also to translate them for the public. Public acceptance of a landscape after changes are made depends, in part, on the ability of forestry professionals to establish a credible dialogue about the visual effects of the proposed changes through time.

The fundamental objectives of visual simulations are understanding, credibility, and lack of bias (Sheppard 1989). Simulations should clearly convey information about a particular scene without misleading or intending to elicit a certain response from a viewer (Bergen et al. 1998).

For years, users of the FVS have used visual simulations produced by the Stand Visualization System (SVS) (McGaughey 1997, 2000) to better understand and communicate the results of their FVS stand projections (McGaughey 2002). This visualization system has now been expanded to the landscape scale: the EnVision Environmental Visualization System by McGaughey (2001, in press). Computer-based landscape simulations have become an important tool for previewing the visual impacts of land-use decisions. The simulations can show how stand and landscape conditions change with management activities, natural disturbances, and growth over time (Figure 1).

EnVision is a full-featured image-generating system for stand- and landscape-scale projects (McGaughey 2001). It builds on many of the concepts used to develop preceding visualization systems, such as Vantage Point (Bergen et al. 1998), Stand Visualization System (SVS), and UVIEW (Ager and McGaughey 1997). EnVision is based on a digital terrain model that defines the ground surface. Color and texture maps show ground surface characteristics and can create synthetic aerial photos or highlight areas of interest, such as roads, streams, or observation points. The program uses individual tree lists and can expand stand tables into individual trees within polygons. Three-dimensional objects range from simple stick trees to photo-realistic icons representing trees and other vegetation (McGaughey 2001).

Figure 1. EnVision (McGaughey 2001) illustrates pre-treatment (1a) and post-treatment (1b) fire risk by depicting live trees by species, dead trees, accumulated understory vegetation, and down fuels.

EnVision can work directly with existing forest inventory data. The minimum data required to use EnVision are a digital terrain model to represent the ground surface, a polygon overlay that delineates individual stands, and stand inventory data or SVS files to represent each stand on the landscape (McGaughey 2001). Size of the landscape area is immaterial. EnVision can visualize a whole landscape or plots less than 0.5 hectare (ha).

Basic components of an EnVision project include a digital terrain model to define the ground surface; color, and texture maps to define ground surface characteristics; and groups of objects (McGaughey 2001). Scene definitions include background imagery used to add clouds and distant landscape features, model components (e.g., terrain models and polygon overlays), viewpoint and camera characteristics, and foreground imagery used to provide high detail in the image foreground. EnVision models individual light sources, including a simulated sun position and atmospheric effects such as fog and haze. Images are rendered by using a geometrically correct camera model, making it possible to match real photographs taken from known viewpoints to simulated scenes.

EnVision is one of the most sophisticated computer simulation tools available for illustrating stand- and landscape-scale projects. EnVision can help land managers and others visualize the effects of different management treatments over time. McGaughey (2002) illustrated this by displaying fuel conditions corresponding to stand development over time (Figure 1). This ability helps people understand the effect of proposed treatment designs and thereby improve forest planning and consideration of management options. As a result, the system shows particular promise for enhancing public communication and increasing public involvement in making informed forest management decisions.

3. Successional Pathways Models

The previous three landscape models (PPE, Prognosis-EI, LMS) are all based on stand models driven by individual tree dynamics. In them, the landscape is populated as a collection of stands, each of which comprises a list of sample trees. Individual tree stand models need not be the basis for a landscape projection, however. The following landscape models use only discrete classes to describe the vegetation in a given location; individual tree information is not used. The area of land in each vegetation class is the fundamental modeling unit. Dynamics enters with the specification of transition rates or probabilities of moving from one vegetation class to another. Essentially, these are successional pathway models that use transition probabilities to move plants between vegetation classes or fuel characteristic classes. Although several of these focus heavily on modeling fire effects across the landscape, they all model silvicultural manipulation of the vegetation and the resulting effect of management actions (or no management) on succession and dynamics through time.

FETM. The Fire Effects Tradeoff Model is a landscape-scale, strategic planning model that simulates changes in vegetation composition over time in response to various human-caused and natural disturbances (Air Sciences 2002). It is designed to simulate the long-term tradeoffs between wildland fire and various fuel treatment alternatives over large areas of the landscape encompassing diverse environmental conditions, natural fire regimes, and land management policies (Air Sciences 2002). FETM is a stochastic, dynamic, nonspatial model that simulates annual changes in vegetation composition, fire behavior, fuel consumption, and emissions across large areas (i.e., greater than half a million ha) and over long periods of time (i.e., hundreds of years). The processes modeled are succession, harvest, fuel treatment, and fire; outputs include fire emission for four weather classes (Barrett 2001). Because the FETM model is capable of assessing the long-term consequences of land management practices, it can provide the information needed to identify the economic and environmental tradeoffs between fuel treatment alternatives and wildland fire, assess the long-range impacts to ecosystems, and assist in identifying specific vegetation types to target for fuel reduction (Air Sciences 2002). FETM is distinguished from other vegetation dynamics models by its link between vegetation composition (fuel loading) and wildland fire behavior and effects. It was first applied to a 485 000 ha landscape in northeastern Oregon.

FETM uses a discrete-time transition matrix modeling approach to annually move acres between vegetation states over time (Schaaf et al. 1998). Vegetation composition is described by 206 fuel characteristic classes. A fuel characteristic class is a unique fuel group that is quantitatively described but keyed to management-related characteristics such as vegetation type (e.g., ponderosa pine [*Pinus ponderosa* Dougl. ex Laws.]), age class (e.g., mature), loading class (e.g., moderate), and management activity class (e.g., precommercial thinning followed by lopping and scattering slash). Various disturbances may be accounted for in the model, including timber harvesting and other forms of fuel removal (e.g., firewood collection), mechanical fuel treatments, prescribed fire, wildland fire, windthrow, insects, disease, and natural succession (Air Sciences 2002).

LANDSUM. LANDscape SUccession Model (Keane et al. 1997) and its predecessor CRBSUM (Keane et al. 1996) were developed as research tools to investigate landscape fire succession modeling, but the models can be used as management tools (Barrett 2001). The processes modeled are succession, harvest, disease, and fire; the classification system used is structural stages and cover types within potential vegetation types. In addition to classified vegetation, outputs can include summaries of land area affected by processes by year, harvest area, and if accompanied by a user-supplied volume table, harvest volume. LANDSUM is stochastic and spatial, although the attributes of neighbors do not directly affect a given polygon. The earlier raster version of the model, CRBSUM, was applied to very large landscapes as part of the Interior Columbia River Basin Assessment. The first application of the polygon-based LANDSUM model was applied to nested landscapes of 89 000 ha and 23 000 ha within central Idaho.

SIMPPLLE. SIMulating vegetative Patterns and Processes at Landscape scaLEs (Chew 1995) is a knowledge-based expert system. It was designed as a management tool to understand how processes and vegetation interact to affect landscape change (Barrett 2001). The processes modeled are succession, harvest, disease, insects, and fire; the classification system used is current species composition, potential vegetation, stand density, and structure. In addition to classified vegetation, outputs include maps and charts of processes. SIMPPLLE is stochastic and spatial (polygon-based). Two variants, the Upper Clark Fork and the Headwaters of the Missouri, have been used in assessments of landscapes of 3000 to 30 000 ha. Knowledge about vegetative change is represented in the form of all possible states that may exist (pathways) but a transition matrix is not used (Weise and Kimberlin 1999). Each plant community is evaluated separately to determine what process may be responsible for vegetative change. The probabilities of processes occurring for each community, such as wildland fire, insects, and disease are derived from a combination of existing hazard rating systems, observed frequency of occurrence, and expert opinion (Jones and Chew 2000).

MAGIS. Multi-resource Analysis and Geographic Information System is a modeling system for integrating ecological and social information and scheduling management treatments spatially and temporally for a landscape (Weise and Kimberlin 1999). It also computes the effects from the schedule of treatments. Management practices that can be accommodated include alternative silvicultural methods, logging methods, and prescribed burning. Various relationship types are available for addressing management issues: outputs (e.g., wildlife habitat indices, sediment yields, harvest volumes), hectares with specified characteristics (e.g., area in a specific stand structure class), kilometers of network with specific characteristics (e.g., length of road closing), costs, and net revenues. These relationships can be calculated for the entire planning area, or subcomponents such as individual drainages. In addition, there are spatial relationships for addressing wildlife security areas, corridors, and activities on adjacent treatment units (Weise and Kimberlin 1999).

SIMPPLLE and MAGIS are executed as separate entities that share information (Weise and Kimberlin 1999). The process begins by using SIMPPLLE simulations to identify the frequency and locations of disturbance, particularly fire, for a no-action alternative. The results from SIMPPLLE help formulate treatment priorities as input into MAGIS. MAGIS is used to develop a schedule of treatments in space and time to achieve management objectives, including impacts on various resource values. The resulting schedule of activities is then used in SIMPPLLE to verify the treatment effects on disturbance occurrence (Weise and Kimberlin 1999).

VDDT. Vegetation Dynamic Development Tool (Beukema and Kurz 1998) was designed to facilitate understanding of vegetation change through successional pathway modeling (Barrett 2001). The processes modeled include succession, harvest, disease, insects, and fire; the classification system used is cover type and structural stage. The primary output is area in each vegetation class. VDDT was developed to support the Interior Columbia River Basin Assessment. VDDT models typically apply to potential vegetation types. For each potential vegetation type, succession classes are defined according to the cover type and structural stage combinations that exist. In the absence of disturbance, vegetation moves (grows) from one successional class to the next. This time-dependent progression is referred to as a successional pathway. Natural or human-caused disturbances also affect vegetation. In VDDT, disturbances are defined for each successional class according to type (e.g., wildland fire, harvest), successional class destination, probability of occurrence, and the relative ages for which each probability applies (Weise and Kimberlin 1999). For each year of the simulation, VDDT cycles through the landscape, generates a random number and determines if each landscape unit merely ages one more year, moves to a new succession class, or experiences a disturbance.

VDDT models have been built for several areas outside the Columbia Basin (Weise and Kimberlin 1999). The Siuslaw National Forest in the Coast Range of Oregon has developed models that apply to both public and private lands in northwestern Oregon. Past studies have assumed that private lands are harvested at some fixed rotation age (e.g., 50 years). In the Siuslaw model, private land harvest probabilities vary according to the class of owner and stand age. This results in a more realistic projection of the condition of intermingled private land (Weise and Kimberlin 1999). The Siskiyou (southwestern Oregon) and Shasta-Trinity (northern California) National Forests are using VDDT to project vegetative conditions in areas including wildlife reserves to depict the consequence of active intervention vs. passive management.

TELSA. Tool for Exploratory Landscape Scenario Analyses is a spatially explicit model that incorporates the forest succession simulator VDDT, stochastic natural disturbances, user specified management actions, and a diverse range of spatial and nonspatial indicators to evaluate landscape scenario projections (Klenner et al. 2000). It is designed to simulate up to

250 000-ha landscape units and has been applied to simulate the dynamics of a 116 000-ha landscape. At the core of the process is a tool that simulates multiple scenarios, each characterized by different assumptions about management actions and natural disturbances. Multiple management scenarios, and multiple simulations of the model within a scenario are used to assess the effects of stochastic natural disturbances and user-defined management actions in user-defined combinations. Because wildfires and other natural disturbance events that affect vegetation dynamics are inherently unpredictable, the model can generate multiple stochastic simulations of each scenario to provide estimates of the mean, range, and variability of the selected performance indicators. This approach allows both the average response and the variability within and between scenarios to be examined. TELSA imposes a user-defined set of natural disturbance agents onto the landscape, including wildfire, several defoliators, bark beetles, and root disease (Klenner et al. 2000).

LANDIS. The LAND Information System model simulates ecological dynamics including forest succession, disturbance, seed dispersal, species establishment, and fire and wind disturbance and their interactions (Mladenoff and He 1999). The purpose of LANDIS is to simulate long-term changes (greater than 100 years) in patterns of forest vegetation across large landscapes while maintaining reasonable realism in important ecological processes and their spatial interactions (Gustafson et al. 2000). The model is not designed to simulate fine-scale processes operating within single stands or to develop operational plans for small groups of stands. Rather, LANDIS is a tool to examine the large-scale, long-term impacts of forest disturbance by wind, fire, and harvest across landscapes from several hundred to several hundred thousand hectares in extent (Gustafson et al. 2000). The model operates on a raster map or grid where each cell contains information on the tree species and their 10-year age cohorts present (species–age list), but not information about the number or size of individual stems. The model simulates differential reproduction, dispersal, and succession patterns by species and incorporates effects of disturbance and environmental heterogeneity across the landscape.

Based on a watershed management case study in the Missouri Ozark Mountains, Gustafson et al. (2000) conclude that LANDIS is well suited for exploring detailed questions about the interaction of current stand conditions, economic forces, and management strategies through time. The authors also conclude that LANDIS is able to simulate succession as a consequence of disturbance by vegetation management, providing insight into changes in both spatial pattern and forest composition produced by management alternatives.

3.1. Discussion: Successional Pathways Models

The SIMPPLLE and MAGIS models are spatial and designed to provide decision support at the project or watershed level for treatment type and sequencing on the landscape (Jones and Chew 2000), whereas FETM and VDDT are nonspatial, larger-scale planning models intended to support forest-level decisions. FETM has a more detailed prescribed and wildland fire component than VDDT to emphasize the tradeoff of fire management actions. VDDT is a more general model than FETM and includes a large array of other disturbance factors presently not available in FETM (Weise et al., in press). LANDSUM is a spatially explicit, deterministic vegetation dynamics simulation model in which disturbance is treated as a stochastic process (Keane et al. 1997). FETM, LANDSUM, SIMPPLLE, and VDDT are all stochastic models with some deterministic components.

FETM, LANDSUM, SIMPPLLE, and VDDT can be used in the general process of understanding vegetation change for forest landscapes (Barrett 2001). The four models only partially overlap, as FETM and VDDT are specialized toward fuel modeling and pathway development, respectively. The strengths of these models lie in the comparison of strongly differentiated alternatives, rather than in prediction. If planners (or the public) are choosing between alternatives of no management and aggressive treatment, these models may still capture the general consequences well enough to aid in decision-making (Barrett 2001).

Two spatial models, LANDSUM and SIMPPLLE, are intended to be aids for understanding the range of possible future vegetative conditions and disturbances, rather than to portray the future landscape that will occur (Barrett 2001). Users should be careful how they interpret maps in this situation because maps tend to convey the idea that particular conditions occur in particular places (Barrett 2001). Developers and users of these models need to stress that any individual future landscape depicted is representative, and that the probabilities and pathways have not been validated (Barrett 2001). Thematic maps showing predicted probabilities may be more appropriate than maps of individual future landscapes.

Weise et al. (in press) present results of the initial testing of these four models on the Bitterroot National Forest in western Montana. Three fuel treatments were simulated over 50 years: (1) no action, (2) prescribed burning in ponderosa pine and Douglas-fir (*Pseudotsuga menziesii* (Mirb.) Franco) habitat types at two different rates, and (3) timber harvesting that returns the stand to a reproduction stage. Simulation results for all the models suggested that the area of Douglas-fir and ponderosa pine would not be appreciably altered by the fuel treatments (Weise et al., in press). Most models suggested the harvest treatment would reduce area burned by wildfire and smoke emissions; some models suggested the prescribed fire treatment would reduce wildfire acreage.

All models suggested that the harvest treatment would reduce acreage of pole-size trees; some models suggested the fire treatment would increase acreage of sapling-size trees.

4. Landscape Optimization Models and Tools

The use of linear programming (LP) to support decisionmaking on forest harvesting and management began in the late 1960s (Weintraub et al. 2000). Timber RAM (Navon 1971) was widely used for long-range forest planning on timber harvests; environmental concerns were not treated explicitly. FORPLAN (Jones et al. 1986) corrected this deficiency by considering environmental issues such as water sedimentation, wildlife, and erosion in an explicit form. It was an outgrowth of the Model II formulation of Johnson and Scheurman (1977). Still, the problem remained that these nonspatial solutions could not necessarily be identified on a map and implemented. The management problem became spatial with inclusion of road building (access for timber harvesting and fire protection) in the analysis. Weintraub and Navon (1976) integrated road building and land management by using mixed integer linear programming (MIP) models.

In the past decade, environmental concerns regarding the protection of wildlife, scenic beauty, reduction in water sedimentation, and erosion have required greater spatial sensitivity in planning models. Constraints that prevent adjacent stands from being clearcut within a given time period are commonly mandated; they are called "adjacency constraints." Once harvesting large tracts of land was prohibited, spatial adjacency constraints needed to be considered for each unit of land in each period by the planning model (McDill and Braze 2000). Each harvesting decision was represented by an integer variable in MIP, which allowed spatial relationships to be expressed by using adjacency constraints. This greatly increased the dimensionality and difficulty of the forest management problem, and required new classes of optimization algorithms. Additional spatial constraints are needed to restrict harvesting near rivers or roads, or to preserve forested travel corridors for wildlife (connectivity constraints). Solving the mathematical formulation of these spatial planning problems is quite difficult, but sometimes possible by strengthening the MIP formulations. Exact and approximate solutions to exact model representations include branch and bound, and Lagrangian relaxation schemes (Weintraub et al. 2000).

Heuristic approaches have proven quite successful (e.g., Bettinger et al. 1996 *et seq.*). They include random search methods, and several local search methods such as Tabu search (Bettinger et al. 1997, 1998c), genetic algorithms (Boston and Bettinger 2002), dynamic programming (Hoganson and Borges 1998), and simulated annealing (Weintraub et al. 2000). One widely used system is SNAP, the Scheduling and Network Analysis Program (including

SNAP II+ and III; see Sessions and Sessions 1991). SNAP is designed to assist in the scheduling and transportation planning for forest harvest areas (Sessions and Sessions 1991). The solution approach is based on several heuristics applied in sequence: random search for the adjacency condition, a Steiner problem for the habitat connection, and a shortest path algorithm for the road-building decisions (Weintraub et al. 2000). It can schedule the harvest for up to 50 time periods where costs, several species, alternative destinations, non-adjacency requirements, and transportation systems are considered (Mowrer 1997). SNAP combines pattern generation and network analysis to find feasible solutions consisting of units that are and are not selected for harvest. Both even- and uneven-age management can be modeled. In addition to normal non-adjacency rules, SNAP can aggregate units during pattern generation to form "super polygons" subject to maximum size of disturbance limits. Also units may be excluded from harvest, and wildlife corridors may be created by connecting sets of polygons that conform to the eligible seral stages defined by the user. SNAP III is capable of doing 5,000 polygons, 10,000 road links, 20,000 stream links, 50 time periods, 100 polygon attributes, and 250 seral stages. The model developers find the system user-friendly and solutions to be quite acceptable. As with heuristics for solving the adjacency problem, this approach is limited in handling additional constraints (Weintraub et al. 2000).

Harvest scheduling models and other decision support systems are important tools for exploring forest resource conflicts (Boyland 2002). A wide variety of tools are available, each with their own strengths and weaknesses. The formulation itself often determines the strengths and weaknesses of a given approach. Often the strengths and weaknesses are simply two sides of the same issue (Boyland 2002): optimization finds good solutions, but perhaps too good to successfully implement; simulations are simple and easily understood, but perhaps too simple to deal with all the complexities of scheduling and planning. Boyland (2002) finds that understanding the tool formulation gives insight into how models can mislead, and what areas of inquiry will produce reliable information.

5. Regional and National Models

In the United States, both legislative mandates and strategic planning require periodic assessments of the forest resource at both a regional and national scale (e.g., Haynes et al. 1995). Such assessments involve long-range projections of 50 years or more. This requires a large-scale forest resource model that can begin with the latest inventory data, aggregate them into appropriate classes for the analysis, and then simulate growth dynamics into the future for each class. A projection system is needed to test the implications of various timber management strategies and analyze policies directed toward future timber supplies at a national scale.

Macro-economists have developed forest sector models that attempt to quantify the tradeoffs between resources (Kallio et al. 1987, Nabuurs et al. 1998). Such exercises often focus on two-dimensional joint production functions and are usually only for market-driven outputs. Our compatible forest management problem is at least 5-dimensional (the interaction of wood production with wildlife, aquatics, biodiversity, and social values). Furthermore, forest sector modelers have traditionally abstracted the output from all forest ecosystem processes into simple averages. Below, I examine one such model that has great flexibility for representing the forest resource in great detail at either the regional or national scale:

ATLAS, the Aggregate TimberLand Assessment System, is a planning and policy tool for projecting the national (or regional) timber inventory into the future (Mills and Kincaid 1992). It was developed by the USDA Forest Service to address broad policy questions related to future timber supplies mandated by the 1974 Renewable Resources Planning Act (RPA). The RPA objective is to compile a comprehensive assessment of the current and future (50-year) national timber situation. The ATLAS model simulates growth, harvest, and regeneration of timber stands. It was first used in the Southern Timber Supply Study (USDA Forest Service 1988). An important design criterion is a link to Forest Inventory and Analysis data, the national forest inventory system for the United States. The ATLAS model was developed to replace TRAS (Larson and Goforth 1974), the timber model used for assessments before 1989. Investigative projections can be designed to test the implications of various timber management strategies or to analyze policies directed toward future timber supplies. Components of change are dependent on the passing of time (periods) and the movement of acres among age classes. The basic requirements for conducting a simulation are a starting inventory, a yield table (empirical or model-based), and estimates of future harvest. The model outputs are designed to provide a complete accounting of inventory, growth, area, and harvest for each period by age class and management intensity. Options allow for period-specific detailed reports and all-period summaries.

The supply of timber at any point in time is modeled, in part, as a function of timber inventory levels, stumpage prices, and the amount of harvest (Haynes et al. 1995). Projections require models of forest growth, and assumptions on private timberland management, timberland area change, forest type transition, the efficiency of harvest utilization, harvest flows from public timberlands, and an array of market forces. The inventory is national, and is condensed into an RPA timber database of over 124,000 plot summary records representing approximately 200 million ha of timberland (Mills 1990). Timberland inventories are aggregated into strata by age class, growth, yield, and management characteristics. Strata are projected through time with the use of yield tables and density change functions. Simulations can differ in size and complexity from a single stratum to over 10,000 strata. To run the system, the user must provide files containing the starting area and volume; growth, yield, management alternatives, and area change parameters; and the desired harvest volumes and reporting parameters (Mills 1990). ATLAS does not have harvest optimization options; harvest must be specified by the user (Mills and Kincaid 1992).

In the RPA modeling process, ATLAS is linked to the economic model TAMM (Adams and Haynes 1980). Thus, it is a forest sector model. The resulting projections of timber inventories, harvests, stumpage prices, wood products consumption and production, and employment in the forest products sector are the result of interaction between biological and economic assumptions, and the ensuing equilibrium between the timber supply and the demand for stumpage. A major attribute of the ATLAS model is that it can simulate the number of acres shifting into different classes of management intensities and compute the resulting changes in volume yields based on alternative assumptions about future land management activities (Haynes et al. 1995).

The projection mechanism computes growth by using current stocking density, the yield table standard, and the density change function (Haynes et al. 1995). Each cell in the starting inventory has an independent yield function. Projected cells reflect net volume of the yield table standard, typically inventory growing stock. Private harvests are generated through an interaction between ATLAS and TAMM that results in an equilibrium solution between supply and demand in both product and stumpage markets.

In principle, ATLAS is not strictly an even-age model (Haynes et al. 1995). When inputs are aggregated at the regional level, age classes function much like growth classes. Growth and harvest take place across this range of classes. The ATLAS harvest mechanism can account for both partial harvests and commercial thinning. Final harvest also can be proportioned across a range of age classes: the treated area may be regenerated in alternative management levels, assumed to change timber type, or withdrawn from the timberland base (Haynes et al. 1995).

6. Global Vegetation Models

For completeness, the logical extreme of "large-scale area" is global. Global vegetation models are large-scale tessellation models that predict the occurrence of a specific vegetation zone or biome for each point on the landscape or globe (e.g., Prentice et al. 1992, Tchebakova et al. 1993). Because they link the suitability of specific types of vegetation with climatic and environmental conditions at a location, they predict potential vegetation. Most have been developed in the last decade to address questions of vegetation distribution under climate change. All are driven by climatic variables, and some include moisture stress and soil properties. Thus, these are policy models that are almost exclusively focused on one particular policy question: What are the ramifications of global climate change?

The best known is the Biome model of Prentice et al. (1992), which was used to predict the vegetation component under the numerous climate change scenarios examined by the Intergovernmental Panel on Climate Change (IPCC, Watson et al. 1995). Primary driving variables are the mean temperature of the coldest month, growing days over 5°C, and a drought index incorporating the seasonality of precipitation and the available water capacity of the soil (Prentice et al. 1992). Other vegetation distribution models include the hierarchical suite of models based on the bioclimatological theories of Budyko (1974): the global Budyko model of Tchebakova et al. (1993), the subcontinental Siberian model of Tchebakova et al. (1994), and the regional Sayan Mountain model of Monserud and Tchebakova (1996). These models predict vegetation as a function of net radiation balance, precipitation, and a dryness index. Because of the broad scales involved, it is difficult to incorporate any management effects (but see Leemans and van den Born 1994).

Vegetation distribution models predict broad areas of vegetation by biome or type, and are thus not applicable as stand or watershed management tools. They operate at a hierarchical level far above the units of management (individual trees and stands). They also predict potential vegetation, and do not begin with the vegetation that is actually present at a location.

7. Discussion

Consider the mathematical properties and problems of generalizing a yield function or prediction from the fine scale (tree or stand) to much broader spatial scales (watershed or landscape, and province or region). At the fine scale (stand), units are the same as inventoried (e.g., a tree list of diameters, species, heights). At broad scales such as a region, units are usually expressed as means on a per-area basis. This is necessitated both by incomplete fine-scale data (e.g., a list of all trees in Oregon does not exist) and by obvious considerations of efficiency and cost. However, three problems are introduced by working with means, two of which can introduce bias. First, all information has been lost except for the measure of central tendency (mean). Second, most stand-level yield functions are curvilinear, and a curvilinear function of the mean of a sample is not equal to the mean of the same curvilinear function of the sample. Both of these effects can bias the estimation of regional volume totals from a sample of tree diameters. Third, working with means provides inadequate information about sample- and model-based uncertainty, both of which are central to prudent investment and decisionmaking.

Yield functions at the fine scale (stand) are commonly evaluated by the statistical properties of their predictions relative to accepted standards, such as permanent plots. Calculating residuals to evaluate a regional model is rather difficult because the units have been abstracted to means from the beginning of the analysis. Therefore, the statistical evaluation of regional models is not straightforward. Furthermore, the models are put to different uses with differing demands for accuracy.

It is common for yield functions to be used outside their original context and applied to a range of other conditions, and even other scales. However, it is not necessarily true that a tree growth function will generalize to provide unbiased estimates at larger scales, such as a province or region. How can such a generalization be statistically justified? Furthermore, what strategies might be employed to assure quality predictions if bias might be present? This is an established problem in sampling theory, where data from different sources (and with different precision) must be integrated into an overall estimate (e.g., the population total). One suggestion (A. Robinson, University of Idaho, personal communication, 2000) is that a hierarchical protocol might be appropriate, with several models at different scales bound together by using techniques based on double-sampling. To illustrate, in phase 1 a regional model such as ATLAS would provide "cheap" estimates quickly and efficiently for broad areas. In phase 2, a stand model such as FVS would use "expensive" tree list and stand data to provide a second, nested estimate for a subset of areas. Any bias that accrued from the entire regional (ATLAS) coverage could then be estimated and corrected. Predictions for this subset could be calibrated with field measurements by using the FVS localization protocols, thus enabling contributions of information from a third layer of data.

One alternative to this double-sampling strategy is a bias correction by simulation. This would follow the lines of the tree-tripling swindle used in FVS (Stage 1973). It is also possible to derive bias corrections from theoretical considerations (e.g., error propagation techniques that involve Taylor Series expansions of the functional form).

8. A Management Example Evaluating Environmental Indicators

Finally, I present an example of a modeling tool for addressing compatible forest management questions at the watershed or landscape scale. For some years now, public sentiment has grown in support of nontimber values in forest management planning (Greenough et al. 1999b). This presents a problem for planners. Although they have sophisticated quantitative tools to project impacts on forest trees, they have had no similar means to account for environmental impacts.

The Prognosis EI model is designed to assist forest management planning by providing environmental impact projections that are comparable to, and linked with the available timber projections (Greenough et al. 1999a,b). The

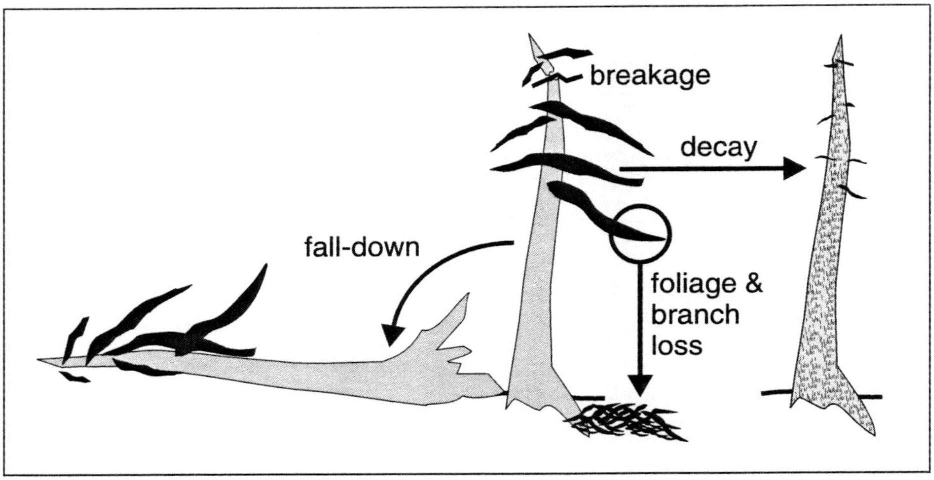

Figure 2: Snag dynamics simulated in Prognosis EI (from Greenough 1999b). By using a model component of the Fire and Fuels model, Prognosis EI captures the processes of snag breakage, decay, foliage and branch loss, and fall-down. Down snag material is added to litter, duff, and woody debris pools on the ground. Its subsequent decay is modeled as losses to these pools.

Prognosis EI watershed model is designed to provide quantitative environmental impact projections, simulations of stand and watershed development, and expected timber flows under a range of management scenarios defined by the user. Prognosis EI is a linked set of models—including FVS and its root disease and fire model extensions (Figure 2), and the new environmental indicators model—all operating within the Parallel Processing Extension of Crookston and Stage (1991).

Snag and woody debris values in Prognosis EI are modeled dynamically by using the FVS Fire and Fuels Extension (FFE; Beukema et al. 2000). As shown in Figure 2, this model simulates the crown loss, breakage, decay and fall-down of the dead trees produced in Prognosis stand projections (Greenough et al. (1999b). The model also simulates the subsequent decay of the different sizes of woody material on the ground, including debris created by management actions.

Thus, PPE coordinates the simulation of management operations on hundreds of stands simultaneously across a watershed, whereas the Environmental Indicators module determines environmental effects for each scenario. As a result, Prognosis EI can provide detailed, credible timber and environmental forecasts for numerous alternatives. In addition, the system is designed to be transportable to other regions.

Greenough et al. (1999a) demonstrated the model with a case study that compared 21 watershed management regimes in the West Arm Demonstration Forest (Kootenay Lake, British Columbia). Five priority management goals guided the selection of alternatives:

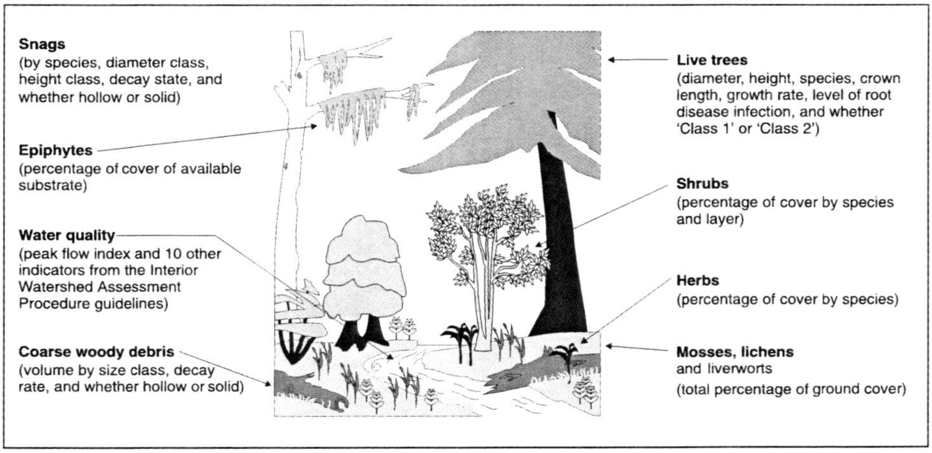

Figure 3. Stand-level and nonspatial indicators included in Prognosis EI (from Greenough 1999b).

1. Obtain a greater merchantable harvest than forecast for the baseline regimes
2. Increase the average area of prime ungulate winter range
3. Maintain a continuous supply of prime grizzly bear (*Ursus arctos*) summer foraging habitat
4. Preserve the condition of the three designated viewsheds
5. Minimize impacts on water quality

The key to the analysis is the development of appropriate and relevant environmental indicators. The West Arm Demonstration Forest analysis used the following indicators (Greenough et al. 1999a):

Stand structural indicators (Figure 3):

a. Overstory characteristics from FVS, including stand height, maximum diameter at breast height (dbh), basal area, species composition, and both total and merchantable standing volume
b. Canopy cover (for all trees, for trees greater than 5 meters (m) tall, and for non-deciduous trees only), number of canopy layers, and developmental stages
c. Understory condition, including percentage of cover by species by layer in the shrub and herb layers, and total percentage of cover in the moss, lichen, and epiphytic layers
d. Volume of coarse woody debris, by species class, diameter class, and hollow or solid state
e. Number of snags, by species, dbh, current height, decay class, and hollow or solid state

f. Seral stage
g. A summary measure of structural diversity based on number of canopy layers, shrub cover, volume of coarse woody debris and volume of snags

Wildlife indicators:

a. Three habitat quality measures for pileated woodpeckers (*Dryocopus pileatus*): winter foraging habitat quality (which depends on proximity to suitable roost trees), availability of drumming trees, and nesting habitat quality
b. Two habitat quality measures for bats: (*Laisurus* spp.) the amount of high-contrast edge available for foraging, and quality of roosting habitat

Landscape spatial indicators (Figure 4):

a. Patch-size distribution across the watershed, based on stand age
b. Location and amount of old growth in the watershed

Water quality indicators:

a. Equivalent clearcut area in each of three elevation bands
b. Length of streams bordered by three levels of canopy cover
c. Indicators (11 of the 13) defined in the Interior Watershed Assessment Procedure Guidebook (BC Ministry of Forests and BC Environment 1995)

Visual quality indicators (Figure 4):

a. Alteration of designated viewsheds, and resulting status (preservation, retention, partial retention, modification) of each viewshed

Timber values:

a. Mean annual increment (m^3 ha^{-1} yr^{-1})
b. Total and merchantable harvest volumes
c. Diameter and species distribution of harvested timber

After analyzing 21 watershed management regimes, the resulting "balanced" management regime succeeded in meeting the first three goals (Greenough et al. 1999a). At the same time, this regime maintained a "retention" visual quality designation in all viewsheds, and reduced the equivalent clearcut area (of concern for water quality) in two out of three elevation bands in the watershed. Visualization was produced with SVS (McGaughey 1997, 1998).

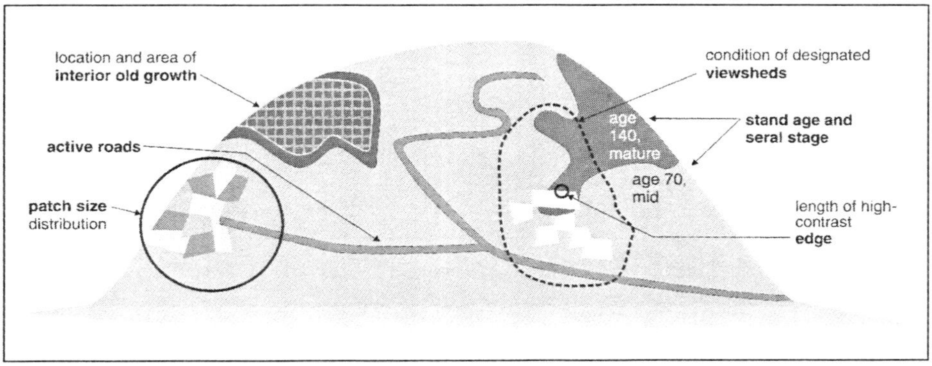

Figure 4. Spatial indicators included in Prognosis EI (from Greenough 1999b).

As expected, there is not one ideal management regime that can simultaneously maximize all values. This watershed-wide exercise demonstrates that a detailed model like Prognosis EI can provide quantitative indicator projections useful in weighing tradeoffs and increasing compatibility among the various forest resources. This case study indicates that Prognosis EI is capable of providing quantitative impact projections for a wide range of timber and environmental indicators simultaneously across a watershed (Greenough et al. 1999b).

This exercise goes to the heart of compatible forest management: it illustrates a tool that allows the manager to specify a range of desired goals or conditions, and an objective means for comparing the competing management strategies by directly examining environmental indicators relevant to the management objectives.

9. References

Adams, D.M., and Haynes, R.W. 1980. The 1980 timber assessment market model: structure, projections, and policy simulations. Forest Science Monograph 22. *Forest Science*, 26(3): 64 p.

Ager, A., and McGaughey, R.J. 1997. UTOOLS: Microcomputer software for spatial analysis and landscape visualization. PNW-GTR-397. Portland, OR: U.S. Department of Agriculture, Forest Service, Pacific Northwest Research Station.

Air Sciences. 2002. Fire Effects Tradeoff Model: FETM 4 user manual. Denver, CO: Air Sciences, Inc. 115 p. plus appendices.

Aplet, G.H., Johnson, N., Olson, J.T., and Sample, V.A. 1993. Defining sustainable forestry. Washington, DC: Island Press.

Barrett, T.M. 1997. Voronoi tessellation methods to delineate harvest units for spatial forest planning. *Canadian Journal of Forest Research*, 27: 903-910.

Barrett, T.M. 2001. Models of vegetation change for landscape planning: a comparison of FETM, LANDSUM, SIMPPLLE, and VDDT. General Technical Report RMRS-GTR-76-WWW. Ogden, UT: U.S. Department of Agriculture, Forest Service, Rocky Mountain Research Station. 14 p.

Barrett, T.M., Jones, J.G., and Wakimoto, R.H. 2000. Adapting forest planning decision support systems for prescribed fire treatments. In: L.F. Neuenschwander, and K.C. Ryan (Eds.), Proceedings of the Joint Fire Science conference and workshop, crossing the millennium: integrating spatial technologies and ecological principles for a new age in fire management. (Vol. 1, pp. 12-17). Moscow, ID: University of Idaho.

BC Ministry of Forests, and BC Environment. 1995. Interior watershed assessment procedure guidebook (IWAP): Level I analysis. Forest practices code of British Columbia. Victoria, BC: British Columbia Ministry of Forests and British Columbia Ministry of Environment. 82 p.

Bergen, S.D., McGaughey, RJ., and Fridley, J.L. 1998. Data-driven simulation, dimensional accuracy and realism in a landscape visualization tool. *Landscape and Urban Planning*, 40(4): 283-293 .

Bettinger, P., Boston, K., and Sessions, J. 1999. Combinatorial optimization of elk habitat effectiveness and timber harvest volume. *Environmental Modeling and Assessment*, 4: 143-153.

Bettinger, P., Johnson, K.N., and Sessions, J. 1996. Forest planning in an Oregon case study: defining the problem and attempting to meet goals with a spatial-analysis technique. *Environmental Management*, 20(4): 565-577.

Bettinger, P., Johnson, K.N., and Sessions, J. 1998a. Evaluating the association between alternative measures of cumulative watershed effects on a forested watershed in eastern Oregon. *Western Journal of Applied Forestry*, 13(1): 15-22.

Bettinger, P., Johnson, K.N., and Sessions, J. 1998b. Improving aquatic habitat conditions over time while producing wood products: an examination of options. *Journal of American Water Resources Association*, 34(4): 1-17.

Bettinger, P., Sessions, J., and Boston, K. 1997. Using TABU search to schedule timber harvests subject to spatial wildlife goals for big game. *Ecological Modelling*, 94: 111-123.

Bettinger, P., Sessions, J., and Johnson, K.N. 1998c. Ensuring the compatibility of aquatic habitat and commodity production goals in eastern Oregon with a TABU search procedure. *Forest Science*, 44(1): 96-112.

Beukema, S.J., Greenough, J.A., Robinson, D.C.E., Kurz, W.A., Smith, E.L., and Eav, B.B. 1997. The Westwide Pine Beetle Model: a spatially-explicit contagion model. In: R. Teck, M. Moeur, and J. Adams (Comps.), Proceedings: Forest Vegetation Simulator conference (pp. 126-130). General Technical Report INT-GTR-373. Ogden, UT: U.S. Department of Agriculture, Forest Service, Intermountain Research Station.

Beukema, S.J., and Kurz, W.A. 1998. Vegetation dynamics development tool: user's guide, version 3.0. Vancouver, BC: ESSA Technologies Ltd. 104 p.

Beukema, S.J., Reinhardt, E.D., Kurz, W.A., Crookston, N.L. 2000. An overview of the Fire and Fuels Extension to the Forest Vegetation Simulator. In: L.F. Neuenschwander, and K.C. Ryan (Tech. eds.), Proceedings of the Joint Fire Science conference and workshop—crossing the millennium: integrating spatial technologies and ecological principles for a new age in fire management. (Vol. 2, pp. 80-85). Moscow, ID: University of Idaho and the International Association of Wildland Fire.

Boston, K., and Bettinger, P. 2002. Combining Tabu search and genetic algorithm heuristic techniques to solve spatial harvest scheduling problems. *Forest Science*, 48: 35-46.

Boyland, M. 2002. Simulation and optimization in harvest scheduling models. ATLAS/SIM FOR project extension report. Vancouver, BC: University of British Columbia. 10 p.

Budyko, M.I. 1974. Climate and life. New York: Academic Press. 508 p.

Camenson, D., Sleavin, K., and Greer, K. 1995. Spectrum: an analytical tool for building natural resource management models. In: R. Päivinen, L. Roihuvuo, and M. Siitonen (Eds.), Proceedings. Large-scale forestry scenario models: experiences and requirements (5: 133-142). Joensuu, Finland: European Forestry Institute.

Ceder, K.R., and Marzluff, J.M. 2002. Linking tools of forest and wildlife managers: wildlife habitat evaluation using the landscape management system. In: N.L. Crookston, and R.N. Havis (Comps.), Second Forest Vegetation Simulator conference proceedings (pp. 200-208). RMRS-P-25. Ogden, UT: U.S. Department of Agriculture, Forest Service, Rocky Mountain Research Station.

Chew, J.D. 1995. Development of a system for simulating vegetative patterns and processes at landscape scales. Missoula, MT: University of Montana. 182 p. Ph.D. Dissertation.

Church, R.L., Murray, A.T., and Figueroa, M.A. 1995. Regional Ecosystems and Land Management Decision Support System (RELMdss) user's manual. Santa Barbara, CA: Department of Geography, University of California.

Crookston, N.L. 1990. User's guide to the Event Monitor: part of Prognosis Model version 6. General Technical Report INT-275. Ogden, UT: U.S. Department of Agriculture, Forest Service, Intermountain Research Station. 21 p.

Crookston, N.L., and Havis, R.N. (Comps.). 2002. Second Forest Vegetation Simulator (FVS) conference proceedings RMRS-P-25. Ogden, UT: U.S. Department of Agriculture, Forest Service, Rocky Mountain Research Station. 208 p.

Crookston, N.L., and Stage, A.R. 1991. User's guide to the Parallel Processing Extension of the Prognosis Model. General Technical Report INT-281. Moscow, ID: U.S. Department of Agriculture, Forest Service, Intermountain Forest and Range Experiment Station.

Faber, B.G., Watts, R., Hautaluaoma, J.E., Knutson, J., Wallace, W.W., and Wallace, L. 1994. A groupware-enabled GIS. In: GIS applications in natural resources 2. Fort Collins, CO: GIS World Books.

Greenough, J.A., Beukema, S.J., Robinson, D.C.E., Kurz, W.A., Densmore, N., Winter, R., and Snowdon, B. 2002. Prognosis EI: a detailed watershed-level environmental indicators model. In: N.L. Crookston, and R.N. Havis (Comps.), Second Forest Vegetation Simulator conference proceedings (pp. 122-125). RMRS-P-25. Ogden, UT: U.S. Department of Agriculture, Forest Service, Rocky Mountain Research Station.

Greenough, J.A., Robinson, D.C.E., Beukema, S.J., and Kurz, W.A. 1999a. Prognosis environmental indicators model: WADF case study methodology. Vancouver, BC: ESSA Technologies Ltd. 24 p.

Greenough, J.A., Robinson, D.C.E., Kurz, W.A., Beukema, S.J., Densmore, N., Winter, R., and Snowdon, B. 1999b. Use of the prognosis EI model in balancing timber and environmental values at the watershed-level. Vancouver, BC: ESSA Technologies Ltd. 30 p.

Gustafson, E.J., Shifley, S.R., Mladenoff, D.J., Nimerfro, K.K., and He, H.S. 2000. Spatial simulation of forest succession and timber harvesting using LANDIS. *Canadian Journal of Forest Research*, 30: 32–43.

Hann, D.W., Hester, A.S., and Olsen, C.L. 1997. ORGANON User's Manual. Corvallis. OR: Oregon State University, College of Forestry.

Haynes, R.W., Adams, D.M., and Mills, J.R. 1995. The 1993 RPA timber assessment update. General Technical Report RM-259. Fort Collins, CO: U.S. Department of Agriculture, Forest Service, Rocky Mountain Forest and Range Experiment Station. 66 p.

Haynes, R.W., and Monserud, R.A. 2002. A basis for understanding compatibility among wood production and other forest values. General Technical Report PNW-GTR-529, Portland, OR: U.S. Department of Agriculture, Forest Service, Pacific Northwest Research Station. 46 p.

Hoganson, H.M., and Borges, J.G. 1998. Using dynamic programming and overlapping subproblems to address adjacency in large harvest scheduling problems. *Forest Science*, 44(4): 526-538.

Johnson, K.N., and Scheurman, H.L. 1977. Techniques for prescribing optimal timber harvest and investment under different objectives—discussion and synthesis. *Forest Science Monograph*, 18: 31.

Jones, J.G., and Chew, J.D. 2000. Applying simulation and optimization to evaluate the effectiveness of fuel treatments for different fuel conditions at landscape scales. In: L.F. Neuenschwander, and K.C. Ryan (Eds.), Proceedings from the joint fire science conference and workshop (Vol. 2, pp. 89-96). Moscow, ID: University of Idaho.

Jones, J.G., Hyde, J.F.C., III, and Meacham, M.L. 1986. Four analytical approaches for integrating land management and transportation planning on forest lands. Research Paper INT-361. Ogden, UT: U.S. Department of Agriculture, Forest Service, Intermountain Research Station. 33 p.

Kallio, M., Dykstra, D.P., and Binkley, C.S. 1987. The global forest sector: an analytical perspective. New York: Wiley & Sons. 703 p.

Keane, R.E., Long, D.G., Basford, D., and Levesque, B.A. 1997. Simulating vegetation dynamics across multiple scales to assess alternative management strategies: proceedings for GIS 1997. Fort Collins, CO: GIS World Books. 640 p.

Keane, R.E., Long, D.G., Menakis, J.P., Hann, W.J., and Bevins, C.D. 1996. Simulating coarse-scale dynamics using the Columbia River Basin Succession Model—CRBSUM. General Technical Report INT-GTR-340. Ogden, UT: U.S. Department of Agriculture, Forest Service, Intermountain Research Station. 50 p

Klenner, W., Kurz, W.A., and Beukema, S.J. 2000. Habitat patterns in forested landscapes: management practices and the uncertainty associated with natural disturbances. *Computers and Electronics in Agriculture*, 27: 243-262.

Laacke, R.J. 1995. Building a decision support system for ecosystem management: KLEMS experience. *AI Applications*, 9(3): 115-127.

Larson, R.W., and Goforth, M.H. 1974. TRAS: a timber volume projection model. Technical Bulletin 1508. Washington, DC: U.S. Department of Agriculture, Forest Service. 15 p.

Leemans, R., and van den Born, G.J. 1994. Determining the potential global distribution of natural vegetation, crops and agricultural productivity. *Water, Air and Soil Pollution*, 76: 133-161.

Lippke, B., and Oliver, C.D. 1993a. How can management for wildlife habitat, biodiversity, and other values be most cost-effective? *Journal of Forestry*, 91: 14-18.

Lippke, B., and Oliver, C. 1993b. An economic tradeoff system for ecosystem management. In: M.E. Jensen, and P.S. Bourgeron (Tech eds.), Ecosystem management: principles and applications. (Vol. 2., pp. 337-345). General Technical Report PNW-GTR-318. Portland, OR: U.S. Department of Agriculture, Forest Service, Pacific Northwest Research Station.

McCarter, J.B. 1997. Integrating forest inventory, growth and yield, and computer visualization into a landscape management system. In: R. Teck, M. Moeur, and J. Adams (Comps.), Proceedings of the forest vegetation simulator conference (pp. 159-167). General Technical Report INT-GTR-373. Ogden, UT: U.S. Department of Agriculture, Forest Service, Intermountain Research Station.

McCarter, J.B., Wilson, J.S., Baker, P.J., Moffett, J.L., and Oliver, C.D. 1998. Landscape management through integration of existing tools and emerging technologies. *Journal of Forestry*, 96(6): 17-23.

McDill, M.E., and Braze, J. 2000. Comparing adjacency constraint formulations for randomly generated forest planning problems with four age-class distributions. *Forest Science*, 46(3): 423-436.

McGaughey, R.J. 1997. Visualizing forest stand dynamics using the stand visualization system. In: Proceedings of the 1997 ACSMIASPRS annual convention and exposition (Vol. 4, 248-257). Bethesda, MD: American Society for Photogrammetry and Remote Sensing.

McGaughey, R.J. 1998. Techniques for visualizing the appearance of forestry operations. *Journal of Forestry*, 96(6): 9-14.

McGaughey, R.J. 2000. Stand visualization system. Retrieved April 4, 2003 from http://www.fs.fed.us/pnw/svs.

McGaughey, R.J. 2001. The EnVision environmental visualization system. Retrieved April 4, 2003 from http://www.fs.fed.us/pnw/envision.

McGaughey, R.J. 2002. Creating visual simulations of fuel conditions predicted by the fire and fuels extension to the forest vegetation simulator. In: N.L. Crookston, and R.N. Havis (Comps.), Second Forest Vegetation Simulator conference proceedings (pp. 8-13). RMRS-P-25. Ogden, UT: U.S. Department of Agriculture, Forest Service, Rocky Mountain Research Station.

McGaughey, R.J. [In press]. Seeing the forest and the trees: visualizing stand and landscape conditions. In: Proceedings of Views From The Ridge: considerations for planning at the landscape level. Portland, OR: Western Forestry and Conservation Association and U.S. Department of Agriculture, Forest Service, Pacific Northwest Research Station.

Miller, R.E., and Blair, P.D. 1985. Input-output analysis: foundations and extensions. Englewood Cliffs, NJ: Prentice-Hall, Inc.

Mills, J.R. 1990. Developing ATLAS growth parameters from forest inventory plots. In: V.J. LaBau, and T. Cunia (Tech. eds.), State-of-the-art methodology of forest inventory: a symposium proceedings (pp. 112-118). General Technical Report PNW-GTR-263. Portland, OR: U.S. Department of Agriculture, Forest Service, Pacific Northwest Research Station.

Mills, J.R., and Kincaid, J.C. 1992. The aggregate timberland assessment system—ATLAS: a comprehensive timber projection model. General Technical Report PNW-GTR-281. Portland, OR: U.S. Department of Agriculture, Forest Service, Pacific Northwest Research Station. 160 p.

Mladenoff, D.J., and Baker, W.L. (Eds.). 1999. Spatial modeling of forest landscape change: approaches and applications. Cambridge, UK: Cambridge University Press.

Mladenoff, D.J., and He, H.S. 1999. Design and behavior of LANDIS, an object-oriented model of forest landscape disturbance and succession. In: D.J. Mladenoff, and W.L. Baker (Eds.), Spatial modeling of forest landscape change: approaches and applications. Cambridge, UK: Cambridge University Press.

Monserud, R.A., and Tchebakova, N. 1996. A vegetation model for the Sayan Mountains, Southern Siberia. *Canadian Journal of Forest Research*, 26: 1055-1068.

Mowrer, H.T. (Tech. comp.). 1997. Decision support systems for ecosystem management: an evaluation of existing systems. General Technical Report RM-GTR-296. Fort Collins, CO: U.S. Department of Agriculture, Forest Service, Rocky Mountain Forest and Range Experiment Station. 154 p.

Nabuurs, G.J., Nuutinen, T., Bartelink, H., and Korhonen, M. (Eds.). 1998. Forest scenario modeling for ecosystem management at landscape level. EFI Proc. No. 19. Joensuu, Finland: European Forestry Institute. 382 p.

Navon, D.I. 1971. Timber RAM: A long-range planning method for commercial timber lands under multiple-use management. Research Paper PSW-70. Berkeley, CA: U.S. Department of Agriculture, Forest Service, Pacific Southwest Research Station.

Prentice, I.C., Cramer, W., Harrison, S.P., Leemans, R., Monserud, R.A., and Solomon, A.M. 1992. A Global Biome Model based on plant physiology and dominance, soil properties and climate. *Journal of Biogeography*, 19: 117-134.

Rauscher, H.M. 1999. Ecosystem management decision support for federal forests in the United States: a review. *Forest Ecology and Management*, 114: 173-197.

Schaaf, M.D., Wiitala, M., Carlton, D., Snell, K., and Ottmar, R. 1998. Modeling the trade-offs between prescribed fire and wildfire emissions in forest and range land ecosystems. In: Proceedings, 3rd International conference on forest fire research and 14th conference on fire and forest meteorology (Vol. 2, pp. 1673-1685). Luso, Portugal: University of Coimbra.

Sessions, J., and Sessions, J. 1991. Tactical forest planning using SNAP II. In: Proceedings, Challenges and solutions for forestry operations in the nineties (pp. 12-16). Nanaimo, BC: Council on Forest Engineering.

Sheppard, S.R.J. 1989. Visual simulation: a user's guide for architects, engineers, and planners. New York: Van Nostrand Reinhold.

Stage, A.R. 1973. Prognosis model for stand development. Research Paper INT-137. Ogden, UT: U.S. Department of Agriculture, Forest Service, Intermountain Forest and Range Experiment Station.

Tchebakova, N., Monserud, R., Leemans, R., and Golovanov, S. 1993. A global vegetation model based on the climatological approach of Budyko. *Journal of Biogeography*, 20: 129-144.

Tchebakova, N.M., Monserud, R.A., and Nazimova, D. 1994. A Siberian vegetation model based on climatic parameters. *Canadian Journal of Forest Research*, 24: 1597-1607.

Teck, R., Moeur, M., and Adams, J. (Eds.). 1997. Proceedings: Forest Vegetation Simulator conference. General Technical Report INT-GTR-373. Ogden, UT: U.S. Department of Agriculture, Forest Service, Intermountain Research Station. 222 p.

Twery, M.J., Rauscher, H.M., Bennett, D.J., Thomasma, S.A., Stout, S.L., Palmer, J.F., Hoffman, R.E., DeCalesta, D.S., Gustafson, E., Cleveland, H., Grove, J.M., Nute, D., Kim, G., and Kollasch, R.P. 2000. NED-1: integrated analyses of forest stewardship decisions. *Computers and Electronics in Agriculture*, 27: 167-193.

U.S. Department of Agriculture, Forest Service. 1988. The South's fourth forest: alternatives for the future. Forest Resources Report 24. Washington, DC. 512 p.

Watson, R.T., Zinyowera, M.C., and Moss, R.H. (Eds.). 1995. Climate change 1995: impacts, adaptations and mitigation of climate change: scientific-technical analyses. Contribution of working group II to the second assessment of the intergovernmental panel on climate change. Cambridge, UK: Cambridge University Press. 878 p.

Weintraub, A., Church, R.L., Murray, A.T., and Guignard, M. 2000. Forest management models and combinatorial algorithms: analysis of state of the art. *Annals of Operations Research*, 96: 271-285.

Weintraub, A., and Navon, D. 1976. A forest management planning model integrating silvicultural and transportation activities. *Management Science*, 22(12): 1299-1309.

Weise, D.R, and Kimberlin, R. 1999. A risk based comparison of potential fuel treatment tradeoff models. Retrieved April 4, 2003 from: http://www.rfl.psw.fs.fed.us/jfs/index.html.

Weise, D.R., Kimberlin, R., Arbaugh, M., Chew, J., Jones, G., Merzenich, J., Wiitala, M., Keane, R., Schaaf, M., and Van Wagtendonk, J. [In press]. Comparing potential fuel treatment trade-off models: initial results. In: G.J. Arthaud, and T.M. Barrett (Eds.), Systems analysis in forest resources, proceedings of the eighth symposium. Managing Forest Ecosystems. Dordrecht, The Netherlands: Kluwer Academic Publishers.

Williams, S.B., Roschke, D.J., and Holtfrerich, D.R. 1995. Designing configurable decision-support software: Lessons learned. *AI Applications*, 9(3):103-114.

Wood, D.B., and Dewhurst, M. 1998. A decision support system for the Menominee Legacy Forest. *Journal of Forestry*, 96(11): 28-32.

Wykoff, W.R., Crookston, N.L., and Stage, A.R. 1982. User's guide to the stand prognosis model. General Technical Report GTR-INT-133. Ogden, UT: U.S. Department of Agriculture, Forest Service, Intermountain Forest and Range Experiment Station.

Appendix 1. Key to Model Acronyms or Names, with References

ArcForest. . . . . Forest management decision support system (ESRI Canada Ltd., Victoria, BC)
ARGIS. . . . . . . Groupware-enabled GIS (Faber et al. 1994)
ATLAS. . . . . . . Aggregate TimberLand Assessment System (Mills and Kincaid 1992)
Biome. . . . . . . Global Biome model (Prentice et al. 1992)
Budyko. . . . . . . Global Budyko model (Tchebakova et al. 1993)
CRBSUM. . . . . Columbia River Basin SUccession Model (Keane et al. 1996)
EnVision. . . . . . EnVision environmental visualization system (McGaughey 2001)
FETM . . . . Fire Effects Tradeoff Model (Air Sciences 2002)
FFE. . . . . Fire and Fuels Extention of the Forest Vegetation Simulator (Beukema et al. 2000)
FORPLAN. . . . FORest PLANning model (Jones et al. 1986)
FVS . . . . Forest Vegetation Simulator (Teck et al. 1997; Crookston and Havis 2002)
IMPLAN. . . . . . IMpact analysis for PLANning economic impact model (Miller and Blair 1985)
INFORMS. . . . . INtegrated FOrest Resource Management System (Williams et al. 1995)
KLEMS. . . . . . . Klamath Landscape Ecosystem Management System (Laacke 1995)
LANDIS . . . . . . LANdscape DIsturbance and Succession (Mladenoff and He 1999, Gustafson et al. 2000)
LANDSUM. . . . LANDscape SUccession Model (Keane et al. 1997)
LMS . . . . Landscape Management System (McCarter 1997)
MAGIS. . . . . . . Multi-resource Analysis and Geographic Information System (Weise and Kimberlin 1999)
Model II. . . . . . Forest management linear programming formulation (Johnson and Scheurman 1977)
NED. . . . . NorthEast Decision model (versions 1 & 2) (Twery et al. 2000)
ORGANON. . . . ORegon Growth ANalysis and projectiON model (Hann et al. 1997)
PPE . . . . Parallel Processor Extension of FVS (Crookston and Stage 1991)
Prognosis-EI. . . Prognosis-Environmental Indicators (Greenough et al. 1999a,b; 2002)

RELMdss. . . . . Regional Ecosystem and Land Management decision support system (Church et al. 1995)
Sayan. . . . . Sayan Mountain vegetation model (Monserud and Tchebakova (1996)
Siberia. . . . . Siberian vegetation model (Tchebakova et al. 1994)
SIMPPLLE. . . . SIMulating vegetative Patterns and Processes at Landscape scaLEs (Chew 1995)
SNAP (II+, III)..Scheduling and Network Analysis Program (Sessions and Sessions 1991)
SPECTRUM. . . Decision support software for ecosystem management (Camenson et al. 1995)
SVS. . . . . Stand Visualization System (McGaughey 1997)
TAMM. . . . . . . Timber Assessment Market Model (Adams and Haynes 1980)
TEAMS. . . . . . . Terrestrial Ecosystem Analysis and Modeling System (Wood and Dewhurst 1998)
TELSA. . . . . . . Tool for Exploratory Landscape Scenario Analyses (Klenner et al. 2000)
Timber RAM. . Timber Resource Allocation Model (Navon 1971)
TRAS. . . . . Timber Resource Analysis System (Larson and Goforth 1974)
UTOOLS. . . . . . Landscape visualization tools (Ager and McGaughey 1997)
UVIEW. . . . . . . Landscape visualization tools (Ager and McGaughey 1997)
Vantage Point. . 3D landscape visualization tool (Bergen et al. 1998)
VDDT. . . . . Vegetation Dynamic Development Tool (Beukema and Kurz 1998)
WPBM. . . . . Westwide Pine Beetle Model (FVS extention) (Beukema et al. 1997)

Section 3—Larger Scales

Time and space form the four dimensions of our world. Scientists frequently describe phenomena as occurring at "multiple scales over time and space," but too often they are vague about exactly what scales they mean. The three chapters in Section 3 consider issues of scale and examine compatible management at landscape and regional scales.

In Chapter 8, Spies and Johnson identify the major components of scale from an ecological perspective, examine how scale influences perceptions of ecological condition, and finally, examine how the effects of land ownership and stand-level actions can be studied at the regional scale. The scale of observation influences any assessment of compatibility between forest commodities and biological diversity. In the Oregon Coast Range, their study uses a set of models to estimate forest structure and diversity with enough spatial and ecological resolution to handle many different fine- and coarse-scale ecological indicators. Results show that broad-scale changes are just now emerging from accumulated decades of stand-level management actions. A mixture of fine-scale indicators (species) and broad-scale indicators (forest structure, disturbance regimes) may provide a more robust assessment of ecological conditions.

Issues of scale are closely linked to all other sections in this book. New stand-level actions such as variable retention will change landscape patterns slowly. Thus it may take a long time for the new silvicultural treatments described in Section 2 to achieve particular outcomes. In any assessment, decisions about scale are a type of assumption and should be specified. Scale also affects social aspects of compatible management; some conflicts over forest management may reflect differences in people's scale of observation.

The landscape scale is a critical bridge between the regional scale, such as the range of the northern spotted owl or other wide-ranging species, and the forest-stand scale, at which trees are cut or planted and other silvicultural actions are taken. In Chapter 9, the authors examine approaches to forest landscape management with the perspectives of both scientists and managers. For a case study in the Blue River watershed, western Cascade Range of Oregon, six landscape change scenarios were modeled for a 100-year period. Lessons from Blue River are relevant for other forests when the findings are scaled to local conditions. Historical landscape conditions are often used as a reference for aspects of management such as sustaining native wildlife species and ecological processes. In relation to social aspects, the Blue River case study includes multiple learning approaches, especially adaptive management, as a way to search for compatible management together with the public instead of in conflict with the public.

In Chapter 10, the authors study the causes and the consequences of widespread management intensification on the forested landscape and in the wood processing industry, and the implications for compatible management. The authors discuss the history of management regimes in the states of Oregon and Washington, a highly productive region that has produced 20 to 30% of annual U.S. softwood harvest since the early 1950s. Over the past 50 years, private forests have become concentrated in age classes younger than 60 years, with federal forests containing a higher proportion of older age classes. The transition to managed stands that produce relatively uniform logs is already well advanced.

Several implications for future markets may affect future timber management regimes in the region and affect compatibility. For example, evolving forest products markets affect landowners' choices of forest management regimes. Higher timber prices provide incentives for more intensive management. If prices drop for the primary grades produced in second-growth stands, however, landowners have less incentive to grow trees in long rotations or adopt other practices to produce high-quality logs.

Decisions made at the scale of stands and individual landowners may have unpredictable consequences at the scale of regions and centuries. In the forests of Oregon and Washington west of the Cascade Range crest, the historical actions of all forest land owners have resulted in a current situation where forests are about evenly distributed across age classes and seral stages, though conditions differ by ownership. Although this regional result was unintended, it offers opportunities for compatible management in the sense that log supplies can be sustained while wildlife habitats are preserved and other forest resources and values are maintained.

Chapter 8

The Importance of Scale in Assessing the Compatibility of Forest Commodities and Biodiversity

Thomas A. Spies
USDA Forest Service, Pacific Northwest Research Station,
3200 SW Jefferson Way, Corvallis, OR 97331, USA
K. Norman Johnson
Department of Forest Resources, College of Forestry,
Oregon State University, Corvallis, OR 97331, USA

1. Introduction

The idea that forest commodities and biological diversity can be simultaneously produced from the same area in a socially acceptable manner (Haynes et al. 2002) is common in natural resources management, but is infrequently tested in natural resources science. This concept is related to the principle of multiple use, which was encoded in the Multiple-Use Sustained Yield Act of 1960 and endorsed in the National Forest Management Act of 1976. It is clear from the heated controversies and changes in forest management in recent decades that managers of the national forests have found it extremely difficult to achieve a socially acceptable mix of commodity production and biodiversity conditions. Although the idea of compatible uses is prevalent, systematic scientific approaches rarely have been used to test particular management approaches at any spatial scale (Haynes et al. 2002). Field experiments have only recently begun to evaluate consequences of alternative stand-level management activities on both commodities and measures of biological diversity (Monserud 2002). While information from silvicultural experiments is certainly needed, finding compatible mixes of different forest uses requires a multiscale perspective. Ecological, social, and economic patterns and

R.A. Monserud, R.W. Haynes and A.C. Johnson (eds.)
Compatible Forest Management, 211-235

processes occur at different scales within spatial, temporal, and phenomenological hierarchies. What may appear compatible at one scale may not be at another. Unfortunately, our scientific understanding of compatibility problems viewed from a multiscale perspective is not well developed. For example, we lack examples demonstrating the effects of regional coarse-scale information on decisionmaking in ecosystem management at landscape and stand scales.

A better understanding of how scale influences compatibility is needed to effectively manage for diverse forest values. Because scale has different meanings and can be an abstract concept, concrete examples of scale effects are needed. The goals of this paper are to (1) identify the major components of scale from an ecological perspective, (2) examine how scale influences perceptions of ecological condition, and (3) examine how ownership and stand-level management actions can be expanded to the regional scale in terms of biodiversity and timber production. The essential question that we address is: How does scale influence measures and perceptions of the compatibility between forest biodiversity and commodities?

We emphasize assessments of ecological and socioeconomic conditions that use spatial information. Not all assessments require spatial information. For example, national and regional assessments of economic and environmental conditions under the Resources Planning Act of 1974 are based on aggregating information from forest inventory plots. In recent years, regional and landscape assessments have emphasized spatial information both to deal with ecological and socioeconomic relationships that are inherently spatial, and to better communicate complex information to general audiences (Johnson et al. 1999). As these spatial assessments have become more common, scale effects have become more apparent, and the need to understand the role of scale in assessing compatibility has become more urgent.

2. Dimensions of Scale

Scale has many different meanings, ranging from the geographer's cartographic scale (i.e., the ratio of distance on a map to distance on the ground) to the ecosystem scientist's concept of levels of scale in an ecological hierarchy (e.g., cells to organisms to biosphere) (O'Neill et al. 1986, Schneider 2001, Dungan et al. 2002). Scale concepts also have relevance to both ecological and socioeconomic sciences and management (Haynes et al. 1998). Typically, in natural resources sciences, scale concepts relate to spatial and temporal characteristics such as *extent*, the area or length of time of a study or management action, or *resolution*, the grain or finest unit of area or time that is measured or recognized. Many ecological and socioeconomic phenomena can be mapped and then characterized in terms of both extent and resolution. One general spatial ecological hierarchy is: patches—landscapes/watersheds—provinces—regions—continents—globe. The elements of this hierarchy can be defined in various ways, including the structure and composition of vegetation, soils, topography, climate, and disturbance regimes. As one moves up this ladder, the spatial dimensions of each level increase but the exact sizes will vary with the user, except for the continents and globe, which are spatially well defined. A parallel, spatially defined socioeconomic hierarchy in forest management is: forest cutting unit—forest districts—ownership types—counties—states—nations—globe.

Scale can also refer to phenomena that are not readily mapped or defined in spatial or temporal terms. For example, ecological hierarchies, such as organism—population—community—ecosystem are defined more by levels of complexity and interaction than by particular spatial or temporal characteristics (O'Neill et al. 1986). In many cases, complexity hierarchies can be approximated with spatial or temporal units, but they are generally open systems (Waring and Schlesinger 1985) that depend on complex flows of biological or physical materials. Boundaries drawn around these units on a map are typically done for a particular study or management objective and are largely artificial. A parallel example of a socioeconomic complexity hierarchy that is difficult to define spatially is person—families—social networks—organizations—common markets. In all these cases, each level is nested in the next higher level.

An important scale question in natural resources management is, how do natural or management actions at one level in a scale hierarchy affect ecological or socioeconomic conditions at other levels? This question deals with *scaling* or the multiple scale aspect of the scale problems. Scaling effects are particularly relevant to the concept of compatibility because actions that may be compatible at one scale may not be at another. If the natural resource manager better understands how scale affects the interactions between potentially conflicting uses of the forest, then he or she may be able to more effectively produce the diverse goods and services from forests that society demands.

Ideally, management and research should consider all scales in decision or analysis but this is generally not feasible. Hierarchy theory (Allen and Starr 1988) provides some guidance for dealing with this problem. According to this concept, phenomena of interest can be treated as in a nested hierarchy, in which a focal scale (e.g., a particular watershed or landscape) for analysis is composed of a lower level of finer scale patterns and processes and is also nested in a broader scale (e.g., a subregion or region), which may constrain the range of condition or behaviors of elements at the finer scale. Thus, if we are interested in the dynamics of forest landscape, we should consider the behavior of stands or patches of trees at the next finer scale and the climatic or political context of the landscape at the next broader scale. Hierarchy theory provides a general framework for thinking about scale, but in practice it can be difficult to know which finer scale or coarse-scale elements to include in an analysis or decision.

Decisions about scale are frequently determined by the nature of the problem. In some cases, the selection of the scales of analysis (either in terms of extent or resolution) emerges from a particular ecological question such as, how do we maintain a population of an endangered species? Or, what influences water temperature in a particular stream? In these cases, information is needed about the kind of forests used for habitat, size of area required for raising offspring, the dispersal distance, the characteristics of individual fine-scale habitat elements such as nest trees or germination sites, the location of streams, riparian vegetation, and geology. In other cases, forest policies provide direction on scale. For example the Northwest Forest Plan (FEMAT 1993) specified the extent of the area (range of the northern spotted owl (*Strix occidentalis caurina*)) and some of the decision criteria (e.g., species viability) that indirectly set some bounds on scale and resolution. After the range of scales is narrowed by scientific and political considerations, other factors will come into play; these include constraints set by resources, analytical capacity, and information availability.

3. Influence of Scale on Ecological Conditions

Because ecosystems and social systems are open and hierarchical, our perception depends on our view. In this section we examine some examples of how spatial and organizational scales influence measures of ecological condition. We draw on specific examples from our work in the Oregon Coast Range.

3.1. Spatial Extent

The spatial extent of the analysis area will influence the estimate of conditions within an area. Most ecological indicators are influenced by the extent over which they are observed. For example, the number of species in an area, the proportion of different habitat or ownership types, and the historical range of variability are all dependent on the extent of the analysis area.

This phenomenon is illustrated by Wimberly et al. (2000) in a study of the historical range of variability in amount of old forest in the Oregon Coast Range. Historical range of variability is an indicator for recognizing that biodiversity of an area is not static and the dynamics of habitats and species are actually the source of long-term diversity and productivity of ecosystems (Landres et al. 1999). Using the mean of a forest condition (e.g., amount of habitat or population level of a species) to set goals for a landscape will be of limited value in characterizing the condition of dynamic ecosystems. Instead, the variation of those conditions may be a more suitable indicator than the mean. The extent of time used to estimate this range is ideally long enough to encompass natural variation in disturbances experienced by species and

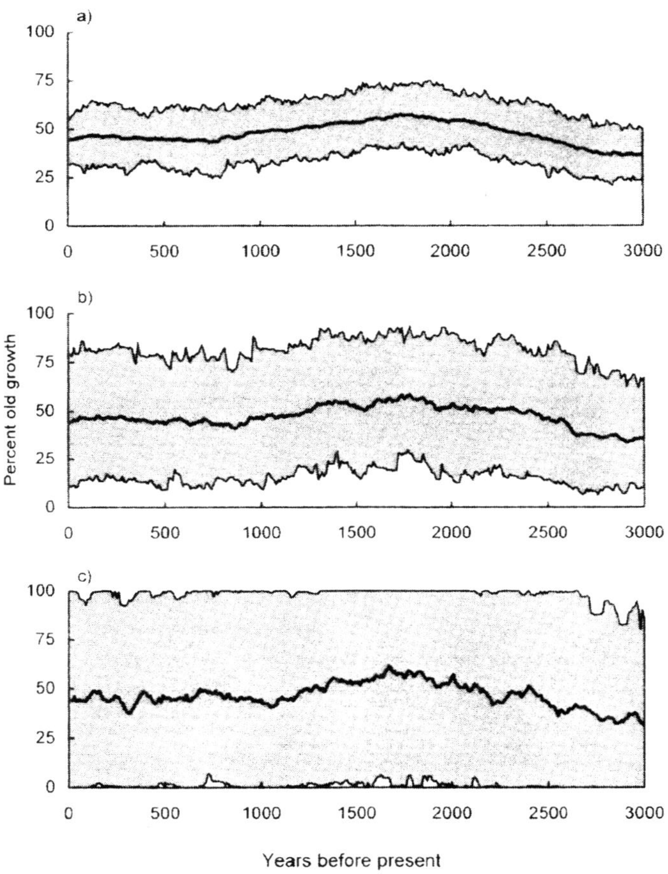

Figure 1. Simulated temporal variability of percentage of old-growth forest over the past 3,000 years in the Oregon Coast Range, summarized at three spatial scales: (a) province (2 265 000 ha), (b) national forest (300 000 ha), and (c) old-growth reserve (40 000 ha). Solid line represents the mean and the shaded area is bounded by the 5% and 95% quantiles (from Wimberly et al. 2000).

processes of an area within a period of similar climate. This may be on the order of 100s to 1,000s of years.

Wimberly et al. (2000) estimated the historical range of variability (HRV) in the Oregon Coast Range by using dendrological and paleoecological information and a spatial simulation model. In this area, the natural fire regime of the last 1,000 years was characterized by relatively large fires, a large proportion of which burned with high severity at frequencies ranging from 75 to more than 300 years, depending on the environment. They estimate the HRV (defined as the range of conditions that would be encountered 90% of the time) for forest age classes at three different spatial extents: a forest reserve of 40 000 (hectares) ha, a national forest of 300 000 ha, and the entire province, 2 265 000 ha (Figure 1). They found that the HRV of old-growth forests (more than 200 years old) for areas the size of reserves and even national forests was so large (0 to 100%; and 10 to 90% of the area, respectively) that the measure had limited use as an indicator at these scales. At the province level however, the HRV in coverage of late successional forest conditions was about 25 to 75%, a range that was more sensitive, and thus more useful to management than a range in which any state was possible.

3.2. Spatial Resolution

All spatial assessments such as those done by the Forest Ecosystem Management Assessment Team (FEMAT 1993) and the Interior Columbia River Basin Ecosystem Management Project (ICBEMP) (USDA 1996) have a minimum mapping area or grain of analysis that can affect the nature of the assessment and its utility. For example, in FEMAT, the vegetation layers were mapped to a resolution of about 16 ha. In ICBEMP, the resolution was about 1 km or 100 ha for the entire area, although finer resolution analyses were done for subsets of the regions in both assessments. Relatively large minimum assessment units can still be valuable for regional assessments but some indicators of ecological condition cannot be evaluated at this scale. An alternative to mapping finer scale attributes is estimating the distributions of conditions within pixels or minimum mapping units. This can be done if data are available to make predictions at finer scales based on coarser scale patterns, such as geology, ownership, and forest type.

In some cases, the spatial location of fine-scale features may be important and force the assessment to use smaller minimum mapping units. For example, fine-scale information is frequently sought for questions about riparian management. Forest and stream interactions occur over relatively small areas, and the spatial position of streams, forests, landslides, and roads can make a difference to the behavior of the system. Stream gradients and junction angles can be used to predict how debris flows through drainage networks and influences riparian habitat (Miller et al., in preparation). To make this prediction it is important to know the pattern of slope gradients across an area. Digital elevation models (DEM) can be used, but if the spatial resolution is too coarse, the models will not accurately estimate the distribution of slope in a watershed, which is an important determinant of landslide risk (Miller et al., in preparation). For example, Clarke and Burnett (in press) found that 10-meter (m) DEMs were far superior to 30-m DEMs for characterizing hydrological patterns and aquatic habitat potential (Figure 2). The least agreement between the two models occurs in slope classes greater than 65%, which are the slopes with

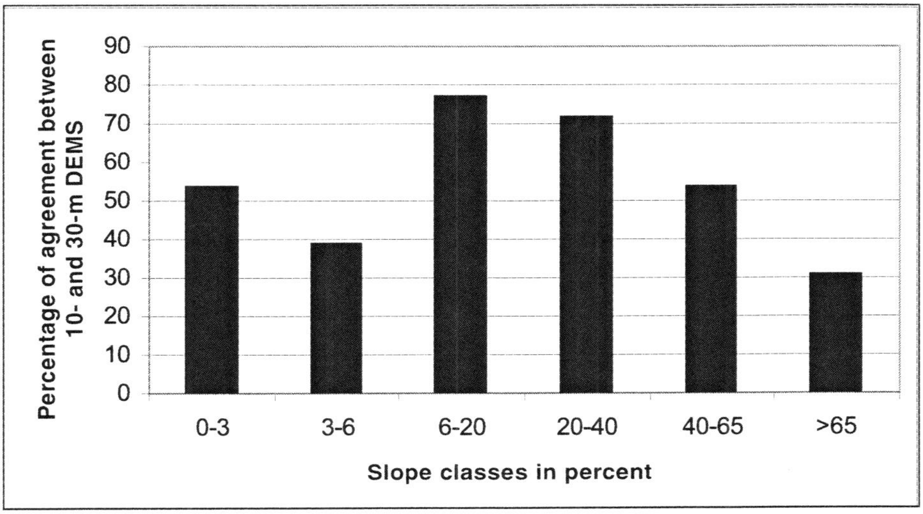

Figure 2. Percentage of agreement between area estimates of slope classes based on 10-meter DEM and 30-meter DEM from nine 7.5-minute quadrangles from the Oregon Coast Range (based on Clarke and Burnett, in press).
Note: DEM = digital elevation model.

the greatest probability of land sliding. Consequently, an assessment of landslide risk based on 30-m DEMs will greatly underestimate risk in steep mountainous areas.

Resolution is also important from a social perspective. When people view maps they often want to "see themselves in the landscape" to effectively participate in decisions. This often means that coarse resolution maps that can not identify stand level features (roughly at least 1 to 10 ha) are not considered relevant or precise enough for assessment purposes. As assessment and planning for forests has become more a collaborative process involving the public, it is increasingly important that the chosen spatial resolution enables citizens to relate to the representation of the forests being studied.

Despite ecological and social pressures to increase the spatial resolution of regional assessments, the full detail of natural systems can never be completely represented in an analysis. Also, the finer the detail considered, the more likely that an analysis may be overwhelmed with the sheer volume of information available. Choosing the spatial resolution for analysis can be one of the most important decisions made in an assessment, and one that will always be full of compromises. As remote sensing, GIS, and computer technology give us the capability to incorporate increasingly fine scale information, our conceptual and institutional ability to use that information (or to know when not to use it) has to increase as well.

4. Ecological Resolution

Ecological entities exist in a hierarchy of ecological scales ranging from individuals to ecosystems (Noss 1990). Fine resolution entities include species, whereas coarse level entities include vegetation types and disturbance regimes. While coarse level indicators play an important role in assessing and maintaining biological diversity, they do not necessarily capture individual species, particularly those that are rare and endemic. However, it is not possible or even necessary to track every species in an area—species are too numerous, the ecology and dynamics of their populations are too poorly known, and resources are not available to manage very many of them individually (Noss et. al. 1997). Consequently, many have called for a strategy in which high-priority species, such as threatened and endangered species, are combined with coarser resolution information (e.g., vegetation types, landscape pattern, and disturbance regimes) (Noss et al. 1997).

To identify these priority species, some have proposed using focal species, not as indicators of other species, but as part of sets of indicators that collectively provide a picture of different ecosystems or elements of ecosystems (Lambeck 1997, Noss et al. 1997, and the Committee of Scientists 1999). In the range of the northern spotted owl, it is sometimes mentioned as an "umbrella" species for late-successional and old-growth ecosystems. In fact, the range of the owl set the boundaries of the FEMAT (1993) assessment area. Three small mammal species have been proposed to measure stand biocomplexity in Douglas-fir (*Pseudotsuga menziesii* (Mirb.) Franco) forests (Carey 1995, Carey and Harrington 2001). However, Lindenmayer et al. (2002) caution that focal species have the same general limitations as indicator species (e.g., species do not respond in the same way to forest structure or disturbance) and that conservation planning and assessment should not solely rely on this approach, but rather on a wide range of strategies.

We developed a set of focal species for coastal Oregon forest management assessments (Spies et al. 2002c). This set is limited, but illustrates how individual species differ in their use of the environment. Our focal species approach was based on habitat suitability models for individual species (McComb et al. 2002). We focused on habitat quality for (1) northern spotted owl, (2) pileated woodpecker (*Dryocopus pileatus*), and (3) western bluebird (*Sialia mexicana*). In addition, we developed an indicator of old-growth development that represented a forest structure-based approach (Lindenmayer et al. 2002). The first two focal species are related to closed forest conditions and are threatened by loss of older forest habitat and large dead trees. The western bluebird is a species representative of open, early-successional forest types. Populations of the bluebird in western Oregon are listed as "sensitive" because of population declines that are possibly related to competition for nest holes from the introduced European starlings (*Sturnus vulgaris*) (Csuti et al. 1997).

Each of these indicators integrates the environment differently in both space and time. For example, the foraging area for northern spotted owls is 1800 ha, while the territory for western bluebirds is less than a hectare. Pileated woodpeckers have a relatively large foraging area, over 400 ha, but require particular microsites, such as relatively large dead trees for nesting. Old-growth structure and development is primarily a stand-level phenomenon based on live and dead attributes occurring over an area of about 1 ha; the index also has a time dimension (age) because some of the biodiversity of older forests is not a product of structure but of time in which species accumulate and spatial diversity increases. In contrast to old growth, the habitat for western bluebird in this forested region is ephemeral, lasting only 10 to 15 years until conifer canopies close.

Habitat patterns of focal species are used as indicators of a larger ecosystem type. However, the habitat of a single focal species does not always match that of its associated forest ecosystem type. For example, in the Oregon Coast Range the habitat distribution of the northern spotted owl, which has been classified as "closely associated" with old growth (Ruggiero et al. 1991) overlaps 75% of the area of old growth defined by the old-growth index (Figure 3). While 75% might be considered relatively high, if the habitat of the owl were the only criterion used to evaluate old-growth habitat, a significant portion of the existing stands with moderate to high old-growth development would not be included. The habitat of the northern spotted owl is distributed differently than that of old growth largely because the owl has specific nest requirements that are a subset of the range of old-growth conditions. A spotted owl uses thousands of meters of forest in different stages of development around its nest site (McComb et al. 2002). This example illustrates the need to use multiple focal species as well as indicators based on forest structure and dynamics in ecological assessments. Given the high dimensionality of ecosystems (i.e., components of ecological variation relating to a large number of relatively independent factors such as effects of disturbance, physical environment, and dispersal capability of the species), several different types of indicators are needed. Theoretically, the number needed to completely represent the ecosystem is closely related to the number of species, interactions among species, number of habitats, variety of disturbance regimes, and variety of physical environments. Complete enumeration, however, is not possible or practical in a management context.

5. Expanding Scale from Stands to Regions

Perhaps the most interesting scaling problems involve estimating how conditions or actions at fine spatial scales affect large areas. For example, a planner may want to extrapolate plot-level inventories to create a vegetation map of larger areas (see Ohmann and Gregory 2002). Policymakers may want to know

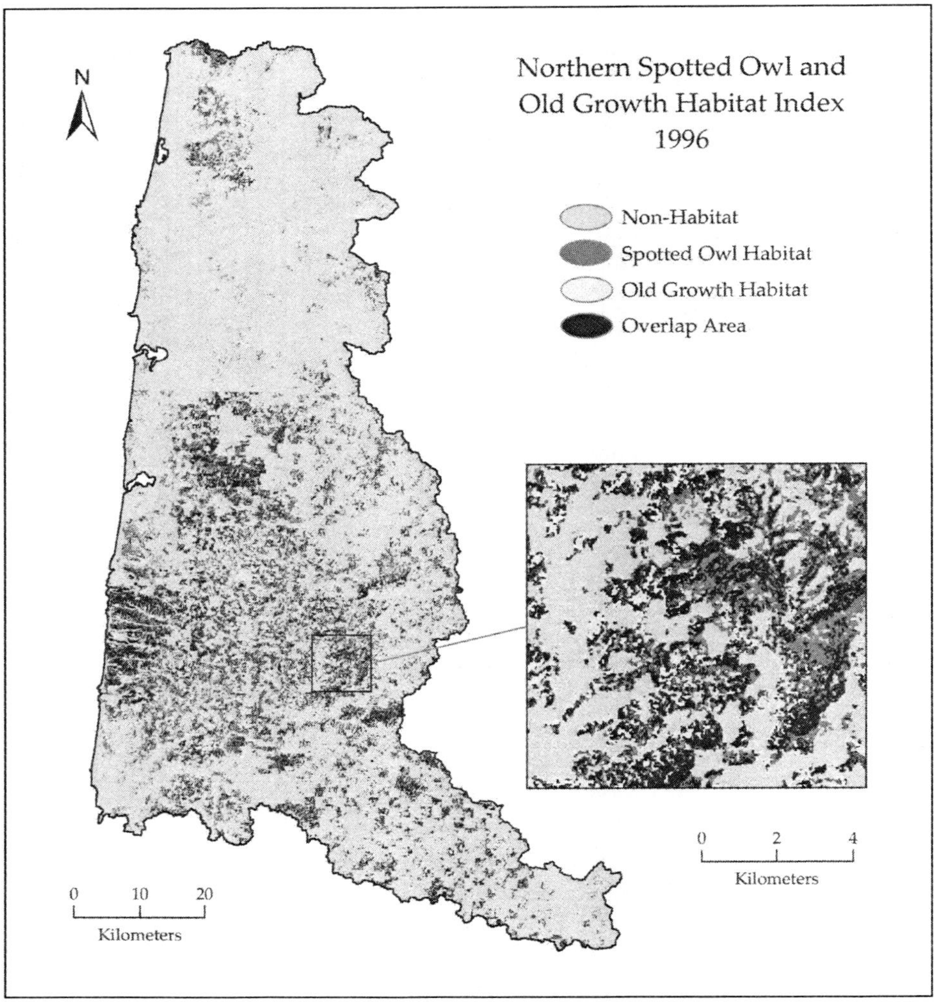

Figure 3. Area of overlap between suitable habitat for northern spotted owls (McComb et al. 2002) and moderate to high old-growth development, and areas where owl and old-growth habitat do not overlap in the central Oregon Coast Range.

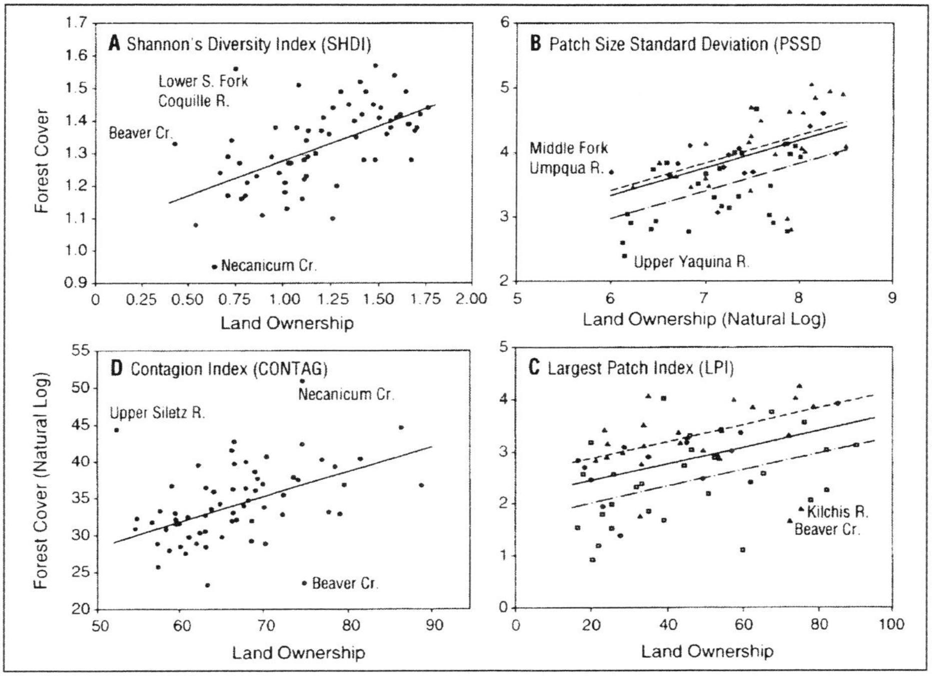

Figure 4. Relationships between landownership and landscape characteristics of forest cover types for 66 fifth-field watersheds in the Oregon Coast Range (from Stanfield et al., in press). Fifth-field watershed (Hydrologic Unit Code 5) are approximately 20 000 to 100 000 ha in size (Seaber et al. 1987). Units for figures are the same for x and y axis; A is unitless, B is hectares, and C and D are percentages.

how the practices of individual landowners affect aggregate conditions of large landscapes or regions. They may also want to know how alternative management approaches at the stand level affect biodiversity and economics over larger areas. In this section we examine these types of scaling problems.

Ownership diversity can lead to ecological diversity if the full range of forest management intensities is present in an area. For example, Stanfield et al. (in press) found that the diversity of forest cover type and spatial patterns could be explained by the underlying ownership pattern (Figure 4). The patterns they observed are a snap-shot of current conditions that are a product of more than two centuries of natural and human actions in the Oregon Coast Range (Wimberly and Ohmann, in review). The study reveals how ownership patterns and landowner behavior can help explain variation in biodiversity in a landscape or region.

Recently developed tools and scientific frameworks are useful for studying how management effects are magnified in space and time. Nonspatial forest planning tools have been used to extrapolate management effects around for many years (Davis et al. 2001, Monserud 2003). However, many ecological

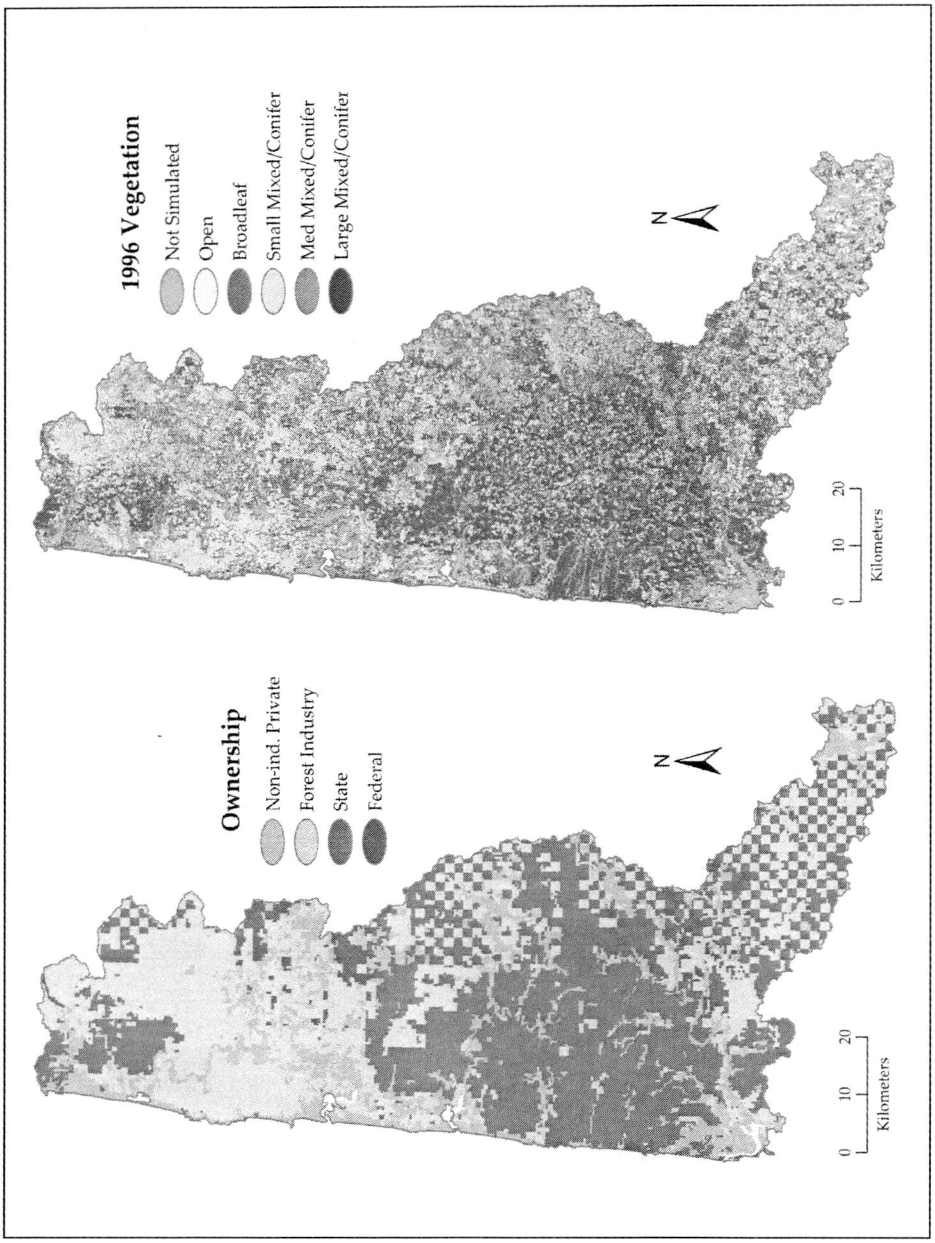

Figure 5. Vegetation and ownership patterns in 1996 across multiple ownerships in the central Oregon Coast Range (from Ohmann and Gregory 2002).

issues require spatial information such as the pattern of riparian zones or the pattern of habitat for wildlife species. New modeling systems are being developed that use spatial information at relatively fine scales to project landscape changes and their consequences (Mladenoff and Baker 1999, Davis et al. 2001). Many management and sustainability questions revolve around how actions at an individual site or stand level affect ecosystem conditions and economic outputs at broader scales. Frequently, this is done by putting forest conditions into several cover types or age classes and then projecting transitions among those classes based on management and disturbance probabilities (Cissel et al. 1999, Pearson et al. 1999). This approach is useful for understanding many landscape-scale effects and processes but it is not well suited for many forestry questions that involve projecting the effects of different silvicultural practices at broad scales. For example, the numerous combinations of initial stand conditions, management practices, and environmental conditions in a region would be difficult to incorporate into an approach based on 10 to 25 forest patch types. The problem is especially complex when landscapes include multiple owners, a situation typically characterized by non-uniform spatial information about vegetation conditions and inadequate specification of management behavior of different ownership types for model projections. In the Coastal Landscape Analysis and Modeling Study (CLAMS) (Ohmann and Gregory 2002, Spies et al. 2002b,c) we developed a set of models that can retain the diversity of initial stand conditions and management strategies to estimate forest structure and compositional diversity with enough spatial and ecological resolution to handle many different fine- and coarse-scale ecological indicators.

CLAMS examines how policies across different forest ownerships in the Oregon Coast Range affect ecological and socioeconomic outputs. The Coast Range is a fitting study area because it contains owners with a wide variety of forest management goals and approaches and patterns of ownership (Table 1, Figure 5). In the CLAMS approach, a set of different spatial simulation models are linked under a conceptual framework in which changes in landscape condition are driven by natural gap disturbances, forest management actions, and land-use changes (Figure 6). The landscape conditions are then used to estimate potential effects on various measures of habitat potential for terrestrial and aquatic organisms (McComb et al. 2002), forest structure and composition, proportion of area in each seral stage, landslides, timber volume, and recreation. The following example illustrates how the CLAMS work can be used to examine the landscape level effects of different forest owner practices and the effects of changing stand-level management practices.

The methods used to simulate landscape change as result of landowner behavior follow those described in Spies et al. (2002b). We used timber harvest volume targets from the Northwest Forest Plan for federal lands (FEMAT 1993). For state lands, we found the maximum sustainable level

Table 1. Forest policies, goals, and strategies dealing with biological diversity in the Oregon Coast Range, by major ownership categories.

Ownership	Policies	Goals[1]	Strategies
USDA Forest Service	NW Forest Plan Forest Plans	Late successional/ old-growth forests, threatened & endangered species, aquatic ecosystems, commodities	Reserves, matrix, green-tree retention, stream buffers, adaptive management areas
USDI Bureau of Land Management	NW Forest Plan Forest Plans	Same as USDA Forest Service	Same as USDA Forest Service but with different matrix prescriptions
State of Oregon	Forest Plans	Healthy forests, indigenous species, abundant timber, threatened & endangered species	"Structure-based" active management, habitat conservation plans
Private industrial	Oregon Forest Practices Act	Growth and harvest of trees, protection of environment and fish and wildlife	Limited retention of individual trees, limited stream-side protection for fish-bearing streams
Private nonindustrial	Oregon Forest Practices Act	Same as private industrial	Same as private industrial

[1] Goals are listed in approximate order of priorities.

(based on current timber volume and expected future growth) subject to achieving certain structural conditions and size distribution of interior habitat patches. For private nonindustrial lands, we used probabilities of clearcutting or partial harvesting from the work of Lettman and Campbell (1997) on historical behavior of nonindustrial private landowners in western Oregon. For forest industry landowners, we simulated the maximum sustainable level for the management intensities they provided. Timber production was calculated as the average volume of timber produced over the 100-year simulation across all ownerships including volume produced by thinning. Our long-term goal is to use the period-by-period harvest targets and management intensities for private forests from the Adams et al. (2001) analysis of economic behavior of private forest landowners in western Oregon.

Stand growth and succession were simulated by using a growth and yield model (ORGANON) (Hann et al. 1995) and an ecological succession model (ZELIG) (Urban et al. 1999) to project the development of forest stands across the entire assessment area. The growth and yield model is used for forest

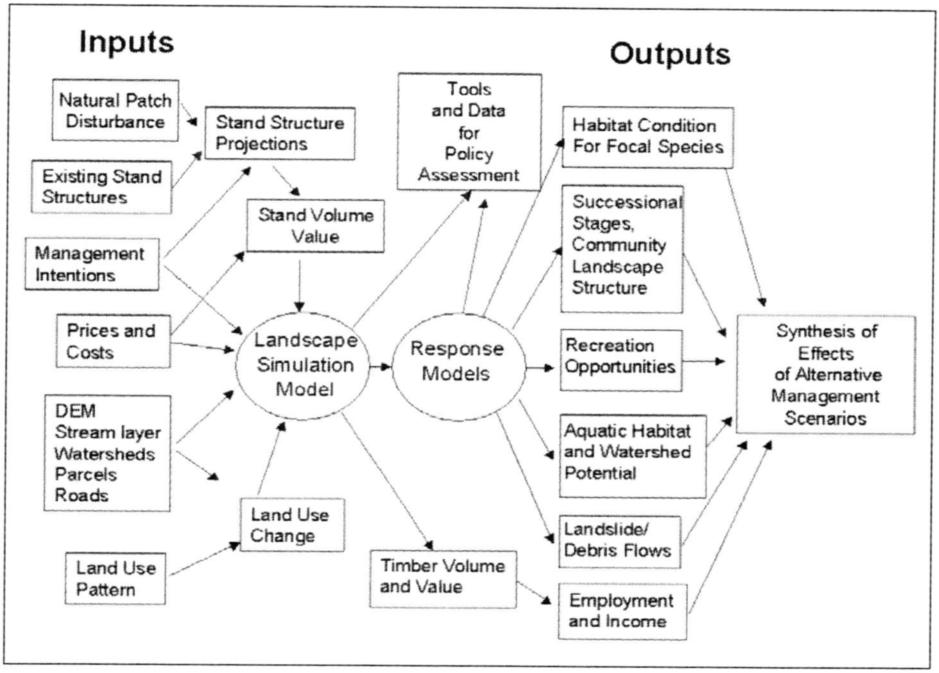

Figure 6. Conceptual model of components and links in CLAMS (Spies et al. 2002a).

management regimes characterized by young stands on short rotations (less than 80 years) and intensive forest management. The ecological succession model is used when stands are grown on longer rotations or allowed to grow without harvest or manipulation until stochastic disturbances occur. ZELIG was calibrated against ORGANON to produce about the same stand structural growth for the first 80 years, a time when mortality is largely density dependent and ingrowth of shade tolerant trees is minimal. The ORGANON model was based on stands less than 80 years old and does not estimate regeneration, ingrowth, and gap mortality that are characteristic of older stands. These models are run nonspatially on every combination of stand types (inventory plots) and management regimes. Results of each projection were then stored in a lookup table, later used in the landscape simulations.

Projections of the future landscape conditions of the Coast Range under current policies indicate that new landscape patterns will emerge over the next 100 years (Figure 7). Specifically, high-contrast landscapes will develop as federal and state agencies implement new management policies that promote the development of mature and old-growth forests to meet goals of providing habitat for late-successional species. These policies were developed largely independently of each other with relatively little effort to examine overall effects, much less coordinate strategies among ownerships. Consequently,

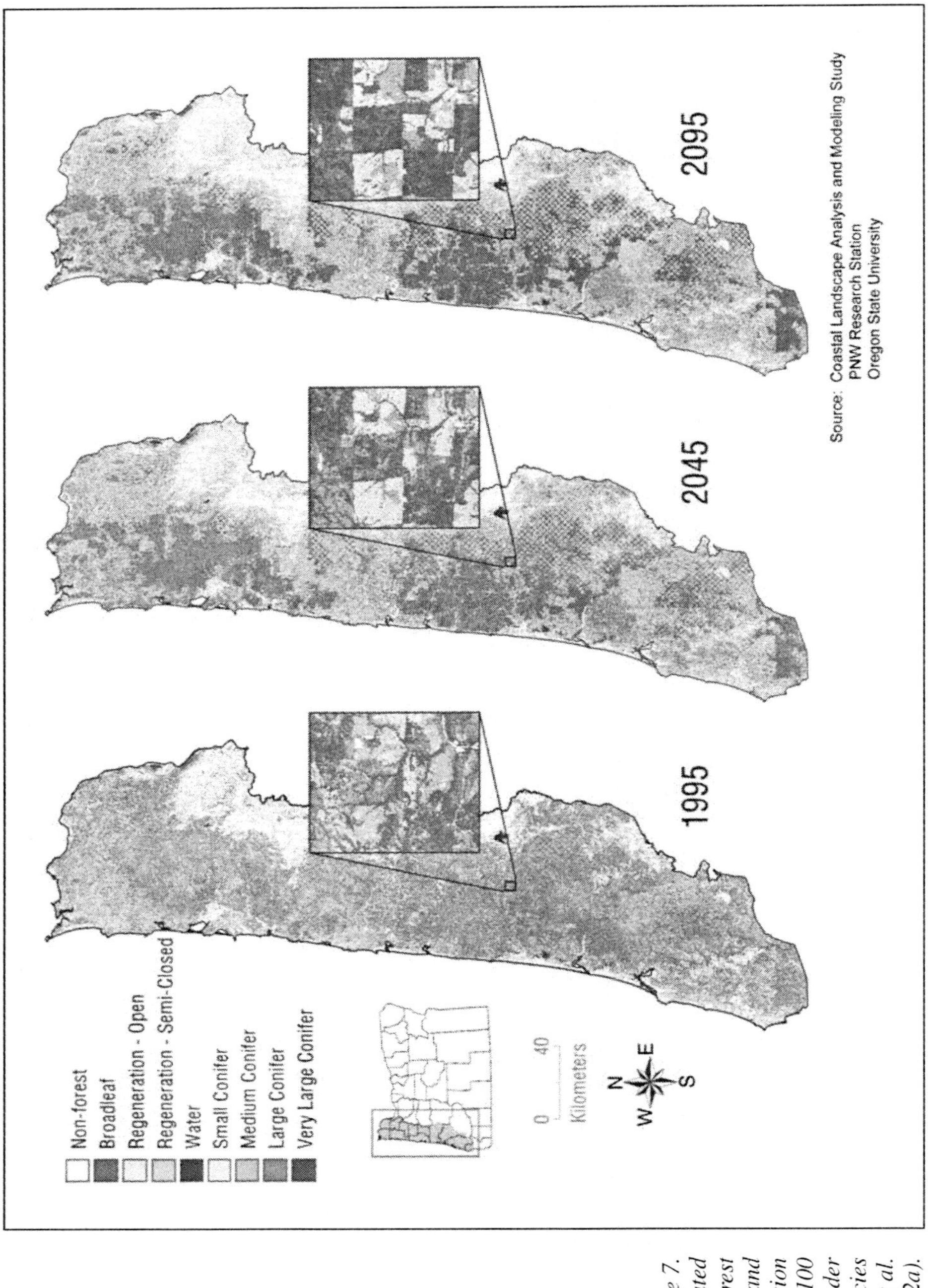

Figure 7. Simulated changes in forest structure and composition classes for 100 years under current policies (Spies et al. 2002a).

Table 2. Characteristics of alternatives used in evaluating ecological and timber production effects (from Spies et al. 2002b).

Alternative	Land ownerships	Description
Current policy	All	Current policies as indicated in federal and state forest plans and the Oregon Forest Practices Act
Increased retention	Private	Retain 5 conifers/ha > 60 cm dbh and 7 conifers/ha > 30 cm dbh
Increased thinning	Federal	Thin from below to 18 m^2/ha in stands less than 40 years old
Long rotation	Private	80-year rotation

some unexpected or unintended trends may occur. For example, the coastal landscape appears to be headed for a condition dominated by two primary states: old, structurally diverse forest and young, relatively uniform conifer plantations. Forests on nonindustrial or state lands may moderate this trend somewhat, but it still appears likely to emerge over time under current policy. This trend will result in sharp edges at ownership boundaries, decline of intermediate-aged forests (e.g., 50 to 150 years), and a decline in young ecologically diverse forests characterized by hardwoods, shrubs, and patches of live and dead large conifers. Although disturbance rates on private lands will be relatively high, particularly on private industrial land, less disturbance will occur in diverse, early-successional stages because private landowners are required to establish well-stocked conifer plantations within a few years after harvest. Federal managers are expected to suppress natural disturbances, such as fire, that reduce the area of older forest. The ecological consequences of these changes are not well understood and therefore, a subject of continuing research.

Forest management occurs at the stand level when foresters cut and plant trees and inhibit the growth of competing shrubs and herbs. We do not completely understand how these changes in silviculture affect ecological and timber production potential across landscapes. A comparison between ecological and timber-production effects at landscape scales for a set of alternative silvicultural scenarios was completed (Table 2) (Spies et al. 2002b) for a 537 700 ha area of the central Coast Range. Major watershed boundaries were used to delineate the study area, which contained a diverse set of ownerships (Figure 7). The LAMPS model (Bettinger and Lennette 2002) was used to simulate 100 years of landscape change under the different scenarios.

A composite biodiversity index was constructed by averaging the mean scores at 100 years into the simulation of four ecological indices: (1) old-growth habitat, (2) northern spotted owl, (3) pileated woodpecker, and (4) western bluebird (Figure 8). The composite index was intended only to illustrate a possible general index of ecological integrity, which has been proposed

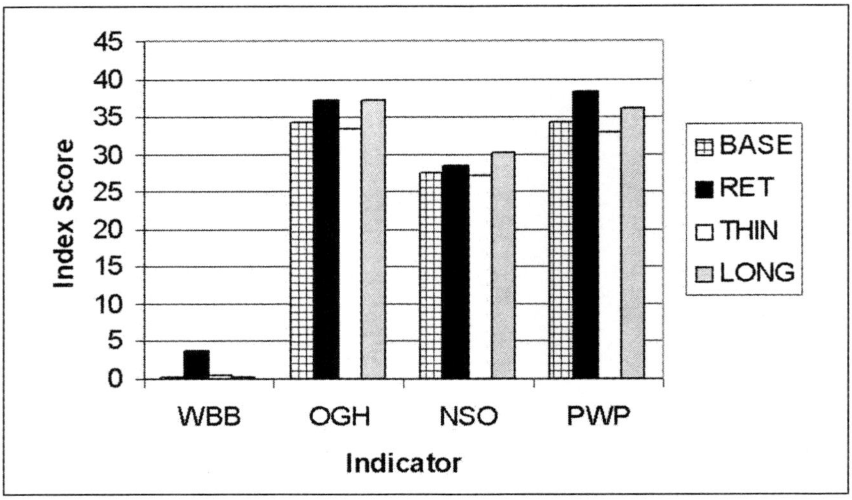

Figure 8. Mean index scores at 100 years for four different ecological indicators and four different scenarios in the central Coast Range. Index scores, which originally ranged from 0 to 1, have been multiplied by 100 to eliminate decimals. Ecological indicators: WBB = western bluebird; OGH = old-growth index; NSO = northern spotted owl; PWP = pileated woodpecker. Scenarios: BASE = base policy; RET = increased retention on private lands; THIN = thinning on federal land; LONG = 80-year minimum rotation on private lands (from Spies et al. 2002b).

as a single measure for the condition of biodiversity in an area (Karr 1991, Quigley et al. 2001). A more complete evaluation of the use of a measure for ecological integrity is needed. For example, indicators could be developed based on an expected landscape-scale distribution of index scores rather than a single mean score or a set of integrity indicators for different major components of an ecosystem.

In general, the absolute mean scores at 100 years of the individual indices differed little among the scenarios (Figure 8). However, for the western bluebird, the relative difference in mean scores was almost 10 times higher for the retention scenario than for any other scenario. This probably resulted from the increase in area of open conditions with snags, which would occur if higher densities of trees were left after clear-cutting on private lands. The retention and long-rotation scenarios resulted in slightly higher mean scores for all the biodiversity metrics when compared with the base policy and the enhanced thinning scenario. These increases in biodiversity metrics resulted from increased numbers of large trees and snags on private lands. It was somewhat unexpected that the thinning scenario did not change the biodiversity indices much over the base scenario. We expected thinning to increase the rate of development of large trees and, therefore, increase the scores of the ecological indices, which are sensitive to this attribute. Densities of large trees did increase faster with thinning, but the effects may not be large enough at 100

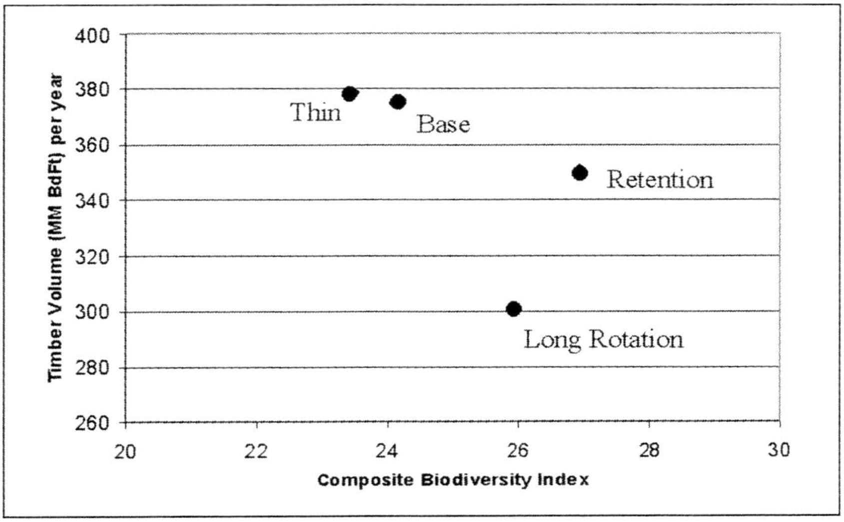

Figure 9. Relationship of timber volume production to the composite biodiversity index at year 100 for four different scenarios (see Figure 8 and Table 2 for explanations) (from Spies et al. 2002b).

years to influence the habitat models. It may be that the thinning regime we simulated did not remove enough biomass. It may also be that the thinning regime reduced amounts of standing dead and down wood, which would reduce the index scores for these measures. Finally, perhaps the thinning regime at a landscape scale was not large enough given the hectares involved. While over 40 500 ha were thinned in this scenario during the first 40 years on federal lands, this amount is only 24 300 ha more than the base policy or only about 5% of the entire central Coast Range area. Further analysis is needed to understand the reasons for the lack of effect of thinning on biodiversity measures at broad scales.

Relatively little difference existed among the scenarios in terms of timber production and the composite index of biodiversity (Figure 9). When compared to the base policy scenario, the greatest change was in the retention and long-rotation scenarios. In terms of the tradeoff between biodiversity and timber production, the retention option proved the more efficient of the two. Under the retention option, the increase in the biodiversity index came at lower cost in terms of reduced timber production. The thinning option differed little from the base policy scenario, showing a slight decline in biodiversity and a slight increase in timber volume. The long-rotation scenario resulted in the greatest decline in timber volume because during the first few decades there would be little or no timber cut that was older than 80 years. For example, during the first 25 years of the simulation, timber volume production per year from private lands would average 3074 cubic meters, whereas during the final

25 years of the study it would average 34 thousand cubic meters. While innovative commercial thinning regimes might offset some of the simulated loss in timber production, there is no doubt the long-rotation scenario significantly reduces private harvest for a number of decades.

The ecological indices are limited in at least two ways. First, they show only trends in habitat quality; this makes it difficult to say what the magnitude of the change means to other ecological criteria that require different types of analyses. However, the indices are useful to indicate direction of trends. A more sophisticated population viability analysis requires much more information (not available for most species) and many more assumptions than the habitat quality analysis we did here. A second limitation is that a single measure of ecological integrity may not change if its components are inversely related to each other. For example, early-successional indices will typically change in the opposite direction of late-successional indices. Consequently, several kinds of ecological integrity measures are needed to provide a more useful characterization of biological diversity.

The measures used to estimate timber production will also influence characterization of tradeoffs. It is relatively easy to measure timber production in terms of volume harvested over time or per time period as we did. However, landowners, especially the forest industry, are typically sensitive to the timing of the harvest or the financial attractiveness of the forest schedule. Most authors (e.g., Adams et al. 2001, Davis et al. 2001) assume that the forest industry attempts to maximize their wealth (net present value). Activities that increase the projected volume of harvest will not always increase net present value and vice versa. Thus, different measures of timber production and assumptions about future interest rates and financial risks are needed to fully evaluate compatibility of timber management and biological diversity from a variety of management viewpoints.

Compatibility is best evaluated by using a set of alternative scenarios. Without prior knowledge of the sensitivity of ecological and socioeconomic indicators to management alternatives, it can be difficult to find a range of scenarios that helps us see the possibilities for tradeoffs. The results show that a wider range of scenarios is needed to fully understand the relation between timber production and ecological conditions. This wide range of scenarios can be explored in a research setting but may prove less possible within a real policy environment where policymakers typically consider only incremental changes to current policies.

The observed lack of strong differences at landscape scales among the scenarios with different stand-level management may be a general phenomenon that occurs when fine-scale actions are extrapolated to broad scales. Stand-level effects can be diluted at broader scales if they occur in a relatively small portion of the total area, such as within one ownership or within one allocation of an ownership. Landscapes can also change slowly because of constraints on future cutting activities or patterns that are imposed by accumulated patterns of managed stands created earlier. In addition, slow change can occur at broad scales simply because forest stands develop slowly following disturbance. Wallin et al. (1994) demonstrated that landscape patterns change slowly, and placement of cutting units early in a rotation can constrain the ability of managers to change landscape patterns in the future. Thus, the temporal extent (100 years) of our analysis may have been too short to observe the full effects of the alternatives.

Human perceptions may also mold our expectations regarding the effects of scaling stand-level activities to landscape scales. When viewing a forest area from the edge of a recent logging unit, the contrast can appear quite dramatic. It is difficult, if not impossible, for the human mind to correctly extrapolate the stand-level effects to large heterogeneous regions in the distant future.

Not all stand-level actions will be diluted at the broader scales. If the ecological process or pattern is restricted to only a small part of the total landscape, such as riparian zones, and the management action is concentrated in those areas, then the impact could be quite large, even though it affects a small portion of the total area. It is important to evaluate each ecological impact for the affected landscape area.

6. Conclusion

The scale of observation will influence the assessment of compatibility between forest commodities and biological diversity. The major hierarchies of scale include space, time, and complexity. Scale must be considered in terms of *extent* (the maximum area, timeframe, or scope of ecological and socioeconomic interactions), and *resolution* (smallest spatial and temporal or biological and ecological units). The process of aggregating fine-resolution information to broad extents is critical for examining variation in compatibility across scales; it is also influenced by decisions about extent and resolution. In this chapter we demonstrated that the extent of observation and the resolution of information will influence measures of ecological condition (e.g., historical range of variability and distribution of steep slopes).

We also demonstrated that decisions made at the scale of stands and individual landowners may have unpredictable consequences at the scale of regions and centuries. For example, the increase in contrast between ownerships and the potential decline in diverse, early-successional forests was not widely appreciated when new forest policies were established by landowners during the 1990s. We also demonstrated that individual fine-scale measures of biodiversity (e.g., habitat for the northern spotted owl) are insufficient to represent the distribution of more complex ecological entities such as old-growth forest ecosystems. At the same time we showed that aggregate measures of biodiversity based on species groups may not be sensitive to the variations of individual species of the group (e.g., western bluebirds). The multidimensionality of ecological communities and ecosystems make it difficult to characterize biodiversity with just one or a few measures. Finally we showed that management influences compatibility (as measured by cost-efficient timber production) by increasing the aggregate habitat score of focal species. The differences among the alternatives were small, however, suggesting that a wider range of management options should be examined.

These findings have several implications for assessing compatibility in forestry. First, because measures and outcomes are influenced by scale and scale hierarchy, it is likely that some past conflicts in forest management have resulted from differences in scale of observation among planners, managers, scientists, modelers, and stakeholders. Some of these conflicts may have been avoided if the effects of scale were better appreciated and publicized. Second, in future assessments, it will be important to explicitly include scale in the analysis and conclusions. Decisions about scale are a type of assumption, reference point, or boundary condition that must be clarified when presenting the results of an assessment. Third, because assessment results can be strongly influenced by decisions about scale, a wide range of spatial, temporal, and complexity scales need to be examined when considering compatibility. Just as the image in a microscope or telescope changes with eyepieces of different magnifications, the image from our assessments changes with different scales of observation and measurement. Our analysis implies that it is not possible to achieve compatible uses by managing at one level. We cannot hope to embrace all the scales of nature and human systems in one assessment; however, we can do a better job of highlighting changing perceptions of compatibility as related to scale. It may be possible to identify compatibility outcomes that are relatively similar across a range of scales. Such outcomes may be more socially acceptable than those that are strongly influenced by the scale of observation.

The use of spatial models and tools, such as those used with CLAMS, can help identify the consequences of decisions about scale and the magnitude of effects of alternative approaches to management. If the ecological resolution is too coarse, such as with community types or indices that combine very different species or processes, the tradeoff analysis may be insensitive to important ecological differences. On the other hand, if the scale of resolution is too fine (e.g., species) assessments of compatibility will concentrate on a relatively small subset of the total species pool and may miss important dimensions of ecological variation. Consequently, a mixture of species and broad-scale indicators (e.g., forest structure, disturbance regimes) may provide a more robust assessment of ecological conditions than assessments that are confined to one type of indicator. The CLAMS work in the Coast Range indicates that large multi-ownership regions can absorb considerable fine-scale change before major changes emerge. Once changes accumulate over a large area, however, it may take a long time to restore the area or achieve particular outcomes. Multiscale modeling approaches help us anticipate how fine-scale management actions will be expressed at broad scales. Currently, we are flying blind when we try to assess effects of actions at multiple scales. Spatial simulation models and tradeoff analysis help us visualize patterns over larger scales of space and time. These models cannot predict the future but they can help identify unintended effects of fragmented policy decisions and identify the level of risks or scales of impacts associated with particular actions.

7. References

Adams, D.M., Schillinger, R.A., Latta, G., and Van Nalts, A. 2001. Timber harvest projections for private lands in western Oregon. Report prepared for the Oregon Department of Forestry, Salem, Oregon. Corvallis, OR: Oregon State University. 70 p.

Allen, T.F.H., and Starr, T.B. 1988. Hierarchy: perspectives for ecological complexity. Chicago: University of Chicago Press.

Bettinger, P., and Lennette, M. 2002. Landscape management and policy simulator (LAMPS). Version 1.1. User's Manual. Corvallis, OR: Oregon State University, College of Forestry. 98 p.

Carey, A.B. 1995. Sciurids in Pacific Northwest managed and old-growth forests. *Ecological Applications*, 5: 648-661.

Carey, A.B., and Harrington, C.A. 2001. Small mammals in young forests: implications for management for sustainability. *Forest Ecology and Management*, 154: 289-309.

Cissel, J.H., Swanson, F.J., and Weisberg, P.J. 1999. Landscape management using historical fire regimes: Blue River, Oregon. *Ecological Applications*, 9(4): 1217-1231.

Clarke, S., and Burnett, K. [In press]. Comparison of digital elevation models for aquatic data development. *ISPRS Journal of Photogrammetry*.

Committee of Scientists. 1999. Sustaining the people's lands: recommendations for stewardship of the national forests and grasslands into the next century. Washington, DC: U.S. Department of Agriculture. Retrieved May 31, 2002 from http://www.fs.fed.us/forum/nepa/rule/cosreport.shtml.

Csuti, B., Kimerling, A.J., O'Neil, T.A., Schaughnessy, M.M., Gaines, E.P., and Huso, M.M.P. 1997. The atlas of Oregon wildlife. Corvallis, OR: Oregon State University Press. 492 p.

Davis, L., Johnson, K.N., Bettinger, P., and Howard, T. 2001. Forest management (4th ed.). New York: McGraw Hill. 804 p.

Dungan, J.L., Perry, J.N., Dale, M.R.T., Legendre, P., Citron-Pousty, S., Fortin, M.J., Jakomulska, A., Miriti, M., and Rosenberg, M.S. 2002. A balanced view of scale in spatial statistical analysis. *Ecography*, 25: 626-640.

Forest Ecosystem Management Assessment Team [FEMAT]. 1993. Forest ecosystem management: an ecological, economic, and social assessment. Portland, OR: U.S. Department of Agriculture; U.S. Department of the Interior [and others]. [Irregular pagination].

Hann, D.W., Hester, A.S., and Olsen, C.L. 1995. ORGANON user's manual: edition 5.0. Corvallis, OR: Department of Forest Resources, Oregon State University. 127 p.

Haynes, R.W., Graham, R.T., and Quiqley, T.M. 1998. A framework for ecosystem management in the Interior Columbia Basin. *Journal of Forestry*, 96(1): 4-9.

Haynes, R.W., Monserud, R.A., and Johnson, A.C. 2002. Emergent results from the wood compatibility initiative. In: A.C. Johnson, R.W. Haynes, and R.A. Monserud (Eds.), Congruent management of multiple resources: proceedings from the wood compatibility initiative workshop (pp. 1-9). General Technical Report PNW GTR-563. Portland, OR: U.S. Department of Agriculture, Forest Service, Pacific Northwest Research Station.

Johnson, K.N., Swanson, F., Herring, M., and Greene, S. (Eds.) 1999. Bioregional assessments: science at the crossroads of management and policy. Washington, DC: Island Press. 398 p.

Karr, J.R. 1991. Biological integrity: a long-neglected aspect of water resource management. *Ecological Applications*, 1: 66-84.

Lambeck, R.J. 1997. Focal species define landscape requirements for nature conservation. *Conservation Biology*, 11: 849-956.

Landres, P.B., Morgan, P., and Swanson, F.J. 1999. Overview of the use of natural variability concepts in managing ecological systems. *Ecological Applications*, 9(4): 1179-1188.

Lettman, G., and Campbell, D. 1997. Timber harvesting practices on private forest land in western Oregon. Salem, OR: Oregon Department of Forestry. 47 p.

Lindenmayer, D.B., Manning, A.D., Smith, P.L., Possingham, H.P., Fisher, J., Oliver, I., and McCarthy, M.A. 2002. The focal-species approach and landscape restoration: a critique. *Conservation Biology*, 16(2): 338-345.

McComb, W.C., McGrath, M., Spies, T.A., and Vesely, D. 2002. Models for mapping potential habitat at landscape scales: an example using northern spotted owls. *Forest Science*, 48(2): 203-216.

Miller, D.J., Burnett, K., and Christiansen, K. [In prep.]. A simple empirical model to estimate debris flow delivery to stream channels. Unpublished manuscript. On file with: Tom Spies, Pacific Northwest Research Station, 3200 SW Jefferson Way, Corvallis, OR 97331.

Mladenoff, D.J., and Baker, W.L. (Eds.) 1999. Spatial modeling of forest landscape change: approaches and applications. Cambridge, UK: Cambridge University Press.

Monserud, R.A. 2002. Large-scale management experiments in the moist maritime forests of the Pacific Northwest. *Landscape and Urban Planning*, 59: 159-180.

Monserud, R.A. 2003. Modeling landscape management. Chapter 7. In: R.A. Monserud, R.W. Haynes, and A.C. Johnson (Eds.), Compatible forest management. Dordrecht, The Netherlands: Kluwer Academic Publishers.

Noss, R. 1990. Indicators for monitoring biodiversity: a hierarchical approach. *Conservation Biology*, 4: 355-364.

Noss, R.F., O'Connell, M.A., and Murphy, D.D. 1997. The science of conservation planning: habitat conservation under the Endangered Species Act. Washington, DC: Island Press. 246 p.

Ohmann, J.L., and Gregory, M.J. 2002. Predictive mapping of forest composition and structure with direct gradient analysis and nearest neighbor imputation in the coastal province of Oregon, USA. *Canadian Journal of Forest Research*, 32: 725-741.

O'Neill, R.V., DeAngelis, D.L., Waide, J.B., and Allen, T.F.H. 1986. A hierarchical concept of ecosystems. Princeton, NJ: Princeton University Press.

Pearson, S.M., Turner, M.G., and Drake, J.B. 1999. Landscape change and habitat availability in the southern Appalachian Highlands and Olympic Peninsula. *Ecological Applications*, 9(4): 1288-1304.

Quigley, T.M., Haynes, R.M., and Hann, W.J. 2001. Estimating ecological integrity in the interior Columbia River basin. *Forest Ecology and Management*, 153: 161-178.

Ruggiero, L.F., Jones, L.C., and Aubry, K.B. 1991. Plant and animal habitat associations in Douglas-fir forests of the Pacific Northwest: an overview. In: L.F. Ruggiero, K.B. Aubry, A.B. Carey, and M.H. Huff (Tech. coords.), Wildlife and vegetation of unmanaged Douglas-fir forests (pp. 447-462). General Technical Report PNW-GTR-285. Portland, OR: U.S. Department of Agriculture, Forest Service, Pacific Northwest Research Station.

Schneider, D. 2001. The rise of the concept of scale in ecology. *BioScience*, 51(7): 545-553.

Seaber, P.R., Kapinos, F.P., and G.L. Knapp. 1987. Hydrologic unit maps. Water Supply Paper 2294. U.S. Geologic Survey. 63 p.

Spies, T. A., Johnson, K.N., Bettinger, P., McComb, W.C., Ohmann, J.L., and Reeves, G.H. 2002a. Challenges to integrating wood production, biodiversity, and other socioeconomic values at broad scales: an example from Coastal Oregon. In: A.C. Johnson, R.W. Haynes, and R.A. Monserud (Eds.), Congruent management of multiple resources: proceedings from the wood compatibility initiative workshop (pp. 13-25). General Technical Report PNW GTR-563. Portland, OR: U.S. Department of Agriculture, Forest Service, Pacific Northwest Research Station.

Spies, T.A., Johnson, K.N., Reeves, G., Bettinger, P., McGrath M.T., Pabst, R., Burnett, K., and Olsen, K. 2002b. An evaluation of tradeoffs between wood production and ecological integrity in the Oregon Coast Range. In: A.C. Johnson, R.W. Haynes, and R.A. Monserud (Eds.), Congruent management of multiple resources: proceedings from the wood compatibility initiative workshop (pp. 111-119). General Technical Report PNW GTR-563. Portland, OR: U.S. Department of Agriculture, Forest Service, Pacific Northwest Research Station.

Spies, T.A., Reeves, G.H., and Burnett, K.M. 2002c. Assessing the ecological consequences of forest policies in a multi-ownership province in Oregon. In: J. Liu, and W.W. Taylor (Eds.), Integrating landscape ecology into natural resource management (pp. 179-207), Cambridge, UK: Cambridge University Press.

Stanfield, B.J., Bliss, J.C., and Spies, T.A. [In press]. Landownership and landscape structure: a spatial analysis of 66 Oregon Coast Range watersheds. *Landscape Ecology*.

Urban, D.L., Acevedo, M.F., and Garman, S.L. 1999. Scaling fine-scale processes to large-scale patterns using models derived from models: meta-models. In: D.J. Mladenoff, and W. Baker (Eds.), Spatial modeling of forest landscape change: approaches and applications (pp. 70-98). Cambridge, UK: Cambridge University Press.

U.S. Department of Agriculture, Forest Service [USDA]. 1996. Status of the interior Columbia Basin: summary of scientific findings. General Technical Report PNW-GTR-385. Portland, OR: U.S. Department of Agriculture, Forest Service; U.S. Department of the Interior, Bureau of Land Management. 144 p.

Wallin, D.O., Swanson, F.J., and Marks, B. 1994. Landscape pattern response to changes in pattern generation rules: land-use legacies in forestry. *Ecological Applications*, 4(3): 569-580.

Waring, R.H., and Schlesinger, W.H. 1985. Forest ecosystems: concepts and management. Orlando, FL: Academic Press, Inc. 340 p.

Wimberly, M.A., and Ohmann, J.L. [In review]. A multi-scale assessment of land cover dynamics in the Oregon Coast Range from 1936-1996. *Landscape Ecology*.

Wimberly, M.A., Spies, T.A., Long, C.J., and Whitlock, C. 2000. Simulating historical variability in the amount of old forests in the Oregon Coast Range. *Conservation Biology*, 14(1): 167-180.

Chapter 9

Landscape Management: Diversity of Approaches and Points of Comparison

Frederick J. Swanson
USDA Forest Service, Pacific Northwest Research Station, 3200 SW Jefferson Way, Corvallis, OR 97331, USA
John H. Cissel
USDI Bureau of Land Management, 3200 SW Jefferson Way, Corvallis, OR 97331, USA
Allison Reger
USDA Forest Service, Willamette National Forest, Box 10607, Eugene, OR 97440, USA

1. Introduction

Forest scientists and managers have long considered landscapes important entities to be studied and managed, although only in the past two decades has the term "landscape" become widely adopted in these fields (Harris 1984, Franklin and Forman 1987, Szaro et al. 1999). Contemporary forest management issues at landscape scales have been framed as concerns about cumulative watershed effects, species viability, fire hazard, and ecosystem health or integrity (Szaro et al. 1999). Because many issues emerge at the landscape scale, it is an essential scale for addressing compatibility and tradeoffs among land use objectives when policy decisions are made.

The type of forest landscapes considered in this chapter (and in many landscape ecology studies) are areas composed of various physical and biotic features as well as socioeconomic units. For example, a landscape may contain patches of forest with different structures and compositions, and there may be different ownerships or land allocations. Landscape studies commonly concern areas ranging from 5000 to 50 000 hectares (ha) and time scales of decades to centuries. These scales are broad enough to encompass many of the coarse-scale temporal and spatial dynamics of ecosystems.

R.A. Monserud, R.W. Haynes and A.C. Johnson (eds.)
Compatible Forest Management, 237-266

Landscapes are a useful scale for management planning for various reasons. Many issues benefit from examination at scales larger than the traditional forest stand or patch. For example, species that require habitat that includes multiple patch types, dispersal of organisms across multiple patches, movement of disturbances such as fire, and watershed processes such as sediment input to streams are all suited to study at the landscape scale. More generally, federal land managers must provide a mixture of habitats so populations of native species remain viable; this requires a landscape large enough to assess the balance of habitat elements across space and time. Relatively small landscapes, such as those considered in this chapter, offer a way to address some of these challenges. These landscapes are large enough for analysts and planners to see broad patterns, assess cumulative effects, and plan for a balance of habitats over space and time. On the other hand, these landscapes are small enough that analysts and planners can retain the site specificity of their data, use the full precision of the primary data, and still comprehend the big picture.

Both broad-scale ecosystem science and management of natural resources commonly consider the world as operating in hierarchies of nested scales—biological, geophysical, sociopolitical, and institutional. A common property of hierarchical systems (O'Neill et al. 1986) is that specific issues may emerge at a particular scale; therefore up- or down-scaling cannot be accomplished as a simple, additive process. As a result, analyses of hierarchical systems must be attentive to multiple scales simultaneously to achieve scientific and management objectives. In forest landscape management, the landscape scale is a critical bridge between the regional scale, such as the range of the northern spotted owl (*Strix occidentalis caurina*) or other wide-ranging species, and the forest-stand scale, at which vegetation management is conducted.

The diversity of forest management approaches represented on the land today, even within a single region, provides a fertile ground for study of landscape dynamics. This diversity reflects specific management objectives, such as recreational use of Congressionally-designated wilderness areas or intensive production of wood fiber from industrial lands. Consequently, existing management approaches span a great spectrum of disturbance frequency, severity, and spatial pattern—from 40-year cutting rotation to no-cut, and fire suppression with disturbance patch sizes ranging from small and uniform to highly variable. Real landscapes are commonly complicated by a mixture of ownerships and land allocations that exhibit different types and stages of management.

The objective of this chapter is to bring science and management perspectives together and examine aspects of diverse approaches to forest landscape management in a region where forestry issues have been hotly contested for several decades. This chapter addresses (1) lessons from a modeling exercise that examined diverse landscape management approaches and selected socioeconomic and ecological consequences, (2) issues of compatibility among forest uses for different objectives, and (3) general perspectives concerning current and future landscape management. We draw from our experiences in developing and implementing different management systems and working collaboratively at the science-management-policy interface. This discussion focuses on the effects of different management systems on forest and watershed processes and features. It is based on examples from Oregon and Washington, where a broad range of management approaches has been used. The range of management systems considered here includes many that have little relevance to private forest management, such as a no-cut and no-fire treatment and very long rotations. However, we believe that the general points concerning landscape planning and management are applicable in diverse settings when the findings are scaled to local conditions.

2. Modeling Alternative Landscape Management Strategies

2.1. Modeling Approach

Modeling future landscape patterns under alternative management approaches is useful for comparing the effects of these approaches in terms of different societal expectations of forest lands, and for examining the compatibility among forest uses and desired conditions. We modeled a single landscape under a wide variety of management rules to produce a set of maps depicting landscape change over time (landscape-change scenarios). These modeled landscape-change scenarios extend over a 600-year period and represent the broad range of stages in forest development that are matters of policy concern in the region. A common landscape is used for the different scenarios so the effects of landscape change can be revealed in the simulations without the potentially confounding influence of topography or other features that would arise if we used different study areas.

These simulations were developed for the 17 500-ha Blue River watershed in western Cascade Range of Oregon, exclusive of the H.J. Andrews Experimental Forest and contained within the Willamette National Forest and the Central Cascades Adaptive Management Area (Figure 1). This steep, mountainous area ranges in elevation from about 300 to over 1600 meters (m) and is dominated by Douglas-fir/western hemlock (*Pseudotsuga menziesii* (Mirb.) Franco/*Tsuga heterophylla* (Raf.) Sarg) forest at lower and middle elevations and Pacific silver (*Abies amabilis* Dougl. ex Forbes) forest at higher elevations. Wildfire over the past 500 plus years and forest cutting since about 1950 have created a complex pattern of forest age classes across the current landscape (Figure 1). This area is broadly representative of steep, federal forest lands west of the crest of the Cascades in the Douglas-fir region (western Oregon and Washington).

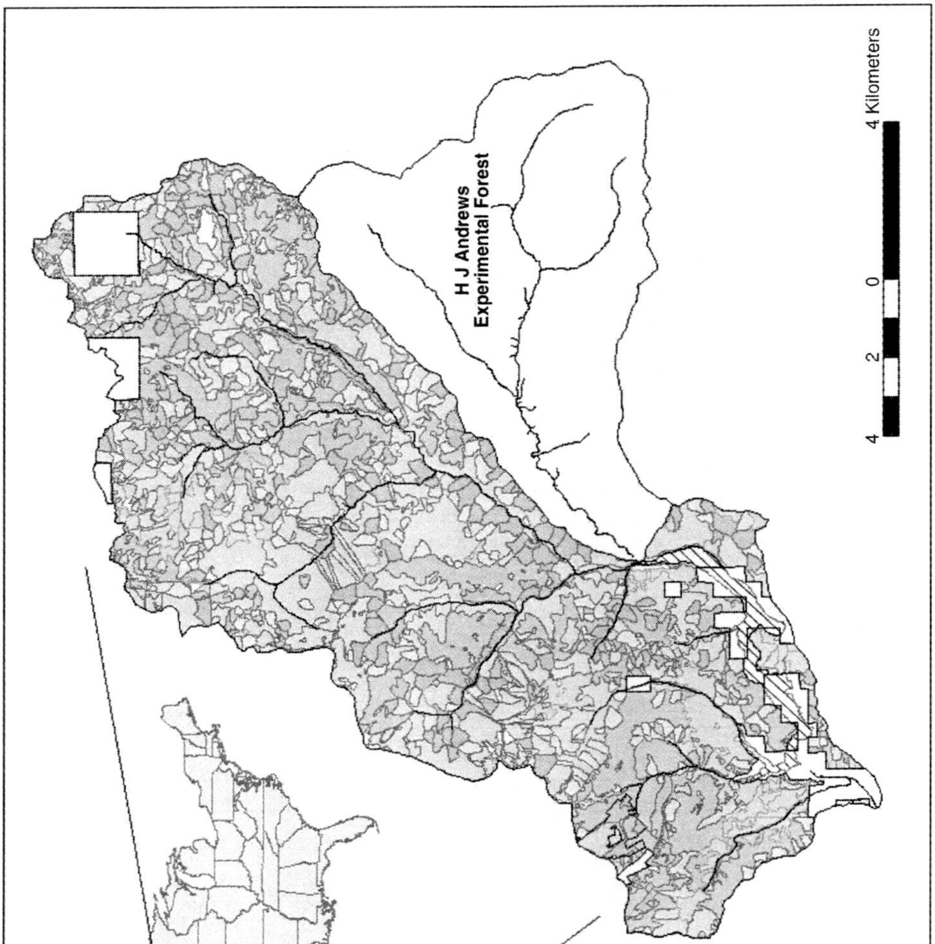

Figure 1. Location of the Blue River, Oregon, landscape study area and current vegetation age classes.

We modeled six scenarios for the Blue River watershed by using the TELSA (Tool for Exploratory Landscape Scenario Analysis) modeling system (ESSA Technologies Ltd. 1999). TELSA is a patch-based, simulation model that employs successional pathways, management regimes, and disturbance regimes defined by the user at levels of temporal and spatial resolution, also controlled by the user. TELSA is both temporally and spatially specific and has numerous methods for controlling or constraining the distribution of management activities and disturbance regimes across space and time. We used 10-year time steps and a minimum spatial resolution of 0.1 ha, although most polygons were substantially larger.

We examined the landscape structure in terms of different forest age classes, mean size of forest patches, and density of the edge between forest and areas of open canopy less than 40 years old. These measures of landscape structure were used because they relate to evaluations of habitat and hydrology effects as well as vulnerability to disturbance, such as windthrow at edges of forest openings (Gratkowski 1956, Sinton et al. 2000). Edge environments are also distinctive in terms of microclimatic effects of the canopy opening penetrating into the adjacent forest (Chen et al. 1995).

We then estimated the consequences of each of these scenarios at year 100 in terms of wood production, extent of early- and late-seral vegetation habitat as indicators of biodiversity protection, and response of annual and summer low streamflow. We used the year 100 for comparisons because most scenarios developed rather stable proportions of the different age classes, and most legacies of earlier landscape patterns (Wallin et al. 1994) were erased. Also, the legacy of forest structure from the current landscape condition, notably dead wood, was largely lost through decomposition and other processes. The amount of carbon stored on the landscape was not compared at year 100, but was estimated based on steady-state landscape conditions in simulation analyses by Harmon and Marks (2002), outlined below. Note that we make simple estimates, requiring major assumptions, which are outlined below, to comparatively analyze scenarios at a coarse resolution.

2.2. Modeled Landscape-Change Scenarios

The modeled scenarios portray a wide range of management approaches and intensities (Table 1). For each scenario involving harvest, we defined the land base subject to timber removal, the type and intensity of harvest prescribed for different zones, and spatial constraints or scheduling rules applicable to the scenario. No harvest was simulated in two of the scenarios, and one scenario simulated historical wildfire regimes.

Several landscape features were simulated in common across scenarios (Table 1). Nonforest vegetation types and lands unsuited for timber harvest because of low productivity and sensitive soils (as determined by the

Table 1. Selected characteristics of modeled landscape-change scenarios.

	Landscape-change scenarios					
Landscape characteristics	Intensive plantation	1990 forest plan	NWFP	AMA	Wildfire	Succession
Cutting rotation (years)	40	80	80	100, 180, 260	119-224[1]	--[2]
Retention (% live canopy)	0	0	15	50, 30, 15	30 (avg.)	--
Reserves (% area)	16	23	42	27	0	0
Fire	No	No	No	No	Yes	No
Context: regional conservation strategy	No	No	Yes	Yes	No	No

[1] Mean fire return interval.
[2] Not applicable.
Note: NWFP = Northwest Forest Plan, AMA = adaptive management area.

Willamette National Forest) were excluded from timber harvest in all scenarios, but might have been subject to burning in the wildfire scenario. Also, 40-ha reserves were defined around all spotted owl nest sites in all scenarios, but these areas were subject to fire. Fire was ignored as a disturbance process in all scenarios except the wildfire scenario. We recognize that some wildfire will occur in the future even with aggressive fire suppression, but excluded it from the modeled scenarios to better contrast tradeoffs and compatibilities.

A plantation forestry scenario, here termed the *intensive plantation scenario*, was intended to represent the highest rate of change, and is common on some private timberlands of the western Cascade Range west of Blue River and elsewhere in the Douglas-fir region. The scenario used a 40-year cutting rotation with aggregated harvest patches, 0% retention of live trees at the time of cutting, and minimal use of reserves for species protection. Riparian reserves (30 m on the main stem of Blue River and 18 m on other streams with significant use by fish for spawning, rearing, or migration) and maximum cutting unit size (64 ha) approximate current Oregon State Practices rules (Table 1). Note that forest practice rules change over time and that actual practices may deviate from regulations.

The *1990 forest plan scenario*, based on the Willamette National Forest plan finalized in that year (Willamette National Forest 1990), represented federal forest land management pre-dating the Northwest Forest Plan (NWFP). This scenario used an 80-year rotation length, dispersed harvest patches, 0% retention of live trees in most cutting units, and limited use of reserves. Special management areas for geologically unique areas and riparian reserves along fish-bearing streams were adopted from the Willamette Forest plan, as were constraints on timber harvest unit dispersion and size (24 ha maximum) of cutting units. The forest planning effort also included constraints to limit the extent of open canopy in areas of rain-on-snow effects, but no regional conservation strategy was involved.

The *Northwest Forest Plan scenario* (USDA and USDI 1994) was based on an 80-year rotation length, extensive riparian reserves approximately 55 m on each side of all non-fish-bearing streams (twice that along fish-bearing streams), and 15% retention of live trees in cutting units. Special management areas and constraints on dispersion and cutting unit size were the same as in the 1990 forest plan scenario. The NWFP sets a strong regional context for work in any included landscape. For example, a network of large late-successional reserves provides prime habitat for old-growth-associated species, including northern spotted owl. It also prescribes management of matrix lands where cutting may occur (such as land in this scenario) with the intent of providing dispersal habitat between late-successional reserves.

The *adaptive management scenario* (AMA) refers to the landscape plan in early stages of implementation in the Central Cascades AMA in western Oregon. The scenario was developed by incorporating interpretations of the historical wildfire disturbance regime to set cutting frequency, severity, and spatial patterns (Cissel et al. 1999). The resulting management approach involved three landscape areas with harvest rotation lengths of 100, 180, and 260 years with retention levels of 50, 30, and 15% of canopy cover, respectively. It also had aquatic reserves and landscape blocks designated to schedule harvest over time and space. This plan significantly reduced the extent of riparian reserves as designated in the Northwest Forest Plan, but added aquatic reserves in certain headwater areas (Cissel et al. 1999). A 40-year sequencing of harvest activity among subbasins was intended to reflect historical wildfire disturbance patterns more closely than in other scenarios. The AMA scenario is set within the regional conservation strategy of the Northwest Forest Plan, so it is strongly linked to management plans in surrounding areas.

The *wildfire scenario* was intended to represent the historical disturbance regime interpreted through reconstruction of fire frequency and severity by using tree-ring dating (Weisberg 1998, Cissel et al. 1999) and data on fire-size distributions from Pennington (2002). Fire frequency in the model ranged from 119 to 244 years. The scenario represented future landscape change as if historical wildfire frequency and spatial pattern were acting on the existing landscape, and into the future. Thus, the landscape change involved no harvest, no reserves, mixed disturbance severity (mean survival of approximately 30% live tree canopy), and no fire suppression. The TELSA model simulates fire by stochastically determining ignitions from the fire frequency distribution, growing a fire across the landscape to the extent that fuel conditions permit and up to the maximum size constrained by the fire size distribution. The upper size limits of successive fires, therefore, may decline.

The *succession-only scenario* represented no disturbance by fire, harvest, or other processes, so landscape change occurred only in response to forest growth over time.

2.3. Methods for Modeling and Assessing Landscape-Change Scenarios

Forest landscapes simulated in the TELSA model are described in terms of forested patch types defined by five age classes: 1 to 20, 21 to 40, 41 to 80, 81 to 200, and 201 plus years. Because many different age-class delineations are used in relevant studies (see Appendix 1, Chapter 1), we refer to age in years rather than by age-class name. In the modeling exercise, each age class was further split into subclasses based on the mean cover of live overstory canopy projected to remain after the initiating disturbance. Data from the Willamette National Forest, Blue River Ranger District (now McKenzie River Ranger District) were used to describe existing conditions. Each of these age classes has distinctive implications in the practice of forestry and in terms of ecological and hydrological effects. The area in the old forest classification, for example, is critically important ecologically because of the habitats and processes it supports, its limited extent in the region, and the length of time required to create more old forest. Many species are linked to old forest habitat; live and dead carbon stores are at a maximum and hydrologic processes are well-buffered in old forests. The youngest age class (less than 20 years old) is distinguished by having the lowest levels of carbon stored in live biomass (assuming no carryover from the previous stand (Harmon et al. 1986)) and minimal biotic control on rain-on-snow peak flow events. The youngest age class provides distinctive habitat important to many species (Johnson and O'Neil 2001). Middle-age classes can fill many of the same ecological roles as old forest, and will eventually grow into the oldest age class, if not disturbed. Wood quantity, quality, and economic incentives for managing for these different age classes vary greatly among ownerships.

2.3.1. Assessing landscape structure—Landscape metrics for the scenarios were calculated by using FRAGSTATS (McGarigal and Marks 1995) to analyze vector maps of existing and future landscape structure. We also used an edge-contrast matrix to represent relative edge contrast among all possible edge types in the landscape to calculate edge density (Cissel et al. 1999 p. 1223). The variety and abundance of patch types, patch size, location of patches, and edge density were selected as key indicators of landscape function.

2.3.2. Assessing timber volume—Prescriptions for thinning and regeneration harvest differed among scenarios. Volume estimates for previously unharvested stands were based on yield tables developed from empirical, stand inventory data stratified by timber type, and on modeling results from the Willamette National Forest plan (Willamette National Forest 1990). Estimates of timber volume yield for managed stands were based on stand inventory data and modeling studies summarized by Garman et al. (in press). Timber harvest values were based on three thinning intensities under the canopy retention levels specified in the scenarios. The prescription for the intensive plantation scenario, for example, consisted of no thinning and a 40-year rotation with 0% retention, whereas a portion of the AMA scenario consisted of three thinnings: a moderate thin at 40 years, a heavy thin at 60 years, and a light thin at 80 years. Note that there is much variation in the type and timing of thinning practices within an individual ownership, across ownerships, and through time, depending on factors such as market conditions and conflict over forest management.

2.3.3. Assessing carbon stores—Estimates of the carbon stored in the simulated landscape are based on results of studies by Harmon and Marks (2002) using their STANDCARB model, which is based on field, remote sensing, and modeling studies of carbon dynamics in the Blue River area and across the region (Harmon et al. 1986, Smithwick et al. 2002). STANDCARB tracks amounts of carbon stored in above- and below-ground organisms, including soil, of a simulated forest landscape that changes under different utilization and rotation lengths in management regimes, and different frequencies and severities of wildfire regimes. Regimes with longer rotations and greater retention level of live and dead vegetation result in more carbon stored in the landscape. Harmon and Marks (2002) simulate a particular regime until the landscape contains a uniform distribution of age classes. They then report the area-weighted amount of carbon stored as a percentage of the maximum stored in old forest for the forest type analyzed. The managed landscape-change scenarios in this chapter contain some no-cut reserves, which we assume have the maximum estimated carbon stores for the area. Thus, we computed total carbon storage in the landscape as an area-weighted average for reserve and managed areas by using values for the managed areas interpreted from Harmon and Marks (2002 Figure 8).

Following Harmon and Marks (2002) and Pennington (2002), we estimated the maximum store of carbon in the modeled landscape as 830 megagrams (Mg) C ha^{-1}, which is well within the range observed in forest plots in the Oregon Cascades (Smithwick et al. 2002). Based on Harmon and Marks (2002), we estimated that the managed part of a landscape produced by the intensive plantation scenario stored 20% of the maximum carbon storage. Managed areas of the 1990 forest plan landscape stored 40% of the estimated maximum (80-year rotation point on the high utilization-high severity line in Figure 8 of Harmon and Marks (2002)). Managed areas in the NWFP scenario also had an 80-year rotation but higher retention levels; 70% of the maximum potential storage was estimated to be stored across the landscape. Each of the three combinations of rotation length and retention level of the AMA scenario stored about 90% of the maximum, so we used this figure for the entire managed area in this scenario. Assuming a wildfire regime averaging moderate severity, we estimated the wildfire scenario would yield a landscape storing 70% of the maximum (Harmon and Marks 2002, Table 3).

2.3.4. Assessing hydrologic effects—Long-term records for small, experimental watersheds in and adjacent to the H.J. Andrews Forest have been used to evaluate and debate the response of peak, low, and annual flows to forest cutting and regrowth, and to a lesser extent, roads (Jones and Grant 1996, Thomas and Megahan 1998, Beschta et al. 2000, Jones 2000, Post and Jones 2001). The low frequency of large peak flows and scarcity of large basins with good flow records lead to small sample sizes and many confounding factors, which are problematic for analyses. Therefore, we do not address peak flow response to the modeled scenarios. To compare scenarios, we assumed that annual water yield was increased by 40% from the 1 to 20 year age class and by 25% from the 21 to 40 age class relative to older forest. For summer low flows we assumed that flow was reduced by 15% for the 21 to 40 year age class relative to all others. These values are based on observations of long-term records for the 10-ha Watershed 10 and other experimental watersheds at the H.J. Andrews Forest (J.A. Jones, Oregon State University, Corvallis, Oregon, personal communication, January 2003). Note that the Blue River modeling study covers a large basin, whereas the experimental watersheds from which these estimates are drawn are all 100 ha or less. We expect that the dominant watershed processes operating in the small watersheds also affect large watersheds, but larger watersheds do not exhibit this magnitude of response to younger age classes because they contain a wide range of age classes.

2.3.5. Assessing biodiversity effects—Many measures of biodiversity have been used to characterize richness, evenness, and other properties of fauna and flora in an area. Intensive field surveys have enumerated many species in many groups of organisms in the Blue River area.[1] These studies reveal strong tendencies for some species to be closely associated with early seral forest habitat and non-forest conditions, whereas others species are associated with late seral habitat, and yet others are generalists (Hansen et al. 1991, Halpern and Spies 1995, Johnson and O'Neil 2001). Hansen et al. (1991) argue that although old-growth forests have been the focus of biodiversity protection issues in the Douglas-fir region and, indeed globally, young forests with standing and down woody debris characteristic of natural, post-disturbance stands may be a critical, limited habitat type in some areas of intensive plantation management.

[1] See http:www.fsl.orst.edu/lter/data and look under data catalog/biodiversity for species lists.

We consider biodiversity in terms of two simple habitat indicators: the extent of young and old forest habitat provided by each scenario. These are distinctive, strongly contrasting habitat conditions; several scenarios differed dramatically in the extent of these habitat types, and provision of a range of habitat conditions is part of a coarse-filter approach to habitat management (Hunter 2002). Additional factors that may affect species include the arrangement of habitat patches within the modeled landscape and the state of the surrounding landscape, which can be expressed as the regional context of the management plan. We consider arrangement in terms of the extent of reserves, which is the direct product of the conservation strategy, and edge density and mean patch size, which are byproducts of the strategy.

2.4. Results

2.4.1. Modeled landscape age-class composition and structure—The composition of the landscape, expressed as the extent of forest vegetation in different age classes produced by each scenario, is an important reference point for judging landscape performance. The proportion of the landscape in each age class represents, at a coarse level, the mix of habitat types, the capacity to store carbon, and the potential for the landscape to either ameliorate or amplify hydrologic processes.

Several features were common to model runs for each scenario. Each scenario began with the vegetation age-class distribution and arrangement existing in the Blue River landscape as of 1995 (Figure 1). The distribution of age classes (Figure 1, year 0 in Figure 2) and landscape structures, as of 1995, include plantations resulting from roughly three decades of patch clearcutting in a landscape with age classes dating from wildfires that occurred in the 1500s and 1800s (Weisberg 1998). Projections of future age-class distributions across the landscape reflect a transition from the current landscape to an age-class distribution resulting from the rules for the scenario (Figures 2 through 7). This transition occurs over the maximum rotation length used in each scenario involving timber harvest; very young stands in reserves in 1995 may take up to 200 years to reach the oldest age class considered here. Forestland in a reserve status, which is effectively the entire landscape for the succession scenario and parts of the managed landscape scenarios, is not subject to management disturbance; therefore, it simply grows into older age classes over the course of the model runs.

We present the landscape-change scenarios in sequence from the highest disturbance frequency and severity to the lowest. Four scenarios have strong management components, which are modeled in a completely deterministic manner, thus restricting variability. The wildfire scenario, which is strongly stochastic, can be viewed as a reference system for a disturbance regime that historically maintained a range of habitat conditions that sustained native

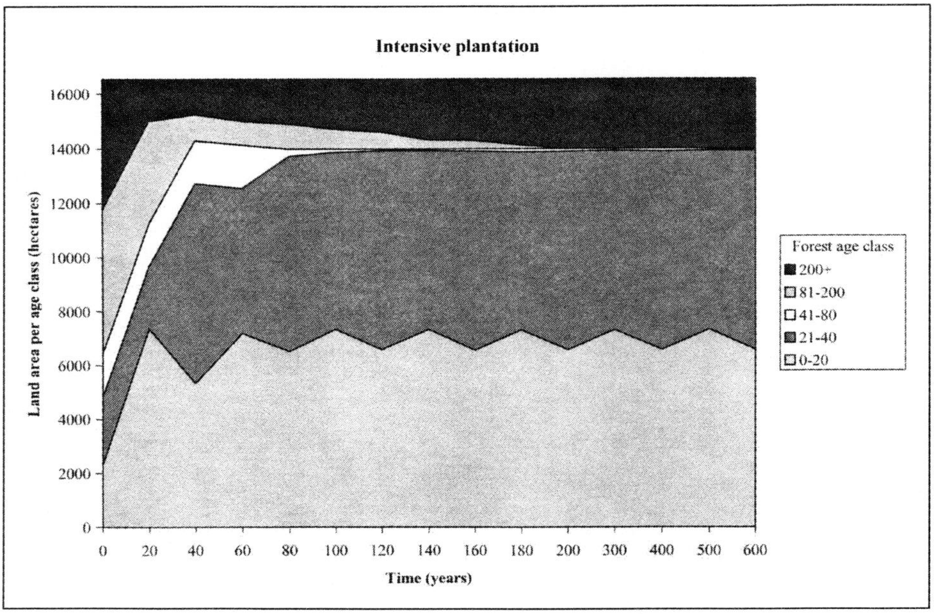

Figure 2. Distribution of forest vegetation age classes over the 600-year modeling timeframe based on the intensive plantation scenario (see Table 1 and text). Note that scale on time axis changes at year 200.

species in the landscape (Engstrom et al. 1999). The succession scenario, which is deterministic but gradually changing, may be considered both a reference condition and a management scenario because wilderness areas and other reserves commonly are managed in this manner. The current condition of the landscape reflects wildfire and about four decades of management similar to the 1990 forest plan, however, with fewer constraints than pertained to federal forestry in the pre-1990 period.

The *intensive plantation scenario* produced a landscape dominated (83% of area) by stands less than 40 years old as a consequence of short (40 years) rotations and limited area in reserves (Figure 2). Within 100 years the 41 to 200 year age classes were nearly eliminated from the landscape. Much of the reserve area in this scenario represented core area reserves for northern spotted owls, which likely would be much smaller on actual intensive plantation forestry lands, so the extent of young age classes in such lands may be higher. This scenario resulted in an initial decrease in mean patch size, characteristic of four of the six scenarios, and increased thereafter (Figure 3), amounting to a nearly 50% increase by year 100. Edge density declined by approximately 18% by year 100 (Figure 4), which in part, reflected the preponderance of very young forest across the landscape.

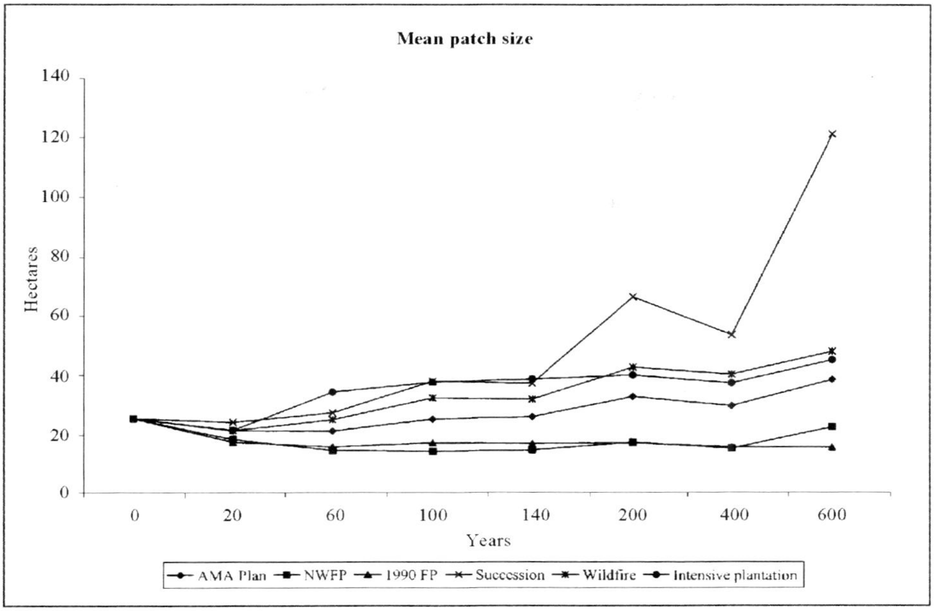

Figure 3. Mean size of vegetation patches in landscapes simulated over 600 years.

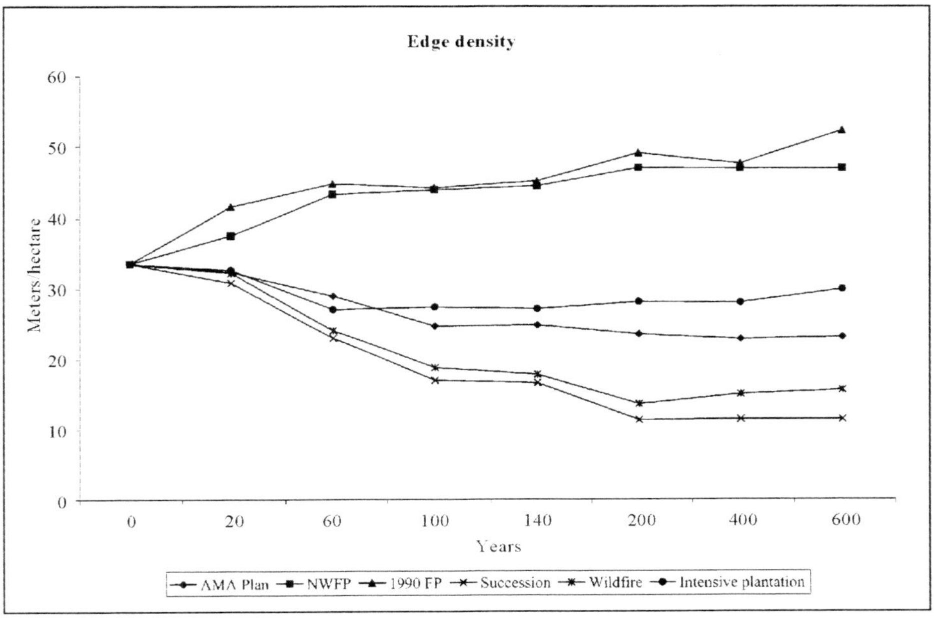

Figure 4. Density of edges between forest and open canopy areas less than 40 years in age.

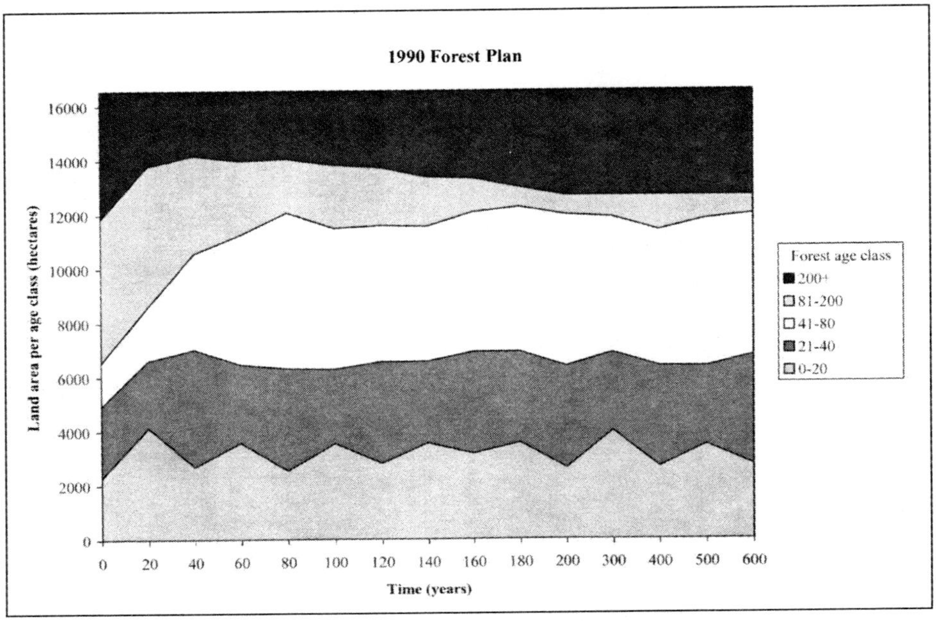

Figure 5. Distribution of forest vegetation age classes over the 600-year modeling timeframe based on the 1990 forest plan scenario (see Table 1 and text).

The *1990 forest plan scenario* yielded a landscape with extensive stands in age classes up to 80 years, the main rotation length used (Figure 5). The mature age class had a limited extent by year 180 of the simulation, producing a bifurcation in landscape structure with much of the landscape in stands either older than 200 years (22%) or less than 80 years (74%). Mean patch size under this scenario declined below the present size, and edge density increased about 30% by year 100.

The *NWFP scenario* also became dominated by stands less than 80 years (Figure 6). The mature age class was truncated to less than 4% of the landscape by year 200, and the landscape contained either old (44% of area greater than 200 years) or young (52% of area less than 80 years) forest age classes. This bifurcation of age class distributions over the landscape was expressed in absence of middle-age classes. The extensive network of riparian reserves in this scenario reduced mean patch size and increased edge density to levels similar to the 1990 forest plan.

The mix of rotation lengths used in the *AMA scenario* yielded a substantial area in the mature age class (Figure 7), unlike the other management scenarios. One consequence was that the AMA scenario produced the smallest area of forest less than 80 years old of the four active management scenarios. Mean patch size declined slightly over the first century and then grew slightly. Edge density declined 25% by year 100. Compared with other management scenarios,

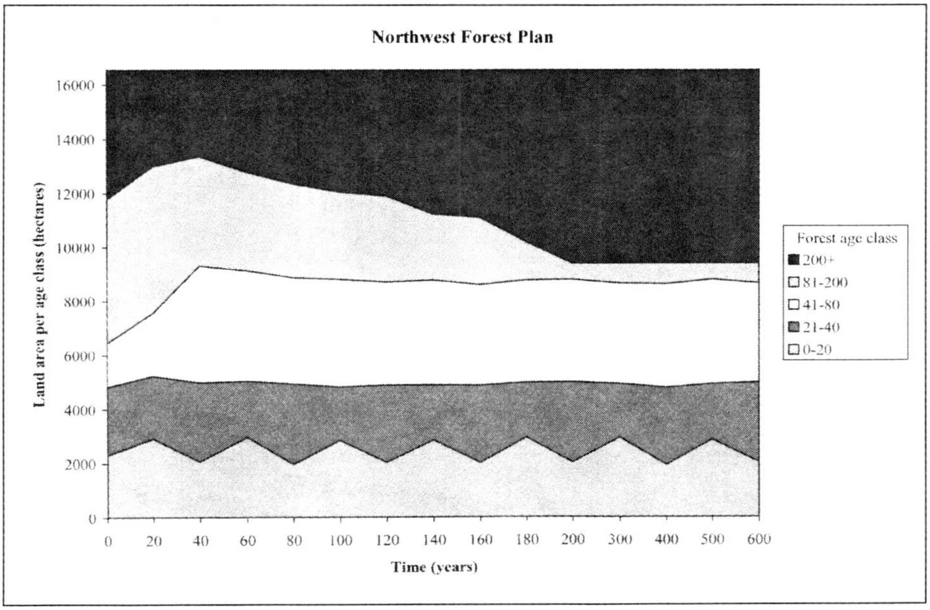

Figure 6. Distribution of forest vegetation age classes over the 600-year modeling timeframe based on the Northwest Forest Plan scenario (see Table 1 and text).

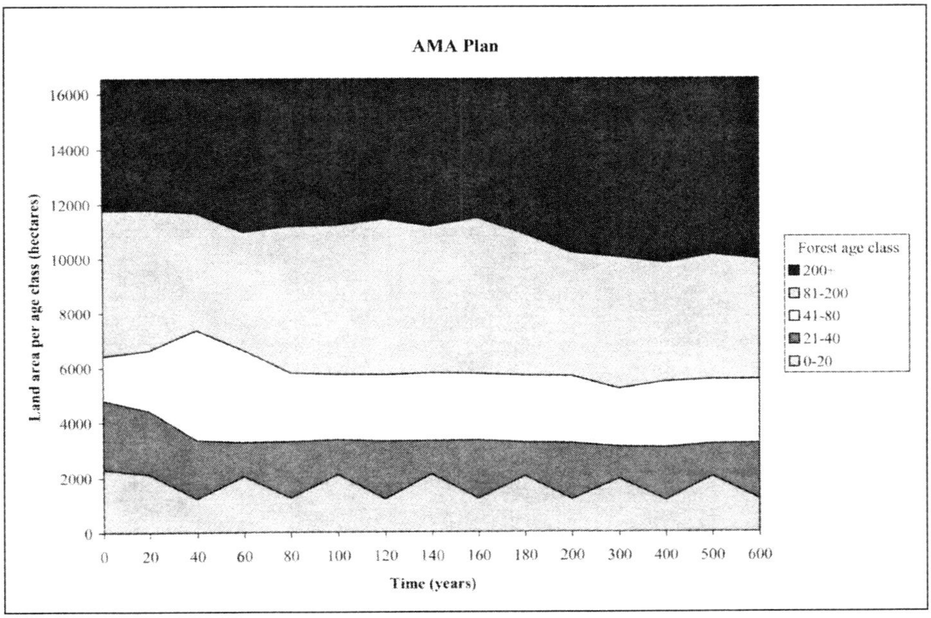

Figure 7. Distribution of forest vegetation age classes over the 600-year modeling timeframe based on the adaptive management area plan scenario (see Table 1 and text).

the AMA plan had more area in forest older than 80 years, larger patches, and more large, live trees in the young age classes. Somewhat surprising was the rapid equilibration of the AMA plan landscape despite the long rotations involved. This resulted, in part, because the current landscape, which involves a mixture of past clearcutting and wildfire-regenerated stands, is not very different in age-class distribution from that produced under this management scenario.

The *wildfire scenario* created the most complex distribution of age classes over time because of the variability in size and timing of fire disturbance (Figure 8). The youngest age classes (less than 40 years) can cover zero hectares after a 40-year period of no disturbance. On the other hand, a period of extensive fire can reduce the extent of old forest to less than 20% of the landscape. The mature age class appeared to be relatively stable across this simulation period, perhaps reflecting loss to fire and growth to the old age class balanced by growth of younger age classes. The wildfire scenario produced a modest increase in mean patch size and a substantial decrease in edge density, reaching a 45% reduction by year 100.

The *succession scenario* created a landscape that progressively lost the younger age classes (Figure 9) because no disturbances were planned and fire suppression was assumed to be completely effective. As stands aged, they entered older age classes and younger age classes gradually disappeared. The succession scenario sets an upper limit to mean patch size because patches grow together as edges become less distinct with the aging of adjacent stands. Mean patch size eventually reached 121 ha in this scenario, which in part represented the large number of small, persistent, nonforest patches scattered across the landscape. Edge density decreased accordingly.

2.4.2. Comparison of landscape composition and structure among scenarios—We compared the landscape-change scenarios in terms of distributions of age classes (Figures 2, 5 through 9), patch sizes (Figure 3), and edge density (Figure 4). A more complete comparison of the NWFP and AMA scenarios is provided in Cissel et al. (1999).

All management scenarios sustained classes up to age 40; the intensive plantation scenario maximized the area in these age classes. The 81 to 200 age class was nearly eliminated in the intensive plantation, 1990 forest plan, and NWFP scenarios, producing a landscape that was divided into old and young age classes with a broad gap in the middle age class. The ultimate extent of forest over 200 years was simply a matter of reserve area in each scenario, except in the AMA scenario where areas with a 260-year rotation contained some older, "managed" forest. The succession-only scenario was distinguished from the others by the absence of disturbance, which resulted in the younger age classes successively disappearing as stands aged and entered the next age class.

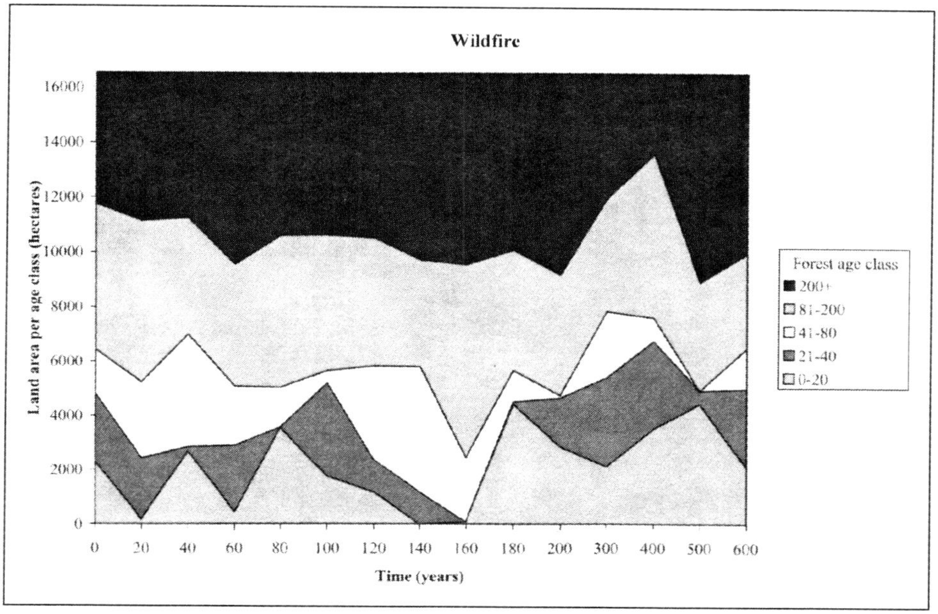

Figure 8. Distribution of forest vegetation age classes over the 600-year modeling timeframe based on the wildfire scenario (see Table 1 and text).

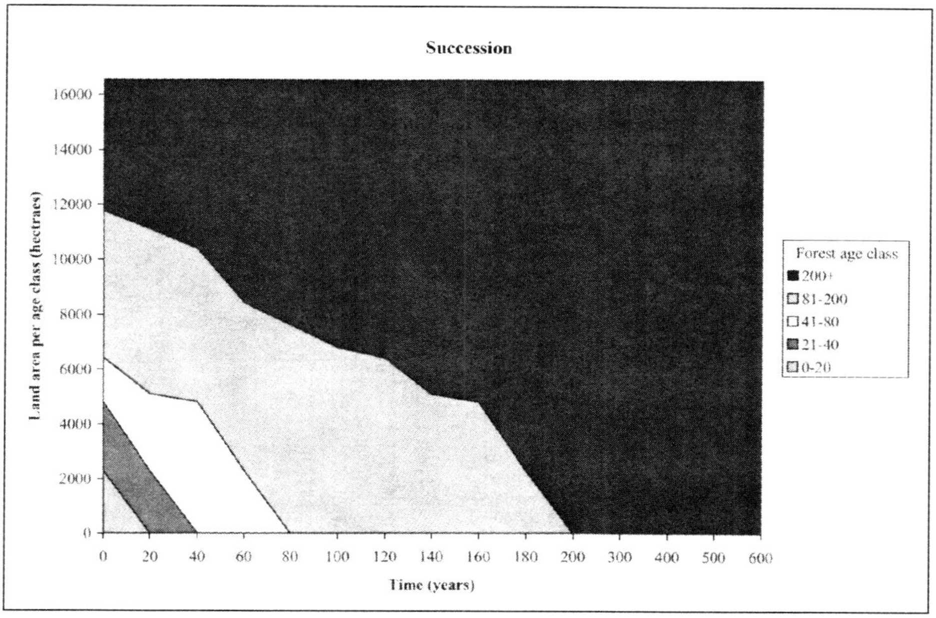

Figure 9. Distribution of forest vegetation age classes over the 600-year modeling timeframe based on the succession scenario (see Table 1 and text).

Table 2. Estimated properties and outputs of scenarios at year 100 for wood yield, extent of young and old forest habitat, annual water yield, and summer low stream flow. Carbon storage is estimated for the landscape once the scenario is fully implemented, based on modeling of Harmon and Marks (2002).

	Landscape-change scenarios					
Landscape properties	**Intensive plantation**	**1990 forest plan**	**NWFP**	**AMA**	**Wildfire**	**Succession**
Timber yield (10^6 m^3 y^{-1})	1.47	1.13	0.70	0.47	0	0
Carbon stored (Mg C ha^{-1})	272	447	686	769	581	830
Habitat (% of area):						
0-20 years	44	21	17	13	11	0
200+ years	11	16	27	32	36	59
Streamflow (proportion of old forest flow):						
Annual yield	1.28	1.13	1.10	1.07	1.09	1.0
Summer low flow	0.94	0.98	0.98	0.99	0.97	1.0

Note: NWFP = Northwest Forest Plan, AMA = adaptive management area, Mg C= megagrams of carbon.

The wildfire scenario resembled the management scenarios because it sustained a mixture of age classes over time. However, the irregular occurrence of fire, in contrast with regular cutting in management scenarios, produced greater temporal variability in the wildfire scenario. Wildfire was expected to produce a large range of patch sizes, high spatial variability, substantial retention of live trees (mean of 30% canopy cover) in disturbance patches, and abundant dead wood. The broad mature age class (81 to 200 years) showed less variability than the three youngest age classes, which spanned only 20 or 40 years each, simply as an artifact of length of the defined age class.

Mean patch sizes of the intensive plantation and AMA plan scenarios generally followed that of the wildfire scenario, exhibiting a slight upward trend after a small initial drop. These three scenarios eventually reached a mean patch size that was 32 to 40% of the mean patch size in the succession scenario. By design, the AMA plan followed the wildfire scenario with greater fidelity than the other scenarios (Figures 3, 4, 7, 8; Table 2). The fit was strongest for age-class distribution and less so for edge density and patch size because the large size of some wildfire patches was not considered socially feasible in a management scenario, even one as adventuresome as the AMA plan. The NWFP and 1990 forest plan scenarios followed a similar relatively level trend line. Mean patch size generally stayed within the range of 12 to 14% of that of the succession scenario over time for these scenarios, with the highest point (19%) reached after 600 years in the NWFP scenario. The timber harvesting in these two scenarios was done in smaller, widely spaces patches because of extensive riparian reserves and 80-year rotations. The net effect of the NWFP and 1990 forest plan scenarios was a highly fragmented landscape with high edge density, hence high vulnerability to windthrow and other edge-related phenomena.

2.4.3. Timber yield comparisons—Timber volume extracted, a traditional measure of wood production from a landscape, is one measure of economic value. The timber volume estimated for these scenarios reflects the vagaries of harvest scheduling models, including the representation of existing age classes in the model. Because harvest levels in this model were regulated to sustain the area harvested over time, volume fluctuations result. In particular, higher volumes resulted in the short term because many of the stands over 80 years carried greater volume than their replacements. This trend was most noticeable in the scenario that harvested on the shortest rotation (intensive plantation scenario), and was not significant for the AMA plan where very long rotations were used.

The volume produced by each scenario reflects the area available for timber harvest, the frequency of harvest, and the intensity of harvest (Table 1). Year 100 results are representative of future decades, and with one exception, reflect the relative differences among scenarios for the first century as well. Not surprisingly, the intensive plantation scenario produced the greatest annual amount (1.47×10^6 cubic meters in year 100), followed by the 1990 forest plan (1.13×10^6 cubic meters in year 100, or 77% of the intensive plantation scenario), the Northwest Forest Plan (0.70×10^6 cubic meters in year 100, or 48% of the intensive plantation scenario), and the AMA plan (0.47×10^6 cubic meters in year 100, or 32% of the intensive plantation scenario).

2.4.4. Carbon stores comparisons—Estimates of stored carbon for the scenarios were based on the final implementation of the landscape-change scenarios rather than carbon storage in landscape simulated at year 100 because the work of Harmon and Marks (2002) makes such an estimate possible and lag effects in stored carbon complicate the use of a year 100 analysis point. The estimated carbon stores on the landscape scenario ranged from 272 Mg C ha^{-1} in the intensive plantation scenario to 830 Mg C ha^{-1} in the succession landscape by the end of the simulations (Table 2). Both the extent of reserved area and the intensity of stand treatments in managed areas (rotation length and retention level) strongly affected carbon storage values for landscape scenarios as a whole. The reserve areas, which ranged from 16% of the intensive plantation scenario landscape to 42% of NWFP, were assumed to accumulate the maximum carbon storage for this landscape (830 Mg C ha^{-1}). The proportion of maximum storage sustained in managed parts of the simulated landscapes ranged from 20% for the high-utilization, 40-year rotation of the intensive plantation scenario to 90% for the high-retention, long-rotation AMA scenario. The wildfire scenario sustained lower carbon stores than either the NWFP or AMA scenarios as a result of the assumptions that no fire occurs in either reserve or managed parts of these scenarios, and that periodic wildfire suppresses carbon accumulation by burning it off and initiating young stands.

2.4.5. Biodiversity comparisons—The scenarios provided dramatically different extents of early (1 to 20 years) and late (80 plus years) seral habitat (Table 2). The intensive plantation scenario emphasized young stands, and the succession scenario was dominated by old forest. We looked to the wildfire scenario for an indication of the historical distribution of age classes and found early-seral stands over 11% and late seral stands over 36% of the landscape. These proportions are very similar to the AMA plan, but deviate by several hundred percent for young and old age classes of the intensive plantation scenario. The NWFP had about 50% more early and 25% less late seral forest than the wildfire scenario.

These management scenarios also differed in landscape pattern, retention levels of live trees in harvest areas, habitat distribution in time and space over the planning area, and regional context—all of which may strongly influence biological diversity. Landscape patterns in the NWFP scenario contained the unnatural aspect of extensive riparian reserves along all streams, which produced a landscape with high edge density and restricted size of individual opening. These conditions may favor some species but not others. Northern spotted owls, for example, may be more prone to predation by great horned owls *(Bubo virginianus)* in an edge-rich landscape, but riparian amphibian species that favor old-growth forest conditions may benefit by an extensive, old-forest riparian network. The AMA scenario was designed to provide patch size distributions and level of connectivity similar to the wildfire scenario. The wildlife scenario may provide the full spectrum of landscape conditions that native species have occupied over the past few millennia. Reserve systems of the NWFP and AMA scenarios differed with respect to riparian and aquatic systems. The AMA scenario had more diversity, which may benefit amphibian species. Landscape patterns of the other management scenarios involved less extensive reserve systems (unless we consider the succession-only scenario as a single, big reserve), which may prove detrimental to biodiversity, if we later learn that reserves are critical components of managed landscapes that sustain species.

Retention levels of live trees in harvest areas differed among some scenarios, with the AMA scenario providing the most diverse set of conditions. Field studies in this area reveal that bird communities shift in composition across a range of retention levels (Hansen et al. 1995), suggesting that a variety of stand structures and compositions may benefit from this type of habitat diversity. Other management scenarios provided a more restricted range of within-stand habitat complexity, which may limit habitat for some species.

Habitat distribution in time and space over the planning area varied substantially across the scenarios. Management scenarios underwent transitions from the current conditions to the simulated steady-state landscape patterns produced by the scenario. Some distinctive landscape structures emerged over the transition, such as the spatial bifurcation of the NWFP landscape with old forest near streams and stands less than 80 years in upslope areas. The wildfire and AMA scenarios, on the other hand, produced greater diversity of spatial patterns and forest age classes, with likely benefits to some species. In contrast to the modeled steady-state conditions of the management scenarios, wildfire produced irregular patterns in time and space. The complex role of wildfire in sustaining and also threatening native species in the landscape is worthy of further consideration.

The regional context of landscape analyses may influence the effects of a particular landscape on protection of biological diversity. Conservation strategies for some species, such as the northern spotted owl in the Douglas-fir region, involve broad-scale plans with networks of large reserves. The NWFP and AMA landscape-change scenarios in this study occur in matrix land nested among large reserves in the nearly 10 million ha, regional NWFP. Consequently, these scenarios may do a better job of protecting biodiversity in the region as a whole than the other management scenarios, which do not have those broader conservation elements.

2.4.6. Hydrology comparisons—The assessment of hydrologic effects of the different modeled scenarios is made in terms of annual flow and summer low-flow. We based this evaluation on extent of the 1 to 20 and 21 to 40 year age classes, which may exhibit increases in these flow parameters, based on long-term, experimental watershed studies in the nearby H.J. Andrews Experimental Forest. We estimated that annual flows would be approximately 10% higher under most management scenarios relative to the landscape fully forested in stands older than 40 years (the succession scenario) (Table 2). The one exception was the intensive plantation scenario, which was projected to have nearly 30% higher annual streamflow because of its extensive area of young forest. Annual water yield under the intensive plantation scenario was estimated as 17% higher than that of the wildfire scenario, and the other scenarios involving forest disturbance differ by only a small percent. Summer low-flow values were about 6% lower in the intensive plantation scenario than in the succession scenario and only 3% lower than in the wildfire scenario (Table 2). All other scenarios differed by only 1 to 2%.

As is the case with the biodiversity assessment, the effects of vegetation type and pattern on streamflow are poorly known. Recent studies in Watershed 1 on the H.J. Andrews Experimental Forest, for example, suggest that a vigorous riparian stand of red alder (*Alnus rubra* Bong.) can reduce summer low-flows more than a conifer-dominated riparian zone (Bond et al. 2002). The extent of alder-dominated riparian zones is, in part, a reflection of recent disturbance by management activities and flooding. Thus, riparian areas can have potential hydrologic effects disproportionate to their extent across a watershed, and they are sensitive to species shifts and disturbance. We have not attempted to account for these phenomena because their magnitude and geographic extent are unknown.

In most cases, the magnitude of effects on annual and summer low-flows were small relative to the succession scenario and even more similar to flows under the wildfire scenario.

2.4.7. Comparing scenarios: compatibility perspectives—Assessment of selected effects of the different landscape-change scenarios revealed tradeoffs between wood production and landscape properties that involve positive and negative, as well as weak and strong interactions. The amount of carbon stored in the landscape relates directly to the extent of older forest (80 plus years) with its associated high pools of live and dead carbon, and to rotation length and density of retained, live trees. For example, in the AMA scenario the long rotations produced a 68% lower timber yield, 2.8 times higher carbon stores, and nearly 3 times more old-forest habitat than the intensive plantation scenario. On the other hand, early seral habitat (1 to 20 years) is directly correlated with the rate of timber yield; it increases by a factor of 3.38 while timber yield increases by 3.13 in the intensive plantation scenario. The streamflow factors we examined were less sensitive to change in forest age-class structure, varying by less than 20% in the case of annual flow from small watersheds and less than a few percentage points for summer low-flow across the range of modeled management scenarios.

In narrow terms of compatibility between commodity and ecological objectives, wood that goes into the human use sector is not retained on the landscape where it can perform ecological functions. Currently, we cannot define with certainty specific thresholds of wood extraction that lead to unacceptable ecosystem consequences; current debate about species viability under different habitat conditions and management regulations is a prime example of attempts to identify such thresholds. Although we recognize this uncertainty, tradeoffs can be managed in several ways to maximize benefits. Modifications of stand-scale treatments can benefit ecological functions and carbon stores. For example, Harmon and Marks (2002) suggest there are substantial gains in carbon stores on the landscape when live and dead trees are retained within cutting units, and certain bird (Hansen et al. 1995) and lichen (Berryman 2002) species typical of older forest can be sustained in managed sites with retention of standing live trees.

Note that factors in addition to those considered in this analysis are at issue when considering tradeoffs between forest commodity extraction and ecological values of forests. Technical issues that we avoided in this modeling study include interactions of natural disturbance processes, such as wildfire and windstorms, with landscapes produced by the management plans; effects of roads on many ecosystem components, such as exotic plants and landslides; effects of forestry on floods; and habitat for threatened species. Social issues extend to public enjoyment of native landscapes through recreation and even simple perception of their existence. Operational landscape plans need to consider these interactions and integrate strategies with usable applications. Broader policy issues in regional forest management plans include shifts of wood production among regions, ownership, and nations with attendant social and ecological consequences.

Given the dynamic nature of landscapes in Oregon and Washington, maintaining future management options is an important consideration when comparing management approaches and taking a long-term perspective of compatibility. A detailed comparison of the AMA and NWFP plans led Cissel et al. (1999) to conclude that the more extensive mature (81 to 200 year) age class of the AMA plan may help maintain future options. For example, if shifts in policy dictate that more old forest should be grown and preserved, this can be accommodated more quickly from the mature age (81 to 200 year) age class than from the younger age classes that dominate the NWFP landscape. Furthermore, the mature age class provides many ecological functions characteristic of old forest, and is the source of future old forest.

3. Perspectives on Landscape Management

Reflections on this modeling work and past experiences working with public land management issues shape our perspectives on future land management. Over the years we have interacted with other scientists, managers, and policymakers on hundreds of field tours and workshops; attempted to implement the AMA plan (Cissel et al. 1999); observed the evolving relevance of science findings to forestry policy; and considered the history of change in forestry policy in the region and more broadly. These experiences lead us to make five general points concerning landscape management:

- Take the long view.
- Take a multiscale approach.
- Historical landscape conditions are a useful reference for considering compatibility.
- Landscape change is inevitable.
- Chart the future with multiple learning approaches, especially adaptive management.

These points are relevant to achieving compatibility among societal objectives for forest landscapes.

3.1. The Long View

Forest landscape management planning generally takes a 10- to 20-year view into the future, a small tick mark on the scale for forest development and ecological change. Landscape change reflects interactions among the slow processes of succession, climate change, cumulative effects of chronic disturbances, and legacies of past landscape structure, as well as abrupt changes due to biophysical or social disturbances. The pace of these phenomena influences outcomes. Distinctive attributes of old forest, for example, develop over centuries, and the legacy of dead wood from an old forest may persist for more than a century into the period of intensive forest management. Given the long time scales of forest stand and landscape change, taking the long view helps illuminate potential outcomes that would otherwise remain obscure.

Modeling future landscapes, is a valuable approach to understanding future landscape change. Examining the evolution of landscape structure over time can reveal unintended consequences of simple landscape-change rules. For example, elimination of the mature age class in the Northwest Forest Plan was not foreseen in the FEMAT analysis (FEMAT 1993). In another example, Franklin and Forman (1987) demonstrated some of the potential problems of dispersed patch clearcutting, partially represented in the 1990 forest plan scenario, through a simple, checkerboard conceptual model. Their analysis, which looked ahead several decades after the patch clearcutting system had been in use for only a fraction of the first rotation, was pivotal in framing the forest fragmentation issues of the late 20th century. We believe that similar modeling of future scenarios, such as we have begun here, is critical for understanding the ramifications of concepts across the landscape and across time. Given the slow pace of forest development through time and space, it is essential to model the future for more than one rotation. Critical challenges to carrying out long-term plans include change in private land ownership and political volatility in public lands policy. Regardless of context, the objective of the long view is to better understand concepts and their potential outcomes, and to communicate with others; these are essential ingredients in finding compatible forest uses.

3.2. A Multiscale Approach

Forest management is evolving from a focus on individual stands to a consideration of scales that range from the individual tree to landscapes, and their context at regional and even global scales (e.g., as influenced by commerce and certification) (Lindenmayer and Franklin 2002). This multiscale perspective is relevant to achieving compatibility because different management and ecological objectives must be addressed at appropriate scales. It is impossible to achieve compatibility across all objectives at a single scale, and many issues cannot be accommodated at only one scale. Protection of wide-ranging species, for example, is addressed at the scale of their range, but other species have smaller ranges. Wood supply issues operate in the global marketplace, yet local issues of wood supply are felt in timber-dependent communities. Efforts to efficiently link assessment and planning at multiple scales are still embryonic.

3.3. The Historical Landscape as Reference

Many short-term studies in ecology and hydrology use plots or small watersheds in stable, old-forest areas as controls or reference points. However, longer term and larger landscape views naturally lead to consideration of landscape dynamics. In disturbance-rich areas, such as Oregon and Washington, it is important to consider the role of disturbance regimes in sustaining native species and ecological processes (e.g., nitrogen fixation mediated by species present in particular successional stages). Many scientists and conservationists have argued for using historical disturbance regimes as a reference point to evaluate and chart future management in landscapes where the objective is to sustain native species and processes (Engstrom et al. 1999). As we discussed in the modeling exercise above, absence of disturbance in disturbance-prone landscapes can remove habitat conditions essential to some species.

Landscapes affected by historical disturbance regimes form useful, even essential, reference systems when considering compatibility among land use objectives. This is particularly true in the case of federal lands, which have high requirements for sustaining native species and some expectation for commercial wood production. Maintaining components of the natural disturbance regime is essential for maintaining some species; critical disturbance influences may include frequency, severity, and spatial pattern of disturbance as well as mechanisms, such as heat and sediment deposition. Whether for commodity extraction or ecological objectives, management actions can suppress, maintain, and augment native disturbance processes, or replace them with non-native processes. The consequences of any of these options should be considered in a compatibility assessment.

3.4. Inevitability of Landscape Change

Forest policy has experienced decades of relative stability punctuated by abrupt change, suggesting that surprising shifts in policy are an important component of landscape management. Forest Service management, for example, began with 50 years that emphasized stewardship while minimizing wood production. This was followed by rapid expansion of timber production in the post-World War II period until it was dramatically curtailed around 1990 as emphasis on species conservation began to dominate federal forest management. What does this policy volatility at the multi-decade time scale mean for landscape management? The planned and the realized future landscapes may differ substantially as a result of social and natural events. Policy shifts and natural processes, such as fire, wind, and insect outbreaks, create persistent and convulsive change in the drivers of landscape change and in the landscape itself. Land use legacies may be hard to erase with new policies because some landscape properties have substantial inertia (Wallin et al. 1994, Foster et al. 2003). Managed landscape patterns can form structures that are particularly vulnerable to disturbance by natural processes, such as windthrow at edges of cutting units (Gratkowski 1956, Sinton et al. 2000). Thus, natural processes commonly disrupt the planned course of landscape change.

In another example of the inevitability of landscape change, we note that each of the scenarios depicted here exists in Oregon and Washington, and each has experienced change or may do so in the near future: industrial forestry is pushed politically to reduce the intensity of management and be more spatially explicit in its application; the 1990 forest plan approach was declared illegal in litigation surrounding protection of the northern spotted owl; the Northwest Forest Plan, which revised the 1990 forest plan, has not approached the harvest levels projected; the AMA plan may require modification if future legislation sets an age limit on harvest; the historical wildfire regime has been thoroughly altered by fire suppression and other aspects of forest management; and the succession-only scenario will not be fully manifested on the landscape because of unsuppressible wildfire, windthrow, and other processes.

3.5. Charting the Future with Multiple Learning Approaches

Lessons about landscape change are best garnered by using multiple approaches: modeling, retrospective analysis of real landscapes, and sustained adaptive management in implementation of multiple management systems. Each of these approaches has strengths and weaknesses, but using them in concert builds a strong base for learning.

A critical approach to learning about landscape-change processes and consequences is to capitalize on the surprising diversity of landscape management systems operating within even restricted regions. Cohen et al. (2002) and Spies et al. (2002), for example, used remote sensing to examine pattern and rate of forest landscape change in western Oregon for mixed ownership and management systems. Most landscapes are hybrids of several approaches because of temporal or geographic factors. The current landscape of Blue River study area (Figure 1), for example, is a composite of wildfire and management similar to the 1990 forest plan scenario. Many landscapes, especially those covering larger areas, contain mixtures of land ownerships with associated landscape management approaches and resulting differences in pattern and function (Spies et al. 2002).

This landscape dynamism calls for us to accept our limited ability to predict the future and appreciate the value of taking an adaptive approach to management. Gunderson et al. (1995), for example, argue that failure to operate within adaptive, learning organizations may doom natural resource systems to operate in cycles punctuated by crisis. Formal adaptive management involves incremental learning through designed tests of alternative approaches to meeting management objectives, followed by monitoring and adaptation of future practices (Holling 1978, Walters and Holling 1990). This permits an incremental approach to seeking compatibility among societal expectations of forests.

Whereas explicit, formal adaptive management seems entirely appropriate for charting the future of landscape management systems, it is very difficult to carry out such programs. In Oregon and Washington, a system of ten Adaptive Management Areas designated by the NWFP met with limited success and ultimately lack of funding (Stankey et al. 2003). However, we persist in implementing the AMA landscape management scenario by using an adaptive management approach (Cissel et al. 1999) despite the challenges that have been raised over the years. Some land management agency (USDA Forest Service and Bureau of Land Management) biologists had difficulty appreciating the landscape perspective and chose to focus on achieving maximum species protection through stand-scale restrictions on harvest. Some agency leaders questioned exploring new options to the NWFP while it is in early stages of implementation despite the specific charge from the NWFP to test alternative approaches in AMAs (USDA and USDI 1994). Criticism of the AMA plan from outside the agency took many forms, including "this looks like a shallow excuse to cut more trees," and "this looks like another excuse to cut fewer trees." Others have viewed the testing of new ideas about landscape management with great interest and enthusiasm. The full spectrum of views is expressed in public discussions taking place in numerous field and workshop forums with the net effect of advancing collective understanding of the social and ecological dimensions of landscape management. A critical component of this process has been the close partnership of research and management, which is leading to new approaches to management with good credibility in both management and science arenas.

In summary, landscape science and management are at a critical juncture. Management and policy issues raise difficult challenges to which science can, at best, give only limited help. The new questions require greater integration across more disciplines than ever before, and new approaches to science beyond traditional, reductionist approaches (Holling 1995). Science needs to broaden its scales and degree of integration to address the challenge; however, science leads to no single, best, sustainable approach to maximizing compatibility among objectives, in part because the standards and objectives, both social and scientific, evolve over time. Ultimately, compatibility is a social choice.

4. Acknowledgments

This work has been supported in part by funding from the Wood Compatibility Initiative of the USDA Forest Service, Pacific Northwest Research Station, and Willamette National Forest. This is a contribution of the H.J. Andrews Experimental Forest and Long-Term Ecological Research program funded by the National Science Foundation. We thank M. Harmon, R. Haynes, A. Johnson, J. Jones, J. Mayo, R. Monserud, and D. Pennington for helpful data, ideas, discussions, and reviews. We thank Tere deSilva for GIS products.

5. References

Berryman, S.D. 2002. Epiphytic macrolichens in relation to forest management and topography in a western Oregon watershed. Corvallis, OR: Oregon State University. Ph.D. dissertation. 153 p.

Beschta, R.L., Pyles, M.R., Skaugset, A.E., and Surfleet, C.G. 2000. Peakflow responses to forest practices in the western Cascades of Oregon, USA. *Journal of Hydrology*, 233: 102-120.

Bond, B.J., Jones, J.A., Moore, G., Phillips, P., Post, D., and McDonnell, J.J. 2002. The zone of vegetation influence on baseflow revealed by diel patterns of streamflow and vegetation water use in a headwater basin. *Hydrological Processes*, 16: 1671-1677.

Chen, J., Franklin, J.F., and Spies, T.A. 1995. Growing-season microclimatic gradients from clearcut edges into old-growth Douglas-fir forests. *Ecological Applications*, 5(1): 74-86.

Cissel, J.H., Swanson, F.J., and Weisberg, P.J. 1999. Landscape management using historical fire regimes: Blue River, Oregon. *Ecological Applications*, 9(4): 1217-1231.

Cohen, W.B., Spies, T.A., Alig, R.J., Oetter, D.R., Maiersperger, T.K., and Fiorella, M. 2002. Characterizing 23 years (1972-95) of stand replacement disturbance in western Oregon forests with Landsat imagery. *Ecosystems*, 5: 122-137

Engstrom, R.T., Gilbert, S., Hunter, M.L., Jr., Merriwether, D., Nowacki, G.J., and Spencer, P. 1999. Practical applications of disturbance ecology to natural resource management. In: R.C. Szaro, N.C. Johnson, W.T. Sexton, and A.J. Malk (Eds.), Ecological stewardship: a common reference for ecosystem management (pp. 313-330). Oxford, UK: Elsevier Sciences Ltd.

ESSA Technologies LTD. 1999. TELSA: Tool for Exploratory Landscape Scenario Analyses, User's Guide Version 1.0. Vancouver, BC: ESSA Technologies Ltd. 164 p.

Forest Ecosystem Management Assessment Team [FEMAT]. 1993. Forest ecosystem management: an ecological, economic, and social assessment. Portland, OR: U.S. Department of Agriculture, Department of the Interior [and others]. [irregular pagination].

Foster, D., Swanson, F., Aber, J., Burke, I., Brokaw, N., Tilman, D., and Knapp, A. 2003. The importance of land-use legacies to ecology and conservation. *BioScience*, 53(1): 77-88.

Franklin, J.F., and Forman, R.T.T.F. 1987. Creating landscape patterns by cutting: ecological consequences and principles. *Landscape Ecology*, 1(1): 5-18.

Garman, S.L., Cissel, J.H., and Mayo, J.H. [In press]. Accelerating development of late-successional conditions in young managed Douglas-fir stands: a simulation study. General Technical Report PNW-GTR-557. Portland, OR: U.S. Department of Agriculture, Forest Service, Pacific Northwest Research Station.

Gratkowski, H.J. 1956. Windthrow around staggered settings in old-growth Douglas-fir. *Forest Science,* 2(1): 60-74.

Gunderson, L.H., Holling, C.S., and Light, S.S. 1995. Barriers and bridges to renewal of ecosystems and institutions. New York: Columbia University Press.

Halpern, C.B., and Spies, T.A. 1995. Plant species diversity in natural and managed forests of the Pacific Northwest. *Ecological Applications,* 5(4): 913-934.

Hansen, A.J., McComb, W.C., Vega, R., Raphael, M.G., and Hunter, M. 1995. Bird habitat relationships in natural and managed forests in the western Cascades of Oregon. *Ecological Applications,* 5(3): 555-569.

Hansen, A.J., Spies, T.A., Swanson, F.J., and Ohmann, J.L. 1991. Conserving biodiversity in managed forests: lessons from natural forests. *BioScience*, 41(6): 382-392.

Harmon, M.E., Franklin, J.F., Swanson, F.J., Sollins, P., Gregory, S.V., Lattin, J.D., Anderson, N.H., Cline, S.P., Aumen, N.G., Sedell, J.R., Lienkaemper, G.W., Cromack, K., Jr., and Cummins, K.W. 1986. Ecology of coarse woody debris in temperate ecosystems. In: A. MacFadyen, and E.D. Ford (Eds.), Advances in ecological research (Vol. 15, pp. 133-302). London: Academic Press.

Harmon, M.E., and Marks, B. 2002. Effects of silvicultural practices on carbon stores in Douglas-fir—western hemlock forests in the Pacific Northwest, USA: results from a simulation model. *Canadian Journal Forest Research*, 32: 863-877.

Harris, L.D. 1984. The fragmented forest. Chicago: University of Chicago Press. 211 p.

Holling, C.S. 1978. Adaptive environmental assessment and management. London: John Wiley and Sons.

Holling C.S. 1995. What barriers?chapterWhat bridges? In: L.H. Gunderson, C.S. Holling, and S.S. Light (Eds.), Barriers and bridges to the renewal of ecosystems and institutions, (pp. 3-34). New York: Columbia University Press.

Hunter, M.L., Jr. 2002. Fundamentals of conservation biology (2nd ed.). Abingdon, England: Blackwell Science.

Johnson, D.H., and O'Neil, T.A. 2001. Wildlife-habitat relationships in Oregon and Washington. Corvallis, OR: Oregon State University Press. 736 p.

Jones, J.A. 2000. Hydrologic processes and peak discharge response to forest removal, regrowth, and roads in 10 small experimental basins, western Cascades, Oregon. *Water Resources Research,* 36(9): 2621-2642.

Jones, J.A., and Grant, G.E. 1996. Peak flow responses to clear-cutting and roads in small and large basins, western Cascades, Oregon. *Water Resources Research*, 32(4): 959-974.

Lindenmayer, D.B., and Franklin, J.F. 2002. Conserving forest biodiversity: a comprehensive multiscaled approach. Washington, DC: Island Press. 351 p.

McGarigal, K., and Marks, B.J. 1995. FRAGSTATS: spatial pattern analysis program for quantifying landscape structure. General Technical Report PNW-GTR-351. Portland, OR: U.S. Department of Agriculture, Forest Service, Pacific Northwest Research Station. 122 p.

O'Neill, R.V., DeAngelis, D.L., Waide, J.B., and Allen, T.F.H. 1986. A hierarchical concept of ecosystems. Monographs in population biology 23. Princeton, NJ: Princeton University Press. 253 p.

Pennington, D.D. 2002. Structural and functional comparison of human-impacted and natural forest landscapes in the western Cascades of Oregon. Corvallis, OR: Oregon State University. Ph.D. dissertation. 263 p.

Post, D.A., and Jones, J.A. 2001. Hydrologic regimes of forested, mountainous, headwater basins in New Hampshire, North Carolina, Oregon, and Puerto Rico. *Advances in Water Resources*, 24: 1195-1210.

Sinton, D.S., Jones, J.A., Ohmann, J.L., and Swanson, F.J. 2000. Windthrow disturbance, forest composition, and structure in the Bull Run Basin, Oregon. *Ecology*, 81(9): 2539-2556.

Smithwick, E.A., Harmon, M.E., Remillard, S.M., and Acker, S.A. 2002. Potential upper bounds of carbon stores in the Pacific Northwest. *Ecological Applications*, 12(5): 1303-1317.

Spies, T.A., Reeves, G.H., Burnett, K.M., McComb, W.C., Johnson, K.N., Grant, G., Ohmann, J.L., Garman, S.L., and Bettinger, P. 2002. Assessing the ecological consequences of forest policies in a multi-ownership province in Oregon. In: J. Liu, and W.W. Taylor (Eds.), Integrating landscape ecology into natural resource management (pp. 179-207). New York: Cambridge University Press.

Stankey, G.H., Bormann, B.T., Ryan, C., Shindler, B., Sturtevant, V., Clark, R.N., and Philpot, C. 2003. Adaptive management and the Northwest Forest Plan. *Journal of Forestry*, 101(1): 40-51.

Szaro, R.C., Johnson, N.C., Sexton, W.T., and Malk, A.J. (Eds.). 1999. Ecological stewardship: a common reference for ecosystem management. Oxford, UK: Elsevier Sciences Ltd.

Thomas, R.B., and Megahan, W.F. 1998. Peak flow response to clear-cutting and roads in small and large basins, western Cascades, Oregon: a second opinion. *Water Resources Research*, 34(12): 3393-3403.

U.S. Department of Agriculture, Forest Service; U.S. Department of the Interior, Bureau of Land Management. 1994. Record of decision for amendments to Forest Service and Bureau of Land Management planning documents within the range of the northern spotted owl. [Place of publication unknown]. 74 p. [plus Attachment A: standards and guidelines].

Wallin, D.O., Swanson, F.J., and Marks, B. 1994. Landscape pattern response to changes in pattern generation rules: land-use legacies in forestry. *Ecological Applications*, 4(3): 569-580.

Walters, C.J., and Holling, C.S. 1990. Large-scale management experiments and learning by doing. *Ecology*, 71(6): 2060-2068.

Weisberg, P.J. 1998. Fire history, fire regimes, and development of forest structure in the central western Oregon Cascade. Corvallis, OR: Oregon State University. 256 p. Ph.D. dissertation.

Willamette National Forest. 1990. Land and resource management plan. Eugene, OR: U.S. Department of Agriculture, Forest Service, Willamette National Forest.

Chapter 10

Contemporary Management Regimes in the Pacific Northwest: Balancing Biophysical and Economic Concerns

Richard W. Haynes
Pacific Northwest Research Station, 620 SW Main St., Suite 400, Portland, OR 97205, USA
Darius M. Adams
Department of Forest Resources, College of Forestry, Oregon State University, Corvallis, OR 97331, USA
John R. Mills
Pacific Northwest Research Station, 620 SW Main St., Suite 400, Portland, OR 97205, USA

1. Introduction

For almost a century timberland in Oregon and Washington (especially in western Oregon and Washington) has been managed with increasing intensity to support a forest products sector that has grown in both size and complexity. Various forest management regimes evolved as landowner objectives, regulatory conditions, and the markets for (and utilization of) harvested timber changed and silvicultural knowledge (including growth and yield information) advanced. The region is characterized by diverse ownerships (Table 1) that have led to a patchwork mosaic of management regimes. Unlike other U.S. regions, however, these ownerships tend to consist of large and relatively contiguous blocks of timberland, leading to an interest in landscape scale management approaches within individual ownerships. This chapter discusses the evolving mix of management regimes, the conditions that influence their adoption, expected trends in forest management, and the effects of these trends on timber harvest potential.

R.A. Monserud, R.W. Haynes and A.C. Johnson (eds.)
Compatible Forest Management, 267-296

Table 1. Area of forestland and timberland in Oregon and Washington, by ownership and region, 1953-97, with projections to 2050.

	Historical					Projections				
Ownership and region	**1953**	**1963**	**1997**	**1987**	**1997**	**2010**	**2020**	**2030**	**2040**	**2050**
	Million hectares									
Timberland:										
Oregon and Washington										
Public	9.3	9.5	9.1	8.9	9.6	9.6	9.6	9.6	9.6	9.6
Forest industry	3.6	3.8	4.0	3.9	3.7	3.6	3.6	3.6	3.6	3.6
NIPF	5.2	4.7	4.0	3.5	3.4	3.3	3.2	3.2	3.2	3.2
Total	18.2	18.0	17.0	16.3	16.7	16.5	16.4	16.3	16.3	16.3
United States										
Public	58.9	59.2	56.0	53.6	59.1	59.1	59.1	59.1	59.1	59.1
Forest industry	23.9	24.9	27.9	28.5	27.1	26.9	26.8	26.6	26.4	26.3
NIPF	123.2	124.5	115.5	114.8	117.7	117.2	116.2	115.1	113.8	112.6
Total	206.0	208.5	199.4	196.9	204.0	203.3	202.1	200.8	199.3	198.0

Note: Data may not add to totals because of rounding.
Source: Haynes 2003.

1.1. Changing Forest Resource Conditions

The states of Oregon and Washington are a unique and highly productive forest region. Encompassing less than 10% of U.S. timberland (Smith et al. 2001 Table 11 and current document Table 1), the region has produced 20 to 30% (Haynes 2003 Table 13) of annual U.S. softwood harvest since the early 1950s. Following World War II, these states became the focus of relatively intensive forest management efforts as they transitioned from an old-growth to a young-growth oriented forest sector.

Basic inventory statistics are shown in Table 2. In general, the historical reductions in total inventory reflect the transition from old to young growth (Appendix 1, Chapter 1), land lost to nonforest uses, and the reclassification of public timberlands, which made some timberlands unavailable for harvest. On private lands, softwood inventory fell by nearly 34% between the early 1950s and late 1990s. This decline resulted from harvest of old growth (mostly on industry land) and conversion of timber stands to other uses (primarily on nonindustrial private (NIPF) ownerships). This transition was accompanied by a doubling in net annual growth and a shift in the age-class structure of the forest, as shown in Figure 1. Two trends are evident: timber has become concentrated in the younger age classes (the proportion of the private inventory over 160 years dropped from 15 to less than 1% over the 50-year period), and the ownership of various age classes have become more concentrated. For example, a higher proportion of the younger age classes are on forest industry lands.

Table 2. Softwood removals, harvest, growth, and inventory for the Douglas-fir region and United States timberlands,[1] 1952-97, with projections to 2050.

Item and owner	Historical							Projections				
	1952	1962	1970	1976	1986	1991	1997	2010	2020	2030	2040	2050
						Million cubic meters						
Douglas-fir region												
National Forest:												
Removals	10	16	15	15	15	8	2	1	2	2	2	3
Harvest	10	17	14	14	19	8	2	1	2	2	2	3
Net annual growth	5	6	7	6	9	9	22	24	22	20	19	18
Inventory	1348	1351	1288	1249	952	952	1456	1754	1967	2161	2334	2493
Other public:												
Removals	4	8	10	12	12	6	5	5	5	5	5	5
Harvest	4	8	10	12	12	7	5	5	5	5	5	5
Net annual growth	5	9	10	11	14	11	14	14	15	16	17	18
Inventory	569	561	556	543	555	459	545	662	769	885	1009	1139
Forest industry:												
Removals	33	26	36	37	35	26	19	28	28	26	27	29
Harvest	35	28	35	36	35	31	18	28	29	27	27	29
Net annual growth	10	11	13	17	26	23	23	27	27	28	30	31
Inventory	927	776	673	623	570	508	500	487	482	487	554	591
Nonindustrial private:												
Removals	9	6	7	6	6	11	9	9	10	12	14	15
Harvest	9	6	7	6	7	12	8	9	10	12	14	15
Net annual growth	8	9	10	10	12	12	11	11	12	14	15	13
Inventory	269	270	292	240	288	316	293	285	308	346	368	357

Table 2. Softwood removals, harvest, growth, and inventory for the Douglas-fir region and United States timberlands,[1] 1952-97, with projections to 2050. (continued)

	Historical							Projections				
Item and owner	1952	1962	1970	1976	1986	1991	1997	2010	2020	2030	2040	2050
						Million cubic meters						
Total												
Douglas-fir region												
Removals	56	55	69	70	67	50	34	43	45	45	48	51
Harvest	59	58	65	68	72	59	33	44	46	46	48	52
Net annual growth	28	34	40	44	61	55	70	77	79	81	84	82
Inventory	3113	2958	2809	2654	2365	2235	2794	3213	3572	3948	4357	4694
United States												
Removals	220	215	260	283	322	308	287	273	305	333	355	379
Harvest	213	207	246	269	320	311	286	279	312	340	363	387
Net annual growth	219	272	321	354	355	339	384	446	460	469	474	464
Inventory	12 232	12 741	13 031	13 228	12 830	12 745	13 707	15 552	17 233	18 724	20 039	21 022

[1] Only timberlands available for management are included in data.
Note: Douglas-fir region includes western Oregon and Washington.
Source: Haynes 2003.

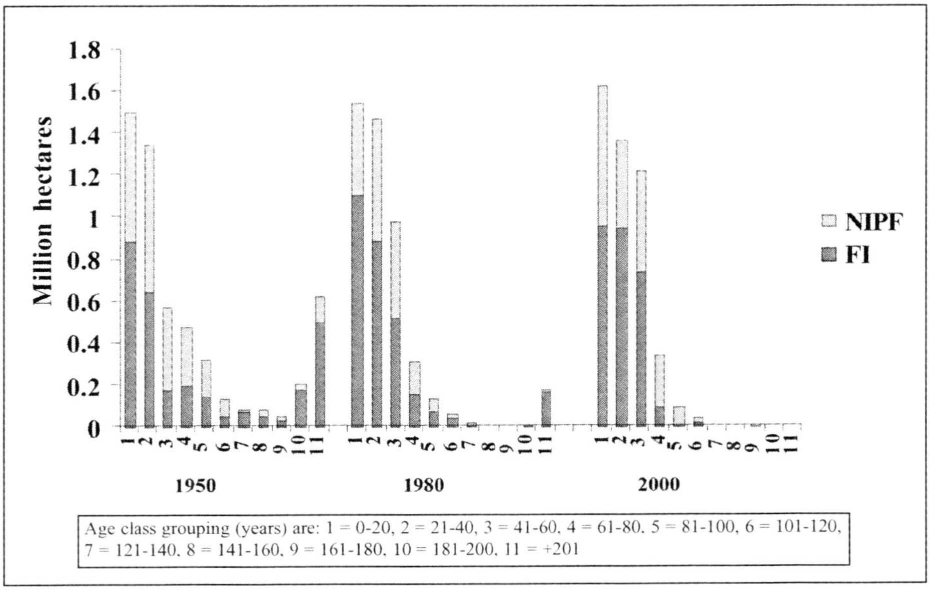

Figure 1. Private inventory by age class for the Douglas-fir region (western Oregon and Washington), 1950, 1980, and 2000. NIPF = nonindustrial private forest, FI = forest industry.

2. Characterization of Current Management Regimes

Over the past 100 years there has been continual speculation among forestry professionals about the "most appropriate" management regimes for lands in Oregon and Washington, though much of the debate has focused on individual practices (see Hummel 2003, Monserud 2003 for review). Despite the debate, there was a tentative consensus about basic practices that comprise the core of management regimes with certain objectives. For example, consensus existed for the conversion of older forests on public lands to managed forests, and for development of regulations addressing forest management practices on private timberlands. During the 1990s, however, broad commonalities in the modes of management practiced in the region began to disappear. Three major forces were at work. Timber production became a lower management priority for public agencies, whereas stand manipulation for nontimber purposes was elevated in importance. A wave of corporate mergers and acquisitions concentrated the ownership of industrial land in a few firms. At the same time, the role and importance of timberlands in the asset portfolios of forest products firms underwent major revisions, and large blocks of timberland were sold to investment companies that did not own processing facilities. Finally, the numbers of NIPF landowners began to rise. This led to increased parcelization of holdings, and more variation in the types of owners and their cultural and economic links to the forest.

Thus, to understand the current and prospective structure of forest management in the region, it is critical to recognize that landowners are becoming more diverse and often have diverse objectives. Contemporary forest management is about expanding owner choices. Management regimes are becoming more flexible both in their application and their outcomes.

2.1. The Evolving Mix of Stand-Level Management Regimes on Private Lands

It was not until the early 1960s that a systematic long-term appraisal of the prospects for development of forest industries and forest resources in Oregon and Washington was undertaken. Timber Trends in western Oregon and western Washington (USDA FS 1963) was a unique study because it analyzed the economic factors expected to influence timber output. It argued that timber output was a function of land use, expected prices, management practices and the rates of time preference of the various owners (their effective discount rates). Landowners with a high discount rate would prefer to realize the value of forests sooner and adopt shorter rotations than those with lower discount rates.

The Timber Trends study used three different management intensities to represent the set of management practices adopted on various portions of the land base. The first intensity, called "exploitive," represented management adopted by owners who operated as if they had a high interest rate and preferred little active timber management. The second management intensity, "intermediate," assumed that landowners operated as if their interest rate was 6% (in nominal terms). Rotation lengths with this intensity were assumed to be shorter (60 years) than with more intensive management. The third management intensity, "conservative," assumed landowners operated with a lower interest rate, favoring longer rotations and more intermediate treatments. The exploitive management intensity involved frequent harvest and a low-cost regeneration scheme. The intermediate intensity was a plant, thin, and harvest regime, differing (based on site and owner) in regeneration methods and the number of thinnings. The conservative regime included site preparation, planting (PL), fertilization, precommercial thin (PCT), commercial thinning (CT) and final harvests. All final harvests were assumed to be clearcuts. There were differences in the number of thinnings based on site and owner. As private forests in Oregon and Washington transitioned from old growth to second growth, it was estimated that 16, 22, and 62% of the private land would ultimately be managed in each of three intensity classes, respectively.

In the years since the Timber Trends analysis, various studies (WDNR 1975; Beuter and others 1976; Sessions 1991; Adams et al. 1992, 2002; Haynes 2003) have examined the timber supply prospects for private forestlands in the Douglas-fir region (western Oregon and Washington). In each case, following the lead of the Timber Trends study, efforts were made to identify the range of management practices applied to the current private forest inventory. Information on industrial management practices was obtained, in some instances, by surveying landowners. For NIPF owners, and in some cases for industrial owners as well, estimates were developed by experts. Six studies were done and provide us a view of current management practices (or management intentions) at three time points: the early 1970s, the late 1980s and the late 1990s. A summary of these results for six categories of management regimes is shown in Table 3. Detail for the full 23 categories found in the original reports is shown in Appendix 1.

Table 3 shows how industrial owners have adopted practices to ensure regeneration, and often some form of commercial thinning. The proportion of thinning has varied over time, with some of the more recent surveys showing a decline. This variation may reflect conditions in forest products markets at the times of the surveys. Practices other than commercial thinning also appear to have gained in importance; these include precommercial thinning, fertilization, partial cutting regimes, and use of genetically improved planting stock. In this case, genetic improvement involves selection of seed source or use of seed orchards derived from phenotypically superior parents, not the use of genetically modified organisms.

It is important to note that use of genetically improved stock has been employed so long and so widely in Oregon and Washington, that it may be impossible to distinguish between improved and unimproved planting stock in some cases. Fewer forestland owners are inclined to pursue highly complex systems such as the "conservative" regime (plant/PCT/fertilize/CT), which the Timber Trends Study projected to involve more than 60% of the private land base. These systems appear to have been replaced by simpler one or two treatment regimes (for example, plant/PCT or plant/PCT/fertilize). In practice, the regime actually applied by an owner can vary simply by postponing or abandoning a specific treatment. Although the owner may intend to practice a plant/PCT/CT regime, the planting may be the only part of the regime that is actually carried out.

Table 3 shows that management on NIPF lands appears largely restricted to securing regeneration with limited use of other treatments. The earlier western Oregon studies make it clear that NIPF lands were in relatively poor management condition (low conifer restocking after harvest) at the time of the surveys (compare Beuter et al. 1976 and Sessions 1991). More recent studies (Adams et al. 2002) rely on plot-level data and individual tree models that reflect actual stocking conditions. They also show a large fraction of the land base in some form of partial cutting or selection management. This data does not reveal if NIPF owners have a long-term objective of developing a multi-age selection system or if they are simply engaging in a sequence of heavy thinnings to postpone clearcutting.

Table 3. Summary of private land base allocation to management regimes from six studies of the Douglas-fir region.[1]

	Forest industry owners					
Region and period	**WOR[2] EARLY 1970s**	**WWA[3] EARLY 1970s**	**WOR LATE 1980s**	**WWA LATE 1980s**	**WOR LATE 1990s**	**PNWW[4] LATE 1990s**
Study	**Beuter et al. (1976)**	**WDNR (1975)**	**Sessions (1991)**	**Adams et al. (1992)**	**Adams et al. (2002)**	**Haynes (2003)**
Regime	*Percentage of land base*					
No genetic improvements regenerate only	43	10	28	25	51	17
No genetic improvements regenerate + commercial thinning + other[5]	41	80	50	21	25	17
No genetic improvements regenerate + other (non-commercial thinning)	0	10	13	47	14	12
Partial cut	0	0	0	0	6	2
Genetic regeneration + other	0	0	0	3	0	47
Conversion / rehab[6]	16	0	9	0	0	0
Reserved	0	0	0	4	3	6
Total	100	100	100	100	100	100
Harvest (H) or minimum harvest (MH) age, assumed (A) or established from survey (S)		$50^{H,A}$	$25^{M,A}$	$45\text{-}60^{M,A}$	$50\text{-}57^{H,S}$	$43\text{-}55^{M,S}$

Table 3. Summary of private land base allocation to management regimes from six studies of the Douglas-fir region.[1] (continued)

	Nonindustrial private forest owners					
Region and period	**WOR[2] EARLY 1970s**	**WWA[3] EARLY 1970s**	**WOR LATE 1980s**	**WWA LATE 1980s**	**WOR LATE 1990s**	**PNWW[4] LATE 1990s**
Study	**Beuter et al. (1976)**	**WDNR (1975)**	**Sessions (1991)**	**Adams et al. (1992)**	**Adams et al. (2002)**	**Haynes (2003)**
Regime	*Percentage of land base*					
No genetic improvements regenerate only	62	95	NA	98	61	83
No genetic improvements regenerate + commercial thinning + other[5]	3	5	NA	0	1	0
No genetic improvements regenerate + other (non-commercial thinning)	0	0	NA	1	3	7
Partial cut	0	0	NA	0	33	0
Genetic regeneration + other	0	0	NA	1	0	10
Conversion / rehab[6]	35	0	NA	0	0	0
Reserved	0	0	NA	0	3	0
Total	100	100	100	100	100	100
Harvest (H) or minimum harvest (MH) age, assumed (A) or established from survey (S)		50[H,A]	45-65[M,A]	45-60[M,A]	43-49[M,S]	40-60[M,A]

NA = not available.
[1] Douglas-fir region includes western Oregon and Washington.
[2] WOR = western Oregon.
[3] WWA = western Washington.
[4] PNWW = western Oregon and Washington.
[5] Other includes precommerical thinning and fertilization.
[6] Conversion or rehabilitation is replacing an existing understocked or undesired stand with a stand better matched to site, or of higher commercial value.

The columns in Appendix 1 shows the full range of management regimes examined in past timber supply studies and their management intentions surveys. In general, surveys suggest that the regimes to be implemented in Douglas-fir (*Pseudotsuga menziesii* (Mirb.) Franco) and western hemlock (*Tsuga heterophylla* (Raf.) Sarg) forest types cover a wider range of options than those for other coniferous types or hardwoods. Except on the most productive sites, management for western hemlock, other coniferous types, or hardwoods is usually restricted to planting or planting and early stocking control (PCT).

Rotation age or age of harvest is one of the most scrutinized elements of the timber management regime. Harvest age establishes the rate of return received by the owner on all forest management investments (regeneration thinning, and taxes and administrative fees). Rotation age also establishes the maximum volume attained by the stand, the characteristics of wildlife habitat afforded by the stand over its life, and the time patterns of atmospheric carbon uptake and release. Yet as important as rotation is in both management decisions and in assessing the broader social and environmental impacts of management, relatively little is known (or objectively measured) about actual harvest age behavior on private lands in Oregon and Washington. There are three types of data that can be used to examine the rotation decision:

- Rotation age intentions derived from management surveys and timber supply studies—these indicate what owners say they are going to do.
- Optimal economic rotations derived from yield, price and cost data for sample stands—these indicate what owners should do if their sole objective were to maximize their investment returns.
- Measurements of the ages of harvested timber from periodic forest surveys—these indicate what owners have actually done in managing their timber.

As shown in Table 3, five of the six timber supply studies that described management regimes also provide information on anticipated or intended rotation ages. In most cases these are "minimum harvest ages," the youngest age at which owners would consider harvesting a stand (not average anticipated harvest ages), and they commonly differ by site, private owner, and management regime. For the most part, available estimates were developed from expert opinion rather than surveys of actual owners or their agents. Nonetheless, they provide a starting point for considering harvest-age behavior. With the exception of the Sessions (1991) study, the range of ages is similar for both industrial and nonindustrial owners, 40 to 65 years depending on site and regime.

An alternative, and more prescriptive, view of rotation age can be derived by considering returns on investments in tree planting and timber growing. We examine a simplified case of a landowner whose sole objective for land holding

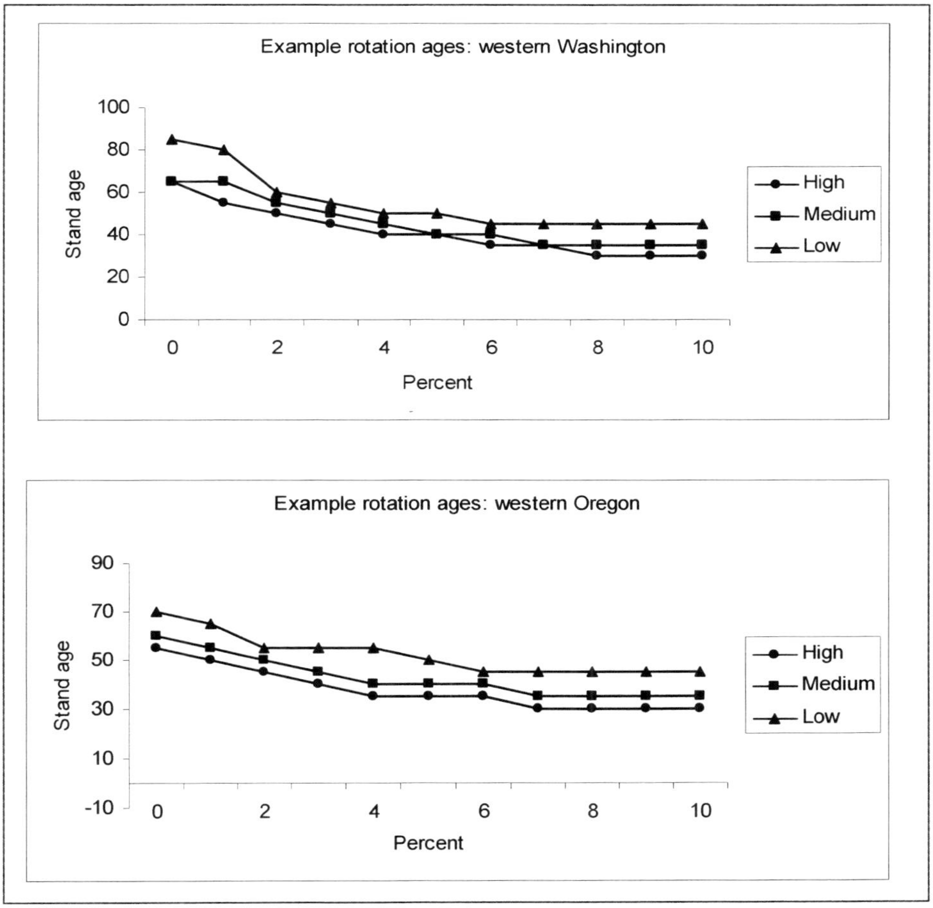

Figure 2. Relation of rotation age and interest rate for sample stands in western Oregon and western Washington by site group (low, medium, high).

is to maximize wealth. For a single timber stand (ignoring any possible interaction with other stands in the same ownership), meeting this objective would involve setting the harvest age to maximize the present value of all future net returns from the land. This value is commonly called the "soil expectation" or "land expectation" value, or simply the "bare land" value. Because the value considers all future net returns, the time horizon of the discounting is infinite. In a mathematical sense, the present value of *all* future net returns (the soil expectation value or SEV) from a simple plant and harvest management regime can be expressed as:

$$\text{SEV} = \frac{\text{PRICE*VOLUME AT ROTATION - PLANTING COST}}{(1+\text{INTEREST RATE})^{\text{ROTATION}} - 1} - \text{PLANTING COST}$$

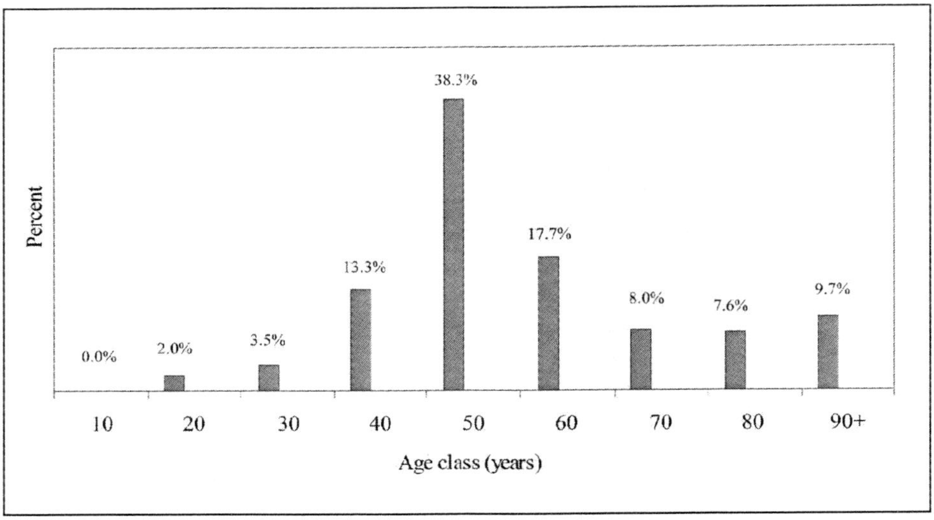

Figure 3. Estimate of the distribution of area harvested by age class on private lands in western Oregon between 1984-86 and 1994.

To illustrate the implications of this approach for rotation age, Figure 2 shows the soil expectation value maximizing rotations for a simple plant and harvest management regime by using yields generated for private lands in western Washington in Adams et al. (1992) and for the western Oregon Cascades by Adams et al. (2002). Since computation of the land value involves discounting, the interest rate plays a key role in determining the best rotation in this criterion. As a result, the rotations in Figure 2 are shown as functions of the discount rate. Higher discount rates mean shorter rotations, so the plots slope downward and to the right in Figure 2. In the customary 4 to 8% range of discount rates, rotations vary from 30 to 55 years across sites in western Oregon and 30 to 50 years in western Washington. The upper bounds on these ranges are similar to the intentions data in Table 3, but the lower bounds are fully a decade shorter. This difference could reflect some bias in the management intentions data (although consider the 25-year minimum age in the Sessions (1991) study). It could also result from simplifications and errors in the land value computations, including failure to (1) adequately capture the rising value of timber as stands age—the longer this increase persists over the life of a stand, the longer the optimal rotations, (2) recognize other returns or costs, and (3) recognize owner adjustments in discount rates to reflect risk.

Finally, information is limited about the stand age at which owners chose to harvest. Lettman and Campbell (1997) examined the areas and volumes harvested by stand age and other attributes in western Oregon between the 1984-86 inventory of private lands and the 1994 update survey. By using their data (from their Figure 12), we estimate the harvest area distribution as shown

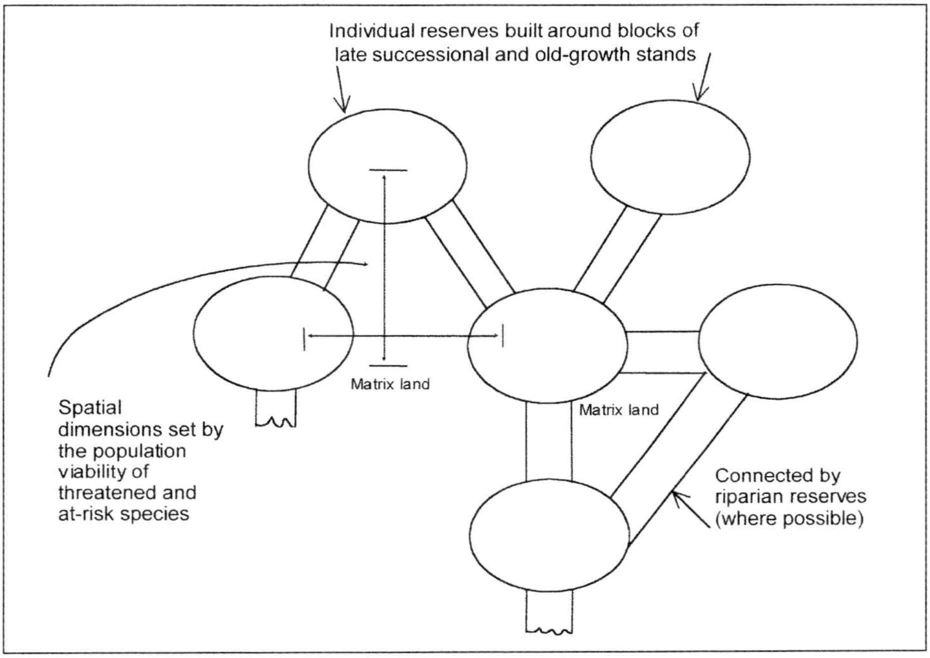

Figure 4. A connected reserve system.

in Figure 3. The area-weighted average harvest age is 48 years from these categorical data, and the modal age (at 38%) is clearly in the 50-year class. The range is from 20 to more than 90 years, but more than two-thirds of the cut comes from the 40- to 60-year groups.

2.2. Federal Land Management in Oregon and Washington

Federal land management in Oregon and Washington since 1993 has been guided by a comprehensive long-term policy (USDA and USDI 1994) for managing habitat for the northern spotted owl (*Strix occidentalis caurina*). This strategy, known as the Northwest Forest Plan (NWFP), was developed from a report prepared by the Forest Ecosystem Management Assessment Team (FEMAT 1993). The NWFP established a connected reserve system (Figure 4) with terrestrial and aquatic components (Haynes and Perez 2001). The federal land base was allocated among late-successional and riparian reserves, matrix lands (all federal lands outside of reserves, withdrawn areas, and managed late-successional areas), and adaptive management areas (AMAs). The AMAs were included because the plan was intended to be an evolutionary strategy that used adaptive management to test and modify assumptions, create opportunities for organizational innovation, and test collaborative approaches.

The land management strategy embedded in the NWFP is based on many of the components common in ecosystem management approaches. The connected reserve system is expected to maintain well-distributed habitat on federal lands for the threatened marbled murrelets (*Brachyramphus marmoratus*) and northern spotted owls. It is likely to reverse habitat degradation for at-risk fish species or stocks. It considered the relations between plant and animal species thought to be closely associated with late-sucessional forests. The design of the connected reserve system also considered its likelihood of long-term persistence. The management regimes that were applied in both the matrix and late successional forests were modified version of those shown in Table 3 but involved extended rotation lengths and partial harvests. This strategy is a departure from the past management expectations. For example, in the 1960s it was assumed that 100% of national forest timberland, after some withdrawal for recreational use, would be managed extensively (USDA FS 1963).

The implementation of the NWFP attempted to balance the economic, environmental, and social challenges facing a broad region. Socioeconomic effects were estimated for different land management strategies and were the basis for extensive public debates (Pipken 1998).

2.3. Impacts of Changing Forest Products Markets

Markets (for both stumpage and finished forest products) have directed the concurrent flows of forest resources to their various uses, and of goods and services to their ultimate consumers. Evolution in these markets has come about as a function of changes in the resource base and changes in the array of manufactured products. For example, western hemlock was not extensively used in lumber manufacturing before the mid 1930s, and it did not become an established export species until the 1970s.

For land managers, understanding change in the demand for various forest products, and assessments of future markets are necessary parts of stewardship considerations. This section summarizes historical and projected trends for both the product and stumpage markets. Much of the discussion is product rather than species specific. Finally, we discuss several implications for future markets that may impact future timber management regimes in the region.

2.3.1. Past market developments**—**The forest products industry in Oregon and Washington developed to take advantage of the abundant Douglas-fir timber. A softwood lumber industry developed quickly in the late 1800s with the advent of mechanical processing and the development of rail transportation systems. These advances lowered processing and delivery costs, allowing forest products from Oregon and Washington to be competitive in numerous world and domestic markets. The industry evolved to include both lumber and plywood mills, and later a pulp and paper industry. From the 1930s through the

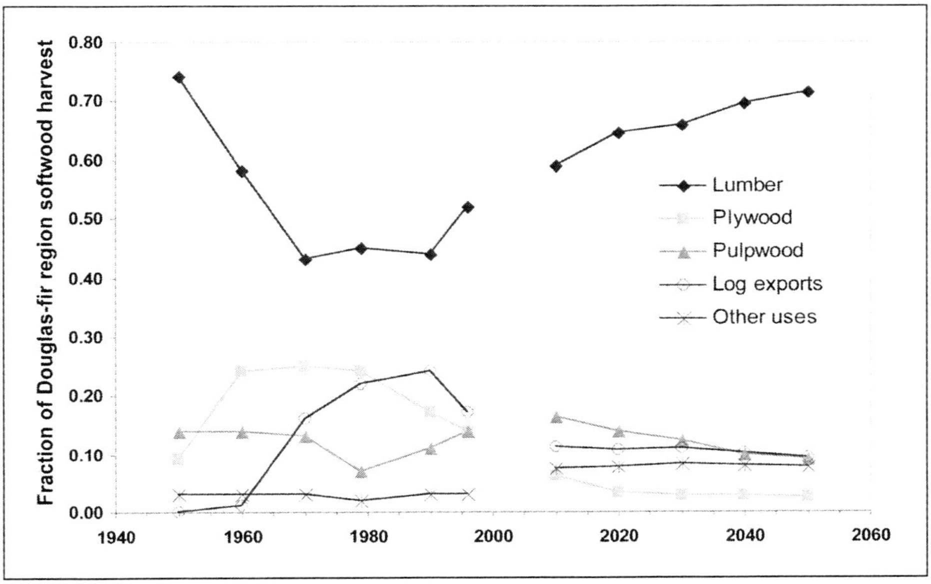

Figure 5. Proportions of the Douglas-fir region (western Oregon and Washington) softwood harvest by product category: history and projections from 2000 RPA Timber Assessment (Haynes 2003).

early 1970s, roughly 40% of the softwood lumber and 65% of the plywood consumed in the United States was produced in these two states. Since then, their share of the market has diminished as lumber and plywood (after 1964) production in the southern United States has increased along with increased lumber imports from Canada, and harvest on public land declined.

Figure 5 illustrates substantial changes in the mix of products produced in the Douglas-fir region during the past five decades. For example, the log export industry started in the early 1960s, peaked in 1989, and now is in decline. The softwood plywood industry, initially based on the high-quality (knot free) large logs available in the region, peaked in the 1960s. With the advent of plywood production in the South (based on smaller and cheaper logs), the introduction of engineered structural panels in the late 1970s, and the decline in peeler log supply from public lands, this industry is also in decline. These products and the softwood lumber industry emphasized relatively high-grade lumber (larger proportions of select, structural and heavy framing grades). This emphasis on high-grade lumber led to price premiums that were reflected in higher stumpage prices. These higher stumpage prices provided incentives for more intensive management of timberlands. For example, the rise of log exports and higher prices for export logs provided landowners with an incentive for growing larger, higher quality logs. But the gradual collapse of the log export market is now changing landowner expectations of economic

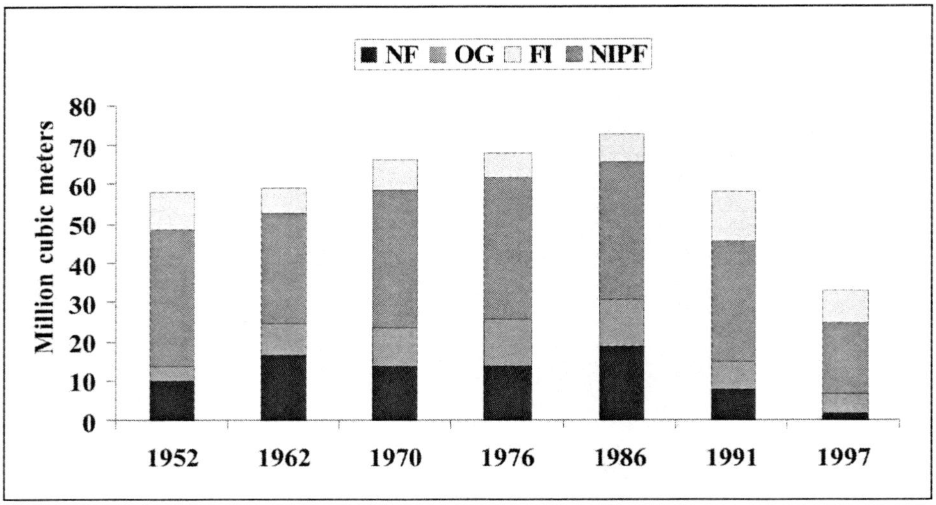

Figure 6. Softwood harvest in the Douglas-fir region (western Oregon and Washington), by owner, 1952-97. NF = national forest, OG = other government, FI = forest industry, NIPF = nonindustrial private forest.

returns from forest management and will eventually change forest management regimes.

From the late 1940s until the late 1980s, timber harvest in the Douglas-fir region increased roughly 25%. As shown in Figure 6, this increase largely came from public timberlands. Since the early 1990s, changes in goals for public land management have led to harvest reductions on public lands and an increased reliance on timber from private ownerships.

A gradual change in the market for logs and stumpage has accompanied the change in domestic lumber and export log markets. Prices of key high-volume log grades have gradually declined, leading to a convergence of prices and reductions in various price premiums. Figure 7 illustrates the sawmill log grades of Douglas-fir in western Oregon. These log grades are ranked from highest to lowest quality: special mill, #2, #3 and #4. In the past, these grades would have gone primarily to sawmills. But with changing technologies in the plywood industry, they are now used in all solid wood sectors although they retained their traditional names. Figure 7 shows the price ratios for these grades by quarter from 1984:1 through 2000:4. The #2 and #3 grades compose the largest volumes of logs in these classes. Note that the price of the #2 grade has declined relative to the #3 price since the late 1980s. The relation of the special mill and #2 grades has been less stable but declined from 1990 to early 1999 and has remained below earlier levels. The price of the #3 grade has risen slowly, though with wide swings, relative to the #4 grade over this same period.

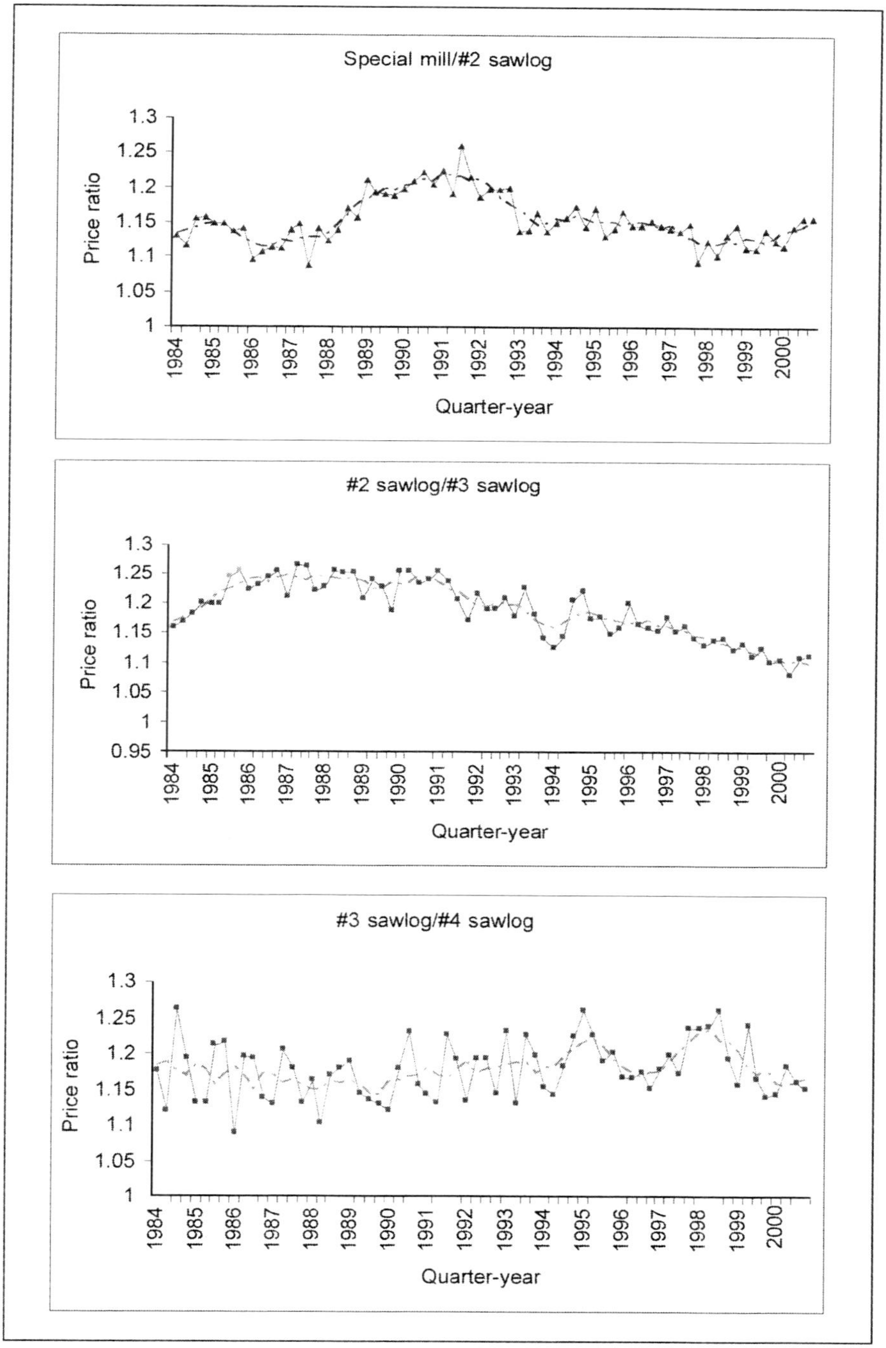

Figure 7. Douglas-fir delivered log price ratios (higher grade/lower grade) in western Oregon from the 1st quarter of 1984 through the 4th quarter of 2000.

Causes of these price trends are not entirely clear. As public timber harvest declined in the 1990s, there were smaller volumes of the higher grades (special mill and #2 grades) available. This led to reductions in the capacity (and number) of mills that processed this timber. These reductions in capacity reduced log demand and consequently prices for these higher grades. At the same time, milling facilities in the region have become more efficient. Logs of a given quality can now produce more lumber and greater proportions of higher (and more valuable) lumber grades. Thus, #3 logs may have become relatively more valuable. Over the past several decades, there also has been a gradual evolution in quality and size of logs and timber available from private lands. Given the relative nature of log grades, the #3 logs produced from timberlands today could be closer to the upper bound of the #3 category whereas the #2s are grouped closer to the lower bound of their category, thus bringing the prices closer together. Finally prices in lumber markets have also changed. Premiums for the highest quality grades decreased as users adopted substitutes, which has led to lower prices for the higher log grades.

Whatever their cause, the implications of the trends shown in Figure 7 for timber growers are significant. With a reduced quality premium for the primary grades produced in second-growth stands, there is less incentive to grow trees in longer rotations or adopt other practices that might produce higher quality logs.

2.3.2. Future market trends**—**The market history for different forest products suggests that care is needed in establishing the objectives of timber management. What, then, are some reasonable expectations for future market activity? Projections of the proportions of harvest by product from the current Resources Planning Act Timber Assessment (RPA, Haynes 2003) are shown in the right-hand portion of Figure 5. This view shows the mix of log markets in the Douglas-fir region returning to conditions similar to the 1950s and 1960s. Options may persist for some utilization in panels, exports, and other products, but softwood lumber output may account for a rising share of production and log use—reaching 70% of the total by 2050. Softwood harvest volumes (Table 2) may rise slightly more than 50% by 2050 compared to 1997, due primarily to expansion on private lands. At the same time, competition from other domestic regions and continued pressure from imports limit growth in product and stumpage prices. Figure 8 shows projections of regional softwood stumpage prices from the RPA (for example, $200 per thousand board feet is equivalent to $35 per cubic meter) (Haynes 2003). A small near-term increase is expected as we transition from existing stands to more managed timberlands. Once this shift is complete (say after 2020 in both the Douglas-fir region and the South), stumpage prices remain at or below those in the South until late in the projection.

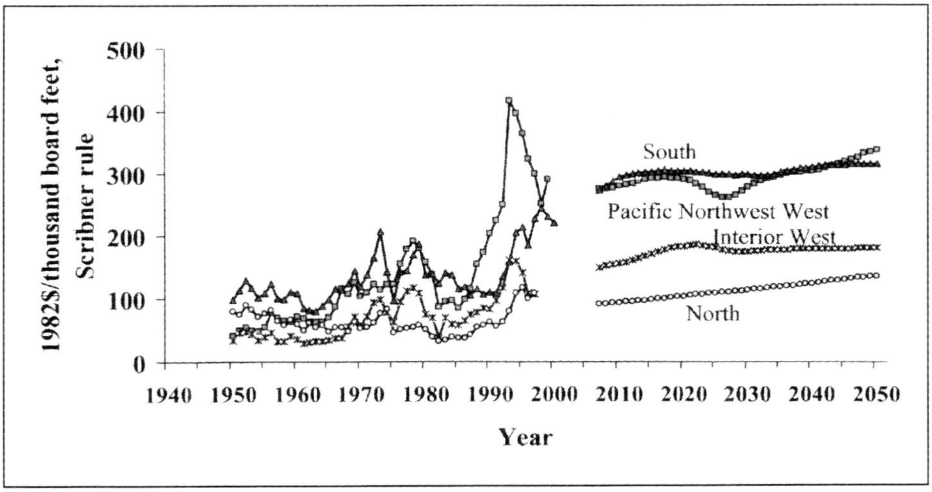

Figure 8. Softwood sawtimber stumpage prices in 1982 U.S. dollars per thousand board feet, Scribner log rule by major U.S. region, with projections to 2050.

These results and recent changes in the mix of the region's industries suggest the Northwest will remain an integrated producer of forest products. However, land management practices will result in a declining range of opportunities for the various products. The future mix of industries, with a growing component of lumber production, implies there will be a greater reliance on domestic markets for commodity grades of softwood lumber than in the past. Embedded in these projections, however, is the loss of the price premium once given to Douglas-fir. This premium, which was based on the generally higher quality of lumber produced from Douglas-fir, and the higher prices received for logs exported to Pacific Rim counties increased all stumpage prices in the Douglas-fir and nearby regions for all landowners. Because of this premium, prices for the Pacific Northwest exceeded other regions, especially during the 1970-90s when the log export market was most active (Figure 8). During this period, stumpage prices averaged 22% higher than prices in the South.

These projections of future markets have implications for future management. First, with the decline in log exports and panel production, softwood lumber will dominate product output in Oregon and Washington as it did in the 1950s and 1960s. The softwood lumber market is expected to remain focused on commodity lumber grades. Second, for Douglas-fir, some of the forces that resulted in various stumpage price premiums have and are expected to disappear. This is the case for some of the export markets for higher quality logs and products. The transition to managed stands that produce relatively uniform logs (both quality and size) is already well-advanced, reducing the opportunities for logs that do not conform to processing standards.

2.4. Impacts of the Changing Regulatory Environment

One of the significant changes in Oregon and Washington over the past 50 years has been the evolution of state and federal regulations that influence the design of forest management practices and control their applications. For the most part, these regulations reflect the manifestation of public concerns about forestlands or forest conditions. The current regulations seek to protect water quality, provide fish and wildlife habitat, protect capital improvements, and ensure that harvested areas are reforested.

Early regulations evolved from public concerns about fire protection and later, restocking of cutover timberland. Significant regulatory events include the Wilderness Act (1964), which permits the designation of wilderness on national forest and other federal lands (such designation removes land from the timberland base); the National Environmental Policy Act (1970), which requires analysis of environmental and economic impacts of significant actions; the Oregon Forest Practices Act (1971), which sets minimum standards for a number of management practices; the Endangered Species Act (1973), which requires the protection of endangered and threatened species and their habitat; the Washington Forest Practices Act (1974), which regulates practices related to growing, harvesting, or processing timber; the National Forest Management Act (1976), which required management plans for each national forest; and the Northwest Forest Plan (1993), a balanced policy for managing federal lands in the range of the northern spotted owl.

Local government, state, and private land in Oregon and Washington is subject to a variety of standards for reforestation, road construction and maintenance, timber harvesting, chemical applications, and slash disposal. The state forest practices acts regulate timber growing and harvesting on, or processing from, all nonfederal forestlands. They encourage the use of best management practices and provide the states with an enforcement mechanism to address noncompliance. In addition, Washington requires habitat conservation plans when forestry operations cover extensive parts of a single watershed. This reflects a move toward landscaped-based plans, emphasis on collaborative efforts by landowners and other stakeholders, and the need to protect public resources.

The current management regimes reflect this regulatory environment but there is constant debate about the possible expansion of various regulations, especially those affecting forest management in riparian areas. In addition, there are concerns about the direct and indirect effects of public policies on land use. For example, Oregon has comprehensive statewide land use planning, which has limited the residential development of forestlands zoned for timber production. Washington, however, struggles with growth management and is seeing conflicts between residential development and timber production along the rural-urban interface.

3. Impact of Harvest and Management Trends on Future Forests

The characteristics of future forests in Oregon and Washington will be the consequence, in part, of decisions made by a diverse group of timberland owners. These forests also will be shaped by the various forest products industries that will attempt to use the species and sizes of timber available, as well as by public concerns that are manifested in regulations to protect nonmarket forest benefits.

3.1. Projections of Private Inventories and Harvests in the Douglas-fir Region

To illustrate the impacts of changing management practices on harvest potential and forest inventory, we examine two volume-flow simulations of future developments on private lands in the Douglas-fir region. The first projection is conducted assuming a continuation of past trends in management on both industrial and nonindustrial lands. The assumptions for shifts in management regimes over time in these projections were derived from the current RPA Timber Assessment (Haynes 2003). These, in turn, were developed in collaboration with the American Forest and Paper Association (AF&PA) and state forestry organizations in Oregon and Washington. The area trends in these states reflect modest declines in nonindustrial private forest ownership resulting from direct conversion of timberland to urban and developed uses, and conversion to replace cropland lost to urban and developed uses. Future area transfers between ownerships are expected to be much smaller than in the past four decades. The most substantial cover type changes are projected to occur on forest industry lands, as more hectares are planted in Douglas-fir.

As noted earlier, the trend in management has generally moved toward intensive practices, and it has occurred more rapidly on industrial lands than on NIPF ownerships. In the second projection, the distribution of land by management regime is frozen in its current form with no trend or change over the projection period. Harvest potential in both simulations is described by a "sequential look-ahead even-flow" projection. These projections were developed by using the ATLAS timber inventory projection model (Mills and Kincaid 1992) and a simple binary search routine that computes both the strict even-flow and sequential look-ahead even-flow schedules. Projections were made for 100 years, though only the first five decades are shown in the figures and tables. The strict even-flow schedule gives the highest periodic harvest volume that can be maintained without decline over the projection period. The sequential even-flow schedule computes the highest strict even-flow for each decade that can be maintained over 5 decades. This schedule is found sequentially for each period in the projection. As a result, the sequential schedule can

Table 4. Definitions of management intensity classes in projections of Douglas-fir region private forests using the RPA Timber Assessment modeling system.

Management intensity class	Description
MI1	Custodial management: natural regeneration, no further treatment until harvest.
MI2	Regeneration by planting. Planted to 998 TPH.
MI3	Adds precommercial thinning. Planted to 998 TPH. Precommercial thinning to 680 TPH at age 15.
MI4	Regeneration with genetically improved planting stock and a fertilization. Planted to 998 TPH. Precommercial thinning to 680 TPH at age 15. Fertilization_high sites: age 20, 30, and 40 with 224 kg nitrogen/ha. Fertilization_medium sites: age 30 and 40 with 224 kg nitrogen/ha.
MI5	MI4 with the addition of a commercial thinning that occurs at age 32 for high sites and age 37 for medium sites. Fertilization occurs, as it does for MI4, within a few years of commercial thinning and about 5 years prior to the earliest final harvest age. A little less than one third of the basal area was removed in each thinning regime (medium and high sites only). Planted to 998 TPH. Precommercial thinning to 680 TPH at age 15. Fertilization-high sites: age 20, 30, and 40 with 224 kg nitrogen/ha. Fertilization-medium sites: age 30 and 40 with 224 kg nitrogen/ha. Commercial thinning: set to age 30 for highly productive sites, age 35 for medium productive sites.

Note: The Douglas-fir region includes western Oregon and Washington. TPH = trees per hectare.

move up and down over time as the levels and growth of the inventory change. Details of the strict and sequential even-flow schedules and binary search can be found in Davis and Johnson (1987).

This projection finds the highest harvest volume in each decade that can be sustained over the ensuing four decades. Thus the cut shown for each period can be maintained without reduction for five periods–an interval called the "look-ahead period." This type of projection is not linked to markets, but is governed by the basic biological productivity of the forest and its inventory. With everything else in the two simulations identical, differences between the "management trend" and "no management trend" projections provide a measure of the effects of continuation of past trends toward more intensive management.

Management regimes are described by five groups of practices (Table 4). Douglas-fir forest types can be managed under all five regimes, whereas western hemlock types (because of their basic silvicultural requirements) can use only the first two forms. The patterns of shift over time in the management trend projection are shown in Figure 9 for industrial and nonindustrial owners. Forest industry lands generally move from lower levels of investment toward

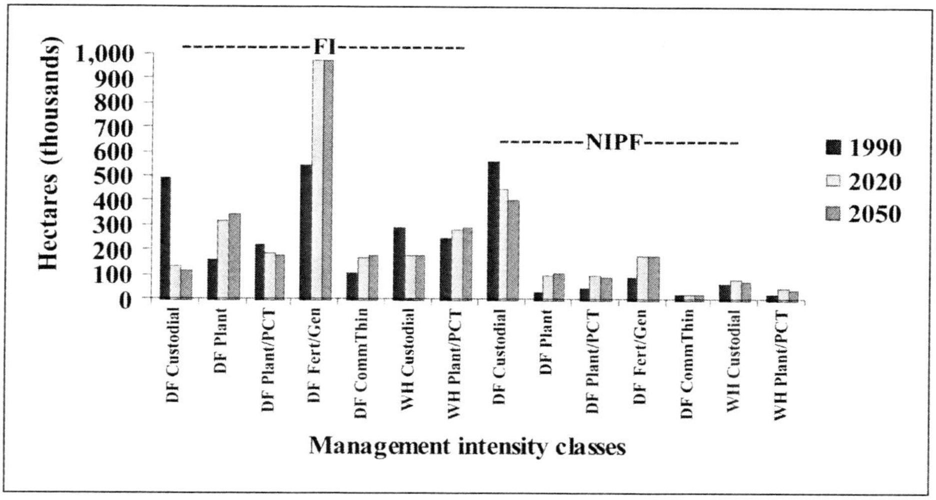

Figure 9. Historical and projected trends in the Douglas-fir region (western Oregon and Washington) forest industry (FI) and nonindustrial (NIPF) trends in management of Douglas-fir and hemlock.

higher levels. As illustrated in Figure 9, area regenerated to the Douglas-fir type is at the more intensive end of the management spectrum. By 2020, the largest shifts in Douglas-fir type are expected to have occurred, with the type gaining over 260 000 hectares (ha). Western hemlock types are projected to lose some 85 000 ha to Douglas-fir during the first 20 years of the projection, and the balance of the remaining western hemlock area is projected to shift toward more intensive management. The nonindustrial private forest ownership shifts more area into Douglas-fir but substantially fewer acres are regenerated under intensive management. By 2020, nonindustrial private forest landowners will hold 211 000 ha compared with industry's 1.1 million ha in higher management intensities. Western hemlock gains area in the nonindustrial private forest ownership.

Projected harvest and inventory levels are shown in Table 5 and Figure 10. With fixed management regimes, potential harvests on both industrial and nonindustrial lands decline steadily. Long-term (50 year) growth is declining and the cut must fall as well. This is clear in the case of industrial owners because inventory falls after 2030 even with the declining harvest. Under a continuation of past management trends, potential harvest is higher than the no-trend case and actually rises for both owners in 2040 and 2050. This reflects the maturation of more intensively managed stands planted in the early part of the projection and the associated increase in growth. The growth increase is large enough that despite higher harvests, inventories rise above the no-management-trend case on both ownerships in the last two periods of the projection.

Table 5. Projected harvest from industrial and nonindustrial (NIPF) ownerships in the Douglas-fir region using sequential constant look-ahead method with and without management intensity class shifts.

	Sequential even-flow with management intensity class shifting								
	Inventory summary				**Harvest summary**				
	Industry		**NIPF**		**Industry**		**NIPF**		
Year	**Softwood**	**Hardwood**	**Softwood**	**Hardwood**	**Softwood**	**Hardwood**	**Softwood**	**Hardwood**	**Total**
	Million cubic meters				*Million cubic meters per decade*				
1990	509.4	89.8	280.7	134.4	273.2	37.4	121.2	52.0	483.9
2000	486.6	81.2	260.6	121.6	276.6	32.6	121.7	51.4	482.3
2010	475.7	78.1	245.8	114.8	275.6	32.1	122.2	49.8	479.7
2020	473.0	77.8	237.0	109.9	271.3	35.2	121.7	49.3	477.5
2030	492.2	77.3	245.9	107.8	270.4	35.3	122.0	48.4	476.2
2040	530.4	77.4	263.4	105.7	310.5	37.7	132.5	47.5	528.1
2050	528.3	74.8	259.8	105.0	311.2	36.1	130.5	48.1	525.9
2060	521.2	74.1	252.2	104.5	309.3	36.5	128.7	48.2	522.7
2070	518.0	73.2	246.1	103.1	308.1	37.0	127.5	47.7	520.3
2080	516.6	71.6	243.3	102.0	307.2	36.5	127.5	47.4	518.6
2090	515.5	70.4	243.6	100.5	307.6	35.8	127.4	47.3	518.2
2100	513.2	69.7	241.6	99.2					

	Sequential even-flow with no management intensity class shifting								
	Inventory summary				**Harvest summary**				
	Industry		**NIPF**		**Industry**		**NIPF**		
Year	**Softwood**	**Hardwood**	**Softwood**	**Hardwood**	**Softwood**	**Hardwood**	**Softwood**	**Hardwood**	**Total**
	Million cubic meters				*Million cubic meters per decade*				
1990	509.4	89.8	280.7	134.4	270.5	40.5	118.6	55.7	485.2
2000	501.1	83.6	266.1	124.6	267.7	37.1	117.1	54.5	476.3
2010	509.8	80.9	257.0	121.1	265.0	35.5	116.2	53.4	470.1
2020	526.9	82.1	251.0	117.1	258.9	36.7	114.1	52.6	462.3
2030	531.4	81.3	246.4	113.3	251.7	37.0	112.3	51.5	452.5
2040	525.3	80.1	241.7	112.4	249.1	37.0	111.3	50.9	448.4
2050	515.0	79.5	235.8	111.6	244.1	37.1	107.9	51.4	440.5
2060	504.6	78.4	231.1	111.8	239.1	37.2	106.0	51.7	434.0
2070	494.8	77.1	225.4	110.4	236.2	36.6	104.0	51.4	428.3
2080	485.9	76.4	220.8	108.8	234.6	36.3	103.1	51.1	425.0
2090	477.2	76.2	217.6	107.9	231.6	36.3	102.7	51.0	421.6
2100	470.0	75.9	213.9	106.9					

Note: The Douglas-fir region includes only western Oregon and Washington.

In these comparisons it is clear that a continued trend in management intensification increases growth, potential harvest and inventory. The increases come with a lag, however, because the primary cause of increased growth is reforestation with improved stock and fertilization at later stages of the rotation.

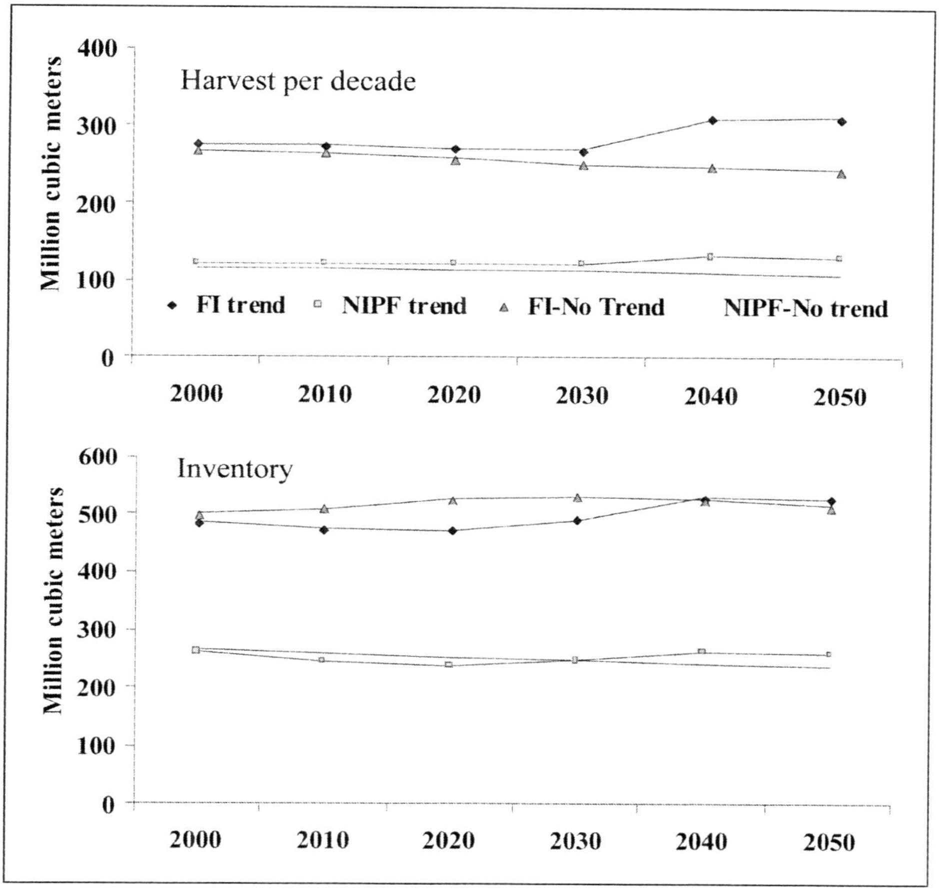

Figure 10. Softwood timber harvest and inventory by owner group for the Douglas-fir region (western Oregon and Washington) with fixed (no trend) and trends in management regimes over projection. FI = forest industry, NIPF = nonindustrial private forest.

3.2. Projections of National Forest Timberlands

Projections of inventories for national forest timberlands were developed in much the same fashion as for private timberlands. The primary difference was the choice of management intensities used for the national forests and the distribution of timberland acres across intensities. National forest timberland area is expected to remain constant over time.

Three management regimes were used for the national forests. Details and development of these regimes are described in Mills and Zhou (2003). The first regime allows a final harvest followed by regeneration, typically by planting. The second regime uses partial harvesting to mimic stands subject to multiple entries. Stand regeneration is accomplished either by underplanting or by favoring natural regeneration. The third regime allows no harvest and mimics stands being held in reserve (often for development into late successional old-growth habitat).

Table 2 shows the projections of national forest inventories for the Douglas-fir region developed by Mills and Zhou (2003). These projections assume that national forest timberlands were assigned to the three management regimes noted in section 2.1. in the proportions 16, 60, and 24%, respectively. Even with such a small area in intensive management (the first regime), inventories rapidly build (they increase by more than 70% between 1997 and 2050) because net growth exceeds removals by a wide margin. Net growth falls after 2010 as larger proportions of the inventory move into the older age classes.

The national forest projections point out a significant shift in expectations for how public lands are managed. In the early 1960s, all available national forest timberland would have been assigned to intensive forest management regimes. This reflected public expectations for efficient management that "has a direct large influence on the well-being of nearly everyone in western Oregon and western Washington" (USDA FS 1963). A generation later public expectations have changed; emphasis is now on habitat conservation set in the context of managing relatively complete ecosystems.

3.3. Implications for the Douglas-fir Region

Within the next 15 years, total regional softwood harvests are expected to return to levels comparable to the early 1990s (Table 2). This will be due almost entirely to harvest expansion on industrial and NIPF lands, despite a loss of nearly 0.4 million ha to other land uses (about 5.7% of the 1997 land base). At the same time, harvest on both private ownerships will be less than growth after the 2010-2020 decade and inventories will rise. In the case of NIPF, inventory will reach levels higher than those observed at any time since 1950. This expansion is a function of (1) an aging private inventory where larger proportions are older than 40 years (see Figure 11) and (2) continued intensification of private management, primarily on industrial ownerships.

Across all ownerships, changes in private timberlands and the gradual aging of inventories on national forest timberlands will lead to a more pronounced bimodal distribution of age classes. As shown in Figure 11, there will be large areas of stands at or below age 60 and at or above age 165. The majority of younger stands (less than 60 years) will be on private land, located typically at lower elevations. Older stands (greater than 80 years) will be on public land typically in higher elevations and headwater areas. These conditions may raise concerns for wildlife species with strong mid-successional habitat requirements.

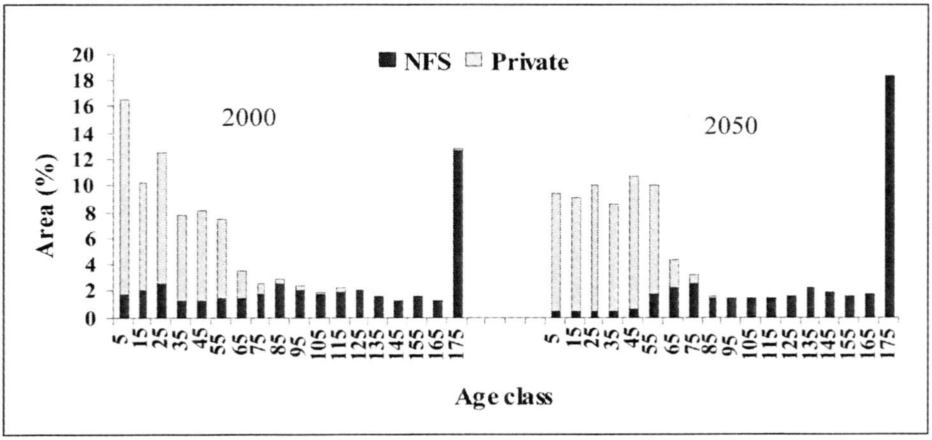

Figure 11. Age class distribution by ownership for softwood forest types on timberland area for the Douglas-fir region (western Oregon and Washington) for 2000, projected for 2050.
Note: NFS = National Forest System.

With stable to rising harvests, the Douglas-fir region will provide higher shares of national softwood removals than in the recent past. In the near term, the region's share will rise sharply from 11.7% (in 1997) to nearly 16% (by 2010) because of lags in the maturation of sawtimber inventory in the South. As southern inventories reach sawlog size and the effects of past management investment are reflected in rising growth, southern output will rise and the Douglas-fir region's share of national removals will decline to about 13.4%. Thus, from a national perspective, a larger portion of U.S. wood requirements will be met by shifting harvest onto more intensively managed private lands (in the South and the Douglas-fir region) while at the same time preserving large areas of older timber on western public lands.

4. Conclusion

Forest management regimes have evolved over the past 50 years in Oregon and Washington. This evolution reveals three lessons that are applicable to forest management everywhere.

The first lesson is the inherent flexibility of management actions within a single rotation. Table 3 shows how management regimes vary from decade to decade, reflecting the expectations of various owners about the financial feasibility of individual actions.

A second lesson is that in spite of this flexibility, the collective actions of all timberland owners are leading to a region which can be generally described as having achieved roughly an area regulated forest. That is, the forests in the Douglas-fir region are about evenly distributed across the age classes (and seral stages). Figure 11 highlights the differences in distribution by ownership; in terms of broad forest conditions these are about equally distributed across the various seral stages. These conditions are expected to remain stable, given the expected market and processing options. These conditions also offer opportunities for compatible management in the sense that flows of timber can be sustained while preserving habitat and the various services derived from different forest structural components.

A third lesson is that a gap exists between management intentions and actual outcomes. Landowners often express a desire to pursue highly complex regimes (plant/PCT/fertilize/CT) involving multiple entries but instead use simpler one or two treatment regimes (for example, plant/PCT or plant/PCT/fertilize). This results from the tendency by owners to postpone or abandon a specific treatment in actual application as objectives or markets change. In spite of a movement towards less complex management regimes in the Douglas-fir region, however, timber inventories continue to increase. These increases reflect improved forest resource conditions and higher net growth resulting from better management practices, including improved planting stock.

5. References

Adams, D., Alig, R., Anderson, D. [and others]. 1992. Future prospects for western Washington's timber supply. CINTRAFOR Special Paper 27. Seattle, WA: College of Forest Resources, University of Washington. 200 p.

Adams, D.M., Schillinger, R.R., Latta, G., and Van Nalts, A. 2002. Timber harvest projections for private land in western Oregon. Research Contribution 37. Corvallis, OR: College of Forestry, Forest Research Laboratory, Oregon State University.

Beuter, J.H., Johnson, K.N., and Scheurman, H.L. 1976. Timber for Oregon's tomorrow: an analysis of reasonably possible occurrences. Research Bulletin 19. Corvallis, OR: Forest Research Laboratory, Oregon State University.

Davis, L.S., and Johnson, N.K. 1987. Forest management. (3rd ed.) New York: McGraw-Hill. 790 p.

Forest Ecosystem Management Assessment Team [FEMAT]. 1993. Forest ecosystem management: an ecological, economic, and social assessment. Portland, OR: U.S. Department of Agriculture; U.S. Department of the Interior [and others]. [Irregular pagination].

Haynes, R.W. (Tech. coord.). 2003. An analysis of the timber situation in the United States: 1952 to 2050. A technical document supporting the 2000 USDA Forest Service RPA assessment. General Technical Report PNW-GTR-560. Portland, OR: U.S. Department of Agriculture, Forest Service, Pacific Northwest Research Station. 254 p.

Haynes, R.W., and Perez, G.E., (Tech. eds.). 2001. Northwest Forest Plan research synthesis. General Technical Report PNW-GTR-498. Portland, OR: U.S. Department of Agriculture, Forest Service, Pacific Northwest Research Station. 130 p.

Hummel, S.S. 2003. Managing structural and compositional diversity with silviculture. Chapter 4. In: R.A. Monserud, R.W. Haynes, and A.C. Johnson (Eds.), Compatible forest management. Dordrecht, The Netherlands: Kluwer Academic Publishers.

Lettman, G.J., and Campbell, D. 1997. Timber harvesting practices on private forest land in western Oregon. Salem, OR: Oregon Department of Forestry. 48 p.

Mills, J.R., and Kincaid, J.C. 1992. The aggregate timberland assessment system-ATLAS: a comprehensive timber projection model. General Technical Report PNW-GTR-281. Portland, OR: U.S. Department of Agriculture, Forest Service, Pacific Northwest Research Station. 66 p.

Mills, J.R., and Zhou, X. 2003. Projecting National Forest inventories for the 2000 RPA timber assessment. General Technical Report PNW-GTR-568. Portland, OR: U.S. Department of Agriculture, Forest Service, Pacific Northwest Research Station. 58 p.

Monserud, R.A. 2003. Modeling stand growth and management. Chapter 6. In: R.A. Monserud, R.W. Haynes, and A.C. Johnson (Eds.), Compatible forest management. Dordrecht, The Netherlands: Kluwer Academic Publishers.

Pipken, J. 1998. The Northwest Forest Plan revisited. [Place of publication unknown]: U.S. Department of the Interior, Office of Policy Analysis. 117 p. On file with: Human and Natural Resources Interactions program, Forestry Sciences Laboratory, 620 SW Main St., Suite 400, Portland, OR 97205. Retrived 1998 from http:///www.doi.gov/nrl/ppa/nwforest/revisit.html.

Sessions, J. (Coord.). 1991 (Revision). Timber for Oregon's tomorrow: The 1989 update. Corvallis, OR: Forest Research Laboratory, Oregon State University.

Smith, W.B., Vissage, J.S., Sheffield, R., and Darr, D.R. 2001. Forest resources of the United States, 1997. General Technical Report NC-219. St. Paul, MN: U.S. Department of Agriculture, Forest Service, North Central Research Station. 109 p.

U.S. Department of Agriculture, Forest Service. 1963. Timber trends in western Oregon and western Washington. Research Paper PNW-5. Portland, OR: Pacific Northwest Forest and Range Experiment Station, Division of Forest Economics Research. 154 p.

U.S. Department of Agriculture, Forest Service; U.S. Department of the Interior, Bureau of Land Management. 1994. Record of decision for amendments to Forest Service and Bureau of Land Management planning documents within the range of the northern spotted owl. [Place of publication unknown]. 74 p. [plus Attachment A: standards and guidelines].

Washington Department of Natural Resources. 1975. Washington forest productivity study, phase 1 report. Prepared for Pacific Northwest Regional Commission. Operations Research Section, Division of Technical Services. Olympia, WA.

Appendix 1. Detail of land base allocated to management regimes from six studies of timber supply in the Douglas-fir region.[1] FI = forest industry, NIPF = nonindustrial private forest landowners.

Region and period	**WOR[2] Early 1970s**		**WWA[3] Early 1970s**		**WOR Late 1980s**		**WWA Late 1980s**		**WOR Late 1990s**		**PNWW[4] Late 1990s**	
Study	**Beuter et al. (1976)**		**WDNR (1975)**		**Sessions (1991)**		**Adams et al. (1992)**		**Adams et al. (2002)**		**Haynes (2003)**	
Owner	**FI**	**NIPF**	**FI**	**NIPF**	**FI**	**NIPF**	**FI**	**NIPF**	**FI**	**NIPF**	**FI**	**NIPF**
Regimes	*Percentage of land base*											
Natural	0	0		90			1	96	35	56	6	78
NAT/F								16			2	
NAT/F/CT											3	
NAT/PCT											5	
NAT/PCT/F									6	2	4	
NAT/PCT/CT											1	
NAT/CT									4	1	2	
NAT/PCT/CT/F									5	0	4	
Selective									6	33	2	
Plant	43	62	10	5	28		24	2	16	4	11	5
PL/PCT					13		31	1				7
PL/PCT/F			10						9	0		
PL/CT	19	3			5				6	0	8	
PL/PCT/CT	16		35	5	6		4					
PL/PCT/CT/F	6		45		39		17		9	0		
G/PL/PCT											5	
G/PL/F										7		
G/PL/F/CT											11	
G/PL/PCT/F							1	1			6	8
G/PL/PCT/CT							1	0			10	
G/PL/PCT/CT/F							1				8	
Conversion	16	35			9							
Reserved			0	0			4		3	3	6	
Total	100	100	100	100	100		100	100	100	100	100	100

[1] The Douglas-fir region includes only western Oregon and Washington.
[2] WOR = western Oregon.
[3] WWA = western Washington.
[4] PNWW = western Oregon and Washington.
Note: NAT = natural, F = fertilization, CT = commercial thinning, PCT = precommercial thinning (stocking control), PL = planted, G = planted with genetically improved stock.

Section 4—Aspects of Management

A forest with many healthy resources not only has greater ecological integrity and resilience, but it also is likely to meet more human desires. Managing for many resources, however, is far more challenging than managing for one or two goals. In this section, the authors examine other aspects of forest management and their relations to compatibility.

In Chapter 11, the authors explain ways that forest management can affect wood quality and the implications for compatible management. In the Pacific Northwest, clear wood with tight grain patterns milled from old-growth trees was replaced gradually by wood from managed second-growth stands. In some cases this is now being replaced by wood from intensively managed plantations. The authors explore the consequences of these trends for wood quality by using simulations for three broad management strategies. The results indicate substantial differences in wood value among the three strategies. This reflects an opportunity to diversify because no single management regime will meet all of society's needs (an area discussed further in Section 5). Diverse management regimes may help achieve compatibility at a landscape scale, in terms of diverse forests and diverse wood qualities.

Silvicultural treatments discussed in Section 2 can alter wood characteristics and thus affect wood product potential and the landowner's return on investment. Stand density management, branch pruning, fertilization, and genetic selection of planting stock can all alter wood properties.

In Chapters 12 and 13, the authors examine nontimber forest products such as native mosses, lichens, ferns, herbs, shrubs, and edible mushrooms. Over 200 species are harvested from public and private lands in the Pacific Northwest for commercial, personal, and traditional purposes. Chapter 12 focuses on the ecology of and management for nontimber forest products in the understory. Chapter 13 focuses on access, which involves physical and legal restrictions. Both chapters discuss a case study on salal, an understory shrub in the Pacific Northwest widely used as a floral green in the florist industry.

In Chapter 12, the authors discuss the effects of forest management practices on understory forest resources. Activities such as thinning that benefit one shrub or herb may be detrimental to another. With site-specific and careful management, however, managers can produce understory resources with commercial value and maintain an economically viable overstory tree crop.

Timber and understory forest resources can be a synergistic mix, providing a cash flow from a maturing forest stand. Biological diversity can be a hedge against future market uncertainties and can help position the landowner to provide possible new products such as conservation credits, carbon credits, and green certification. The compatible management of understory forest resources is jointly a social, biological, and ecological issue.

In Chapter 13, Alexander and Fight discuss access management to non-timber forest resources. Access rights are essentially management choices that determine who can harvest, where, how much, and how long. The social ecology of some resources, such as huckleberries and certain mushroom species, is at least as complicated as the biophysical ecology. Some policy solutions may have unintended and possibly negative social and ecological consequences. Any access management for these products must be flexible enough to fit the diverse products and people who collect the resources, and also respectful of traditional and other personal uses. Permits and contracts can be designed to reward harvesters for stewardship actions such as deferred harvest of young plants.

People are interested in conserving wildlife and in conserving overall biological diversity. In Chapter 14, Carey discusses compatibility and wildlife and management. Public attention is shifting to biodiversity in second-growth forests. Managing for wildlife requires managing multiple ecological processes over the long term. Some species, such as salmon and Roosevelt elk, can be indicators of total landscape functions because of the species' wide ranges and dependence on a number of habitats throughout their life histories.

Results from several studies suggest that joint, efficient production of many forest values, including wildlife and social acceptability, is possible. Experimental treatments in second-growth forests have proven successful at increasing biodiversity. In a modeling study for lands in western Washington, an alternative to intensive management on 40-year rotations emerged. The alternative would manage for biodiversity, including wildlife, wood, and water, and would include harvest, legacy retention, planting, and variable-density thinnings on rotations that alternated between 70 and 130 years. The modeled loss of net present value between the two alternatives was slight.

Chapter 11

Managing for Wood Quality

R. James Barbour
USDA Forest Service, Pacific Northwest Research Station,
620 SW Main St., Suite 400, Portland, OR 97205, USA
David D. Marshall
USDA Forest Service, Pacific Northwest Research Station
3625 93rd Ave., Olympia, WA 98512, USA
Eini C. Lowell
USDA Forest Service, Pacific Northwest Research Station,
620 SW Main St., Suite 400, Portland, OR 97205, USA

1. Introduction

During the 20th century, the Pacific Northwest (western Oregon and Washington, coastal British Columbia, and southeastern Alaska) produced some of the highest quality timber in North America. Forests were old and trees were large; wood products from this region were easily distinguished from those made from timber produced in other regions in North America. In some cases, the timber had unique visual characteristics; its slow growth produced clear wood and tight grain patterns and its superior mechanical properties were prized. The sheer size of the trees made it possible to manufacture extremely large pieces. Early in the century, Douglas-fir trees (*Pseudotsuga menziesii* (Mirb.) Franco) (at the time also known as Oregon pine) that were tall and cylindrical possessed a stiffness that made them valuable as ships' masts (Allin 1995). Later, these same characteristics made this species important as electric power transmission poles. Long spans, made possible by the high strength and stiffness of Douglas-fir, allowed architects to create distinctive open interior spaces. Several mills specialized in beams up to 185 feet (57 meters (m)) long and several feet deep. The fine grain and infrequent knots in large old trees permitted manufacture of clear, straight-grained moulding and millwork that was valued around the world in door and window manufacturing. The high strength-to-weight ratio of Sitka spruce (*Picea*

R.A. Monserud, R.W. Haynes and A.C. Johnson (eds.)
Compatible Forest Management, 299-336

sitchensis (Bong.) Carr.) was attractive to early airplane manufacturers and to musical instrument craftsmen who valued the tonal quality derived from its straight, even grain (Burns and Honkala 1990). The decay resistance and dimensional stability of western redcedar (*Thuja plicata* Donn ex D. Don) made it ideal for siding, roof shingles and shakes, and other exterior uses where both durability and appearance were important (DeBell et al. 1999).

Early timber operations in the Pacific Northwest were limited to areas near the waterways that provided for transport of large raw logs and finished lumber. Because felling was done by hand, these harvests also focused on the highest value trees. Harvest of the region's forests was more like a mining operation where the resource was seen as endless. In the late 1880s, railroads made much more timberland accessible. The 1930s brought both the chain saw and widespread use of trucks for hauling logs. This further expanded timber harvest because logging roads were easier and cheaper to build than railroad spurs. Once log transportation systems became established, clearcuts with natural regeneration became the dominant harvesting system. Only limited resources were devoted to investigating other silvicultural systems (Curtis 1998). About this same time, researchers produced the first yield tables for the region (McArdle and Meyer 1930), demonstrating the productive potential of the region and further stimulating research on topics such as regenerating stands after harvest and controlling wildfire. By the 1950s, however, the forests of the Pacific Northwest were no longer seen as an endless resource.

By the late 1970s, large, old trees were gone from most private timberland and much of the easily accessible public land base, but the wood from second-growth stands that replaced old-growth stands still had highly desirable properties. Stands 70 to more than 100 years old were widespread, and trees were large in comparison to those grown in other timber producing regions. Manufacture of wide and long pieces with mechanical properties as good or better than those available elsewhere was common. These trees even contained enough clear wood to make production of high-value appearance grades of lumber or veneer possible.

Technological changes during this time began to erode markets for long, wide, solid wood products. For example, the development of the softwood plywood industry following World War II contributed to a change in the product mix produced at sawmills as softwood plywood was substituted for sheathing lumber in residential construction. The development of synthetic adhesives in the 1950s and 1960s greatly improved performance, particularly durability and moisture resistance of laminated products. Later, glulam beams replaced large solid sawn beams. In the 1970s and 1980s, wood I-beams and engineered trusses eliminated much of the need for even, moderately wide (greater than 8 inches (20 centimeter (cm))) or long (greater than 20 feet (6 m)) solid sawn lumber. Development of laminated veneer lumber (LVL) and parallel strand lumber (PSL) further reduced the need for large pieces (McKeever 2002).

Today it is possible to manufacture engineered wood products that can substitute for all types of structural solid wood and panel products. Some of these products also have appearance characteristics that make them desirable for architectural applications.

Land ownership and forest management patterns in the Pacific Northwest are also unique among U.S. timber producing regions (Table 1 Chapter 10). More than half of Pacific Northwest forestland is managed by public agencies (federal and state) and much of the private timberland is held in large tracts. This general pattern is prevalent throughout the West. Outside of the Pacific Northwest, nonindustrial private forests are the dominant form of private land ownership.

Through the 1970s, timber production was the primary objective in managing most forestland. Forest management began to shift from a reliance on natural regeneration and the harvest of second-growth stands to intensively managing plantations. This was an important decision in terms of wood quality because it meant that by the end of the 20th century a shift would occur from harvesting and processing managed natural young growth to plantation-grown wood. Research helped develop effective methods for establishing plantations of Douglas-fir and maintaining their rapid growth. Emphasis was placed on developing better methods of weed control, stocking control, and fertilization along with programs in genetic manipulation to increase volume growth. Clearcutting remained the most popular system for harvesting and regenerating stands because it was economically efficient and promoted the reestablishment of Douglas-fir, which grows rapidly in full sunlight. This resulted in a forest landscape where young even-aged, Douglas-fir plantations were the primary forest type. Given the relatively high returns, many landowners have practiced intensive forest management, which leads to younger, smaller, more quickly grown trees (Table 3 Chapter 10) managed for increased wood volume production. Federal forest management policies beginning in the 1990s placed more emphasis on protection and enhancement of biological diversity than on timber production (Haynes et al. 2003b). State land managers are also seeing increased pressure to consider more than just timber yields as criteria for management. The use of extended rotations and thinnings are seen as one way to achieve these goals on public lands. These changes in public timber management policy have contributed to the acceleration toward the production of younger, more intensely managed timber on private land. The longer term impacts of these management changes are still unknown. With a move toward management of broader scale ecosystems on federal and state lands and more intense management on private lands, we can anticipate further changes in the characteristics of logs produced from forestlands in the Pacific Northwest. The consequence, in terms of these management trends on wood quality, will be explored in this chapter by using currently available tree growth and wood quality models, as well as information from empirical wood product recovery studies.

2. Key Wood Quality Attributes

Wood quality is generally defined as a measure of the characteristics of wood that influence the properties of the products manufactured from it (Haygreen and Bowyer 1996). These wood properties (Megraw 1986) can be divided into two groups: microscopic and macroscopic. Microsocopic properties include, but are not limited to, chemical composition, fiber size (length and diameter), cell type, cell wall structure (e.g., cell wall thickness, microfibril angle), permeability to treating chemicals, and dimensional stability. Macroscopic properties include the types of characteristics that are generally considered in lumber or veneer grading systems or are important in determining volume recovery. Some examples are: knot size, frequency, and distribution; grain orientation; grain pattern; stem form; stem straightness; reaction wood (wood produced in response to stimulus, usually a bending or leaning of the tree); color and texture. Links to growing conditions and life history have been established or postulated for most of these properties but the relation between wood properties and silviculture is much easier to demonstrate for macroscopic than for microscopic properties.

2.1. Microscopic Properties

Within the wood cell wall, there are several characteristics that influence wood quality. Each cell has a primary and secondary wall, and the secondary wall is composed of three layers. The middle, or S2, layer is generally the thickest and within this layer are cellulose strands (Panshin and DeZeeuw 1980). The orientation of these strands in respect to the long axis of the cell (microfibril angle) plays an important role in the strength and shrinkage properties of wood. Increasing microfibril angle and a thicker S2 layer cause greater longitudinal shrinkage. When longitudinal shrinkage varies across the width or thickness of a board, warp can occur when the board is dried (Lamb and Wengert 1987).

Cells differ within an annual ring. Each annual ring is made up of early (spring) wood and late (summer) wood with a transition zone in between. The earlywood typically has large-diameter cells, thinner cell walls, and a lower specific gravity (dry wood mass per unit volume) while the latewood has smaller diameter cells, thicker cell walls, and higher specific gravity. The proportion of latewood within an annual ring is not necessarily the same for rings of equal width. This is important for softwoods as there is a relation between the proportion of latewood and specific gravity. Specific gravity is important in wood products as a measure of physical and mechanical properties (Haygreen and Bowyer 1996). As a general rule, the wood with a higher specific gravity is used in applications where strength is important. Within a tree, there are patterns in specific gravity variations that can relate to both growth rate and age (Jozsa et al. 1989, Cave and Walker 1994).

The trend to growing trees faster with emphasis on volume production has led to concerns about juvenile wood. Juvenile wood consists of those rings closest to the pith that were produced under strong influence from the apical meristem and vigorous crown branches (Megraw 1986). Both microscopic and macroscopic properties in juvenile wood create differences between wood formed under the influence of the actively growing crown and wood formed in the part of the stem below the live crown (mature wood). There is long-standing theory that wood properties change at the base of the live crown (Larson 1969, DiLucca 1989).

Several theories have been proposed to explain why wood formed within the vigorously growing crown is different from wood formed lower in the stem. Two of the most widely accepted theories are that physical changes in the vascular cambium are (1) due to repeated cellular division, which causes changes in the characteristics of the cambial initials that translate into changing wood properties (Bannan, 1956, 1962, 1964, 1966, 1970) and (2) response to hormonal influence based on distance from, and vigor of, the live crown (Larson 1962, 1963, 1969).

The best documented example of the first theory is the development of spiral grain in conifers. Bannan (1966) developed retrospective histological techniques to demonstrate how the increase in girth of the vascular cambium through anticlinal division of cambial initials and the subsequent apical intrusive growth alters the vertical orientation of the initials. This gradual slanting of the vascular cambium with respect to the long axis of the stem results in the formation of spiral grain. Bannan explains this transition as mainly a physical phenomenon.

Larson (1962, 1963, 1969), on the other hand, explains the differences in properties between juvenile wood and mature wood in terms of physiological processes that change in response to hormone and nutrient availability at various locations along the stem. Larson makes a compelling case that portions of the vascular cambium located within the live crown are exposed to higher levels of growth hormones than those lower in the stem. Subsequently, the vascular cambium closest to the actively growing branch tips, where these compounds arise, produces xylem with different characteristics than what is produced lower in the stem. Larson used experimental methods to establish a link between cell size and hormone production by meristematic tissue. He also showed that cell wall thickness was related to the quantity of nutrients produced by foliage. These two findings lead to the logical conclusion that proximity to vigorously growing branches is important in determining the microscopic properties of xylem cells that form the stem wood. In addition, because the crown is where the live branches are, and the heartwood has not yet formed in this youngest part of the stem, many of the macroscopic properties are also different higher in the stem than below the live crown.

The group of characteristics composing juvenile wood varies from species to species, but most softwood species display a radial gradient in physical properties (Sanio 1872, Trendelenburg 1936). The concern with juvenile wood is that when compared to mature wood, the mechanical properties differ and it can be less dimensionally stable because of radial gradients in longitudinal shrinkage (Senft et al. 1985, Forest Products Research Society 1986, 1987). It is generally believed that wood properties change as each growth ring is added until a point that is roughly correlated with the recession of the live crown at any given height. After crown recession, wood properties become relatively constant as subsequent growth rings are added. The point at which the transition from juvenile wood to mature wood formation occurs, and whether mature wood properties are in fact constant, are discussed in the wood science literature (Saucier 1990, Sauter et al. 1999, Koubaa et al. 2000, Gartner et al. 2002, Yang 2002). There is, however, widespread agreement that the wood formed within the vigorously growing crown is different than that formed below the crown base (Megraw 1985, 1986; Johnson 1986; Forest Products Research Society 1986; Jozsa et al. 1989; Cave and Walker 1994).

Juvenile wood content is expressed as a percentage of the total volume (Bendtsen and Senft 1986, Briggs and Smith 1986) based on the transition point between juvenile and mature wood production. Wood technologists have attempted for years to find a consistent age where the transition from juvenile wood to mature wood production occurs, and they have generally failed. It is usually possible to say where the demarcation is for a particular property for a given height in any tree, but it is difficult to generalize among trees. DiLucca (1989) was one of the first to explore this problem for Douglas-fir. He used segmented regression modeling to try to identify the transition age. He found that for a subsample of the trees used by Kellogg (1989) for Forintek's Douglas-Fir Task Force study, the transition age was about 20 rings from the pith at breast height. Coincidentally, this age also corresponded to the age at which crown recession occurred in these trees. Later Fahey et al. (1991), following the methods of Kellogg (1989) in a similar study for young, fast-growing Douglas-fir from western Oregon and Washington, also used the "20 ring from the pith" rule to define the transition from juvenile wood to mature wood. This number was incorporated into the ORGANON (ORegon Growth ANalysis and projectiON) model (Maguire et al. 1991, Hann et al. 1997) wood quality module along with an alternative provision to use the base of the live crown as the transition point from juvenile wood to mature wood. Given the available data, it is not really possible to say which method provides a more accurate estimate of the amount of juvenile wood. However, the two methods provide very different results. We do not resolve the question of whether a fixed number of rings or a physiological event, such as crown lift, is a better indicator of the end of the juvenile period although the idea of a physiological trigger is more consistent with the theory that growth conditions influence wood quality.

As the crown lifts, there is a gradual change in microscopic properties whereas below the base of the live crown, microscopic properties are generally assumed to become relatively constant as the tree ages. This could take a decade or more but the trend is toward more uniform properties over time.

2.2. Macroscopic Properties

Changes in macroscopic properties affect characteristics that influence the suitability of a tree for the manufacturing of specific wood products.

Ring width is often used in the log and lumber grading rules as a measure of quality. Ring width alone has little meaning as an indicator of properties (Jozsa and Middleton 1994). In softwoods, the proportion of latewood is the more important factor. Slow-growing trees have narrow rings whereas wide rings are associated with faster growth. Rate of growth is influenced by several factors including site quality (moisture, nutrients) and competition (stand density). Trees with wide annual rings may create machining and finishing problems for products manufactured from them (Fahey and Willits 1995).

Grain can refer to either the direction of the fibers in the finished product (e.g., vertical, flat, or cross) or the size of the wood fibers and corresponding growth rate (coarse, medium, or fine). Rapid growth rates produce coarse grain. Rapid growth may also result in increased longitudinal shrinkage and subsequent warping when products are dried. Fine-grain material (slow grown) is often prized for its appearance.

Compression wood is found in softwoods and is related to tree vigor and angle of tree lean (Timmel 1986, Jozsa and Middleton 1994). Eccentric stem shape and offset pith are visual clues to the presence of compression wood. Its importance from a solid wood quality perspective is that it has reduced strength, can fail unexpectedly, and is not dimensionally stable. Furthermore the shrinkage can be 10 to 20 times greater than normal wood. Compression wood fibers also have high lignin contents and do not refine well, making them undesirable for pulp and paper manufacture.

Heartwood formation occurs primarily below the live crown. The heartwood of trees is made up of non-functioning, dead cells. In contrast, the sapwood in trees conducts water, nutrients, and photosynthates and usually has a higher moisture content than the heartwood. The parenchyma cells of the sapwood, both in rays and longitudinal orientation, are physiologically active. Extractives (various organic substances) produced upon the death of parenchyma, will often darken heartwood and make it more difficult to dry or treat with preservatives. In some species, the resulting changes in color and permeability are important in determining wood product potential. These same extractives can also make the heartwood more durable.

Knots are a major factor in determining wood quality, both from a visual and a structural standpoint. The most important macroscopic change is the death and shedding of branches that result first in dead (black or loose) knots and later in the formation of clear or knot-free wood. Following death of a branch, the influence of the branch on mechanical properties changes and the branch begins to deteriorate. The change in the contribution to mechanical properties happens because the connection between the branch and the stem's vascular cambium is broken. The dead branch no longer lays down new xylem tissue and as a result, a new grain pattern develops around the branch. Instead of being integrated with the branch, the grain of the stem begins to "flow" around the now dead branch. The branch essentially becomes a plugged hole in the stem. Initially the branch is held tightly in the stem wood but as it deteriorates it loosens. Eventually, when the branch is shed, it becomes a hole. Finally, the end of the branch is overgrown or occluded. Each of these stages is recognized in the grading rules for lumber (WWPA 1998) and veneer. The zone of the stem between the base of the live crown and the point where branch stubs begin to occlude is an area of declining quality. This is an important but rarely recognized factor in describing the link between silviculture and wood quality. Quality declines because in most lumber grading rules a live knot (red or tight) is considered higher quality than a dead knot (black or loose), which is considered higher quality than a knothole.

Once the branch stub becomes occluded, clear wood is produced. If there is sufficient clear wood to recover as products, this material has exceptionally high economic value. Typically clear lumber or veneer has three to five times the value of good-quality wood that contains knots or defects such as pitch pockets or stain. As a general rule, the lowest quality wood is produced near the pith and the highest quality wood is produced near the bark.

3. Silviculture Affects Wood Quality Attributes

Whereas wood quality characteristics are inherent to a species, silvicultural treatments can alter wood characteristics (see Table 1) and thus affect wood product potential. Ultimately, most silvicultural treatments that are designed to enhance growth manipulate crown size or vigor and thus form the link to wood characteristics. As a result, practices that change stand structural conditions and tree growth are believed to alter wood properties and may unintentionally affect wood quality. For example, practices that alter crown length and the amount of the juvenile core, the area of branch occlusion below the live crown, and the size and rate of growth of the mature wood zone may influence the suitability of the wood for manufacturing certain products.

Managing stand density (usually expressed in terms of numbers of trees or amount of basal area per unit area) is a common practice of forest managers because it is the major factor influencing the development of individual trees

Table 1. Effects that various cultural practices may have on wood characteristics (adapted from Briggs and Smith 1986).

Wood characteristic	**Wider spacing**	**Precommercial thinning**	**Commercial thinning**	**Pruning**	**Fertilization**
Knots	Larger, sound	Larger, sound	Larger, sound	Removed	Larger, sound
Rings	Wide, uniform	Wide, abrupt change	Wide, abrupt change	Narrowed	Wide, abrupt change
Juvenile wood	Enlarged all logs	Enlarged most logs	Enlarged upper logs	Reduced	Enlarged
Compression wood	Increase	Increase	Increase	Unknown	Increase
Taper	Increase	Increase	Increase	Decrease	Increase or unchanged
Spiral grain	Increase	Increase	Increase	Unknown	Increase
Crookedness	More prevalent	Reduced	Reduced	No effect	No effect? (snow, ice)
Specific gravity	Little change in young trees, since age effect dominates (may be an effect with extreme treatment)		± 5 to 10%	Increase	Decrease 5 to 15% (except severely deficient sites)
Fibril angle	Same as above	Same as above	Increase	Decrease	Increase
Cell dimensions:					
Length	Same as above	Same as above	Small decrease	Small increase	Small decrease
Diameter	Same as above	Same as above	Increase	Decrease	Increase
Wall thickness	Same as above	Same as above	Decrease	Increase	Decrease

and stands (Marshall and Curtis 2002, Hummel 2003). For young Douglas-fir, low densities can reduce the amount of total stand growth because trees do not fully occupy the site throughout the rotation. There is less growing stock (i.e., fewer trees) to produce new wood per unit area. However, lower densities also mean less competition for individual trees and, therefore, can cause greater individual tree growth, especially in diameter. Stand density can be managed through initial planting spacing and subsequent thinning. The level and development of stand thinning can significantly impact live crown length and width, the timing of crown closure, the amount of stem taper, the size of branches, and the diameter of the juvenile core. Lower stand densities can delay crown closure, and therefore, alter branch size and increase juvenile wood content.

The timing of thinning can determine crown length and influence the distribution of dead knots and juvenile wood along the stem.

Available information suggests that a potential decline in the mechanical properties of young-growth Douglas-fir as compared to old-growth or unmanaged second-growth is a concern (Barret and Kellogg 1989, Stevens and Barbour 2000). The problem will probably be more important for western hemlock (*Tsuga heterophylla* (Raf.) Sarg.) (Middleton and Munro 2001). The limited information on warp in lumber from young trees of these two species also suggests that dimensional stability will be more of a problem for young hemlock than for young Douglas-fir.

Pruning live branches will accelerate live crown recession and hasten the production of clear wood that should reduce the period of juvenile wood production. Kotok (1951) found the natural rate of pruning to a stem height of 17 ft (5 m) on a Douglas-fir tree was 77 years. Management regimes that emphasize growth (wider growth rings) also produce large branches and subsequently large knots (Briggs and Smith 1986). Pruning is a way to limit the quality-reducing effect of wide spacing (Fight et al. 1995). Many other considerations are tied in with pruning and some of these (e.g., environmental and policy concerns, wood supply and projections for clear wood, economics and management) are covered in Hanley et al. (1995).

Fertilizer is used to improve growth by adding limiting nutrients to increase productivity. This increase in growth can also improve crown vigor, which in turn increases branch size and changes the properties of juvenile wood. Decreases in wood density (specific gravity) have been observed (Erickson and Lambert 1958, Siddiqui et al. 1972, Parker et al. 1976) with fertilization. The effects of fertilization last from 3 to 5 years (Briggs and Smith 1986, Jozsa and Brix 1989), and there is less effect on the juvenile wood than on the mature wood. Although the idea that stand fertilization will change the properties of juvenile wood and pruning will reduce its duration is attractive, no conclusive evidence has been offered.

Genetic manipulation of planting stock can also alter wood properties. The major thrust of early research was on increased volume growth (Silen 1966, Wahlgren and Schumann 1975, Silen and Wheat 1979, Kellogg 1982, Adams and Howe 1984). Breeding programs have been developed to increase or maintain specific gravity, improve stem straightness, reduce knot size, and reduce reaction wood content. Tree breeding programs in the Pacific Northwest select for growth rate and wood density. Growth rate is under low to moderate genetic control; 10 to 30% of the within-stand variation is controlled by genetics (Stonecypher et al. 1996, Johnson et al. 1997, King et al. 1988). Wood density is controlled even more; over 50% of the variation in a stand is a function of the genetics of the trees (Bastien et al. 1985; King et al. 1988; Loo-Dinkins and Gonazlez 1991; Vargas-Hernandez and Adams 1991, 1992; St.Clair 1994). Both traits need to be examined in a breeding program because there can be a negative adverse genetic correlation between the two traits (Bastien et al. 1985, King et al. 1988, Vargus-Hernadez and Adams 1991, St.Clair 1994). There has been little compelling evidence that genetic manipulation, by itself, adversely affects wood properties in the major softwood species grown in the Pacific Northwest to the degree that they can no longer compete in structural wood product markets.

Management practices (e.g., wide spacings) that promote long-lived vigorous branches almost always detract from wood product value. Vigorous branches lead to knots that detract from functionality and therefore, value in both solid sawn lumber and structural veneer. In widely spaced Douglas-fir, a phenomenon known as multiple flushing can cause trees to develop numerous relatively small branches whose quantity, rather than size, are of concern (Figure 1a). These trees also tend to have large branches at steep angles (Figure 1b) that will ultimately result in large solitary knots. In fast growing Douglas-fir from New Zealand, nodal swelling associated with branch whorls was recognized as a source of cross grain sufficient to cause a substantial downgrade in solid sawn lumber (Whiteside 1978). The problem was associated with every branch whorl and therefore, could not be trimmed out of lumber. Knots also cause problems in fiber-based products and paper; they are difficult to digest and refine because their mechanical and chemical properties differ compared to stem wood. An abundance of knots in young trees is a quality issue that technology has not yet overcome in structural uses of solid wood.

There has been enough interest in wood quality issues that several regional symposia have been held on the topic. These include Douglas-fir: Stand Management for the Future (Oliver et al. 1986), Juvenile Wood: What Does it Mean to Forest Management and Forest Products? (Forest Products Research Society 1986), and Forest Pruning and Wood Quality (Hanley et al. 1995). Silviculture and wood quality was considered at the High Quality Forestry Workshop: The Idea of Long Rotations (Weigand et al. 1994). There also have been reports that detail resource characteristics. For example, Kellogg (1989) addressed the management and product value of second-growth Douglas-fir and Middleton and Munro (2001) analyzed the relation of stand density management in second-growth western hemlock to product yields and attributes. In addition, the University of Washington, Seattle, offered workshops on the relation between forest management and wood quality in the 1990s.

4. Expected Effects of Management Trends on Wood Quality

The changes in management practices brought about by the Northwest Forest Plan on federal lands (Haynes and Perez 2001) and shifting regulations for state and private lands altered silvicultural practices substantially during the 1990s. Management on federal land has become less active and focused on structural and compositional diversity (Hummel 2003). Management on private

Figure 1. Tree exhibiting multiple flushing (a) and ramicorn branching (b).

timberlands, however, continues to emphasize timber production and increasing attention is given to protecting riparian and any upland areas that might provide habitat for species of concern. Management practices on state timberland falls somewhere between federal and private practices. In general, the question of how to manage all forested lands remains unsettled but changing regulations arising from public concerns for maintenance of aquatic and terrestrial biological diversity and aesthetics are likely to continue to force changes in silvicultural practices for some time.

In the Pacific Northwest, the general trend for timber supply is toward a future where most timber produced in the region will come from private timberland and will be much younger than it has been in the past (Haynes et al. 2003a). Younger timber is likely for several reasons. It is economically expedient to use short rotations and much of the timberland is owned by a forest industry with an interest in maximizing economic return. Harvest of younger (e.g., 40 to 50 years) stands is perceived as less risky because they are less frequently used by threatened or endangered species. Publicly traded companies that maintain large areas of "mature" timberland face the possibility of corporate takeovers by groups who will liquidate their older timber to finance the takeover. Finally, there is the risk that the wood processing infrastructure will no longer be capable of handling large logs from older stands (Barbour et al. 2002). This latter possibility might force both large and small nonindustrial landowners to favor younger stands with smaller trees.

The move toward younger stands affects wood quality three ways: (1) reduction in tree size, (2) reduction in clear wood without pruning because of insufficient time for branch shedding, and (3) reduction in physical properties (particularly strength, stiffness, and dimensional stability). The suitability of such wood for certain uses is affected because younger, longer crowned trees have a higher proportion of juvenile wood. The wood from these younger trees are increasingly used for structural products, fiber-based products, or engineered products designed to compensate for changing raw material properties.

Smaller trees eliminate the possibility of manufacturing the large timbers for which the Pacific Northwest region was once renowned, although technological changes have already eliminated a large part of this market. There are still specialty uses for the large, stiff timbers and very long or wide pieces, but in most commodity markets, engineered wood products, and in some cases alternative materials, have been substituted for these products (Eastin et al. 1996). As a result, the changes in quality associated with tree size have, for the most part, become irrelevant once trees are larger than about 10 to 12 inches (25-30 cm) diameter at breast height (dbh). In the Pacific Northwest, lack of processing capability for large trees actually results in a discount for logs larger than about 24 inches (60 cm) diameter on the large end.

Silvicultural practices that establish stands at relatively wide spacings and use thinning to keep them growing quickly also promote a long, live crown.

This virtually eliminates the possibility of recovering clear lumber and veneer. By the early 1990s, products like these were no longer manufactured in coastal Washington and Oregon (Table 9 in Warren 2002) because of insufficient raw material. In timber producing regions like New Zealand and parts of South America, pruning has been used to remove branches so that clear wood can be recovered from young trees. This wood is visually quite different from the high value Select and Shop lumber (WWPA 1988) that was once produced from old-growth Douglas-fir and hemlock in the Pacific Northwest. It has much faster growth rates and generally coarser texture than old growth.

In a recent survey of manufacturers who use clear wood, growth rate was not identified as a quality concern (Eastin et al. 1998). New technologies that use veneer, photographic, or other faux surface treatments to overlay composite or fast-grown clear wood foundations have largely filled the niche for clear finished or stained appearance uses where grain pattern is important. Alternative species (such as poplars (*Populus* spp.)), edge-glued and finger jointed panels, extruded wood plastic composites, and even pure plastics have largely replaced painted uses for clear wood.

5. Relating Wood Quality to Products

Wood product recovery studies designed to trace the processing of wood products (lumber and veneer) from the forest to the final product on a tree-by-tree basis have been conducted by the Ecologically Sustainable Production of Forest Resources (ESP) team and its predecessors at the Pacific Northwest Research Station since the early 1960s. Data for nearly 125 empirical wood product recovery studies are maintained in an electronic database (Stevens and Barbour 2000). In this section, we use these data to illustrate how product quality has varied by tree age and species.

Lumber grade distributions for visually graded lumber from Douglas-fir (data for Douglas-fir come from two studies, Fahey et al. 1991, and an unpublished study archived in the ESP Wood Quality Database (Stevens and Barbour 2000)) and hemlock trees for two age groups; less than 40 years old and from 40 to about 60 years old (Figures 2a and 2b). There are substantial differences in the distributions both between the two species and within each species. In both species, the proportion of lumber in the Select Structural (SS) and Number 1 (No. 1) grades is much higher in the older trees. In Douglas-fir (Figure 2a), most of the difference between the older and younger trees is caused by a shift from lumber graded as No. 2 into the two higher grades. In hemlock (Figure 2b), the largest shift in grade proportion is from the two lowest grades (Utility and Economy) into the No. 2 and Better grades. Although the quality of both species improves with age, the quality of Douglas-fir is better than hemlock in both age classes.

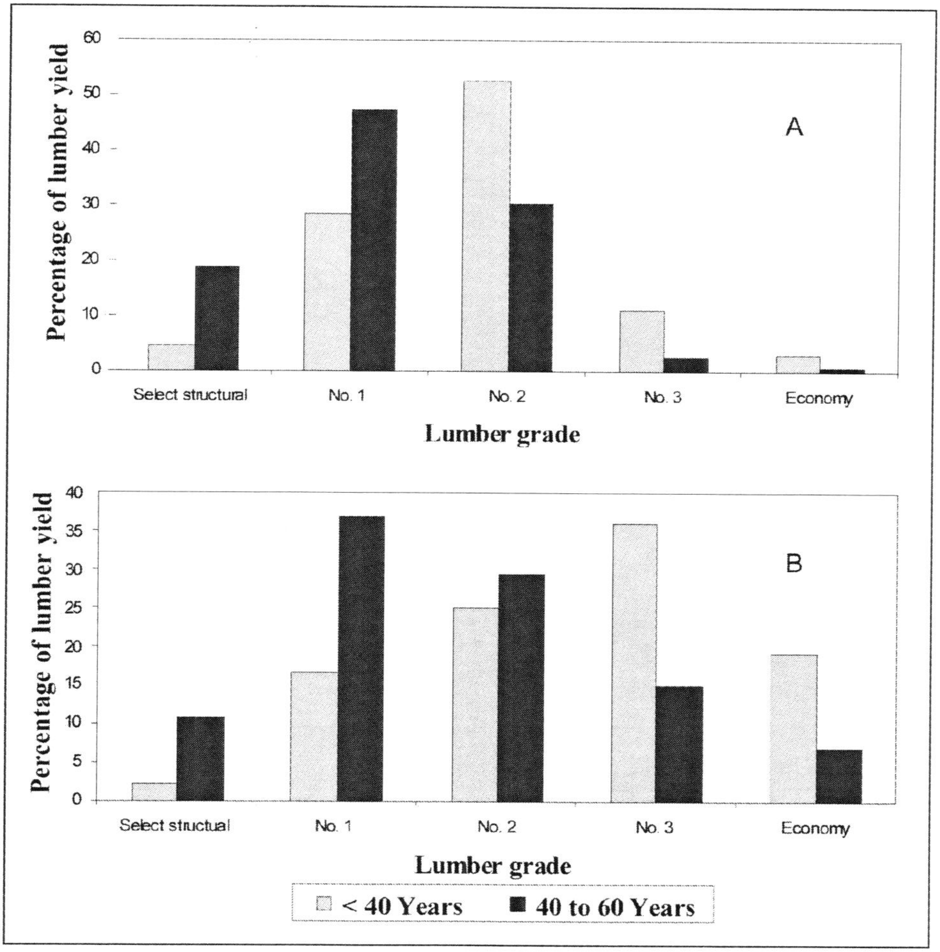

Figure 2. (a) Visual grade lumber yield from young Douglas-fir trees less than 40 years old and from trees 40 to 60 years old. (b) Visual grade lumber yield for young western hemlock trees less than 40 years old and from trees 40 to 60 years old.

In a practical sense the grade shift in hemlock is more important than that in Douglas-fir because only rarely does No. 1 lumber command a premium over No. 2 lumber, but No. 2 always commands a substantial premium over Utility and Economy grade lumber. Most structural lumber is currently sold in grade groups of Standard and Better or No. 2 and Better depending on whether it is graded under Light Framing or Structural Light Framing and Joists and Planks rules (WWPA 2002). In the example given here, the total amount of No. 2 and Better is probably more important than the distribution of grades within the grade group. The proportion of Number 2 and Better lumber is greater for the older trees in both species; the difference is greater in Douglas-fir and is lower for hemlock. Young Douglas-fir had a better grade distribution than older hemlock.

A rule of thumb for sawmills in the western United States is that an efficient mill with a good-quality resource should produce 90% or more No. 2 and Better lumber. The average for the two Douglas-fir studies represented in Figure 2a suggest that this would easily be possible for the 40- to 60-year-old trees, and would be almost possible for trees younger than 40 years. However, the result for hemlock is much different. Only 44% of the lumber volume from the younger trees and 77% of the lumber volume from the older trees was graded as No. 2 and Better. The corresponding results for Douglas-fir were 86% and 96%.

When the same lumber used to construct Figure 2 was tested for machine stress rating, the material from the younger trees of both species performed poorly (Figure 3 a,b). Only 44% of the younger Douglas-fir and less than 2% of the younger hemlock was assigned stress ratings for 1450 fb (extreme fiber stress in bending) or higher. This is important because 1450 fb is essentially the same stress rating that is assigned to No. 2 visually graded lumber for many North American species, although Douglas-fir and southern pine have higher stress ratings (WWPA 1988, National Lumber Grades Authority 2003). Because the distributions for visually and machine graded lumber came from the same sample trees, this result suggests that a large amount of the lumber graded as No. 2 and Better under the visual rules did not actually have mechanical properties required by those grades. This was more important in hemlock, where almost none of the lumber met the required minimum stress limits. Even in Douglas-fir, less than half of the lumber met the minimums. The situation was no better with the older hemlock. It did improve somewhat with older Douglas-fir where almost 78% of the lumber was rated as greater than 1450 fb. Even in this case, the amount of lumber that was rated in the higher stress classes was fairly low.

The data used to construct Figures 2 and 3 came from trees that were selected for rapid growth. Little or no information on stand history was available. It is reasonable to draw the general inference that the quality of younger trees, as expressed by visual lumber grade or lumber bending properties, is likely to be lower than for older trees, and that one or two decades of extra growth can make considerable differences in quality. The results for mechanically rated lumber suggest that the influence of juvenile wood is still strongly felt in 40-year-old Douglas-fir trees and even to some extent in 60-year-old trees.

For Douglas-fir, it is possible to explore the relations between stand management and quality more fully by using simulation because a reasonably reliable simulation model is available for this species (Hann et al. 1997). The wood quality module of ORGANON uses various submodels for branching and juvenile wood that were developed from large empirical data sets (Maguire et al. 1991). There are no similar simulation tools for hemlock at this

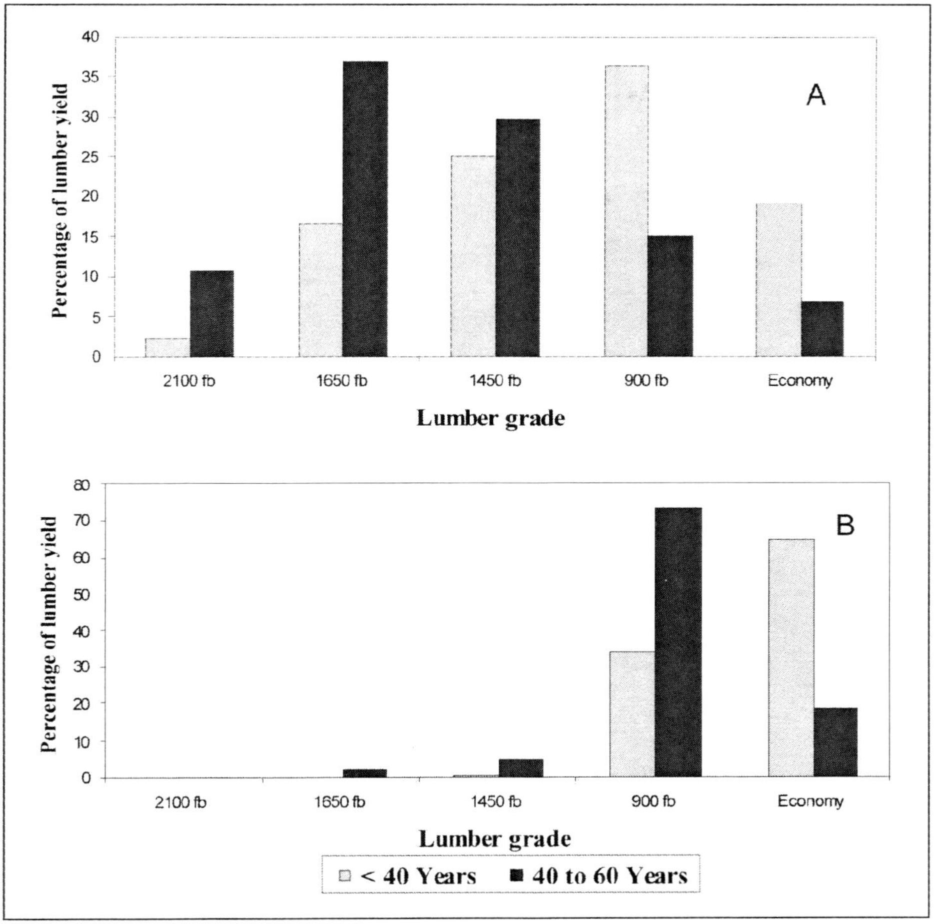

Figure 3. (a) Machine-stress-rated lumber grade yield for young Douglas-fir trees less than 40 years old and from trees 40 to 60 years old. (b) Machine-stress-rated lumber grade yield for young western hemlock trees less than 40 years old and from trees 40-60 years old.
Note: fb = extreme fiber stress in bending.

time, although more research is being devoted to this species and the development of wood quality models (Stevens and Barbour 2000; Middleton and Munro 2001; Singleton et al., in press).

6. Possible Future Management Strategies

Although timber production and clearcutting once dominated the region, society is demanding a greater range of commodity and noncommodity products and outcomes from a decreasing forestland base. Contemporary forest management in the Pacific Northwest falls into three broad strategies: (1)

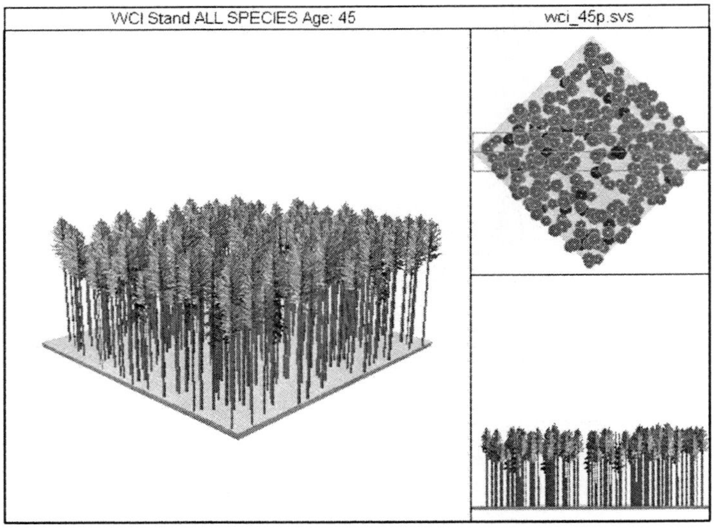

Figure 4. Simulated stand for the Timber Production regime at rotation age 45 years from the Stand Visualization System (McGaughy 1997).

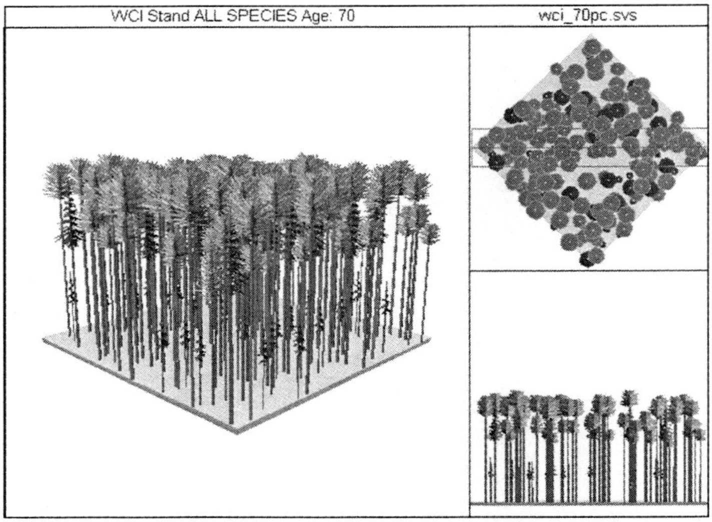

Figure 5. Simulated stand for the Extended Rotation regime at rotation age 70 years from the Stand Visualization System (McGaughy 1997).

timber and fiber production, (2) extended rotations for production of wood and other values, and (3) management for biodiversity objectives. Timber is the primary objective of the first, but demonstrating of the contribution of other nontimber outputs is becoming increasingly important in terms of public good will toward forestry (Curtis et al. 1998). The second balances timber and non-

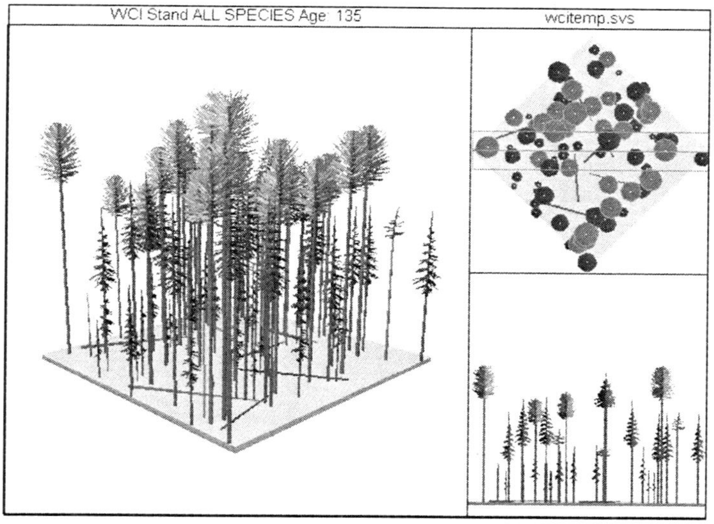

Figure 6. Simulated stand for the Biodiversity regime at rotation age 135 years from the Stand Visualization System (McGaughy 1997).

timber outputs, whereas under the third strategy, timber production is possible as a byproduct of managing for other values. The lines between these three strategies are, of course, blurred in actual practice, and they continue to evolve as the political climate and economic conditions change.

The rest of this section discusses inferences for wood quality associated with these three general management strategies that are defined below. They were simulated by using the Stand Management Cooperative version of the ORGANON (Hann et al. 1997, Monserud 2003) growth model for western Oregon and Washington (Figures 4, 5, 6). These inferences are developed from simulations based on available data and they illustrate differences in wood characteristics that may result from different types of management.

(1) Timber and Fiber Production. The primary objective on these timberlands is to produce an economically efficient crop while meeting forest practices rules regulating harvesting, planting, and other forest management activities. Oregon, Washington, Alaska, and British Columbia all have forest practices rules that require protecting riparian areas during harvesting and management activities, prompting regeneration, and leaving some residual standing trees and down wood (Alaska Division of Forestry, Department of Natural Resources 2000, Oregon Department of Forestry 2001, Washington Department of Natural Resources 2002, British Columbia Ministry of Forests 2003). Silvicultural regimes on land primarily managed for timber production typically begin with planting about 400 trees

per acre (1000 trees per hectare (ha)) after a clearcut harvest. To insure tree survival and encourage rapid stand establishment and tree growth, the young plantation may be treated to reduce competing vegetation. Early stand density may be controlled with a precommercial thinning, especially if natural regeneration is abundant, which can reduce growth or survival of the planted crop trees. As the trees begin to reach merchantable size, a commercial thinning may be applied to reduce competition and to salvage anticipated mortality. Fertilization is sometimes used to increase growth. Stands are typically harvested by clearcutting by age 50, but shorter rotations of 40 years are becoming more common. Rotation ages will likely decrease over time as long as a target tree size of 14 to 20 inches (35-50 cm) at breast height can be achieved. The silvicultural methods for effective establishment and rapid growth of Douglas-fir are well developed (Curtis et al.1998) and are represented in the management of millions of acres of industrial land in western Oregon and Washington (see Table 1 Chapter 10). The ORGANON simulation used in the timber and fiber strategy was a 45-year rotation with a precommerical thinning at age 15 to 330 trees per acre (800 trees per ha).

(2) Extended Rotations for Production of Wood and Other Values. Given that Douglas-fir stands can maintain high growth rates in volume and value for extended periods (in excess of 100 years), several researchers have proposed extended rotations (Kuehne 1994, Curtis 1995) based on the premise that frequent thinnings and a final harvest of high-value trees will provide both financial benefits for landowners and improved habitat opportunities. The basis for this approach is a set of studies that show repeated commercial thinning can prolong the period of high productivity (Curtis and Marshall 1993, Curtis 1995). Such a management strategy provides income during the rotation while producing aesthetically desirable large trees. Furthermore, extended rotations reduce the area of land in the highly visible (and considered by some to be unattractive) regeneration stage. Repeated commercial thinning is also expected to produce an understory of shade-tolerant species (e.g., western hemlock), which add to the structural and species diversity of stands (see Curtis and Carey 1996, Curtis 1997 for discussion of these and other potential benefits to extended rotations). Typical rotation ages will range from 70 to 120 years, with regeneration initiated by clearcutting, shelterwood, or patch cuts. These regimes will produce large and potentially high-value trees in both domestic and export markets. However, with the recent collapse of the log export market and growing concern about lack of processing facilities for converting larger logs (greater than 20 to 24 inches (50-60 cm) on the large end), perceptions are changing about the value of large logs (Barbour et al. 2002). These changing perceptions could cause private landowners to rethink the

viability of extended rotations. The Extended Rotation simulated regime is a 70-year rotation with a precommerical thinning at age 15 to 400 trees per acre (1000 trees per ha) and two commercial thinnings: one at age 30 to a stand density index (SDI) of 200 and a second at age 45 to SDI of 230.

(3) Management for Biodiversity. Owing to past management practices, the late seral-stages (Appendix 1, Chapter 1) of forest development occupy much less area than they did when Europeans settled the region. Consequently, the status of organisms and processes associated with late-seral-stage forests is currently a concern. On much of the region's federal lands, the primary management goal has become restoration or maintenance of the ecological integrity of forest ecosystems and resilience of social and economic systems that depend on the forest (Haynes et al. 1996). Current forest management policies are intended to (1) maintain viable populations of native species, (2) support evolutionary processes and adaptation to changing biotic and abiotic conditions within robust ecological communities, (3) account for the multiple spatial and temporal contexts within which ecosystems function, (4) recognize the human sense of "place" or the spiritual value of the forest, and (5) manage to enhance the social and economic resiliency of human communities associated with forests. Although still under development, many aspects of the management systems designed to achieve these policy goals will be applied at the landscape scale and alter existing age class and species distributions and spatial patterns (Crow and Gustafson 1997). At the stand level, management objectives are designed to provide mid- to late-seral characteristics as related to tree size, crown closure, and stand structure (Appendix 1, Chapter 1). Some management regimes include thinning at early ages to low densities to stimulate tree growth and understory development and using rotations of 100 or more years (Barbour et al. 1997). Through legacy retention and intentional management activities, stands can be moved more rapidly toward increased diversity as compared to a no-management or reserve strategy (Carey and Curtis 1996). Most past silvicultural practices were aimed at enhancing the quantity or quality of wood produced. However, Curtis et al. (1998) state that many of these same techniques can be modified to maintain, produce, or restore wildlife habitat, diverse stand structures (including those usually associated with old forests), and visual aesthetics. The biodiversity regime outlined by Carey and Curtis (1996) minimizes or eliminates the competitive exclusion stage through heavier reductions in initial densities with heavier precommercial thinnings, which promotes initiation and retention of an understory. With repeated heavy commercial thinnings, stands are maintained in a more open condition, greater understory diversity develops, and new trees are established by natural regeneration. Keeping a legacy of large trees and down wood from

Table 2. Initial stand conditions for all simulations. Initial stand density index was 88, height of 40 largest trees was 27.4 feet (8.4 meters), crown ratio was 0.74, and the site index was 128 feet (39 m) at 50 years.

Species	Trees		Basal area		Quadratic mean diameter	
	Number/acre	*Number/hectare*	*Square feet /acre*	*Square meter/hectare*	*Inches*	*Centimeters*
Douglas-fir	450	1110	26.9	6.2	3.3	8.4
Western hemlock	100	250	3.5	0.8	2.5	6.3
Total	550	1350	30.4	7.0	3.2	8.1

the previous stand creates snags and coarse wood on the forest floor. By extending the rotation, a fully functional stage of ecosystem development can be achieved. Carey et al. (1999) simulated this biodiversity strategy and compared it to a similar timber and fiber production regime and a no-management preservation regime. They concluded that practicing biodiversity regimes on portions of the landscape (including shorter rotations than used here) could produce economically valuable forest products, achieve ecological benefits, and support 100% of the species occurring in western Washington forests (Carey and Curtis 1996, Carey 2003). This Biodiversity regime simulated a 135-year rotation with a precommerical thinning at age 15 to 295 trees per acre (725 trees per ha) and commercial thinnings at age 30 to SDI of 160, at age 45 to SDI of 175, and at age 65 to SDI of 210, with ingrowth added through the rotation.

7. Simulated Management Results

We simulated these three management strategies to illustrate (Figures 4, 5, 6) how the conditions under which trees develop will influence wood characteristics and ultimately general categories of wood product potential (Barbour and Parry 2001; Barbour et al. 2002; Barbour et al., in press). All simulations were conducted by using ORGANON and the same initial stand conditions (Table 2). By starting with a single set of stand conditions, comparisons can be made among regimes. The outputs of these simulations are in terms of characteristics such as knot size and frequency, tree and log size, and log grade. These characteristics can be translated into wood product yields, specifically lumber and veneer yields (see Kellogg 1989, Fahey et al. 1991). The limited tree age ranges used to develop the underlying models introduces uncertainty when they are used to analyze the wide age and tree size range expected in these management strategies. Instead, we display the results in terms of potential log grade coupled with a direct summary of results from empirical lumber recovery data. We believe this provides a more precise picture of the likely wood product recoveries from the three management strategies.

These three broad scenarios allow managers to exercise wide latitude in designing specific management options. They can potentially result in large differences in ecosystem development such as those shown in Table 3. While all three of the scenarios go through an initial stage of ecosystem initiation (stage 1, Appendix 1, Chapter 1), managing for timber production by using either short or extended rotations has extended the time of competitive exclusion where trees fully occupy the site and there is high stand volume growth, a primary objective of these regimes. Franklin et al. (2002) proposed a modified classification scheme of stand development stages that call this the biomass accumulation/competitive exclusion stage (Appendix 1, Chapter 1), which is the objective in timber production. For the shorter rotation regime, higher densities are maintained to maximize wood production with only limited understory development. By extending the rotation and reducing densities by using multiple thinnings, some understory initiation and development is encouraged and biodiversity is increased. Figures 4, 5, and 6 illustrate the simulated regimes at various "rotation" ages by using the Stand Visualization System (McGaughey 1997).

Results show substantial differences in wood characteristics among the three regimes used to illustrate likely management pathways for different landowners in the Pacific Northwest. A summary of the stand characteristics and volume yields are given in Table 4. As expected, the size of the trees produced under the various regimes increases with total regime age. The average annual volume production is, however, lowest for the Biodiversity regime, which also produces the oldest trees. This occurs because frequent thinning keeps the stand from being fully occupied over much of the rotation.

Yield by tree size on a merchantable volume basis is displayed for the three regimes in Tables 5, 6, and 7. These tables document the shift from smaller trees, generally in the 10 to 16 inch (25-40 cm) dbh range for the Timber Production regime to trees in the 24 to 32 inch (60-80 cm) dbh range for the final harvest from the Biodiversity regime. In this latter simulation, about 25% of the volume is from the thinning at stand age 90 years, and 40% of the volume from the final harvest comes from trees greater than 24 inches (60 cm) dbh. The butt logs (at base of stem) and perhaps the second logs from these trees are larger than those typically processed by sawmills in the region (Barbour et al. 2002); currently their prices are discounted by 15 to 20% as compared to smaller logs. These trees could contain wood with characteristics (e.g., slower growth rates, tight rings, and no or few knots) that would make them desirable for high-value uses (Barbour and Parry 2001).

The three simulated regimes also influence the percentage of volume of juvenile wood in the tree. The two methods for defining juvenile wood (base of live crown and number of rings) and percentage of total volume (includes all thinnings and the final harvest) are illustrated in Figure 7a. The difference between the two methods declines as older trees are brought into the mix.

Table 3. Development of three management regimes in the western hemlock zone forest of western Washington by stages of ecosystem development (Appendix 1, Chapter 1) and potential stand management treatments.

	Timber and fiber production		Extended rotation		Biodiversity	
Decade	**Stage**	**Treatments**	**Stage**	**Treatments**	**Stage**	**Treatments**
1	1: Ecosystem initiation	Plantation management	1: Ecosystem initiation	Plant	1: Ecosystem initiation	Plant
2	1: Ecosystem initiation	Precommercial thin	2/3: Competitive exclusion	Precommercial thin	4: Understory reinitiation	Precommercial thin
3	2/3: Competitive exclusion	Commercial thin and fertilization	2/3: Competitive exclusion	Commercial thin	4: Understory reinitiation	Variable-density thin
4	2/3: Competitive exclusion	None or clearcut	4: Understory reinitiation	None	4: Developed understory	None
5	2/3: Competitive exclusion	Clearcut	4: Understory reinitiation	Commercial thin	4: Developed understory	Coarse woody debris Thin
6			4: Developed understory	None	5: Niche diversification	None
7			4: Developed understory	Commercial thin	5: Niche diversification	Coarse woody debris Thin
8			5: Botanically diverse	None	5: Niche diversification	None
9			5: Botanically diverse	None or harvest	5: Niche diversification	None
10					5: Fully functional	None
11					5: Fully functional	None or harvest
12					5: Fully functional	None or harvest

Table 4. Summary statistics for the three regimes where merchantable cubic foot volume is calculated to a 6-inch top (15 cm) for 32 foot logs (9.75 m) with a minimum top log length of 16 feet (4.9 m).

Regime	Final harvest age	Commercial thinning(s)				Final harvest				Total volume		Average volume per year	
		Volume[1]		QMD[2]		Volume		QMD					
		ft³/ac	*m³/ha*	*in*	*cm*	*ft³/ac*	*m³/ha*	*in*	*cm*	*ft³*	*m³*	*ft³*	*m³*
Timber													
Production	45	0	0	0	0	8,041	565.8	11.8	30.0	8,041	565.8	179	12.6
Extended													
Rotation	70	2,939	206.8	9.9	25.1	10,668	750.6	15.6	39.6	13,607	957.4	194	13.6
Biodiversity	135	5,589	393.2	11.3	28.7	15,342	1079.4	19.9	50.5	20,931	1472.7	155	10.9

[1] ft^3/ac = cubic feet per acre; m^3/ha = cubic meters per hectare.
[2] QMD = quadratic mean diameter.

Table 5. Percentage of merchantable volume cut in each entry by diameter class and species for the Commercial Timber regime.

Diameter class		Percentage of merchantable volume in final harvest	
		Douglas-fir	Western hemlock
Inches	*Centimeters*		
8–10	20–25	15.7	-
10-12	25–30	25.5	-
12–14	30–35	33.9	-
14-16	35–40	5.5	6.0
16–18	40–45	-	13.4
18–20	45–50	-	-

Table 6. Percentage of total volume cut in each entry by diameter class and species for the Extended Rotation regime.

Diameter class		Percentage of merchantable volume for Extended Rotation					
		30 years		45 years		70 years	
		Douglas-fir	Western hemlock	Douglas-fir	Western hemlock	Douglas-fir	Western hemlock
Inches	*Centimeters*						
2-4	5-10	-	-	-	-	-	-
4-6	10-15	-	-	-	-	-	-
6-8	15-20	16.0	-	-	-	-	-
8–10	20-25	54.0	9.0	15.0	-	-	-
10-12	25-30	6.0	15.0	27.0	-	9.7	-
12-14	30-35	-	-	27.0	-	12.5	-
14-16	35-40	-	-	7.0	10.0	18.1	-
16-18	40-45	-	-	-	14.0	19.4	-
18-20	45-50	-	-	-	-	13.2	-
20–22	50-55	-	-	-	-	1.8	19.1
22–24	55-60	-	-	-	-	-	6.1
24–26	60-65	-	-	-	-	-	-

Plotting the results for the two methods for the Extended Rotation and Biodiversity regimes (Figures 7b and 7c) illustrate the large differences between the two methods for young trees and how the two estimates converge for older trees.

Another interesting aspect of the Biodiversity regime is an expected tendency to favor regeneration of western hemlock over Douglas-fir. This is expected to occur because shade-tolerant hemlock regenerates more successfully in partially cut stands than Douglas-fir. On a volume basis, western

Table 7. Percentage of merchantable volume from trees cut in each entry by diameter class and species for the Biodiversity regime.

Diameter class		**Percentage of merchantable volume for Biodiveristy regime**							
		30 years		**55 years**		**90 years**		**135 years**	
		Douglas-fir	**Western hemlock**	**Douglas-fir**	**Western hemlock**	**Douglas-fir**	**Western hemlock**	**Douglas-fir**	**Western hemlock**
Inches	*Centimeters*								
0–2	0-5	-	1.0	-	-	-	-	-	-
2-4	5-10	-	-	-	-	-	-	-	-
4–6	10-15	-	-	-	-	-	-	-	-
6–8	15-20	-	-	-	1.0	-	-	-	-
8-10	20-25	30.0	-	-	-	-	2.0	-	3.0
10-12	25-30	46.0	-	6.0	-	-	1.0	-	-
12-14	30-35	6.0	17.0	16.0	-	-	1.0	-	-
14-16	35-40	-	-	32.0	-	6.0	5.0	-	1.0
16-18	40-45	-	-	25.0	-	13.0	1.0	8.0	6.0
18-20	45-50	-	-	3.0	10.0	17.0	-	6.0	6.0
20–22	50-55	-	-	-	7.0	14.0	-	7.0	7.0
22–24	55-60	-	-	-	-	16.0	-	7.0	10.0
24–26	60-65	-	-	-	-	8.0	14.0	9.0	-
26–28	65-70	-	-	-	-	2.0	-	9.0	-
28–30	70-75	-	-	-	-	-	-	5.0	11.0
30–32	75-80	-	-	-	-	-	-	3.0	0
32–34	80-85	-	-	-	-	-	-	-	-
34–36	85-90	-	-	-	-	-	-	-	-
36–38	90-95	-	-	-	-	-	-	3.0	-

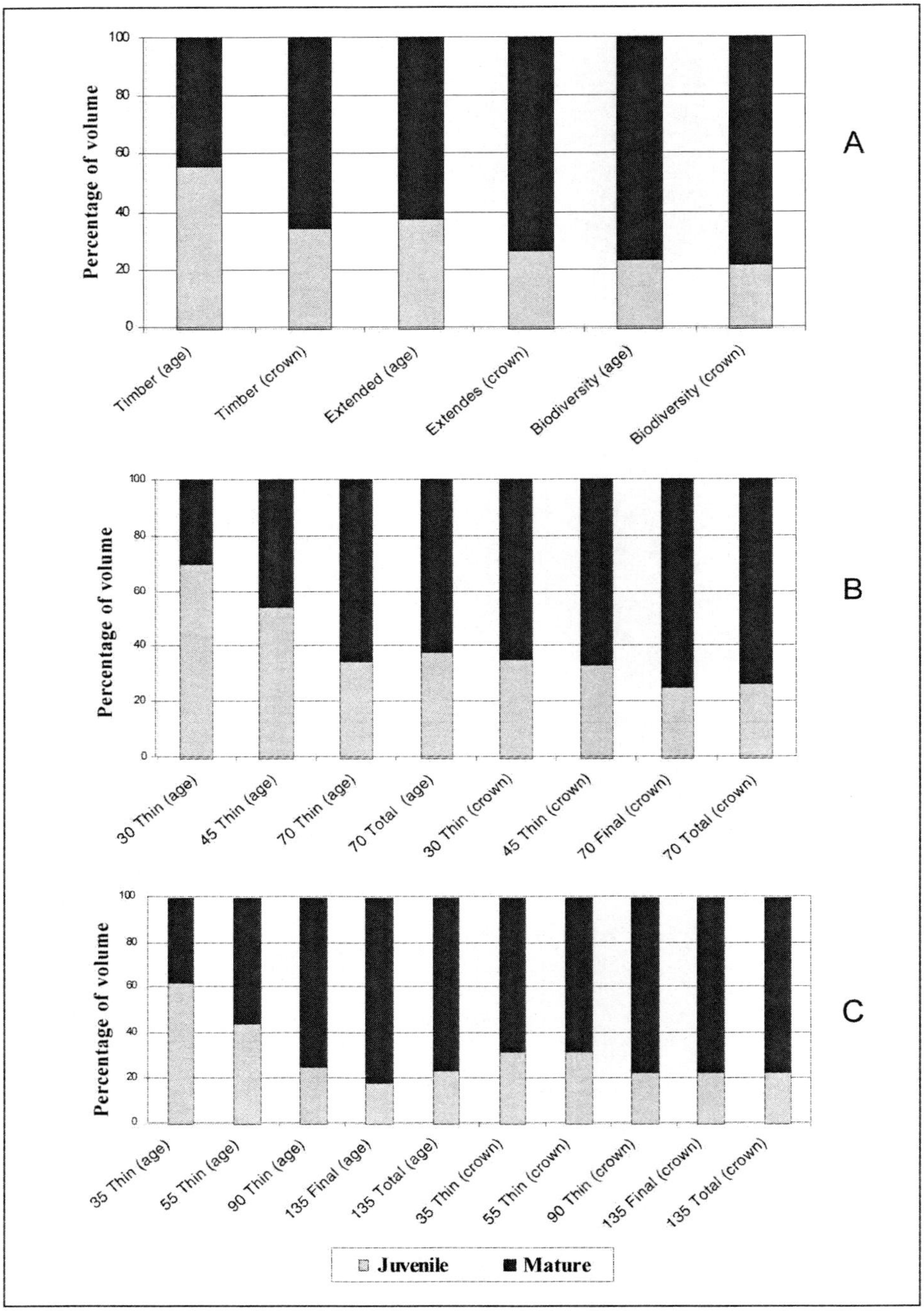

Figure 7. Percentage of total volume in juvenile or mature wood for (a) all thinnings and final harvest (b) Extended Rotation regime and (c) Biodiversity regime. Juvenile wood is determined by age (20 rings from pith) or crown (base of live crown).

Table 8. Criteria used to estimate log grades[1] (from Barbour and Parry 2001).

Grading criteria	Log grade Special Mill	No. 2 Sawmill	No. 3 Sawmill	No.4 Sawmill
Minimum rings per inch outer portion	6	None	None	None
Minimum scaling diameter (inches)	16	12	6	5
Maximum branch diameter (inches)	1.5	2.5	3	None
Minimum length (feet)	17	12	12	12

[1] These criteria follow the Northwest Log Rules Advisory Group (1998) rules except for the branch frequency requirements for Special Mill and net scale requirements for all grades. Failure to include the net scale requirement is most important for No. 4 Sawmill logs because in practice, larger logs with high defect levels are placed in this grade. In this analysis, only small logs or those with large branches were graded as No. 4 Sawmill.

hemlock represents about 20% of the harvest from the timber production regime and 25% of the volume from each of the entries and the final harvest for the Extended Rotation. Under the Biodiversity regime the proportion of hemlock increases from about 20% after the first two commercial thinnings, to 25% after the third thinning, and 45% after the final harvest. In current markets, Douglas-fir receives a 21% premium over hemlock so the increased yield of hemlock in both the Extended Rotation and Biodiversity regimes may present a further disincentive to private landowners who choose to manage their land for older stands. Neither of these potential disincentives to longer rotations, or size and species shift consider the quality of the trees. We are unable to simulate wood quality for hemlock because there is no branch model for this species as there is for Douglas-fir (Maguire et al. 1991).

Log grade yield is a practical indicator of wood quality. It is easily derived for Douglas-fir from simulations using ORGANON (Christensen 1997, Barbour and Parry 2001). The criteria we used to assign log grades are given in Table 8. The existing log grading rules (Northwest Log Rules Advisory Group 1998) consider additional criteria but they generally apply to older trees that contain defects not commonly found in healthy, undamaged, young (less than 200 years old) trees such as those considered here. The current log mix harvested from private land typically consists of No. 2 Sawmill and No. 3 Sawmill logs. Marked deviations from those grades in the simulations would indicate a change in quality. Log grade yields for Douglas-fir from the three simulated regimes are given in Table 9. About a third of the logs recovered under the Bodiversity regime are Special Mill grade (which may be a little conservative because this regime is an extrapolation of the branch models used and more clear wood might be expected at this age than predicted by the models), but otherwise all of the logs from all three regimes were graded either No. 2 Sawmill or No. 3 Sawmill. Based solely on external characteristics, the quality of the logs recovered from the Timber Production and Extended

Table 9. Percentage of total harvested merchantable volume (thinnings plus final harvest) by wood quality characteristic.

Wood quality characteristic		Regime		
		Timber	Extended	Biodiversity
			Percent	
Log grade:				
#3 peeler		0	0	1
Special Mill		0	0	24
#2 sawlog		34	28	35
#3 sawlog		66	72	40
Avg. largest knot:				
Inches	*Centimeters*			
0–0.5	0-1.25	0	0	
0.6–1.0	1.5-2.5	59	56	30
1.1–1.5	2.8-3.8	41	43	57
1.6–2.0	4.1-5.1	0	2	13
2.1–2.5	5.3-6.3	0	0	1
Number of whorls:				
< 16		94	79	77
16		6	15	8
17		1	5	4
18			1	1
19				1
> 19				10

Rotations regimes may be similar to those harvested today; especially those harvested from private land.

The logs from the Biodiversity regime would be expected to have higher quality than those typically available on the market today because of a combination of diameter (at least 16 inches (40 cm) on the small end), knot size, and growth rate (Table 9). Barbour and Parry (2001) found that logs from trees grown under regimes similar to the Extended Rotation and Biodiversity regimes were placed in the Special Mill grade. Some logs, though, will be assigned to lower log grades for various reasons, including the persistence of branches and the number of growth rings per inch. In theory, the Special Mill logs contain at least 65% of their net scale (the portion of the log recoverable as lumber or veneer) in appearance grades of lumber and veneer (Northwest Log Rules Advisory Group 1998). However, it is uncertain if the industry in the coastal Pacific Northwest would take advantage of the potential to manufacture higher value products because of the almost exclusive focus on structural products (Table 9 in Warren 2002).

On the other hand, if ORGANON accurately projects the number and size (Table 9) of branches in Douglas-fir trees managed under the Extended Rotations, then the lumber or veneer grade yields from these regimes should be good regardless of the assigned log grade. Even the No. 2 and No. 3 Sawmill logs would contain relatively few knots, with none of the knots particularly large. Under the Biodiversity regime less than 20% of the log volume has more than 16 branch whorls. This suggests that structural lumber manufactured from these trees could have relatively high yields of higher value structural grades (such as Select Structural). Furthermore, they could have reasonable yields of higher stress level lumber if machine stress rating is used to select the pieces that are best suited for engineered wood products.

8. Discussion

The context for discussing wood quality is set by the relentless interaction among markets for forest products, resource conditions, and wood characteristics. In the Pacific Northwest, these discussions are influenced by a transition (see Haynes et al. 2003a) in the products manufactured from the available resources. In addition to technical and resource changes, changes in consumer preferences, including declining use of appearance grades is influencing notions of wood quality. These changes are focusing the discussion of wood quality on the production of commodity grade structural lumber products, specialty panels (for the domestic and export markets), and a declining proportion of log and cant exports. All but the last of these products can be produced from common log mixes.

Proponents of producing high-quality material are left with declining (but highly valuable) niche markets that utilize unique wood characteristics. There have been several discussions on whether the costs of producing high-quality logs justify the results (Johnson 1986, Fight et al. 1995). These concerns prompt discussions about whether the logging, manufacturing, and product distribution infrastructure will be in place if and when larger trees with higher quality wood are harvested (Barbour et al. 2002).

Uncertain opportunities for profitable conversion raises questions about the emphasis of silvicultural treatments and whether the quantity of wood available for a targeted market may be sufficient to sustain processing facilities. There is also the consideration of removing the material from the woods. Accessibility is a key point, both from the logging standpoint (e.g., roads, topography, and geographic location) and from the market viewpoint (e.g., price and demand).

At the same time, shifting forest management objectives influence the resource available for wood products. Each regime potentially influences wood quality (e.g., species mix, juvenile wood proportion, and knot size), which in turn affects value and feasibility of use. Within current management discussions, overcoming the economic challenges of growing trees more slowly (longer rotations) is a greater problem than developing new technologies to use younger wood. Technology has enabled us to develop substitute wood products from the resource as it now exists by focusing on performance-based standards to develop engineered wood products. We now find ourselves considering the question of what can be done with larger and relatively higher quality trees. Given that the forest products industry is currently using a much different raw material, when will a market (that pays a premium) for trees with old-growth characteristics reappear?

There are still some people, however, who value the aesthetics of solid wood products produced from trees with old-growth characteristics. The presence of various niche markets for old-growth-like products will lead to the development of management regimes that produce trees of the size and quality associated with these old-growth characteristics. These markets will have to be large enough to overcome the long-term investment costs of producing larger and older trees. Finally, the loss of export markets in the mid to late 1990s also has helped shift utilization efforts toward niche markets where producers take advantage of the unique quality characteristics in the wood.

Finally, given its diverse land tenure, the Pacific Northwest will continue to be a region where a variety of management strategies are used to meet societal demands for goods and services from the forest. Some of these strategies, however, will produce trees with limited markets, reducing the financial incentives for their implementation. At the same time, the diversity of management regimes will lead to a variety of tree sizes and other quality attributes that will create opportunities for the intrepid entrepreneur looking for an opportunity in the forest products industry.

9. References

Adams, W.T., and Howe, G. 1984. Measurement strategies for young Douglas-fir progeny tests: study plan. Pacific Northwest Tree Improvement Research Cooperative. Corvallis, OR: Department of Forest Science, College of Forestry, Oregon State University.

Alaska Division of Forestry, Department of Natural Resources. 2000. Alaska Forest Resources and Practices Act. Anchorage, AK. 25 p.

Allin, L.C. 1995. Shipbuilding and shipping in the period of ascendancy. In: R.W. Judd, E.A. Churchill, and J.W. Eastman (Eds.), Maine: the pine tree state. Orono, ME: University of Maine Press.

Bannan, M.W. 1956. Some aspects of the elongation of fusiform cambial cells in *Thuja occidentalis* L. *Canadian Journal of Botany*, 34: 175-196.

Bannan, M.W. 1962. Cambial behaviour with reference to cell length and ring width in *Pinus strobus* L. *Canadian Journal of Botany*, 40: 1057-1062.

Bannan, M.W. 1964. Tracheid size and anticlinal divisions in the cambium of *Pseudotsuga*. *Canadian Journal of Botany*, 42: 603-631.

Bannan, M.W. 1966. Spiral grain and anticlinal divisions in the cambium of conifers. *Canadian Journal of Botany*, 44: 1515-1538.

Bannan, M.W. 1970. A survey of cell length and frequency of multiplicative divisions in the cambium of conifers. *Canadian Journal of Botany*, 48: 1585-1589.

Barbour, R.J., Johnston, S., Hayes, J.P., and Tucker, G.F. 1997. Simulated stand characteristics and wood product yields from Douglas-fir plantations managed for ecosystem objectives. *Forest Ecology and Management*, 91: 205-219.

Barbour, R.J., and Parry, D.L. 2001. Log and lumber grades as indicators of wood quality in 20- to 100- year-old Douglas-fir trees from thinned and unthinned stands. General Technical Report PNW-GTR-510. Portland, OR: U.S. Department of Agriculture, Forest Service, Pacific Northwest Research Station. 22 p.

Barbour, R.J., Marshall, D., Parry, D., and Christensen, G. 2002. Do large trees always have high wood product value? In: A.C. Johnson, R.W. Haynes, and R.A. Monserud (Eds.), Congruent management of multiple resources: proceedings from the compatibility initiative workshop (pp. 135-144). General Technical Report PNW-GTR-563. Portland, OR: U.S. Department of Agriculture, Forest Service, Pacific Northwest Research Station.

Barbour, R.J., Zaborske, R.R., McClellan, M.H., Christian, L. and Golnick, D. [In press]. Young-stand management options and their implications for wood quality and other values. *Landscape and Urban Planning*.

Barret, J.D., and Kellogg, R.M. 1989. Strength and stiffness of dimension lumber. In: R.M. Kellogg (Ed.), Second growth Douglas-fir: its management and conversion for value (pp. 50-58). Special Publication No. SP-32. Vancouver, BC: Forintek Canada Corp.

Bastien, J.C., Roman-Amat, B., and Vonnet, G. 1985. Natural variation of some wood quality traits of coastal Douglas-fir in a French progeny test: Implications on breeding strategy. S2.02.05 –Working Party. Vienna: IUFRO.

Bendtsen, B.A., and Senft, J. 1986. Mechanical and anatomical properties in individual growth rings of plantation-grown eastern cottonwood and loblolly pine. *Wood and Fiber Science*, 18(1): 23-38.

Briggs, D.G., and Smith, R. 1986. Effects of silviculture practices on wood properties of conifers: a review. In: C.D. Oliver, D.P. Hanley, and J.A. Johnson (Eds.), Douglas-fir: stand management for the future (pp. 108-117). Contribution No. 55. Seattle, WA: University of Washington, College of Forest Resources, Institute of Forest Resources.

British Columbia Ministry of Forests. 2003. Forest practices code of British Columbia Act. Chapter 159. Victoria, BC: Queen's Printer.

Burns, R.M., and Honkala, B.H. (Tech. cords.). 1990. Silvics of North America, Vol. 1, Conifers. Agriculture Handbook 654. Washington, DC: U.S. Department of Agriculture, Forest Service. 675 p.

Carey, A.B. 2003. Managing for wildlife: a key component for social acceptance of compatible forest management. Chapter 14. In: R.A. Monserud, R.W. Haynes, and A.C. Johnson (Eds.), Compatible forest management. Dordrecht, The Netherlands: Kluwer Academic Publishers.

Carey, A.B., and Curtis, R.O. 1996. Conservation of biodiversity: a useful paradigm for forest ecosystem management. *Wildlife Society Bulletin*, 24(6): 610-620.

Carey, A.B., Lippke, B.R., and Sessions, J. 1999. Intentional systems management: managing forests for biodiversity. *Journal of Sustainable Forestry*, 9: 83-125.

Cave, I.D., and Walker, J.C.F. 1994. Stiffness of wood in fast-grown plantation softwoods and the influence of microfibril angle. *Forest Products Journal*, 44(5): 43-48.

Crow, T.R., and Gustafson, E.J. 1997. Ecosystem management: managing natural resources in time and space. In: K.A. Kohm and J.F. Franklin (Eds.), Creating a forestry for the 21st century (pp. 215-228). Washington, DC: Island Press.

Christensen, G.A. 1997. Development of Douglas-fir log quality, product yield, and stand value after repeated thinnings in western Oregon. Corvallis, OR: Oregon State University. M.S. thesis. 176 p.

Curtis, R.O. 1995. Extended rotations and culmination age of coast Douglas-fir: old studies speak to new issues. Research Paper PNW-RP-485. Portland, OR: U.S. Department of Agriculture, Forest Service, Pacific Northwest Research Station. 49 p.

Curtis, R.O. 1997. The role of extended rotations. In: K.A. Kohm and J.F. Franklin (Eds.), Creating a forestry for the 21st century (pp. 165-170). Washington, DC: Island Press.

Curtis, R.O. 1998. "Selective cutting" in Douglas-fir: history revisited. *Journal of Forestry*, 96(7): 40-46.

Curtis, R.O., and Carey, A.B. 1996. Timber supply in the Pacific Northwest: managing for economic and ecological values in Douglas-fir forests. *Journal of Forestry*, 94(9): 35-37.

Curtis, R.O., DeBell, D.S., Harrington, C.A., Lavender, D.P., Tappeiner, J.C., and Walstad, J.D. 1998. Silviculture for multiple objectives in the Douglas-fir region. General Technical Report PNW-GTR-435. Portland, OR: U.S. Department of Agriculture, Forest Service, Pacific Northwest Research Station. 123 p.

Curtis, R.O., and Marshall, D.D. 1993. Douglas-fir rotations–time for reappraisal? *Western Journal of Applied Forestry*, 8(3): 81-85.

DeBell, J.D., Morrell, J.J., and Gartner, B.L. 1999. Within-stem variation in tropolone content and decay resistance of second-growth western redcedar. *Forest Science*, 45: 101-107.

DiLucca, C.M. 1989. Juvenile-mature wood transition. In: R.M. Kellogg (Ed.), Second growth Douglas-fir: its management and conversion for value (pp. 23-38). Special Publication No. SP-32. Vancouver, BC: Forintek Canada Corp.

Eastin, I.L., Lane, C.L., Fight, R.D., and Barbour, R.J. 1998. An assessment of the industrial markets for softwood clearwood lumber. *Forest Products Journal*, 48(11/12): 48-54.

Eastin, I.L., Simon, D., and Shook S. 1996. Softwood lumber substitution in the US residential construction industry. CINTRAFOR Working Paper No. 57. Seattle, WA: University of Washington.

Erickson, H.D., and Lambert, G.M.G. 1958. Effects of fertilization and thinning on chemical composition, growth, and specific gravity of young Douglas-fir. *Forest Science*, 4(4): 307-315.

Fahey, T.D., Cahill, J.M., Snellgrove, T.A., and Heath, L.S. 1991. Lumber and veneer recovery from intensively managed young-growth Douglas-fir. Research Paper PNW-RP-437. Portland, OR: U.S. Department of Agriculture, Forest Service, Pacific Northwest Research Station. 25 p.

Fahey, T.D., and Willits, S.A. 1995. Volume and quality of clear wood from pruned trees. In: D.P. Hanley, C.D. Oliver, D.A. Maguire, D.G. Briggs, and R.D. Fight (Eds.), Forest pruning and wood quality of western North American conifers (pp. 115-126). Seattle, WA: College of Forest Resources, University of Washington.

Fight, R.D., Johnston, S., Briggs, D.G., Fahey, T.D., Bolon, N.A., and Cahill, J.M. 1995. How much timber quality can we afford in coast Douglas-fir stands? *Western Journal of Applied Forestry*, 10(1): 12-16.

Franklin, J.F., Spies, T.A., Van Pelt, R., Carey, A.B., Thornburgh, D.A., Berg, D.R., Lindenmayer, D.B., Harman, M.E., Keeton, W.S., Shaw, D.C., Bible, K., and Chen, J. 2002. Disturbances and structural development of natural forest ecosystems with silvicultual implications, using Douglas-fir forests as an example. *Forest Ecology and Management*,155: 399-423.

Forest Products Research Society. 1986. A technical workshop: juvenile wood—What does it mean to forest management and forest products? Proceedings 47309. Madison, WI: Forest Products Research Society. 81 p.

Forest Products Research Society. 1987. Executive summary plenary session IVA: juvenile wood in hardwoods and conifers. Madison, WI: Forest Products Research Society. [not paginated].

Gartner, B.L., North, E.M., Johnson, G.R., and Singleton, R. 2002. Effects of live crown on vertical patterns of wood density and growth in Douglas-fir. *Canadian Journal of Forest Research*, 32(3): 439-447.

Hanley, D.P., Oliver, C.D., Maguire, D.A., Briggs, D.G., and Fight, R.D. (Eds.) 1995. Forest pruning and wood quality of western North American conifers. Seattle, WA: College of Forest Resources, University of Washington. 403 p.

Hann, D.W., Hester, A.S., and Olsen, C.L. 1997. ORGANON user's manual edition 6.0. Corvallis, OR: Oregon State University, Department of Forest Resources. 133 p.

Haygreen, J.G., and Bowyer, J.L. 1996. Wood science and forest products—an introduction (3rd ed.). Ames: IA: State University Press.

Haynes, R.W., Adams, D.M., and Mills, J.R. 2003a. Contemporary management regimes in the Pacific Northwest: balancing biophysical and economic concerns. Chapter 10. In: R.A. Monserud, R.W. Haynes, and A.C. Johnson (Eds.), Compatible forest management. Dordrecht, The Netherlands: Kluwer Academic Publishers.

Haynes, R.W., Graham, R.T., and Quigley, T.M. (Tech. eds.) 1996. A framework for ecosystem management in the interior Columbia Basin and portions of the Klamath and Great Basins. General Technical Report PNW-GTR-374. Portland, OR: U.S. Department of Agriculture, Forest Service, Pacific Northwest Research Station. 66 p.

Haynes, R.W., Monserud, R.A., and Johnson, A.C. 2003b. Compatible forest management: background and context. Chapter 1. In: R.A. Monserud, R.W. Haynes, and A.C. Johnson (Eds.), Compatible forest management. Dordrecht, The Netherlands: Kluwer Academic Publishers.

Haynes, R.W., and Perez, G.E. 2001. Northwest Forest Plan: research synthesis. General Technical Report PNW-GTR-498. Portland, OR: U.S. Department of Agriculture, Forest Service, Pacific Northwest Research Station. 130 p.

Hummel, S.S. 2003. Managing structural and compositional diversity with silviculture. Chapter 4. In: R.A. Monserud, R.W. Haynes, and A.C. Johnson (Eds.), Compatible forest management. Dordrecht, The Netherlands: Kluwer Academic Publishers.

Johnson, G.R., Sniezko, R.A., and Mandel, N.L. 1997. Age trends in Douglas-fir genetic parameters and implications for optimum selection age. *Silvae Genetica*, 46: 349-358.

Johnson, J.A. 1986. Wood quality and its relationship to uses, grades, and prices: past, present, and future. In: C.D. Oliver, D.P. Hanley, and J.A. Johnson (Eds.), Douglas-fir: stand management for the future (143-148). Contribution No. 55. Seattle, WA: University of Washington, College of Forest Resources, Institute of Forest Resources.

Jozsa, L.A., and Brix, H. 1989. The effects of fertilization and thinning on wood quality of a 24-year-old Douglas-fir stand. *Canadian Journal of Forest Research*, 19: 1137-1145.

Jozsa, L.A., and Middleton, G.R. 1994. A discussion of wood quality attributes and their practical implications. Special Publication No. SP-34. Vancouver, BC: Forintek Canada Corp.

Jozsa, L.A., Richards, J., and Johnson, S.G. 1989. Relative density. In: R.M. Kellogg (Ed.), Second growth Douglas-fir: its management and conversion for value (pp. 5-22). Special Publication No. SP-32. Vancouver, BC: Forintek Canada Corp.

Kellogg, R.M. 1982. Coming to grips with wood quality. *Forestry Chronicle*, 58(6): 254-257.

Kellogg, R.M. (Ed.). 1989. Second growth Douglas-fir: its management and conversion for value. Special Publication No. SP-32. Vancouver, BC: Forintek Canada Corp. 173 p.

King, J.N., Yeh, F.C., and Heaman, J.C.H. 1988. Selection of wood density and diameter in controlled crosses of coastal Douglas-fir. *Silvae Genetica*, 37: 152-157.

Kotok, E.S. 1951. Shall we prune to provide peeler logs for the future? *Timberman*, 52(10): 104-109.

Koubaa, A., Zhang, S.Y., Isabel, N., Beaulieu, J., and Bousquet, J. 2000. Phenotypic correlations between juvenile-mature wood density and growth in black spruce. *Wood and Fiber Science*, 32(1): 61-71.

Kuehne, M.J. 1994. High quality forestry: an alternative for management of national forest lands in western Washington. In: J.W. Weigand, R.W. Haynes, and J.L. Mikowski (Comps.), Proceedings, high quality forestry workshop: the idea of long rotations (pp. 211-243 plus appendix). SP-15. Seattle, WA: University of Washington, College of Forest Resources, Center for International Trade in Forest Products.

Lamb, F.M., and Wengert, E.M. 1987. The drying properties of lumber containing juvenile wood. In: Executive summary plenary session IVA: juvenile wood in hardwoods and conifers [not paginated]. Madison, WI: Forest Products Research Society.

Larson, P.R. 1962. A biological approach to wood quality. *TAPPI*, 45(6): 443-448.

Larson, P.R. 1963. Stem form development of forest trees. *Forest Science-Monograph 5*. 31 p.

Larson, P.R. 1969. Wood formation and the concept of wood quality. Bulletin No. 74. New Haven, CT: Yale University, School of Forestry. 54 p.

Loo-Dinkins, J.A., and Gonzalez, J.S. 1991. Genetic control of wood density profile in young Douglas-fir. *Canadian Journal of Forest Research*, 21: 935-939.

Maguire, D.A., Kershaw, J.A., Jr., and Hann, D.W. 1991. Predicting the effects of silvicultural regime on branch size and crown wood core in Douglas-fir. *Forest Science*, 37: 1409-1428.

Marshall, D.D., and Curtis, R.O. 2002. Levels-of-growing-stock cooperative study in Douglas-fir: Report No. 15–Hoskins: 1993-1998. Research Paper PNW-RP-537. Portland, OR. U.S. Department of Agriculture, Forest Service, Pacific Northwest Research Station. 80 p.

McArdle, R.E. and Meyer, W.H. 1930. The yield of Douglas-fir in the Pacific Northwest. Technical Bulletin 201. Washington, DC: U.S. Department of Agriculture, Forest Service. 65 p.

McGaughey, R.J. 1997. Visualizing forest stand dynamics using the stand visualization system. In: Proceedings of the 1997 ACSM/ASPRS annual convention and exposition. (Vol. 4, pp. 248-257). Bethseda, MD: American Society for Photogrammetry and Remote Sensing.

McKeever, D.B. 2002. Domestic market activity in solid wood products in the United States, 1950-1998. General Technical Report PNW-GTR-524. Portland, OR: U.S. Department of Agriculture, Forest Service, Pacific Northwest Research Station. 76 p.

Megraw, R.A. 1985. Wood quality factors in loblolly pine. Atlanta, GA: TAPPI Press.

Megraw, R.A. 1986. Douglas-fir wood properties. In: C.D. Oliver, D.P. Hanley, and J.A. Johnson (Eds.), Douglas-fir: stand management for the future (pp. 81-96). Contribution No. 55. Seattle, WA: University of Washington, College of Forest Resources, Institute of Forest Resources.

Middleton, G.R., and Munro, B.D. (Eds.). 2001. Second-growth western hemlock product yields and attributes related to stand density. Special Publication SP-41. Vancouver, BC: Forintek Canada Corp.

Monserud, R.A. 2003. Modeling stand growth and management. Chapter 6. In: R.A. Monserud, R.W. Haynes, and A.C. Johnson (Eds.), Compatible forest management. Dordrecht, The Netherlands: Kluwer Academic Publishers.

National Lumber Grades Authority. 2003. Standard grading rule for Canadian lumber. New Westminster, BC: National Lumber Grades Authority.

Northwest Log Rules Advisory Group. 1998. Official log scaling and grading rules (8th ed.). [Place of publication unknown]. 48 p.

Oliver, C.D., Hanley, D.P., and Johnson, J.A. (Eds.). 1986. Douglas-fir: stand management for the future. Contribution No. 55. Seattle, WA: College of Forest Resources, University of Washington, Institute of Forest Resources. 388 p.

Oregon Department of Forestry. 2001. Oregon Forest Practices Act ORS Chapter 527. 610 - .770 & .990-.992. Salem, OR. Retrieved April 2003 from http://www.odf.state.or.us.

Panshin, A.J., and de Zeeuw C. 1980. Textbook of wood technology Vol. 1. New York: McGraw-Hill Book Co. 705 p.

Parker, M.L., Hunt, K., Warren, W.G., and Kennedy, R.W. 1976. Effect of thinning and fertilization on intra-ring characteristics and kraft pulp yield on Douglas-fir. *Journal of Applied Polymer Science: Applied Polymer Symposium*, 28: 1075-1086.

Sanio, K. 1872. On the size of the wood cells of the Scotch pine (*Pinus silvestris*). *Jahrbucher fur Wissenschaftliche Botanik*, 8: 401-420.

Saucier, J.R., 1990. Forest Management and Wood Quality. In: J.R. Saucier, and F. Cubbage (Comps.), Proceedings of southern pine plantation wood quality workshop (pp. 47-56). General Technical Report SE-63. Asheville, NC: U.S. Department of Agriculture, Forest Service, Southeastern Forest Experiment Station.

Sauter, U.H., Mutz, R., and Munro, B.D. 1999. Determining juvenile-mature wood transition in Scots pine using latewood density. *Wood and Fiber Science*, 31(4): 416-425.

Senft, J.F., Bendtsen, B.A., and Galligan, W.L. 1985. Weak wood, fast-grown trees make problem lumber. *Journal of Forestry*, 83(8): 476-484.

Siddiqui, K.M., Gladstone, W.T., and Marton, R. 1972. Influence of fertilization on wood and pulp properties of Douglas-fir. In: Proceedings of effect of growth acceleration on the properties of wood (C1-C18). Madison, WI: U.S. Department of Agriculture, Forest Service, Forest Products Laboratory.

Silen, R.R. 1966. A simple progressive tree improvement program for Douglas-fir. Research Note PNW-RN-45. Portland, OR: U.S. Department of Agriculture, Pacific Northwest Research Station. 13 p.

Silen, R.R., and Wheat, J.G. 1979. Progressive tree improvement program in Douglas-fir. *Journal of Forestry*, 77(2): 78-83.

Singleton, R., DeBell, D.S., Marshall, D.D. and Gartner, B.L. [In press]. Eccentricity and fluting in young-growth western hemlock. *Western Journal of Applied Forestry.*

St.Clair, J.B. 1994. Genetic variation in tree structure and its relation to size in Douglas-fir. I. Biomass partitioning, foliage efficiency, stem form and wood density. *Canadian Journal of Forest Research*, 24: 1226-1235.

Stevens, J.A., and Barbour, R.J. 2000. Managing the stands of the future based on the lessons of the past: estimating western timber species product recovery by using historical data. Research Note PNW-RN-528. Portland, OR: U.S. Department of Agriculture, Forest Service, Pacific Northwest Research Station. 11 p.

Stonecypher, R.W., Piesch, R.F., Helland, G.G., Chapman, J.G., and Reno, H.J. 1996. Results from genetic tests of selected parents of Douglas-fir (*Pseudotsuga menziesii* (Mirb.) Franco), in an applied tree improvement program. *Forest Science Monograph*, 32: 1-35.

Timmel, T.E. 1986. Compression wood in gymnosperms. Vol. 1-3. New York: Springer. 2150 p.

Trendelenburg, R. 1936. Variations in the density of important coniferous timbers, due to locality, habitat and differences in individual trees. Oxford: University of Oxford, Imperial Forestry Institute. 10 p.

Vargas-Hernandez, J., and Adams, W.T. 1991. Genetic variation of wood density components in young coastal Douglas-fir: implications for tree breeding. *Canadian Journal of Forest Research*, 21: 1801-1807.

Vargas-Hernandez, J., and Adams, W.T. 1992. Age-age correlations and early selection for wood density in young coastal Douglas-fir. *Forest Science*, 38(2): 467-478.

Wahlgren, H.E., and Schumann, D.R. 1975. Properties of major southern pines. Part 1: wood density survey. Research Paper FPL-176. Madison, WI: U.S. Department of Agriculture, Forest Service, Forest Products Laboratory.

Warren, D.D. 2002. Production, prices, employment, and trade in Northwest forest industries, all quarters 2000. Research Bulletin PNW-RB-236. Portland, OR: U.S. Department of Agriculture, Forest Service, Pacific Northwest Research Station. 171 p.

Washington Department of Natural Resources. 2002. Forest Practices Act. RCW Chapter 76.09. Retrieved April 2003 from http://www.wa.gov/dnr/htdocs/forestpractices/rules/rcw76.09.pdf.

Weigand J.W., Haynes, R.W., and Mikowski, J.L. (Comps.). 1994. Proceedings, high quality forestry workshop: the idea of long rotations. SP-15. Seattle, WA: University of Washington, College of Forest Resources, Center for International Trade in Forest Products. 266 p.

Western Wood Products Association [WWPA]. 1998. Western lumber grading rules 98. Portland, OR. 248 p.

Western Wood Products Association [WWPA]. 2002. Coast F.O.B. price summary. Portland, OR. 8 p.

Whiteside, I.D. 1978. Machine stress grading studies and grading rules for Douglas-fir. In: A review of Douglas fir in New Zealand (pp. 273-287). Symposium No. 15. Rotorua, New Zealand: NZ Forest Service, Forest Research Institute.

Yang, K. 2002. Impact of spacing on juvenile wood and mature wood properties of white spruce (*Picea glauca*). *Taiwan Journal of Forest Science*, 17(1): 13-29.

Chapter 12

Compatible Management of Understory Forest Resources and Timber

Becky K. Kerns
USDA Forest Service, Pacific Northwest Research Station, 3200 SW Jefferson Way, Corvallis, OR 97331, USA
David Pilz
Department of Forest Science, College of Forestry, Oregon States University, Corvallis, OR 97331, USA
Heidi Ballard
Department of Environmental Science, Policy and Management, University of California, 151 Hilgard Hall #3110, Berkeley, CA 94720, USA
Susan J. Alexander
USDA Forest Service, Pacific Northwest Research Station, 3200 SW Jefferson Street, Corvallis, OR 97331, USA

1. Introduction

Many native mosses, lichens, ferns, herbs, shrubs, and fungi are harvested by humans from the understories of Pacific Northwest forests. These understory products are used personally and commercially for decorative, culinary, medicinal, cultural and educational purposes. Understory species harvested for any of these purposes are typically and awkwardly referred to as nontimber or special forest products (Vance et al. 2001, IFCAE 2002, Table 1). Hereafter, we refer to these species simply as understory forest resources and understory products. These species have important ecological roles in forest communities. They contribute to biological diversity and long-term ecosystem productivity (Alaback and Herman 1988, Halpern and Spies 1995), underpin mammalian and avian abundance (Morrison 1982, Carey 1995, Carey and Johnson 1995) and are important aesthetic components of forests. The aesthetic values of flowering trees and shrubs was a major basis for language in the National Forest Management Act requiring the USDA Forest Service to maintain and

R.A. Monserud, R.W. Haynes and A.C. Johnson (eds.)
Compatible Forest Management, 337-381

Table 1. Common Pacific Northwest understory non-tree species with commercial, social, or cultural value that have also been the subject of studies listed in Appendix 1.

Life-form	Common name	Scientific name	Type of use or product
Shrubs and small trees	Vine maple	*Acer circinatum*	Floral, craft
	Dwarf Oregon-grape	*Berberis nervosa*	Floral, medicinal
	Redstem ceanothus	*Ceanothus sanguineus*	Floral, medicinal
	Snowbush	*Ceanothus velutinus*	Floral, medicinal
	Prince's pine	*Chimaphila umbellata*	Floral, medicinal, flavorings
	Beaked hazelnut	*Corylus cornuta*	Food, fiber material
	Salal	*Gaultheria shallon*	Floral, medicinal, ornamental, food
	Thimbleberry	*Rubus parviflorus*	Food
	Salmonberry	*Rubus spectablis*	Food
	Dewberry	*Rubus ursinus*	Food
	Willow	*Salix*	Fiber
	Elderberry	*Sambucus*	Food, medicinal
	Pacific yew	*Taxus brevifolia*	Medicinal, craft
	Alaskan blueberry	*Vaccinium alaskaense*	Food
	Thin leaved huckleberry	*Vaccinium membranaceum*	Food, medicinal
	Evergreen huckleberry	*Vaccinium ovatum*	Floral, food
	Red huckleberry	*Vaccinium parvifolium*	Floral, medicinal, food, ornamental
	Bear-grass	*Xerophyllum tenax*	Floral, fiber
	Cascara	*Rhamnus purshiana*	Medicinal
Ferns	Wood fern	*Dryopteris*	Food
	Western sword fern	*Polystichum munitum*	Food, floral, ornamental
	Bracken fern	*Pteridium aquilinum*	Food, floral, ornamental
Herbs	Pearly everlasting	*Anaphalis margaritacea*	Floral, medicinal
	Oregon goldthread	*Coptis laciniata*	Medicinal, ornamental
	Fireweed	*Epilobium angustifolium*	Food, fiber
	Rattlesnake plantain	*Goodyera oblongifolia*	Medicinal
	St. John's wort	*Hypericum perforatum*[1]	Medicinal
	Stinging nettle	*Urtica dioica*	Medicinal, cosmetic
Lichen	Lung moss	*Lobaria*	Medicinal, floral
	Beard lichen	*Usnea*	Floral, medicinal
Bryophytes	Oregon beaked moss	*Kindbergia oregana*	Floral
	Isothecium moss	*Isothecium myosuroides*	Floral
Fungi	Chanterelles	*Cantharellus* species	Food
	American matsutake	*Tricholoma magnivelare*	Food
	Morels	*Morchella* species	Food

[1] Exotic to region. Source: Vance et al. 2001, IFCAE 2002, Schroeder 2002.

enhance diversity in the forests it manages. Recent increased attention to understory forest resources stems from social and ecological concerns that parallel rising demand for consumptive use of these species as decorative floral products, medicinals, and edible wild foods.

Since World War II, clearcut logging and even-age management have dominated forest practices in Oregon and Washington (Haynes et al. 2003). Objectives were timber production and maintaining vigorous crop trees. The status of the understory in managed stands was generally an unintentional by-product of timber management. In some areas, humans harvest understory species within these management legacies. Today, there is interest in developing alternative silvicultural systems and new forest practices that embrace multiple values, such as fish and wildlife, aesthetic values, biodiversity and forest structural complexity (Curtis and Carey 1996; DeBell et al. 1997; Carey et al. 1999a,b; Franklin et al. 2002; Muir et al. 2002; Hummel 2003). Compatible forest management (defined by Haynes et al. 2003) is management that simultaneously produces wood and at least one other forest value without decreasing any other values. Interest and social demand for compatible forest management is not limited to public lands; private landowners are also finding it necessary to develop compatible management regimes (Loehle et al. 2002). Using forest management practices to enhance commercial understory products while maintaining an economically viable over-story tree crop has become a focus for some forest managers (e.g., Weigand 1998).

The primary objectives of this chapter are to

- *Assess the ecological and biological literature from the understory resource and compatible management perspective*
- *Synthesize this information to provide a synopsis of scientific knowledge*
- *Discuss the compatibility of management of understory forest resources and timber*

We emphasize peer-reviewed empirical studies, widely available literature, and research conducted under the USDA Forest Service Pacific Northwest Research Station, Wood Compatibly Initiative (Haynes et al. 2003). Our geographic focus is western Oregon and Washington, although we include literature from British Columbia and southeastern Alaska. These studies address understory species response from a range of scales and viewpoints. Some use general measures such as species composition, abundance, and diversity while others focus on a single species, species-specific responses, or species reproduction (Appendix 1). This chapter is not an exhaustive literature review; rather we tried to summarize major themes gleaned from empirical research to frame a discussion of management implications for understory resources. Our focus will largely be on non-tree understory vascular plant species and fungi.

We will address these key questions: What are the effects of forest management practices on understory resources? How can timber production and understory resource enhancement be compatible? We will discuss the degree to which managers can manage both for timber and understory forest resources. The first section of this chapter is focused on understory ecology, forest development, and management. Relationships between understory species, biological abundance, and commercial productivity are discussed in the second section. In the last section, we comment on implications of forest management decisions on understory resources, and outline critical issues and future directions. Throughout this chapter we use Latin names for all species with parenthetical reference to common names the first time the species is mentioned. Vascular plant nomenclature is based on Hitchcock and Cronquist (1973).

2. Understory Ecology, Forest Development and Management

The Pacific Northwest (western Oregon and Washington, coastal British Columbia, and southeastern Alaska) forests are characterized by a mild, wet, maritime climate strongly influenced by cyclonic storms that approach from the Pacific Ocean on dominant westerlies. Annual precipitation ranges from 100 to 350 cm, with most moisture falling as rain between October 1 and March 31 (Franklin and Dyrness 1988). In summer, storm tracks shift north and the southern part of the region experiences relatively cool, dry summers. In general, an increase in precipitation and decrease in temperature corresponds with the increase in latitude. Our discussion focuses on the two major vegetation zones described in Haynes et al. 2003 (*Picea sitchensis* (Bong.) Carr. [Sitka spruce], and *Tsuga heterophylla* (Raf.) Sarg. [western hemlock], Franklin and Dyrness 1988), but we discuss pertinent literature from forest types throughout Oregon, Washington, and Alaska (Appendix 1).

In the Pacific Northwest, as in many forested ecosystems, the overstory canopy tree species found in a particular area may vary little, but understory species in the same area often vary widely. Site location, interactions among species, and past disturbance and forest management activities on the site contribute to the diversity of understory species. Species life-form, reproductive strategies, and nutritional modes are also important. Our discussion is limited to herbs, shrubs, bryophytes, lichen and fungi (Table 1). For fungi, we will refer to four categories based on nutritional mode: ectomycorrhizal fungi (EM-fungi) that require healthy host trees, saprobes that decompose coarse woody debris (CWD-saprobes), saprobes that decompose litter layers or soil organic matter (soil-saprobes), and parasites. Specific species and life-form examples will be provided in the following sections.

Understanding the response of understory species to forest management is central to developing compatible approaches for managing understory forest resources. Forest management practices that can affect understory species include the initial effects of disturbance associated with logging and site preparation, and subsequent activities that directly or indirectly alter rates or patterns of succession (Halpern and Spies 1995) (Figure 1). These practices influence understory communities at a variety of temporal and spatial scales within a matrix of other scale-dependent processes. At fine spatial scales, overstory stand structure strongly influences understory species distribution and

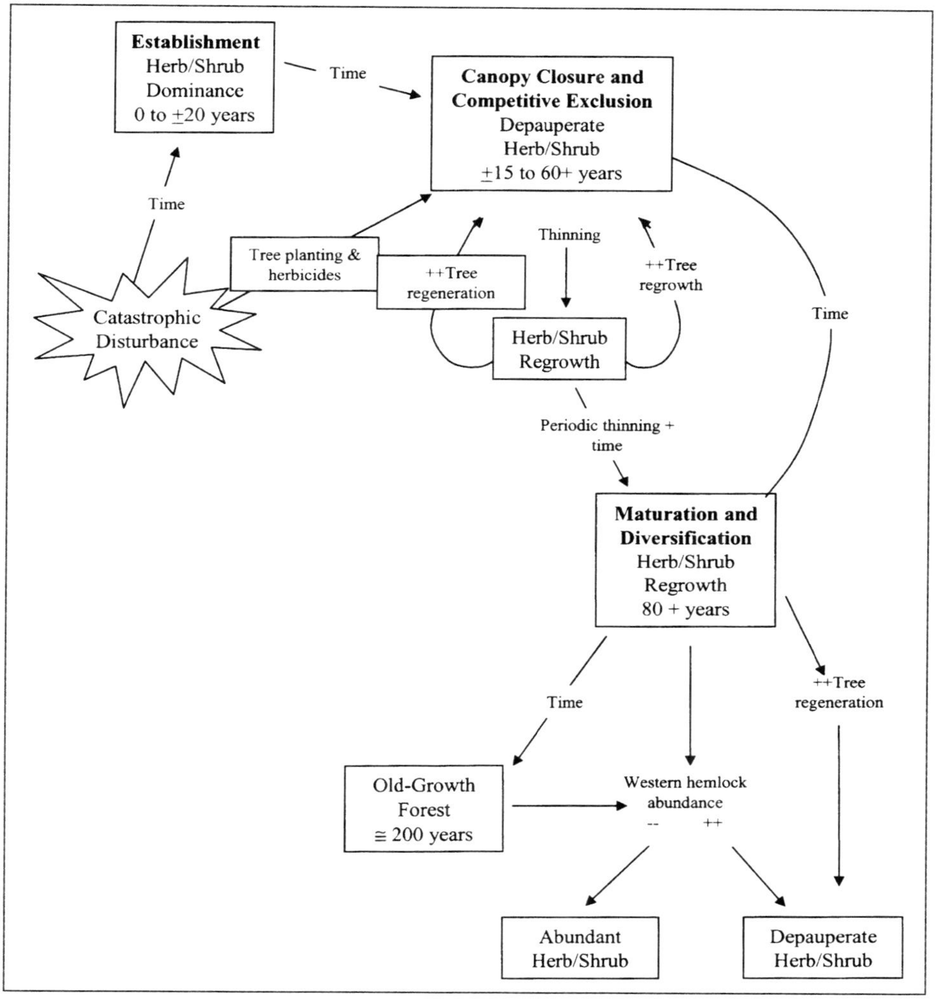

Figure 1. Conceptual model for forest stand development, herb and shrub understory abundance, and forest management practices. Model is illustrated by using a Pseudotsuga menziesii *dominated sere growing in the* Tsuga heterophylla *zone (Franklin and Dyrness 1988).*

abundance by altering microsites, resources, and environmental conditions. Presumably, light availability and shade tolerance of understory plant species are primary determining factors because light is considered a major limiting resource for most photosynthetic understory species in coniferous forests, particularly in wet, mild, low elevation locations (Alaback 1982). Canopy conditions can also dramatically change understory environmental condition.

Many other factors and processes influence understory species distribution and abundance; these include the catastrophic disturbance that originated the stand, the degree of biotic legacies retained following the disturbance, and the rapidity with which trees established on the site and formed a dense canopy. For example, the responses of fungi are largely determined by their mode of nutrition. For these species, carbohydrates and other nutrients are usually gained from live trees or coarse woody debris rather than photosynthesis. Thus, nutritional substrate alterations, rather than light, are key. For bryophytes and lichens, which do photosynthesize, substrates for growth, such as large branches, tree architecture, hardwoods and large shrubs are important. Forest management practices affect understory attributes through a complex matrix of ecological processes that operate at different temporal and spatial scales (Figure 1). These effects differ among species (Appendix 2). To facilitate our discussion, we present six categories of forest developmental stages and management activities (see Appendix 1, Chapter 1): (1) first stage – establishment, (2) second and third stages – closure and competitive exclusion, (3) fourth and fifth stages – maturation and diversification, (4) operational stand thinning, (5) fertilization, and (6) other silvicultural systems.

2.1. First Stage – Establishment

Because of the prevalence of catastrophic disturbances such as stand-replacing fires and even-age management focused on clearcut logging in Pacific Northwest forests, much research concerning understory response has focused on early-successional changes after these disturbances (Appendix 1). After major disturbances, most forest species will temporarily decline and non-forest herbs (ruderals) invade or increase. Ruderal species are generally short-lived and allocate the abundant available resources to reproduction (Appendix 2). On northern Vancouver Island, thick, wet, slowly decomposing forest floors of decayed wood on recently logged and burned sites inhibit growth of non-forest invaders. Such sites are quickly dominated by regrowth of *Gaultheria shallon* Pursh (salal) (Messier and Kimmins 1991). *G. shallon* has also been the subject of numerous scientific studies because of its conspicuous dominance in western Pacific Northwest forests and competition with conifers. This economically important floral green offers potential for compatible management opportunities, which are highlighted in our case study (Box 1).

Despite an early increase in frequency and abundance of non-forest species, 70 to 90% of the species present before the catastrophic disturbance event (residual species) will survive and recover, especially over larger spatial scales (Schoonmaker and McKee 1988, Halpern and Spies 1995). Dyrness (1973) reported that residual shrubs such as *Acer circinatum* Pursh (vine maple) and *Vaccinium parvifolium* Sm. (red huckleberry) can regain dominance after clearcutting in as little as 5 years. Schoonmaker and McKee (1988) suggested that residual species made up 97% of total understory plant-cover 40 years after logging. The regenerative structures of many plants are below ground, ensuring survival. Rapidly resprouting and woody clonal species (e.g., *G. shallon*, *Berberis nervosa* Pursh [dwarf Oregon grape], and *Acer circinatum*) are particularly tolerant of major disturbance. Site preparation can impede recovery of forest understory species, but many species will still rapidly increase in abundance (Knowe et al. 1997). *G. shallon* will even survive and thrive after significant herbicide treatments (Stein 1995). On the other hand, species with regenerative structures close to the soil surface or above ground (e.g., delicate herbs such as *Coptis laciniata* Gray [Oregon goldthread]) can be sensitive to major disturbance, especially fire (Dyrness 1973, Halpern 1989a, Halpern and Spies 1995).

Control of competing vegetation during early conifer establishment was and still is commonly practiced, particularly on private timberlands. Removal or control of competing vegetation is done to alter resource availability during the first stage of forest development to favor crop tree species. However, some research suggests that understory cover may be more beneficial for nutrient cycling rather than the absence of understory growth altogether (Bradley et al. 2000). Rapid nutrient loss due to major vegetation removal and reduced retention of nutrients is a well-documented phenomenon (Likens et al. 1978). However, few studies have examined the impact of understory vegetation control treatments on nutrient retention and availability. In response to public concerns surrounding the use of herbicides, air quality, and maintaining native shrub diversity, federal and state agencies recently have moved away from intensive site preparation methods.

Many prized edible fungi (chanterelles, matsutake, boletes) are ectomycorrhizal (EM), and require living trees for their carbohydrate nutrition. During stand replacement events, edible species typically stop fruiting for one to several decades until a new stand is well established. For instance, chanterelles can start fruiting after 12 to 15 years in productive stands of *Pseudotsuga menziesii* (Mirb.) Franco (Douglas-fir) in western Washington (W. Littke, Weyerhaeuser, Co., Centralia, Washington, personal communication, 1995). In Sweden, where host trees grow more slowly, chanterelles typically start fruiting when the stand age is 40 years (Danell 1994, Pilz et al. 2003). Matsutake commonly start fruiting at stand age 40 (Hosford et al. 1997).

Severity of soil compaction or erosion, loss of litter layers, number and distribution of surviving trees, presence of alternate host species that quickly sprout from roots, and the length of time until EM-fungi recolonize new trees all interact to determine whether mycelial colonies of EM-fungi persist in the soil or die out and regenerate from spores (Amaranthus 1992, Amaranthus and Perry 1994). Trees and EM-fungi exhibit different degrees of specificity in their formation of ectomycorrhizae with each other (Molina and Trappe 1982, Molina et al. 1992). Reforesting the stand with appropriate species of host trees is necessary to restore EM-fungal fruiting regardless of the regenerative or reproductive mechanism that operates.

Tree death, fire, and soil disturbance are known to produce prolific crops of morels for 1 or 2 years in some forest types. The reasons for this response remain speculative because morel life cycles are complex and because multiple species with differing modes of nutrition (soil-saprobic and EM-fungi) likely coexist in any given stand (Weber et al. 1996, Dahlstrom et al. 2000). Coarse woody debris saprobes such as *Laetiporus sulphureus* (chicken of the woods) and parasites such as *Armillariella mellea* (honey mushrooms), among a variety of other harvested species, can also fruit in clearcuts and stands destroyed by fire or wind. If the opening is especially xeric, fruiting of these species might be impaired or delayed relative to fruiting in more humid forests.

2.2. Second and Third Stages – Canopy Closure and Competitive Exclusion

These stages are typified by the dominance of trees and the competitive exclusion of many organisms. Growing space diminishes and tree canopies close, a process that takes about 25 to 35 years in southeastern Alaska, and about 15 to 30 years in the Oregon Coast Range. Speed of canopy closure increases with high site fertility (Alaback 1982). Although a common stand condition today, the competitive exclusion phase was historically rare in many landscapes (Tappeiner et al. 1997). Catastrophic disturbances that remove most or all biological legacies (e.g., clearcut logging and conventional site preparation), dense regeneration, and lack of intermediate disturbances preclude crown-class differentiation and quickly move stands into the canopy closure and competitive exclusion stage (Carey et al. 1999b). Herbaceous and shrub abundance declines, and bryophytes (e.g., mosses) and ferns begin to dominate the understory (Turner and Long 1975, Alaback 1982, Alaback and Herman 1988, Tappeiner and Alaback 1989, Klinka et al. 1996). Flowering and seed production of understory species in young, dense stands can be nonexistent (Tappeiner and Alaback 1989).

The degree to which understory species decline or disappear, and the temporal persistence of this phase of competitive exclusion depend on the density and species of trees, spatial heterogeneity of the site and its vegetative cover, understory species reproductive strategies, and tree regeneration (Spies 1997). Slow growth, clonal development and low rates of seedling survival retard herb and shrub establishment in young stands (Tappeiner and Alaback 1989), but species tolerant of such stress can persist. The decline of understory species will be more pronounced during this stage if the overstory canopy is dominated by *Tsuga heterophylla* rather than *Pseudotsuga menziesii* (Alaback 1982, Stewart 1988). Local extinction of some species could shape longer-term patterns of understory composition, but little research has been devoted to mechanisms by which plants persist through this phase or reestablish once

Box 1. Case Study - *Gaultheria shallon* (Salal)

Gaultheria shallon is a dominant understory shrub on relatively dry and nutrient-poor stands throughout much of the Pacific Northwest (Klinka et al. 1996). This species is also one of the most economically important understory plants in the Pacific Northwest. The floral and Christmas greens market in the Pacific Northwest was estimated at $128.5 million in 1989 (Schlosser et al. 1991). Kerns et al. (in press) estimate that *G. shallon* composes approximately two-thirds of the total wild-harvested floral market in the Pacific Northwest. Peruse a flower shop or grocery store in the United States and Europe, and flower bouquets with its foliage as a backdrop will undoubtedly be found. In the following section, we synthesize literature and present harvest, price, and product information based on our discussions with harvesters, buyers, business owners, and managers in the Olympic Peninsula region of Washington.

Harvest techniques and characteristics. Much of the *G. shallon* sold for use in the fresh floral market is harvested from the Olympic Peninsula and Puget Sound areas of Washington. Extensive floral industries have been in place for almost a century and depend entirely on understory plants harvested from public and private forests.

G. shallon can be harvested commercially July through April; harvesting ceases during the spring growth period only because the new growth is too fragile to survive the picking and shipping process. Harvesting techniques require very little technology: most pickers do not even use a knife. Depending on the length of the most recent new stems, stems from 1 to 3 years old are removed in order to obtain the required minimum length for a commercial purposes—36 to 71 cm, depending on the particular product or grade desired. Only the most recent foliage free of insect and disease damage is considered commercial quality; blemished and older foliage is stripped from the stem. The length of the stem with new green growth determines the quality or the grade (and thus value) of the stem, therefore harvesters seek patches of *G. shallon* with a high density of long new shoots. Bunches of 25 to 35 stems weighing (0.5 to 0.7 kg) are tied with rubber bands and sold to wholesalers for $0.40 - $1.50 per bunch, depending on the quality, commercial grade, and time of year.

Ecology. Because of its conspicuous dominance, competition with conifers, and importance as a forage species, *G. shallon* has been the subject of numerous scientific studies (Appendices 1 and 2). In younger forests with uniform, open canopies, this species can form a dense monoculture, interfere with conifer regeneration and growth and exclude other understory species (Turner and Long 1975, Halpern and Spies 1995). Colonization rates after silvicultural treatments can be rapid—even in areas that have been clearcut and intensively treated (Bunnell 1990, Stein 1995). *G. shallon* typically increases cover via rhizome extension and vegetative expansion but it will regenerate from seed when large amounts of downed and decomposing wood provide suitable conditions (Huffman et al. 1994a). Bunnell (1990) notes that sexual reproduction may only occur at less than 33% stand closure; however, others observe flowering and fruiting under a variety of overstory and light conditions (Huffman et al. 1994b, Harrington et al. 2002).

G. shallon cover is generally inversely related to overstory canopy cover (Long and Turner 1975, Klinka et al. 1996, He and Barclay 2000). Increasing overstory closure and decreasing irradiance reduce *G. shallon* longevity, size, and vigor (Bunnell 1990). Huffman et al. (1994a) find that *G. shallon* rhizome biomass and density, size and expansion rate of individual clonal fragments, aerial stem biomass and density, and total biomass are negatively correlated with overstory density. Rhizome growth is hindered when overstory density is greater than 360 stand density index (Huffman et al. 1994a), and at canopy covers greater than 80 to 85%, *G. shallon* is essentially "shaded out" (Messier et al. 1989, Klinka et al. 1996). With decreasing light, *G. shallon* responds with structural changes expected in more shade-tolerant species such as erect, loosely branched growth pattern and low foliar biomass (Sabhasri 1961, in Klinka et al. 1996). At overstory canopy cover approaching 80%, *G. shallon* was observed to be as tall as 4 meters (m) (Klinka et al. 1996). In dense unthinned stands, it will often occur as small (1m x 1m), sparse, single-species patches.

In thinned and more open stands, *G. shallon* may be abundant (> 70% cover) and from 1 to 1.5 meters tall. Commercial thinning operations in Douglas-fir stands tend to favor establishment of *G. shallon* (Appendix 2). Extensive rhizome mats exist in stands even where cover is low. Any disturbance that breaks these mats will generate new growth (Bunnell 1990). Harvesters note that disturbance associated with thinning can initially decrease the commercial productivity of a stand for 2 to 3 years, but vigorous rhizome and aerial stem growth will resume several years after thinning. Because of shifts in below- and above-ground carbon allocation, simple visual assessments of above-ground *G. shallon* understories within stands may not provide a good measure of the competitive role of *G. shallon* and subsequent response to disturbance (Messier et al. 1989).

Compatible Management Approaches. Commercial production of *G. shallon* offers potential for compatible management opportunities. *G. shallon* has been in high demand relatively consistently for over 30 years (Kerns et al., in press). Some floral greenery buyers suggest that cumulative revenue from stands managed for *G. shallon* production may be equal to or exceed timber revenues at the end of a typical rotation period. More realistically, harvest and sale of *G. shallon* can provide some revenue in precommerical timber stands and for landowners interested in long rotations. Some managers theorize that on some favorable sites, harvest actually increases commercial productivity by stimulating aerial stem and rhizome growth, even at high harvest levels. However, this has not been documented.

Approaches to compatible management must consider not only the species' response to stand development and management found in much of the literature, but issues concerning commercial quality. Because harvesters and buyers only seek undamaged, large-leaved, long-stemmed growth of the past 1 to 3 seasons, commercial productivity of *G. shallon* can differ greatly from biological productivity in any given stand. While studies of understory response to overstory canopy cover and silvicultural practices focus on cover and/or stem density of the species, commercial quality may not be closely associated with these metrics. While *G. shallon* may be present or even abundant at overstory canopy densities ranging from 0 to 80% (Long and Turner 1975, Klinka et al. 1996, He and Barclay 2000), the canopy cover required for commercial-quality foliage is more limited. Moreover, a variety of environmental factors can contribute to insect and pathogen outbreaks that affect commercial productivity in the understory that are neither due to overstory conditions nor reflected in standard cover measures. Presently, new studies are underway specifically examining response of this species to harvest as a floral greenery product and silvicultural management options to enhance commercial productivity and characteristics.

Managers may choose to alter their management practices to actively enhance commercial *G. shallon* production. Such activities may include elimination of herbicide use, and maintaining an overstory canopy cover that promotes production but provides for commercial characteristics. Numerous studies have shown that stand thinning can greatly increase *G. shallon* production (Appendix 2). We estimate that overstory canopy covers in the range of 50 to 70% provide such conditions. However, thinning operations will reduce cover in the short-term (1 to 3 years) due to mechanical destruction.

On the Olympic Peninsula, highly productive *G. shallon* sites frequently occur on lower quality timber sites. In such stands, the period of *G. shallon* productivity may be substantially longer than on higher quality sites. Little management may be required during this period. On other sites, timing of precommercial and commercial thinning can be adjusted, depending on the site characteristics, to lengthen the period of commercial productivity for *G. shallon*. Use of fertilization treatments have been proposed–if fertilization is combined with thinning, both *G. shallon* and timber productivity may be enhanced; however, additional studies are needed. In older extended-rotation stands, fertilization may promote competition-related mortality and the presumed understory regrowth that would occur in gaps.

growing conditions improve (Halpern et al. 1999). Bryophyte and lichen abundance depends on hardwood tree species abundance and presence of host species such as *Acer circinatum* (Peck and McCune 1998).

Not all stands will progress rapidly to tree dominance and overstory stand closure. In the Oregon Coast Range, vegetative growth and replacement of *Rubus spectabilis* Pursh (salmonberry) allows this species to maintain a dense canopy that substantially inhibits regeneration of trees and taller shrubs (Tappeiner et al. 1991, Stein 1995, Knowe et al. 1997). Once persistent cover of *Rubus spectabilis* is established, specifically under *Alnus rubra* Bong. (red alder) or in the open, succession to other tree or shrub communities could be unlikely without severe disturbance or management interference.

In southeastern Alaska, stem exclusion after catastrophic disturbance in conifer stands can persist for 100 years or more (Alaback 1982, Tappeiner and Alaback 1989). Persistent stem exclusion is not typical for a natural disturbance regime characterized by high-frequency, low-magnitude disturbance events and gap-phase replacement. However, recent studies indicate that mixed *Alnus rubra*-conifer stands have species rich, abundant understories (Hanley and Hoel 1996, Wipfli et al. 2003). As part of the Wood Compatibility Initiative, resources associated with a range of tree compositions (pure conifer, to nearly pure *Alnus rubra*) were evaluated in young southeastern Alaska stands. Preliminary results suggest that nearly pure *Alnus rubra* stands host a large biomass of shrubs such as *Vaccinium* spp. and *Rubus spectabilis*, and moderate amounts of *Alnus rubra* are associated with the highest biomass of understory plants considered desirable deer browse species (Deal and Orlikowska 2002, Hennon et al. 2002).

Dense, unthinned conifer stands 20 to 60 years old frequently offer the best opportunities for collecting many species of mushrooms. Ectomycorrhizal networks have been reestablished, litter layers have begun to accumulate, coarse woody debris often remains from the previous stand, and parasitic fungi can persist or be spread by logging operations. The lack of thinning slash and suppressed shrub layers provide conditions for easy walking and spotting mushrooms at a distance. Dense canopies sustain the high humidity levels necessary for mushroom growth on the forest floor.

Additionally, harvester observations suggest that some ectomycorrhizal fungi, such as chanterelles, fruit more abundantly in young to middle-aged stands than in old-growth forests (Love et al. 1998). During stem exclusion, dying trees can provide new habitat for CWD-saprobes, such as important medicinal conks (Stamets and Yao 2002) as well as several prized edible fungi. Mushrooms used for natural dyes (EM-fungi and soil saprobes) can also proliferate in such stands.

2.3. Fourth and Fifth Stages – Maturation and Diversification

Most remaining stands in these stages of forest development in Oregon and Washington are concentrated on federal lands (Bolsinger and Waddell 1993). Many state and private landowners use rotation lengths that preclude the development of older forest stands. Thus, this section is largely pertinent for federal, state, industrial, and private land managers implementing habitat conservations plans or those interested in longer rotations compatibility, conservation of biodiversity, sustainability, and green certification.

During stand maturation and diversification, gradual changes in populations, structures, and vegetative processes can last from 100 to 1,000 years, depending on forest type and disturbance history (Spies 1997). Stands open up as gaps begin to form and release and recruitment of understory species is initiated (Oliver and Larson 1996). Abundance and diversity of vascular plants increase with stage of forest development (Spies 1991, Halpern and Spies 1995). Herbaceous and shrub species that were able to survive in low light conditions may show vigorous response to small-scale disturbances such as windthrow (Tappeiner and Alaback 1989), and well-developed herb and shrub layers may form. In southeastern Alaska, bryophyte biomass peaks at about 140 to160 years after logging and then shrub and herb components increase during the later stages of forest development (Alaback 1982). In the Oregon Coast Range, commercial moss collection areas are generally in stands about 100 years or older with hardwood trees and shrubs (Peck and McCune 1998).

As stands continue to age, levels of resources change and become highly heterogeneous. A dynamic mosaic of patches created by old trees dying, and young trees, herbs, and shrubs filling the gaps, characterizes older *Pseudotsuga menziesii* forests of Oregon and Washington. A wide range of tree size classes, forest floor substrates and microenvironments are available for understory species. Heterogeneous canopies also allow for greater development of epiphyte biomass and diversity. Overstory canopy gaps persist for long periods and have well-developed understories of herbs, shrubs, and small trees

(Franklin and Spies 1991, Carey et al. 1999a). In the western Cascades of Oregon, Stewart (1986) found that in mixed stands, most understory species are more frequent in canopy openings than in other canopy types. However, in old, tall, multi-layered forests, complex understory light conditions exist. Patterns and distribution of forest canopy foliage, not just gap location, have important implications for the distribution and abundance of understory species (Carey et al. 1999a, Van Pelt and Franklin 2000).

Although understory plant species richness and abundance may increase during the understory release phase as denser forests dominated by *Tsuga heterophylla* replace more open *Pseudotsuga menziesii* dominated stands, plant abundance may decline (Halpern and Spies 1995). In older *Pseudotsuga menziesii* stands, well-developed understories may be present by 120 years, but in dense canopied *Tsuga* stands, understory development may be minimal, even at 150 years (Stewart 1988). In much older stands (500 to 830 years), understory shrubs are also more abundant beneath *Pseudotsuga* canopies, than beneath *Abies—Tsuga* canopies (Stewart 1986).

Kerns and Ohmann (in review) found that shrub abundance in the Oregon Coast Range was negatively correlated with the basal area of *Tsuga heterophylla*. The proportion of *Tsuga* and *Pseudotsuga* in the overstory influence conditions in the understory in several ways. *Tsuga* canopies intercept more precipitation compared to *Pseudotsuga* canopies and have high foliar biomass, which produces dense high leaf area canopies and a sparse understory (Alaback 1982, Stewart 1986, Alaback and Herman 1988), a trend that has been observed in *Tsuga* forests elsewhere (Rogers 1980). The wide-spreading crowns of *Tsuga* also create larger gaps than *Pseudotsuga* when they die standing or are windthrown (Stewart 1986).

Patterns of understory abundance in older forests most likely reflect both direct resource limitations and time-dependent responses for which overstory characteristics may be surrogates (McKenzie et al. 2000). For example, some species exhibit slow rates of establishment and growth (e.g., *Taxus brevifolia* Nutt. [Pacific yew]). The absence of epiphytic lichens and bryophytes in younger stands may reflect, in part, their limited dispersal capabilities and slow growth rates.

Smith et al. (2002) document greater EM-fungal productivity in young to mature *Pseudotsuga* stands than in mature to older stands, although species diversity did not change dramatically following canopy closure. However, Love et al. (1998) speculate that in dry years chanterelles might fruit more abundantly in older stands because the well-decomposed coarse woody debris common in older forests retains more soil moisture. The greater abundance of coarse woody debris, along with the broader range of microhabitats, and the larger variety of nutrient and fruiting substrates in old-growth forests likely fosters a greater diversity of edible, medicinal, and decorative fungi than found in more uniform younger stands. Rare and endemic fungal species are also more common in older forests (Moore et al. 2001, Smith et al. 2002); hence such forests are valuable mycological prospecting habitats for pharmaceuticals and superior fungus strains suitable for cultivation.

2.4. Operational Stand Thinning

Operational stand thinning is the purposeful removal of a portion of the trees from a stand. Precommercial thinning (trees are too small for economic use) of trees is often applied to young stands prior to canopy closure to enhance the survival, growth, and the value of residual trees (DeBell et al. 1997, Hummel 2003). Commercial thinning (thinning in older stands) was not widely applied in the Pacific Northwest until recently for a variety of reasons (Curtis et al. 1998). However, the practice is increasing because it provides opportunities to both produce wood and enhance understory resource objectives. For example, public and some private land managers are now developing and using thinning, including variable-density thinning, in combination with extended rotations to promote forest structural complexity and diversity (Franklin and Spies 1991, Curtis and Carey 1996, DeBell et al. 1997, Bailey et al. 1998, Muir et al. 2002). There is increasing interest in using stand thinning to enhance commercial and noncommercial understory productivity while maintaining overstory growth and productivity.

Thinning operations are low- to intermediate-intensity disturbances (e.g., thinning from below) compared to catastrophic disturbances such as clearcut logging and stand-replacing fires. As with gap-phase replacement processes associated with more mature forests, thinning can increase microhabitats, creating new germination sites and small openings in the canopy. Light, water, nutrient availability, and soil temperatures may increase. Therefore, thinning favors species that can (1) rapidly colonize or expand into newly available resources, (2) respond through vegetative propagation, or (3) tolerate newly created environments (Grime 1979, Klinka et al. 1996, Bailey and Tappeiner 1998). If conifer regeneration is limited, understory species respond to stand thinning by increased biomass and cover, particularly for clonal, woody shrub, grass, sedge, nitrogen fixing, and exotic species (Appendix 1). Thomas et al. (1999) found that 85% of all understory species responded positively to thinning in the overstory.

Intensity, stand age at the time of thinning, frequency, uniformity of thinning, and operational disturbance are all important issues to consider when making inferences regarding understory species response. Conventional uniform thinning produces evenly spaced trees with a uniformly distributed, compositionally simple understory. Irregular, or variable-density thinning that increases spatial heterogeneity and other structural attributes can increase variability in understory species composition and distribution (Carey and Johnson 1995, Carey 1995, Curtis et al. 1998, Carey et al. 1999a, Thysell and Carey

2001). Traditional, light commercial thinning may not preclude or move a stand out of the competitive exclusions state (Carey et al. 1999a). Many understory species will benefit from heavier thinning treatments. Understory response to thinning, particularly shrubs, is typically correlated with the amount of canopy removal (Alaback and Herman 1988; Kerns et al., in press). Thomas et al. (1999) note that understory cover values were highest in the intensely thinned stands (final stem density of c. 500 trees/ha). Alaback and Herman (1988) report that understory biomass shifted from bryophytes in unthinned control plots, to herbs in lightly thinned areas, and to shrubs in extremely thinned plots. However, heavy thinning with even spacing can cause stands to become drier through increased wind and sunlight and may result in brushfields (e.g., *G. shallon*) (Carey et al. 1999a). While effects of increased understory abundance during the first several years following thinning can be substantial and prolonged (Thomas et al. 1999, Kerns and Ohmann, in review), a single thinning may not maintain stable herb and shrub populations. Thus, effects of a single thinning may not be apparent (He and Barclay 2000).

While periodic light-to-moderate thinning may encourage herbaceous and shrub growth, issues of soil disturbance, exotic colonization, and economics may prohibit this approach. On the other hand, economic conditions may encourage periodic light-to-moderate thinning depending on local, regional and global conditions. At very low thinning intensities, the initial disturbance may potentially outweigh any increase in resource availability (Thomas et al. 1999). Greater frequency and abundance of exotic species in thinned compared to unthinned and older stands has been reported and these effects may last for decades (Bailey et al. 1998, Thysell and Carey 2000). The degree to which thinning promotes desirable species or encourages the spread of invasive and exotic species depends, in part, on the species of seeds present in the forest floor and soil (Halpern et al. 1999) and the preparation of the seed bed. Based on an analysis of seeds in soil (the soil seed bank) from young, closed-canopy forests, Halpern et al. (1999) suggested that silvicultural thinning will result in limited germination of forest species and will favor recruitment of early seral and exotic species. Training, use of designated skid trails and corridors, proper equipment, and initial heavy thinning that allows for relatively long intervals between subsequent thinning may reduce understory loss and disturbance due to thinning prescriptions (Curtis et al. 1998).

Thinning can prolong the herbaceous or shrub stage and reduce the depauperate understory stage associated with canopy closure (Figure 1). However, recommendations for thinning to maintain or maximize understory development, or overall biodiversity, ideally will consider existing species composition, competitive interactions of the species present and their associated life-history traits, and edaphic factors. Because some species (e.g., *Vaccinium parvifolium* Sm. [red huckleberry], Klinka et al. 1996) are often found on decaying wood, occurrence and distribution of coarse woody debris may also affect understory response to thinning treatments. Thinning treatments can also damage large shrubs, and shrub size is positively correlated to flowering and fruiting (Wender et al., in review). Thinning, particularly heavy thinning, may also reduce diversity and promote dominance by a single aggressive species such as *G. shallon* (Long 1977, Alaback and Herman 1988, He and Barclay 2000, Carey 2002), and effects on other species may be insignificant. Thinning can have untoward effects on the ecosystem process and elements of biodiversity associated with decadence in live standing trees, dead standing trees, and fallen, decaying trees by forestalling or preventing recruitment of these trees.

Heavy thinning also can promote hardwood and conifer regeneration and expansion, particularly *Tsuga*, precluding opportunity for herb and shrub regrowth (Alaback and Herman 1988, Fried et al. 1988, Tappeiner and Alaback 1989, Bailey and Tappeiner 1998). This has been particularly problematic in southeastern Alaska where partially cut stands show no positive herbaceous and shrub response because of conifer regeneration–intense thinning may exacerbate this problem (Alaback 1982, Tappeneir and Alaback 1989, Deal and Farr 1994, Deal 2001). Thinning in mixed-alder, rather than pure conifer stands may mitigate this problem (Hanley and Hoel 1996, Deal and Orlikowska 2002, Hennon et al. 2002).

The effect of stand thinning on the EM-fungal productivity (number or mass of mushrooms ha^{-1} yr^{-1}) depends on changes in carbohydrates available to the fungi, soil disturbances that impact ectomycorrhizal formation and existing fungal mats, and forest floor micro-environmental conditions for mushroom development. For instance, removing tree species that do not form ectomycorrhizae with the EM-fungus in question will not reduce carbohydrate supplies available to that fungus. Similarly, for tree species that do form EM with the fungus, removing fewer or suppressed trees will not reduce carbohydrate supplies as much as removing more or dominant trees. Often several tree species in a stand can form ectomycorrhizae with a given fungus, and the relative benefits that each host tree species provides a particular fungus species remains largely unknown. However, many managed stands are even-aged monocultures and any heavy thinning will have short-term impacts (Carey 2002).

Activities that minimize soil compaction or disturbance help preserve soil environments conducive to ectomycorrhizal formation (Amaranthus et al. 1996). Logging slash can make forest stands hard to walk through and mushrooms difficult to find. Low-intensity broadcast burning of slash in thinned stands is not likely to permanently harm EM-fungi but, to date, most studies of fire effects on ectomycorrhizae have examined high-intensity stand-replacement wildfires. EM-fungi often fruit in the same discrete clumps or patches each year. If the locations of these patches are mapped during sale planning, leave trees can be concentrated around these areas and skidding corridors or slash piles placed to avoid them.

The effect of altered forest floor temperature and humidity regimes on productivity of EM-fungi is less clear. We know of no studies that correlate stand-level productivity of EM-fungi with forest floor conditions. Temperature changes and precipitation often trigger fruiting, and both can be more intense under open canopies. On the other hand, prolonged high humidity needed for mushroom growth is more common under closed canopies. Cumulative seasonal mushroom productivity likely varies by complex interactions of canopy cover, local and regional weather patterns, typical fruiting season, site aspect, litter layer thickness, soil disturbance, and mushroom species. Hence, generalizations based on current knowledge are not useful.

Old, but often-cited research on the Japanese species of matsutake (*Tricholoma matsutake*) suggested that clearing understory brush and non-EM host trees warmed the forest floor, thus inducing matsutake mycelial colonies to grow and fruit more vigorously (Hosford et al. 1997). Significantly, the arboreal EM host species for the matsutake (dominant pines) were not thinned. These reports were the impetus for thinning trials in Oregon matsutake habitat (Weigand 1998) that have yet to be evaluated for efficacy.

Tree death, soil disturbance, and prescribed fire also can produce morel crops after thinning operations, although fewer morels typically fruit than in stand-replacement events, commensurate with the lesser degree of disturbance. Significant opportunities exist for enhanced coproduction of edible and medicinal CWD-saprobic fungi, large-diameter timber, and wildlife habitat by inoculating leave trees, snags, or stumps (Stamets 2000) during thinning operations.

2.5. Fertilization

Nitrogen (N) fertilizers offer substantial economic returns for timber production under certain circumstances and are used widely. The effects on understory plant communities are not well understood. Because broadcast applications of fertilizer tend to produce earlier canopy closure in newly established and very young plantations (DeBell et al. 1997), without stand density management, stem exclusion and concomitant understory plant decline could be accelerated. Studies show that N-fertilization in young stands increases tree growth, accelerates canopy closure, decreases understory light levels, and decreases understory plant abundance, especially species tolerant of low nutrient conditions such as the Ericaceae family (e.g., *Gaultheria* spp., *Vaccinium* spp.) (Stanek et al. 1979, Prescott et al. 1993, Thomas et al. 1999, Bradley et al. 2000). This trend can be more pronounced in unthinned stands (Thomas et al. 1999). Understory plant species abundance tends to remain unchanged or increase in stands that are thinned in combination with fertilization (Stanek et al.

1979, Thomas et al. 1999). In older stands, fertilization may accelerate stand development processes, including competition-related mortality (Miller and Tarrant 1983) that could accelerate gap development and understory regrowth.

While increased tree growth and shading due to fertilization treatments may partially explain understory plant decline, alteration of competitive interactions among species, species-specific nutritional characteristics, and physiology are also important (Prescott et al. 1993, Thomas et al. 1999). Some understory plants may be sensitive to the toxic effects of ammonia volatilization, a problem exacerbated in low light conditions (Thomas et al. 1999). Species tolerant of low N conditions may be unable to make efficient use of nitrate as their major source of N (Hayes 1986), and efficient nitrate assimilators may replace low N tolerant species. Prescott et al. (1993) note that repeated applications of about 400 kg N ha^{-1} yr^{-1} reduced *G. shallon* cover in coastal *Pseudotsuga* forests, but reductions were less pronounced on plots that received sulfur in addition to nitrogen. Moreover, *Symphoricarpos* increased dramatically in some of the N-fertilized stands.

Evidence from Europe suggests that nitrification of forest soils resulting from continuous air pollution may reduce the diversity and inhibit the growth of EM-fungi (Wallenda and Kottke 1998) and reduce the productivity of edible EM mushrooms (Arnolds 1991). Periodic fertilization of commercial timberlands with nitrogen could have a similar effect while new flushes of soluble soil nitrogen are being assimilated. Chanterelle fruiting can decline for several years following nitrogen fertilization and then rebound (W. Littke, Weyerhaeuser Co., Centralia, Washington, personal communication, 1995). The soil saprobe fungal community is too complex to derive generalizations regarding the effects of fertilization on fungal species (Moorhead and Reynolds 1992). The CWD-fungi are not likely to be affected in the short-term because they are decomposing previous substrates, but parasites might respond to changes in host health.

2.6. Other Silvicultural Systems

While clearcut logging and even-age management have dominated forest practices in the Pacific Northwest since World War II, other silvicultural systems are being proposed (see Hummel 2003). Although rationale and names differ (seed-tree, shelterwood, clearcut with reserves, green-tree retention) the common result is that more trees and their associated biological legacies are left on the site. Indeed, it has been suggested that these old terms, developed for the primary purpose of timber production, be replaced with "variable-retention harvest systems." Variable-retention harvest systems are based on the concept that natural disturbances leave residual "legacy" structures which are believed to play important roles in ecosystem function and biological diversity (Franklin et al. 1997). Therefore a variable-retention system leaves trees distributed throughout harvest areas to provide structural diversity (see Hummel 2003). Biological legacies are important for the rapid reestablishment of ecosystems that have high levels of structural, functional, and compositional diversity (Perry and Amaranthus 1997). However, understory species effects are not well documented although establishment of new studies (e.g., DEMO, McKenzie et al. 2000) are underway. Using a retrospective approach, North et al. (1996) found that herb and shrub richness were significantly higher, and shade tolerant plants were more common and abundant in green-tree retention cuts compared to clearcuts, depending on site preparation treatments. Green-tree retention management practices also have the potential to increase the abundance of certain lichens (Peck and McCune 1997). For fungi, leaving more ectomycorrhizal host trees, ericaceous shrubs, and large coarse woody debris could potentially renew EM-fungal fruiting and productivity sooner than in clearcut sites. As with thinning, the extent of this response would also depend on changes in carbohydrates available to the fungi, soil disturbances that impact formation, and forest floor micro-environmental conditions. It is unclear how large or dense leave-tree-patches must be to provide adequate nutrition and appropriate environmental conditions for the fruiting of EM-fungi.

3. Biological vs. Commercial Productivity

Overstory stand conditions and management practices can have important effects on the understory. However, there is limited information quantitatively relating management practices and commercial productivity of the understory. Most past studies measure species cover to quantify abundance (Appendix 1); few studies directly quantify commercial productivity or other attributes of interest for harvesters (some products may not be harvested and sold commercially, such as crafts, but still have associated desirable attributes).

At the stand scale, only a small percentage of understory biological productivity will be of sufficient quality for commercial harvest. Pilz et al. (1999b) suggested that harvesters collected no more than half the American matsutake that fruited in a popular commercial collection area. While some commercial floral greenery species may survive or be abundant after clearcut logging and heavy thinning, commercial characteristics may be lacking. Many fresh foliage products such as *G. shallon* (Box 1) and *Xerophyllum tenax* (Pursh) Nutt. (bear-grass), require shade to develop dark green foliage, a requirement for commercial value (Higgins et al., in press). Light thinning is thought to increase abundance of species used as floral products, but heavy thinning may reduce the amount of dark green foliage (Schlosser et al. 1992); variable-density thinning may provide the best of both desired results (Carey and Curtis 1996).

For some products, such as berries and leaves of certain medicinal species, commercial product abundance and biological abundance may be more closely linked. Some data exist relating flowering and fruiting of understory shrubs to forest development and forest management (Appendix 2). Most species require some light to flower and fruit, and significant fruit production is usually needed in an area to be realistically harvestable. Seed production in clearcuts can be dramatically greater compared to stands in the later stages of forest development (Tappeiner and Alaback 1989). Bunnell (1990) reports that *G. shallon* will not flower with crown closure greater than 33%. However, Huffman et al. (1994b) noted that this species flowered and fruited in as much as 70% sunlight. Berry production for *V. membranaceum* Dougl. ex Torr. (thin leaved huckleberry) tends to decline in closed canopy forests; however, production can be high in certain years (Minore 1984). Meteorological events rather than local conditions may be more important for determining annual *V. membranaceum* berry production (Minore 1978).

4. Management Implications and Considerations

Opportunities for compatible management of understory resources differ widely across forest developmental stages, management approaches, and land classes (Appendix 3). Opportunities will also differ due to site conditions, even across relatively small spatial scales, and that makes management of understory species particularly problematic. According to some studies listed in Appendix 2, certain species may respond in opposite ways to a particular management action or stand condition. Site-specific predictions usually cannot be made from broad general theory, and specific recommendations regarding understory management are often quite complex. Prescriptions to achieve specific landowner objectives must be site-specific (or aggregate site-specific frameworks for larger geographic scales) and based on information such as stand density, age, species composition, history, reproduction and dispersal, species biology, and product attributes. Appendix 2 presents species-specific information that can be used in conjunction with Appendix 3, literature citations, knowledge of species biology, and product attributes, to develop tailored silvicultural prescriptions for landowner objectives. We illustrate use of the information presented in these tables with a specific example below.

A landowner or manager may be interested in managing intensively for diverse commercial values, for example, for both timber and *G. shallon*. *G. shallon* can provide revenue opportunities in young stands for landowners or managers utilizing extended rotation times, or in relatively slow-growing stands. Revenues from understory products, supplemented with thinning revenues, can provide cash flow that allows flexibility in the timing of management decisions and fuller consideration of market conditions; biological diversity can provide economic portfolio diversification (Box 1). Although *G.*

shallon can be prolific in recently clearcut areas, commercial quality will be lacking. Stands in the second and third forest development stages may have sufficient plants with commercial quality leaves, and stand thinning could increase *G. shallon* abundance (Appendix 2). However, Appendix 3 details that heavy thinning can alter leaf color and commercial-quality of many floral greens. Fertilization may decrease the abundance of this species as it is tolerant of low N conditions while competing species capable of efficient nitrate uptake may be enhanced (Appendix 3). Fertilization in young stands may also hasten canopy closure, which could potentially reduce *G. shallon* abundance.

Although the response of fungi to forest management practices relate more to their mode of nutrition and quality of growth substrates than to growth forms, light levels, or reproductive mechanisms, opportunities for coproduction of harvested fungi and timber are varied and numerous. As with plants, optimal coproduction opportunities change as forest stands become established and mature. Long timber rotations on portions of the landscape, density management, snags and coarse woody debris management, intentional tree species selection or planting, and artificial inoculations all have the potential to create a synergistic mix of timber and fungal resources.

A major issue associated with our synthesis and the information presented in Appendix 2 is the lack of quantitative information regarding relationships between biological response or abundance and product attributes, such as commercial productivity (Kerns et al. 2002). Almost all the studies listed in Appendix 2 record biological species response and abundance (general cover). These data are informative for general species response and biodiversity issues. However, information regarding product attributes is rarely provided in these studies. Stands may contain understory species in great abundance, but in a condition or appearance that holds little or no value for humans interested in harvesting them. This issue is relevant for formulating site-specific prescriptions, for managers issuing permits, and for large-geographic-scale assessments. We caution that species-specific inventory data used in regional assessments only documents biological abundance.

Unfortunately, few methodologies currently exist for estimating understory product quantity and quality (Kerns et al. 2002). Natural variability in productivity, particularly for edible fungi, makes it difficult to assign future landscape productivity values (Pilz et al. 1998). Constraints on estimating economic value per unit area are also problematic for landowners and managers making decisions regarding cost-effective, active management for understory forest resources. Assessment of compatible management of wood and understory forest resources would be aided by permanent-plot studies using designs that incorporate appropriate product attributes as response variables. Such studies could then be used to develop models relating biological abundance to commercial characteristics.

Rapidly shifting market conditions also complicate development of frameworks for compatible management of commercial understory forest resources (Kerns et al., in press). Some species highlighted in this chapter may no longer be in demand in the future. In contrast, most management prescriptions take several years to plan and execute, and considerable time may pass before increases in desired species or product abundance are realized. Prediction of future market viability is also problematic. However, products can be actively marketed to stimulate demand, especially if they are widely available due to current local or future forest conditions. Collaboration of landowners, managers, business owners, and harvesters could provide such options for compatible management. Diversification can also help address future uncertainties–diverse kinds of wood (e.g., *Alnus rubra*), qualities of wood, diverse commercial valuable products (e.g., wood and floral greens), conservation credits, carbon credits, green certification; all may or may not be important in the future. Intentional management of spatial heterogeneity (e.g., variable density thinning, legacy retention) to provide a diversity of resources could prove to be an important key to management of wood and understory production. Demand for noncommercial products, especially those with known cultural and recreational associations (e.g., berries and mushrooms) tends to be more consistent over time. It may be easier to design management prescriptions for these species.

The greatest challenge for landowners and managers in generating socially acceptable and beneficial management strategies for understory resources at any geographic scale may be integrating social and ecological information. Both site-specific and aggregate regional and landscape-level forest management actions exist within social and cultural contexts, regardless of landownership. Nontimber forest resource harvesters, buyers, users, and permit data are rarely incorporated into inventory, research, and large-scale forest assessments. However, the USDI Bureau of Land Management is using a method in which "human geographic maps" are generated with GIS for application in a larger process known as the Human Geographic Information Management System. Based on spatially explicit and empirically derived "human resource units," this method shows promise as a tool for integrating social and ecological data to aid in responsive management of multi-jurisdictional, multi-species landscape-level ecosystem projects (Kent and Preister 1999). In the context of managing for understory resources, permit information, and experiential and indigenous knowledge gathered from harvesters, buyers, and business owners could be mapped spatially and incorporated into existing GIS databases. Compatible management approaches that are collaborative and provide local involvement and control may have the best chance for success (Behan 2001). As with many other resource management issues, achieving compatible management of understory forest resources is fundamentally a social, rather than a biological or ecological issue. Given the complex and interconnected biological, ecological, silvicultural, social, and economic context of timber and understory resource management, a conservative general approach to managing forests for uncertain future values is to maintain forest community, age, and structural diversity across the landscape over time.

5. References

Alaback, P.B. 1982. Dynamics of understory biomass in Sitka spruce-western hemlock forests of southeast Alaska. *Ecology*, 63: 1932-1948.

Alaback, P.B., and Herman, F.R. 1988. Long-term response of understory vegetation to stand density in *Picea-Tsuga* forests. *Canadian Journal of Forest Research*, 18: 1522-1530.

Amaranthus, M.P. 1992. Mycorrhizas, forest disturbance and regeneration in the Pacific Northwestern United States. In: D.J. Read, D.H. Lewis, A.H. Fitter, and I.J. Alexander (Eds.), Mycorrhizas in ecosystems (pp. 202-207). Wallingford, UK: CABI.

Amaranthus, M.P., Page-Dumroese, D., Harvey, A., Cazares, E., and Bednar, L.F. 1996. Soil compaction and organic matter affect, conifer seedling nonmycorrhizal and ectomycorrhizal root tip abundance and diversity. Research Paper PNW-RP-494. Portland, OR: U.S. Department of Agriculture, Forest Service, Pacific Northwest Research Station. 12 p.

Amaranthus, M.P., and Perry, D.A. 1994. The functioning of ectomycorrhizal fungi in the field: linkages in space and time. *Plant and Soil*, 159: 133-140.

Amaranthus, M.P., Pilz, D., Moore, A., Abbott, R., and Luoma, D.L. 2000. American matsutake (*Tricholoma magnivelare*) across spatial and temporal scales. In: R.F. Powers, D.L. Hauxwell, and G.M. Nakamura (Tech. coords.), Proceedings of the California Forest Soils Council conference on forest soils biology and forest management. General Technical Report PSW-GTR-178. Albany, CA: U.S. Department of Agriculture, Forest Service, Pacific Southwest Research Station.

Arnolds, E. 1991. Decline of ectomycorrhizal fungi in Europe. *Agriculture, Ecosystem, Environment*, 35: 209-244.

Bailey, J.D., Mayrsohn, C., Doescher, P.S., St. Pierre, E., and Tappeiner, J.C. 1998. Understory vegetation in old and young Douglas-fir forests of western Oregon. *Forest Ecology and Management*, 112: 289-302.

Bailey, J.D., and Tappeiner, J.C. 1998. Effects of stand thinning on structural development in 40- to 100-year old Douglas-fir stands in western Oregon. *Forest Ecology and Management*, 108: 99–113.

Behan, R.W. 2001. Plundered promise: capitalism, politics, and the fate of federal lands. Washington, DC: Island Press.

Bolsinger, C.L., and Waddell, K.L. 1993. Area of old-growth forests in California, Oregon, and Washington. Resource Bulletin PNW-RB-197. Portland, OR: U.S. Department of Agriculture, Forest Service, Pacific Northwest Research Station.

Bradley, R.L., Titus, B.D., Preston, C.M., and Bennett, J. 2000. Improvement of nutritional site quality 13 years after single application of fertilizer N and P on regenerating cedar-hemlock cutovers on northern Vancouver Island, B.C. *Plant and Soil*, 223: 195-206.

Bunnell, F.L. 1990. Reproduction of salal (*Gaultheria shallon*) under forest canopy. *Canadian Journal of Forest Research*, 20: 91-100.

Carey, A.B. 1995. Sciurids in Pacific Northwest managed and old-growth forests. *Ecological Applications*, 5: 648-661.

Carey, A.B. 2002. Globalization of flora: inviting worldwide ecosystem disaster. *Renewable Resources Journal*, 20: 13-17.

Carey, A.B., and Curtis, R.O. 1996. Conservation of biodiversity: a useful paradigm for forest ecosystem management. *Wildlife Society Bulletin*, 24: 61-62.

Carey, A.B., and Johnson, M.L. 1995. Small mammals in managed, naturally young, and old-growth forests. *Ecological Applications*, 5: 336-352.

Carey, A.B., Kershner, J., Biswell, B., and Dominguez De Toledo, L. 1999a. Ecological scale and forest development: squirrels, dietary fungi, and vascular plants in managed and unmanaged forest. *Wildlife Monographs*, 63.

Carey, A.B., Lippke, B.R., and Sessions, J. 1999b. Intentional systems management: managing forests for biodiversity. *Journal of Sustainable Forestry*, 9: 83-125.

Curtis, R.O., and Carey, A.B. 1996. Timber supply in the Pacific Northwest. Managing for economic and ecological values in Douglas-fir forests. *Journal of Forestry*, 94: 5-7; 35-37.

Curtis, R.O., DeBell, D.S., Harrington, C.A., Lavender, D.P., St. Clair, B., Tappeiner, J.C., II, and Walstad, J.D. 1998. Silviculture for multiple objectives in the Douglas-fir region. General Technical Report PNW-GTR-435. Portland, OR: U.S. Department of Agriculture, Forest Service, Pacific Northwest Research Station.

Dahlstrom, J.L., Smith, J.E., and Weber, N.S. 2000. Mycorrhiza-like interaction by *Morchella* with species of the Pinaceae in pure culture synthesis. *Mycorrhiza*, 9: 279-285.

Danell, E. 1994. *Cantharellus cibarius*: mycorrhiza formation and ecology. Acta Universitatis Upsaliensis, Comprehensive summaries of Uppsala dissertations from the faculty of science and technology 35. Uppsala: Swedish University of Agricultural Science.

Deal, R.L. 2001. The effects of partial cutting on forest plant communities of western hemlock–Sitka spruce stands in southeast Alaska. *Canadian Journal of Forest Research*, 31: 1-13.

Deal, R.L., and Farr, W.A. 1994. Composition and development of conifer regeneration in thinned and unthinned natural stands of western hemlock and Sitka spruce in southeast Alaska. *Canadian Journal of Forest Research*, 24: 976-984.

Deal, R.L., and Orlikowska, E.H. 2002. Development of mixed red alder-conifer stands in southeast Alaska. In: A.C. Johnson, R.W Haynes, and R.A. Monserud (Eds.), Congruent management of multiple resources: proceedings of the wood compatibility initiative workshop (pp. 127-132). General Technical Report PNW-GTR-563. Portland, OR: U.S. Department of Agriculture, Forest Service, Pacific Northwest Research Station.

DeBell, D.S., Curtis, R.O., Harrington, C.A., and Tappeiner, J.C. 1997. Shaping stand development through silvicultural practices. In: K.A. Kohm, and J.F. Franklin (Eds.), Creating a forestry for the 21st century (pp. 141-150). Washington, DC: Island Press.

Dyrness, C.T. 1973. Early succession following logging and burning in the western Cascades of Oregon. *Ecology*, 54: 57-69.

Franklin, J.F., Berg, D.R., Thornburgh, D.A., and Tappeiner, J.C. 1997. Alternative silvicultural approaches to timber harvesting: variable retention systems. In: K.A. Kohm, and J.F. Franklin (Eds.), Creating a forestry for the 21st century (pp. 111-139). Washington, DC: Island Press.

Franklin, J.F., and Dyrness, C.T. 1988. Natural vegetation of Oregon and Washington. Corvallis, OR: Oregon State University Press. Originally published by the USDA Forest Service 1973.

Franklin, J.F., and Spies, T.A. 1991. Composition, function, and structure of old-growth Douglas-fir forests. In: L.F. Ruggiero, K.B. Aubry, A.B. Carey, and M.H. Huff (Eds.), Wildlife and vegetation of unmanaged Douglas-fir forests (pp. 71-81). General Technical Report PNW-GTR-285. Portland, OR: U.S. Department of Agriculture, Forest Service, Pacific Northwest Research Station.

Franklin, J.F., Spies, T.A., Van Pelt, R., Carey, A., Thornburgh, D.A., Berg, R., Lindenmayer, D.B., Harmon, M.E., Keeton, W.S., Shaw, D.C., Bible, K., and Chen, J. 2002. Disturbances and structural development of natural forest ecosystems with silvicultural implications, using Douglas-fir as an example. *Forest Ecology and Management*, 155: 399–423.

Fried, J.S., Tappeiner, J.C., II, and Hibbs, D. 1988. Bigleaf maple seedling establishment and early growth in Douglas-fir forests. *Canadian Journal of Forest Research*, 18: 1226-1233.

Grime, J.P. 1979. Plant strategies and vegetation processes. New York: Wiley, Inc.

Halpern, C.B. 1989a. Early successional pattern of forest species: interactions of life history traits and disturbance. *Ecology*, 70: 704-720.

Halpern, C.B. 1989b. Early successional pathways and the resistance and resilience of forest communities. *Ecology*, 69: 1703-1715.

Halpern, C.B., Evans, S., and Nielson, S. 1999. Soil seed banks in young, closed canopy forests of the Olympic Peninsula, Washington: potential contributions to understory reinitiation. *Canadian Journal of Botany*, 77: 922-935.

Halpern, C.B., and Franklin, J.F. 1990. Physiognomic development of *Pseudotsuga* forests in relation to initial structure and disturbance intensity. *Journal of Vegetation Science*, 1: 475-482.

Halpern, C.B., and Spies, T.A. 1995. Plant species diversity in natural and managed forests of the Pacific Northwest. *Ecological Applications*, 5: 913-934.

Hanley, T.A., and Hoel, T. 1996. Species composition of old-growth and riparian Sitka spruce –western hemlock forests in southeastern Alaska. *Canadian Journal of Forest Research*, 26: 1703-1708.

Harrington, C.A., Buermeyer, K.R., Brodie, L.C., and Wender, B.W. 2002. Factors influencing growth and flowering of understory plants in conifer stands in western Washington. In: A.C. Johnson, R.W. Haynes, and R.A. Monserud (Eds.), Congruent management of multiple resources: proceedings of the wood compatibility initiative workshop (pp. 159-168). General Technical Report PNW-GTR-563. Portland, OR: U.S. Department of Agriculture, Forest Service, Pacific Northwest Research Station.

Hayes, R.J. 1986. Mineral nitrogen in the plant-soil system. Toronto: Academic Press.

Haynes, R.W., Monserud, R.A., and Johnson, A.C. 2003. Compatible forest management: background and context. Chapter 1. In: R.A. Monserud, R.W. Haynes, and A.C. Johnson (Eds.), Compatible forest management. Dordrecht, The Netherlands: Kluwer Academic Publishers.

He, F., and Barclay, H.J. 2000. Long-term response of understory plant species to thinning and fertilization in a Douglas-fir plantation on southern Vancouver Island, British Columbia. *Canadian Journal of Forest Research*, 30: 556-572.

Hennon, P., Wipfli, M., Deal, R., Johnson, A.C., De Santo, T., Schultz, M., Hanley, T., Orlikowska, E., Takashi, G., Bryant, M., and Edwards, R. 2002. Mixed alder-conifer forests: managing upland ecosystems in southeast Alaska for wood products, wildlife, and fish. In: A.C. Johnson, R.W Haynes, and R.A. Monserud (Eds.), Congruent management of multiple resources: proceedings of the wood compatibility initiative workshop (pp. 51-58). General Technical Report PNW-GTR-563. Portland, OR: U.S. Department of Agriculture, Forest Service, Pacific Northwest Research Station.

Higgins, S., Blatner, K., Kerns, B.K., and Worthington, A. [In press]. Relationship between *Xerophyllum tenax* and canopy density in the southern Cascades of Washington. *Western Journal of Applied Forestry*.

Hitchcock, C.L., and Cronquist, A. 1973. Flora of the Pacific Northwest. Seattle and London: University of Washington Press.

Hosford, D., Pilz, D., Molina, R., and Amaranthus, M. 1997. Ecology and management of the commercially harvested American matsutake mushroom. General Technical Report PNW-GTR-412. Portland, OR: U.S. Department of Agriculture, Forest Service, Pacific Northwest Research Station.

Huffman, D.W., and Tappeiner, J.C. 1997. Clonal expansion and seedling recruitment of Oregon grape (*Berberis nervosa*) in Douglas-fir (*Pseudotsuga menziesii*) forests: comparisons with salal (*Gaultheria shallon*). *Canadian Journal of Forest Research*, 27: 1788-1793.

Huffman, D.W., Tappeiner, J.C., and Zasada, C. 1994a. Regeneration of salal (*Gaultheria shallon*) in the central Coast Range forests of Oregon. *Canadian Journal of Botany*, 72: 39-51.

Huffman, D.W., Zasada, C., and Tappeiner, J.C. 1994b. Growth and morphology of rhizome cuttings and seedlings of salal (*Gaultheria shallon*): effects of four light intensities. *Canadian Journal of Botany*, 72: 1702-1708.

Hummel, S.S. 2003. Managing structural and compositional diversity with silviculture. Chapter 4. In: R.A. Monserud, R.W. Haynes, and A.C. Johnson (Eds.), Compatible forest management. Dordrecht, The Netherlands: Kluwer Academic Publishers.

IFCAE. 2002. United States nontimber forest products database. Portland, OR: Institute for Culture and Ecology. Retrieved November 20, 2002 from http://www.ifcae.org/ntfp/.

Kent, J.D., and Preister, K. 1999. Methods for development of human geographic boundaries and their uses, Cooperative Agreement No. 1422-P850-A8-0015 between James Kent Associates (JKA) and U.S. Department of Interior, Bureau of Land Management (BLM), Task Order No. 001.

Kerns, B.K., Alexander, S.J., and Bailey, J.D. [In press]. Nontimber forest products and Douglas-fir stand conditions in western Oregon: ecology and economics of huckleberry species. *Economic Botany*.

Kerns, B.K., Liegel, L., Pilz, D., and Alexander, S.J. 2002. Biological inventory and monitoring. In: E.T. Jones, R.T. McLain, and J. Weigand (Eds.), Nontimber forest products in the United States (pp. 237-269). Lawrence, KS: University Press of Kansas.

Kerns, B.K., and Ohmann, J. [In review]. Factors that influence shrub abundance in the coastal Oregon forest landscape.

Klinka, K., Chen, H.Y.H., Wang, Q., and de Montigny, L. 1996. Forest canopies and their influence on understory vegetation in early seral stands on west Vancouver Island. *Northwest Science*, 70: 193-200.

Knowe, S.A., Stein, W.I., and Shainsky, L.J. 1997. Predicting growth response of shrubs to clear-cutting and site preparation in coastal Oregon forests. *Canadian Journal of Forest Research*, 27: 217-226.

Kranabetter, J.M., Trowbridge, T., Macadam, A., McLennan, D., and Friesen, J. 2002. Ecological descriptions of pine mushrooms (*Tricholoma magnivelare*) habitat and estimates of its extent in northwestern British Columbia. *Forest Ecology and Management*, 158: 249-261.

Likens, G., Bormann, F.H., Pierce, R.S., and Reiner, W.A. 1978. Recovery of a deforested ecosystem. *Science*, 19: 492-496.

Loehle, C., MacCracken, J.G., Runde, D., and Hicks, L. 2002. Forest management at landscape scales. *Journal of Forestry*, 11(6): 25-33.

Long, J.N. 1977. Trends in plant species diversity associated with development in a series of *Pseudotsuga menziesii/Gaultheria shallon* stands. *Northwest Science*, 51: 119-130.

Long, J.N., and Turner, J. 1975. Aboveground biomass of understory and overstory in an age sequence of four Douglas-fir stands. *Journal of Applied Ecology*, 12: 179-188.

Love, T., Jones, E., and Liegel, L. 1998. Valuing the temperate rainforest: wild mushrooming on the Olympic Peninsula Biosphere Reserve. In: L.H. Liegel (Comp.), The biological, socioeconomic, and managerial aspects of chanterelle mushroom harvesting: the Olympic Peninsula, Washington State, U.S.A. *AMBIO*, Special Report No. 9: 16-25.

Luoma, D.L., Eberhart, J.L., and Amaranthus, M.P. 1996. Response of ectomycorrhizal fungi to forest management treatments: Sporocarp production. In: C. Azcon-Aguilar, and J.M. Barea (Eds.), *Mycorrhizas* in integrated systems: from genes to plant development. (pp. 553-556). Luxembourg: Office for Official Publications of the European Communities.

McCune, B. 1993. Gradients in epiphyte biomass in three *Pseudotsuga–Tsuga* forests of different ages in western Oregon and Washington. *The Bryologist*, 96: 405-411.

McKenzie, D., and Halpern, C. 1999. Modeling the distribution of shrub species in Pacific Northwest forests. *Forest Ecology and Management*, 114: 293-307.

McKenzie, D., Halpern, C., and Nelson, C.R. 2000. Overstory influences on herb and shrub communities in mature forests of western Washington, U.S.A. *Canadian Journal of Forest Research*, 30: 1-12.

Mead, B.R. 2002. Constancy and cover of plants in the Petersburg and Wrangell Districts, Tongass National Forest and Associated Private and other public lands, southeast Alaska. Research Paper PNW-RP-540. Portland, OR: U.S. Department of Agriculture, Forest Service, Pacific Northwest Research Station.

Messier, C., Honer, T.W., and Kimmins, J.P 1989. Photosynthetic photon flux density, red: far-red ration, and minimum light requirement for survival of *Gaultheria shallon* in western red cedar–western hemlock stands in coastal British Columbia. *Canadian Journal of Forest Research*, 19: 1470-1477.

Messier, C., and Kimmins, J.P. 1991. Above- and below-ground vegetation recovery in recently clearcut and burned sites dominated by *Gaultheria shallon* in British Columbia. *Forest Ecology and Management*, 46: 275-294.

Miller, R.E., and Tarrant, R.F. 1983. Long-term growth response of Douglas-fir to ammonium nitrate fertilizer. *Forest Science*, 29: 127-137.

Minore, D. 1978. Big huckleberry abundance as related to environment and associated vegetation near Mount Adams, Washington. Research Note PNW-322. Portland, OR: U.S. Department of Agriculture, Forest Service, Pacific Northwest Forest and Range Experiment Station.

Minore, D. 1984. *Vaccinium membranaceum* berry production seven years after treatment to reduce overstory tree canopies. *Northwest Science*, 58: 208-212.

Molina, R., Massicotte, H., and Trappe, J.M. 1992. Specificity phenomena in mycorrhizal symbioses: community-ecological consequences and practical implications. In: M.F. Allen (Ed.), Mycorrhizal functioning (pp. 357-423). New York: Chapman and Hall.

Molina, R., and Trappe, J.M. 1982. Patterns of ectomycorrhizal host specificity and potential among Pacific Northwest conifers and fungi. *Forest Science*, 28: 423-458.

Moore, D., Nauta, M.M., Evans, S.E., and Rotheroe, M. (Eds). 2001. Fungal conservation: issues and solutions. A special volume of the British Mycological Society. Cambridge, UK: Cambridge University Press.

Moorhead, D.L., and Reynolds, J.F. 1992. Modeling the contributions of decomposer fungi in nutrient cycling. In: G.C. Carroll, and D.T. Wicklow (Eds.), The fungal community: its organization and role in the ecosystem (2nd ed.) (pp. 691-714). New York: Marcel Dekker, Inc.

Morrison, M.L. 1982. The structure of western warbler assemblages: ecomorphological analysis of the balk-throated gray and hermit warbler. *The Auk*, 99: 503-513.

Muir, P.S., Mattingly, R.L., Tappeiner, J.C., II, Bailey, J.D., Elliot, W.E., Hagar, J.C., Miller, J.C., Peterson, E.B., and Starkey, E.E. 2002. Managing for biodiversity in young Douglas-fir forests of Western Oregon. Biological Science Report USGS/BRD/BSR 2002-0006. Corvallis, OR: U.S. Geological Survey Division. 76 p.

North, M., Chen, J., Smith, G., Krakowiak, L., and Franklin, J. 1996. Initial response of understory plant diversity and overstory tree diameter growth to a green tree retention harvest. *Northwest Science*, 70: 24-35.

O'Dea, M.E., Zasada, J.C., and Tappeiner, J.C. 1995. Vine maple clone growth and reproduction in managed and unmanaged coastal Oregon Douglas-fir forests. *Ecological Applications*, 5: 63-73.

Oliver, C.D., and Larson, B.C. 1996. Forest stand dynamics. New York: John Wiley and Sons, Inc.

Pabst, R.J., and Spies, T.A. 1998. Distribution of herbs and shrubs in relation to landform and canopy cover in riparian forests of coastal Oregon. *Canadian Journal of Botany*, 76: 298-315.

Peck, J.E. 1997. Commercial moss harvest in northwestern Oregon: describing the epiphyte communities. *Northwest Science*, 71: 187-194.

Peck, J.E., Acker, S.A., and Mckee, W.A. 1995. Autoecology of mosses in coniferous forests in the central western Cascades of Oregon. *Northwest Science*, 69: 185-190.

Peck, J.E., and McCune, B. 1997. Remnant trees and canopy lichen communities in western Oregon: a retrospective approach. *Ecological Applications*, 7: 1181-1187.

Peck, J.E., and McCune, B. 1998. Commercial moss harvest in northwestern Oregon: biomass and accumulation of epiphytes. *Biological Conservation*, 86: 299-305.

Perry, D.A., and Amaranthus, M.P. 1997. Disturbance, recovery, and stability. In: K.A. Kohm, and J.F Franklin (Eds.), Creating a forestry for the 21st century (pp. 31-56). Washington, DC: Island Press.

Pilz, D., Mayo, J., and Molina, R. 1999a. Chanterelle mushroom productivity responses to young stand thinning: young stand thinning and diversity study. Corvallis, OR: Oregon State University. Retrieved November 20, 2002 from http://www.fsl.orst.edu/mycology/youngstndthin/Yss.html.

Pilz, D., Molina, R., and Liegel, L. 1998. Biological productivity of chanterelle mushrooms in and near the Olympic Biosphere Reserve. *AMBIO,* Special Report No. 9: 8-13.

Pilz, D., Norvell, L., Danell, E., and Molina, R. 2003. Ecology and management of commercially harvested chanterelle mushrooms. General Technical Report PNW-GTR-567. Portland, OR: U.S. Department of Agriculture, Forest Service, Pacific Northwest Research Station.

Pilz, D., Smith, J., Amaranthus, M.P., Alexander, S., Molina, R., and Luoma, D. 1999b. Managing the commercial harvest of the American matsutake and timber in the southern Oregon Cascade Range. *Journal of Forestry*, 97: 8-15.

Pilz, D., Weber, N.S., Carter, C., Parks, C., and Molina, R. [In review]. Investigations of morel mushroom productivity, ecology, taxonomy, and population genetics following wildfires and tree mortality in northeastern Oregon.

Prescott, C.E., Coward, L.P., Weetman, G.F., and Gessel, S.P. 1993. Effects of repeated nitrogen fertilization on the ericaceous shrub, salal (*Gaultheria shallon*), in two coastal Douglas-fir forests. *Forest Ecology and Management*, 61: 45-60.

Qian, H., Klinka, H., and Sivak, B. 1997. Diversity of the understory vascular vegetation in 40-year-old and old-growth forest stands on Vancouver Island, British Columbia. *Journal of Vegetation Science*, 8: 773-780.

Rogers, R.S. 1980. Hemlock stands from Wisconsin to Nova Scotia: transitions in understory composition along a floristic gradient. *Ecology*, 6: 178-193.

Rominger, E.M., Allen-Johnson, L., and Oldemeyer, J.L. 1994. Arboreal lichen in uncut and partially cut subalpine fir stands in woodland caribou habitat, northern Idaho and south-western British Columbia. *Forest Ecology and Management*, 70: 195-202.

Schlosser, W.E., Blatner, K.A., and Chapman, R.C. 1991. Economic and marketing implications of the special forest products harvest in the Coastal Pacific Northwest. *Western Journal of Applied Forestry*, 6: 67-72.

Schlosser, W.E., Blatner, K.A., and Zamora, B. 1992. Pacific Northwest forest lands potential for floral greenery production. *Northwest Science*, 66: 44-55.

Schoonmaker, P., and Mckee, A. 1988. Species composition and diversity during secondary succession of coniferous forests in the western Cascade Mountains of Oregon. *Forest Science*, 34: 960-979.

Schroder, R. 2002. Contemporary subsistence use of nontimber forest products in Alaska. In: E.T. Jones, R.T. McLain, and J. Weigand (Eds.), Nontimber forest products in the United States (pp. 300-326). Lawrence, KS: University Press of Kansas.

Smith, J.E., Molina, R., Huso, M.M.P., Luoma, D.L., McKay, D., and Castellano, M. 2002. Species richness, abundance, and composition of hypogeous and epigeous ectomycorrhizal sporocarps in young, rotation-age, and old-growth stands of Douglas-fir (*Pseudotsuga menziesii*) in the Cascade Range of Oregon. *Canadian Journal of Botany*, 80: 186-204.

Smith, N.J., and Clark, D.R. 1990. Estimating salal leaf area index and leaf biomass from diffuse light attenuation. *Canadian Journal of Forest Research*, 20: 1265-1270.

Spies, T.A. 1991. Plant species diversity and occurrence in natural, young, mature, and old-growth Douglas-fir stands in western Oregon and Washington. In: L.F. Ruggiero, K.B. Aubry, A.B. Carey, and M.H. Huff (Eds.), Wildlife and vegetation of unmanaged Douglas-fir forests (pp. 111-121). General Technical Report PNW-GTR-285. Portland, OR: U.S. Department of Agriculture, Forest Service, Pacific Northwest Research Station.

Spies, T.A. 1997. Forest stand structure, composition, and function. In: K.A. Kohm, and J.F. Franklin (Eds.), Creating a forestry for the 21st century (pp. 11-30). Washington, DC: Island Press.

Spies, T.A., and Franklin, J.F. 1991. The structure of natural young, mature, and old-growth Douglas-fir forests in Oregon and Washington. In: L.F. Ruggiero, K.B. Aubry, A.B. Carey, and M.H. Huff (Eds.), Wildlife and vegetation of unmanaged Douglas-fir forests (pp. 91-109). General Technical Report PNW-GTR-285. Portland, OR: U.S. Department of Agriculture, Forest Service, Pacific Northwest Research Station.

Stamets, P. 2000. Growing gourmet and medicinal mushrooms. (3rd ed.). Berkeley, CA: Ten Speed Press.

Stamets, P., and Yao, C.D.W. 2002. MycoMedicinals: an informational treatise on mushrooms (3rd ed.). Olympia, WA: MycoMedia Productions.

Stanek, W., Beddow, D., and State, D. 1979. Fertilization and thinning effects on a Douglas-fir ecosystem at Shawnigan Lake on Vancouver Island. Some observations on salal and braken fern undergrowth. Report BS-R-1. Victoria, BC: Environment Canada, Pacific Forest Research Centre.

Stein, W.I. 1995. Ten-year development of Douglas-fir and associated vegetation after different site preparation on Coast Range clearcuts. Research Paper PNW-RP-473. Portland, OR: U.S. Department of Agriculture, Forest Service, Pacific Northwest Research Station.

Stewart, G.H. 1986. Forest development in canopy openings in old-growth Pseudotsuga forest of the western Cascade Range, Oregon. *Canadian Journal of Forest Research*, 16: 558-568.

Stewart, G.H. 1988. The influence of canopy cover on understory development in forests of the western Cascade Range, Oregon, USA. *Vegetatio*, 76: 79-88.

Tappeiner, J.C., and Alaback, P.B. 1989. Early establishment and vegetation growth of understory species in the western hemlock–Sitka spruce forests in southeast Alaska. *Canadian Journal of Botany*, 67: 318-326.

Tappeiner, J.C., Huffmann, D., Marshall, D., Spies, T., and Bailey, J.D. 1997. Density, ages, and growth rates in old-growth and young-growth forests in coastal Oregon. *Canadian Journal of Forest Research*, 27: 638-648.

Tappeiner, J.C., and Zasada, J.C. 1993. Establishment of salmonberry, salal, vine maple, and bigleaf maple seedlings in the coastal forests of Oregon. *Canadian Journal of Forest Research*, 23: 1775-1780.

Tappeiner, J.C., Zasada, J.C., Ryan, P., and Newton, M. 1991. Salmonberry clonal and population structure: the basis for persistent cover. *Ecology*, 72: 609-618.

Thomas, S.C., Halpern, C., Falk, D.A., Liguori, D.A., and Austin, K.A. 1999. Plant diversity in managed forests: understory responses to thinning and fertilization. *Ecological Applications*, 9: 864–879.

Thysell, D.R., and Carey, A.B. 2000. Effects of forest management on understory vegetation: a retrospective study. General Technical Report PNW-GTR-488. Portland, OR: U.S. Department of Agriculture, Forest Service, Pacific Northwest Research Station.

Thysell, D.R., and Carey, A.B. 2001. Manipulation of density of *Pseudotsuga menzieii* canopies: preliminary effects on understory vegetation. *Canadian Journal of Forest Research*, 31(9): 1513-1525.

Turner, J., and Long, J.N. 1975. Accumulation of organic matter in a series of Douglas-fir stands. *Canadian Journal of Forest Research*, 5: 681-690.

Vance, N., Borsting, M., Pilz, D., and Freed, J. 2001. Special forest products. Species information guide for the Pacific Northwest. General Technical Report PNW-GTR-513. Portland, OR: U.S. Department of Agriculture, Forest Service, Pacific Northwest Research Station.

Vance, N., Gray, A., and Haberman, B. 2002. Assessment of western Oregon forest inventory for evaluating commercially important understory plants. In: A.C. Johnson, R.W. Haynes, and R.A. Monserud (Eds.), Congruent management of multiple resources: proceedings of the wood compatibility initiative workshop (pp. 183-190). General Technical Report PNW-GTR-563. Portland, OR: U.S. Department of Agriculture, Forest Service, Pacific Northwest Research Station.

Van Pelt, R., and Franklin, J.F. 2000. Influence of canopy structure on the understory environment in tall, old-growth, conifer forests. *Canadian Journal of Forest Research*, 30: 1231-1245.

Wallenda, T., and Kottke, I. 1998. Nitrogen deposition and ecto-mycorrhiza. *New Phytologist*, 139: 169-187.

Weber, N.S., Pilz, D., and Carter, C. 1996. Morel life histories: beginning to address the unknowns with a case study in the Fremont National Forest near Lakeview, Oregon. In: D. Pilz, and R. Molina (Eds.), Managing forest ecosystems to conserve fungus diversity and sustain wild mushroom harvests (pp. 62-68). General Technical Report GTR-PNW-371. Portland, OR: U.S. Department of Agriculture, Forest Service, Pacific Northwest Research Station.

Weigand, J.F. 1998. Management experiments for high-elevation agroforestry systems jointly producing matsutake mushrooms and high-quality timber in the Cascade Range of southern Oregon. General Technical Report GTR-PNW-424. Portland, OR: U.S. Department of Agriculture, Forest Service, Pacific Northwest Research Station.

Wender, B.W., Harrington, C.A., and Tappeiner J.C. [In review]. Flower and fruit production of understory shrubs in western Oregon and Washington. *Northwest Science*.

Wipfli, M.S., Deal, R.L., Hennon, P.E., Johnson, A.C., Edwards, R.T., De Santo, T.L., Gomi, T., Orlikowska, E.H., Bryant, M.D., Schultz, M.E., LeSage, C., Kimbirauskus, R., and D'Amore, D.V. 2003. Compatible management of red alder-conifer ecosystems in southeastern Alaska. Chapter 3. In: R.A. Monserud, R.W. Haynes, and A.C. Johnson (Eds.), Compatible forest management. Dordrecht, The Netherlands: Kluwer Academic Publishers.

Xiao, G., and Berch, S.M. 1996. Diversity and abundance of ericoid mycorrhizal fungi of *Gaultheria shallon* on forest clearcuts. *Canadian Journal of Botany*, 74: 337-346.

Yarie, J. 1980. The role of understory vegetation in the nutrient cycle of forested ecosystems in the mountain hemlock biogeoclimatic zone. *Ecology*, 6: 1498-1514.

Appendix 1. Studies detailing non-arborescent understory response to forest management practices and other disturbances in Pacific Northwest forests.

Reference	Location[1]	Vegetation[2]	Study type, stand type and forest management treatments[3]	Response variables[4]
Alaback (1982)	SEA	PISI, PISI/TSHE, TSHE/PISI, TSHE/THPL	Retrospective chronosequence of 60 plots in clearcut and forest stands (0 – approximately 400 yrs. old).	Shrub, herb, fern, bryophyte, lichen; functional group & species-specific biological biomass.
Alaback and Herman (1988)	OCR	PISI, TSHE/PSME	6 month & 17 yr. understory response to five experimental thinning treatments at 2 sites, stands thinned at 13 yrs.	Shrub, herb, fern; functional group & species-specific biomass.
Amaranthus et al. (2000)	OCA, SO	TSME/ABSH/PIMO	Descriptive study, fungi relative host specificity, spatial proximity of fungi to potential host trees, & annual productivity variation from 4 years (1992-1995).	Fungi; *T. magnivelare* density & commercial biomass; spatial interactions.
Bailey et al. (1998)	OCR, OCA	TSHE/PSME	Retrospective chronosequence of 80 stands - including 50- 120 yr. old thinned & unthinned pairs (10-23 yrs. since thinned) and 18 older (LSOG) stands.	Shrub, herb, fern; some species-specific frequency & cover; diversity, composition.
Bunnell (1990)	BC	TSHE/PSME	Retrospective variable density (open, dense, and typically stocked) sequence of mature (80-100 yrs.) stands.	Shrub; *G. shallon* regeneration & rate of spread.
Deal (2001)	SEA	TSHE/PSME, PISI	Retrospective sequence of partially cut stands from 18 multi- and all-aged stands.	Shrub, herb, fern, bryophyte, lichen; cover, some species-specific; diversity, composition.
Dyrness (1973)	HJA	TSHE/PSME	Permanent plot time series from watersheds 3,[5] 5 yrs. after experimental clearcut & burn.	Shrub, herb, fern; functional group & species-specific cover.
Halpern (1989a)	HJA	TSHE/PSME	Permanent plot time series from watersheds 1 & 3,[5] 21 yrs. after experimental clearcut & burn.	Shrub, herb, fern; functional group & species-specific (most common) cover.

Appendix 1. Studies detailing non-arborescent understory response to forest management practices and other disturbances in Pacific Northwest forests. (continued)

Reference	Location[1]	Vegetation[2]	Study type, stand type and forest management treatments[3]	Response variables[4]
Halpern (1989b)	HJA	TSHE/PSME	Permanent plot time series from watersheds 1 & 3,[5] 21 yrs. after experimental clearcut & burn.	Shrub, herb, fern; species-specific cover; compositional & gradient analysis.
Halpern and Franklin (1990)	HJA	TSHE/PSME	Permanent plot time series from watersheds 1 & 3,[5] 25 yrs. after experimental clearcut & burn.	Shrub, herb, fern; functional group cover; structure.
Halpern and Spies (1995)	HJA, OCA, OCR, WCA	TSHE/PSME	Permanent plot time series from watersheds 1, 3, & 10,[5] 27 yrs. after experimental clearcut & burn, & retrospective chronosequence from 196 young and older (LSOG) unmanaged stands.[6]	Shrub, herb, fern; some species-specific cover; diversity.
Halpern et al. (1999)	OP	TSHE/PSME, PISI	Soil seed bank study—soil samples collected from three plots from young, pre-commercially thinned forests c. 40-60 yrs. old.	Shrub, herb, fern; species-specific frequency & density of germinant.
Hanley and Hoel (1996)	SEA	TSHE/PSME, ALRU	Retrospective study of two stands from three forest types: LSOG riparian, LSOG upland, and 40-year-old alder riparian.	Shrub, herb, fern, bryophyte, lichen; species-specific biomass.
Harrington et al. (2002)	WOP	TSHE/PSME	Experimental variable density thinning treatments as part of the Olympic Habitat Development Study.	Shrub; functional group & species-specific cover & flowering; diversity.
He and Barclay (2000)	VI	TSHE	27-year response to treatments applied once at 24 years—treatments were three thining levels combined with three N-fertilization levels (a 3 x 3 factorial design) from 51-yr.-old stands in a PSME plantation.	Shrub, herb, fern, bryophyte, lichen; functional group & some species-specific cover (*G. shallon*); diversity.
Higgins et al. (in press)	WCA	ABAM, TSHE	Descriptive study from 20 sites with different tree canopy densities from 2 plant associations.	Herb; *X. tenax* marketability, color, basal area, leaf length & commercial (fresh/dry) biomass.
Huffman et al. (1994a)	OCR	TSHE/PSME	Retrospective study of three sites from six 30–80 yr. old stands.	Shrub; clone growth, morphology & seedling regeneration of *G. shallon*.

Appendix 1. Studies detailing non-arborescent understory response to forest management practices and other disturbances in Pacific Northwest forests. (continued)

Reference	Location[1]	Vegetation[2]	Study type, stand type and forest management treatments[3]	Response variables[4]
Huffman et al. (1994b)	OCR	NA	2-year experimental nursery study using different light treatments to mimic different overstory conditions.	Shrub; *G. shallon* rhizome, aerial stem, seedling, and fruit biological biomass, density, length and height.
Huffman and Tappeiner (1997)	OCR	TSHE/PSME	Retrospective study of ten 35-55 year old stands–five stands thinned at age 5-8 yr., five stands unthinned.	Shrub; regeneration & expansion of *B. nervosa* and *G. shallon*.
Kerns and Ohmann (in review)	OCR	Various	Landscape model using forest inventory (FIA, CVS) & research plot data; multiple forest types, stand ages & management histories.	Shrub; total & deciduous shrub cover.
Kerns et al. (in press)	OCR, OCA	TSHE/PSME	Retrospective chronosequence study of 50 – 120 yr-old thinned & unthinned paired stands (10 - 23 yrs. since thinned) & older (LSOG) stands.[7]	Shrub; cover & frequency of *Vaccinium ovatum*, *V. parvifolium*, & *V. membranaceum*.
Klinka et al. (1996)	VI	THSE	Descriptive study of 157 35-yr-old plantation stands (replanted with various tree species) from 7 locations.	Shrub, herb, fern, bryophyte; functional group & *G. shallon*, *R. spectabilis* cover.
Knowe et al. (1997)	OCR	TSHE/PSME	Experimental study of 4 stands with 6 site preparation treatments (chemical and manual) after clearcut logging (same sites as used in Stein 1995).	Shrub; functional group & *G. shallon*, *R. parvifolius*, *R. spectabilis*; 10-year cover projection models.
Kranabetter et al. (2002)	BC	TSHE/THPL	Descriptive study developing site properties to identify *T. magnivelare* habitat & spatial extent.	Fungi; *T. magnivelare*
Long (1977)	WCA	TSHE/PSME	Retrospective study of four 22-73 yr. old naturally regenerated stands & a 5-yr-old plantation stand.	Shrub, herb, fern, bryophyte; species-specific cover especially *G. shallon*; diversity.

Appendix 1. Studies detailing non-arborescent understory response to forest management practices and other disturbances in Pacific Northwest forests. (continued)

Reference	**Location**[1]	**Vegetation**[2]	**Study type, stand type and forest management treatments**[3]	**Response variables**[4]
Long and Turner (1975)	WCA	TSHE/PSME	Retrospective study of four 22-73 yr. old naturally regenerated stands & a 9 yr. old plantation stand.	Shrub, herb, fern, bryophyte; species-specific biological biomass; diversity.
Luoma et al. (1996)	OCA, WCA, SO	Various. Mostly PSME/TSHE	Experimental study of mature forest thinning densities & leave tree patterns.	Epigeous & hypogeous EM-fungi; some species-specific frequency & biological biomass; diversity.
McCune (1993)	HJA, WRE	TSHE/PSME	Retrospective chronosequence of three stands (95, 145, and >400 years).	Bryophyte, lichen; epiphyte functional group biological biomass.
McKenzie and Halpern (1999)	WCA, OCA	Various	Modeling study using plot-level data from the Area Ecology program, USDA Forest Service.	Shrub, herb; probability of 9 common species occurrence.
McKenzie et al. (2000)	WCA, WPT	TSHE/PSME	Modeling study using four 65-170 yr. old stands from the DEMO study (Demonstration of Ecosystem Management Options).	Shrub, herb, fern; functional group & *A. cincinatum* cover.
Mead (2002)	SEA	Various	Descriptive analysis of plot-level inventory data (Forest Inventory and Analysis Program, USDA Forest Service) from the Stikine area, Tongass National Forest and adjacent lands.	Shrub, herb, fern, bryophyte; species-specific (by plant association) constancy & cover.
Messier et al. (1989)	BC	TSHE, THPL, PISI, ABAM	Retrospective study of three 36 year old variable density stands.	Shrub; *G. shallon* density, height, leaf morphology & relationship between photosynthetic photon flux density & cover.
Messier and Kimmins (1991)	VI	TSHE, THPL, ABAM	Retrospective chronosequence (2, 4, and 8 yrs. old) using six post clear-cut sites.	Shrub, herb; *G. shallon*, *Vaccinium* spp., *E. angustifolium* & *C. canadensis* vertical distribution, cover, leaf area, above & belowground biological biomass.

Appendix 1. Studies detailing non-arborescent understory response to forest management practices and other disturbances in Pacific Northwest forests. (continued)

Reference	**Location[1]**	**Vegetation[2]**	**Study type, stand type and forest management treatments[3]**	**Response variables[4]**
Minore (1984)	WCA	ABAM	Seven year results from five experimental treatments (cut and burn, herbicides, and inoculation) to reduce overstory canopies (4 replicates).	Shrub; *V. membranaceum* fruit production.
North et al. (1996)	WCA	TSHE/PSME/ABAM	Retrospective sequence of clearcut, green-tree retention, patch tree, and no-harvest treatments.	Shrub, herb, fern; species-specific cover; diversity.
O'Dea et al. (1995)	OCR	TSHE/PSME	Retrospective chronosequence from 25, 5-240 yr. old managed and unmanaged stands.	Shrub; *A. circinatum* clone morphology & reproduction.
Pabst and Spies (1998)	OCR	TSHE/PSME, PISI	Descriptive study of 94 sites along 42 streams representing a range in landform and forest characteristics.	Shrub, herb, fern; species-specific cover.
Peck (1997)	OCR, OCA	Not specified	Retrospective study of 10 OCR sites (65-110 years) chosen to represent commercially harvested areas, 10 OCA sites (50-290) chosen to represent hotspots of diversity.	Bryophyte, lichens; harvestable mat & species-specific biomass.
Peck et al. (1995)	HJA	TSHE/PSME, ABAM	Descriptive study from 15 mature and older (LSOG) sites	Bryophyte; species-specific cover & autecology.
Peck and McCune (1997)	WCA	TSHE/PSME, ABAM	Retrospective chronosequence of paired stands (55-120 years), with and without large old (> 300 yrs.) remnant trees (green-tree retention management).	Epiphytic lichens; functional group biological biomass.
Peck and McCune (1998)	OCR, OCA	Not specified	Retrospective study of 10 c. 100 yr. old OCR sites; nine OCA stands spanning a range of ages and management histories.	Bryophyte, lichen; harvestable mat biomass & accumulation rates.
Pilz et al. (1998)	OP	PISI/TSHE/PSME	Retrospective study of young, mature, and older forests, managed and unmanaged multi-ownership stands.	Fungi; *Cantharellus* density & biological biomass (ha^{-1} yr^{-1}).

Appendix 1. Studies detailing non-arborescent understory response to forest management practices and other disturbances in Pacific Northwest forests. (continued)

Reference	**Location[1]**	**Vegetation[2]**	**Study type, stand type and forest management treatments[3]**	**Response variables[4]**
Pilz et al. (1999a)	OCA	PSME/TSHE	Experimental study of 50 yr. old unthinned, moderately thinned, & heavily thinned stands; 1 year pre-thinning & 5 of 7 years post-thinning data.	Fungi; *Cantharellus* density & biological biomass (ha^{-1} yr^{-1}).
Pilz et al. (1999b)	OCA	ABSH/PIMO/PILA/ PICO/PIPO	Descriptive study of stands from different elevations and habitat types.	Fungi; *T. magnivelare* density, biological biomass & commercial productivity (ha^{-1} yr^{-1}).
Pilz et al. (in review)	BM	ABCO/PSME/ PICO/LAOC	Descriptive & experimental study of healthy, insect-killed, and wildfire-burned stands. Commercial harvest vs. no harvest treatments.	Fungi; species-specific *Morchella* density & biomass (ha^{-1} yr^{-1}).
Prescott et al. (1993)	WCA, VI	PSME, PSME/ TSHE/THPL	Experimental study of two sites from stands (45-70 years) that received various fertilization treatments (N, S).	Shrub, herb, fern; species-specific (especially *G. shallon*) cover.
Qian et al. (1997)	VI	TSHE	Retrospective study of 40 yr. old plantation & older (LSOG) stands. Plantation stands from two locations composed of 7 different single-species monocultures.	Shrub, herb, fern; diversity.
Rominger et al. (1994)	BC	ABLA	Retrospective study of two sites with partially cut and uncut stands.	Lichen (arboreal); total functional group & *Alectoria sarmentosa* and *Bryoria* biomass.
Schoonmaker and McKee (1988)	HJA,[8] OCA	TSHE/PSME	Retrospective chronosequence from 20 clearcut, young, and older (LSOG) stands.	Shrub, herb, fern; species-specific & functional group cover; diversity.
Smith and Clark (1990)	VI	THSE	Retrospective sequence of unthinned, thinned, and old (LSOG) stands (13 total) ranging in age from 15-250 years.	Shrub; *G. shallon* leaf area index & leaf biomass estimated from photosynthetically active radiation.

Appendix 1. Studies detailing non-arborescent understory response to forest management practices and other disturbances in Pacific Northwest forests. (continued)

Reference	Location[1]	Vegetation[2]	Study type, stand type and forest management treatments[3]	Response variables[4]
Smith et al. (2002)	HJA, OCA	TSHE/PSME	Retrospective chronosequence of young, mature, and old (LSOG) stands.	Epigeous & hypogeous EM fungi; functional group and some species-specific biomass & frequency; diversity.
Spies (1991)	WCA, OCA, OCR	TSHE/PSME, ABAM	Retrospective chronosequence from 196 young, mature, and old (LSOG) unmanaged stands.[6]	Shrub, herb, fern, bryophyte, lichen; species-specific cover & frequency; diversity.
Spies and Franklin (1991)	WCA, OCA, OCR	TSHE/PSME, ABAM	Retrospective chronosequence from 196 young, mature, and old (LSOG) unmanaged stands.[6]	Shrub, herb, fern, bryophyte, lichen; functional group cover; composition.
Stein (1995)	OCR	TSHE/PSME	Time series from four sites after experimental clearcut logging & six site preparation treatments (chemical & manual).	Shrub, herb, fern; functional group & species-specific cover & height.
Stewart (1988)	OCA	TSHE/PSME – ABAM transition	Retrospective study of three 102–129 yr. old stands, canopy gaps included as a subset variable.	Shrub, herb, fern; common species-specific. cover.
Tappeiner and Alaback (1989)	SEA	TSHE/PISI	Retrospective study of mature, old, young, 4- 6 yr. old stands & laboratory trials.	Shrub, herb; germination, survival & clonal development of 5 species.
Tappeiner and Zasada (1993)	OCR	TSHE/PSME	Retrospective study of clearcut, thinned and unthinnd 40-50 yr. old stands from two sites.	Shrub, herb; *R. spectabilis*, *G. shallon*, & *A. circinatum* seedling emergence & survival.
Tappeiner et al. (1991)	OCR	TSHE/PSME, ALRU	Retrospective study of 20 sites in clearcut, young and mature stands (2 -180 yrs).	*R. spectabilis* clonal architecture & population structure.
Thomas et al. (1999)	WCA, WPT	TSHE/PSME	Experimental thinning & periodic N fertilization treatments from six 21-27 yr. old plantation stands, thinned at 8–11 yrs.	Shrub, herb, fern; species-specific & functional group cover; diversify.

Appendix 1. Studies detailing non-arborescent understory response to forest management practices and other disturbances in Pacific Northwest forests. (continued)

Reference	**Location[1]**	**Vegetation[2]**	**Study type, stand type and forest management treatments[3]**	**Response variables[4]**
Thysell and Carey (2000)	WPT	TSHE/PSME	Retrospective study from 16, 13–ha 55-70 yr. old post-clearcut naturally regenerated stands that were either managed with thinning (thinned at 50–60 years) or legacies (no thinning).	Shrub, herb, fern, bryophyte; functional group & dominant species-specific cover; diversity.
Thysell and Carey (2001)	WPT	TSHE/PSME	Experimental variable density thinning treatments of 16, 13–ha stands.	Shrub, herb, fern, bryophyte; functional group & dominant species-specific cover; diversity.
Turner and Long (1975)	WCA	TSHE/PSME	Retrospective study of four 22-73 yr. old unmanaged stands.	Shrub, herb, fern, bryophyte; total understory & biomass of *G. shallon*.
Vance et al. (2002)	OCR, OCA	Various	Modeling study using western Oregon plot-level inventory data from Forest Inventory and Analysis (FIA).	Shrub, herb, fern; commercial species-specific frequency.
Van Pelt and Franklin (2000)	WCA, WRE, HJA, CCR	TSHE/PSME, ABAM, SESE	Retrospective study of the influence of tree & foliage spatial distribution on the understory environment; 6 stands in 3 yr. old (LSOG) sites.	Shrub, herb; functional group cover.
Wender et al. (in review)	WOP, WPT,	TSHE/PSME	Experimental variable density thinning treatments as part of the Olympic Habitat Development Study.	Shrub; 9 species-specific cover; flowering & diversity.
Xiao and Berch (1996)	VI	TSHE/ABAM, THPL/TSHE	Description of root samples removed from clearcut sites in each vegetation type.	Fungi; Ericoid mycorrhizal fungi isolated and identified from roots of *G. shallon*
Yarie (1980)	BC	TSME, ABAM, CHNO	Retrospective study of the role of understory in forest nutrient cycling from three plant associations from mature and old (average 260 years) forests.	Shrub, herb, fern, bryophyte; species-specific biomass.

Appendix 1. Studies detailing non-arborescent understory response to forest management practices and other disturbances in Pacific Northwest forests. (continued)

[1] **Location:** BC = British Columbia; BM = Blue Mountains, Oregon; CCR = California Coast Range; EW = eastern Washington; HJA = H.J. Andrews Experimental Forest (OCA); OCA = Oregon Cascades; OCR = Oregon Coast Range; OP = Olympic Peninsula; SEA = Southeast Alaska; SO = southern Oregon; VI = Vancouver Island, BC; WCA = Washington Cascades; WOP = Washington Olympic Peninsula; WPT = Washington Puget Trough; WRE = Wind River Experimental Forest.

[2] **Vegetation:** ABAM = *Abies amabilis*; ABCO = *Abies concolor*; ABLA = *Abies lasiocarpa*; ABSH = *Abies x shastensis*; ALRU = *Alnus rubra*; CHNO = *Chamaecyparis nootkatensis*; LAOC = *Larix occidentalis*; PICO= *Pinus contorta*; PILA = *Pinus lambertiana*; PIMO = *Pinus monticola*; PIPO = *Pinus ponderosa*; PISI = *Picea sitchensis*; PSME = *Pseudotsuga menziesii*; SESE = *Sequoia sempervirens*; THPL = *Thuja plicata*; TSHE = *Tsuga heterophylla*; TSME = *Tsuga mertensiana*.

[3] **Study types:** retrospective (area is studied after treatment, rather than before and after), descriptive, experimental, or other. Young stands generally refer to the first through third stages of forest development, mature stands are generally equivalent to the fourth stage, old stands are eqivalent to the fifth stage and may exhibit late-successional old-growth characteristics (LSOGS) (Franklin and Spies 1991) (see Appendix 1, Chapter 1). Classifications of stand age may vary based on author, study area, and forest type.

[4] **Descriptions of response variables** include life-form specificity: functional group, specific-species (listed), species-specific (not listed, > 3 and/or common, dominant or other); abundance measure: biological biomass (may be measured or estimated), commercial biomass or productivity, density, frequency, cover; and other response variables: e.g., diversity (species richness, heterogeneity, Simpson's index, etc.), compositional, or otherwise specified.

[5] Units within the H.J. Andrews Experimental Forest include several watersheds where understory vegetation has been remeasured periodically on permanent plots.

[6] The same 196 stands are used in these papers.

[7] Study used a subset of data presented in Bailey and Tappeiner (1998).

[8] Location within the H.J. Andrews was not specified.

Appendix 2. Common Pacific Northwest understory non-tree species with commercial, social, or cultural value, and associated response to forest management or forest development stage.

Stand type/forest management	Understory forest species response or association
Catastrophic disturbance	Susceptible:
	Chimaphila spp. (Dyrness 1973, Halpern 1989a, Halpern and Spies 1995)
	Goodyera oblongifolia (Dyrness 1973, Halpern 1989a)
	Taxus brevifolia (Schoonmaker and McKee 1988, Halpern 1989a, Halpern and Spies 1995)
	Many bryophyte and lichen species (Rominger et al. 1994)
	Sensitive or slow to recover:
	Coptis laciniata (Halpern 1989a especially with burning)
	Vaccinium membranaceum (Minore 1984, Halpern and Spies 1995 especially with burning)
	Vaccinium parvifolium (Halpern and Spies 1995)
	Xerophyllum tenax (Halpern and Spies 1995)
	Ectomycorrhizal fungi (*Cantharellus, Craterellus, Boletus, Tricholoma magnivelar, Morchella elata*) (Amaranthus 1992, Amaranthus and Perry 1994, Hosford et al. 1997, Smith et al. 2002)
	Initially declines, but recovers:
	Acer circinatum (Dyrness 1973, Schoonmaker and McKee 1988, Halpern 1989a, O'Dea et al. 1995)
	Berberis nervosa (Dyrness 1973, Schoonmaker and McKee 1988, Halpern 1989a)
	Corylus cornuta (Dyrness 1973, Halpern 1989a, Stein 1995)
	Gaultheria shallon (Dyrness 1973, Schoonmaker and McKee 1988, Halpern 1989a, Stein 1995)
	Polystichum munitum (Dyrness 1973, Schoonmaker and McKee 1988, Halpern 1989a, Stein 1995)
	Rubus ursinus (Schoonmaker and McKee 1988)
	Vaccinium membranaceum (Schoonmaker and McKee 1988)
	Vaccinium ovatum (Stein 1995, sensitive to burning)
	Vaccinium parvifolium (Dyrness 1973; Alaback 1982; Stein 1995, sensitive to burning)
	Increases or invades; may be ephemeral:
	Anaphalis margaritacea (Dyrness 1973 especially moist sites, Halpern 1989a, Stein 1995)
	Ceanothus sanguineus (Dyrness 1973 especially on burned, dry sites; Schoonmaker and McKee 1988; Halpern 1989a)
	Ceanothus velutinus (Dyrness 1973 especially on burned, dry sites; Schoonmaker and McKee 1988; Halpern 1989a)
	Dryopteris austriaca (Alaback 1982)
	Epilobium angustifolium (Dyrness 1973, Schoonmaker and McKee 1988, Halpern 1989a, Stein 1995)
	Gaultheria shallon (Messier and Kimmins 1991)
	Hypericum perforatum (Schoonmaker and McKee 1988)
	Pteridium aquilinum (Schoonmaker and McKee 1988, Halpern 1989a, Stein 1995)
	Rhamnus purshiana (Stein 1995)
	Rubus parviflorus (Schoonmaker and McKee 1988, Halpern 1989a, Stein 1995)

Appendix 2. Common Pacific Northwest understory non-tree species with commercial, social, or cultural value, and associated response to forest management or forest development stage (continued).

Stand type/forest management	Understory forest species response or association
	Rubus spectabilis (Alabeck 1982, Tappeiner and Zasada 1993, Stein 1995, North et al. 1996) *Rubus ursinus* (Dyrness 1973; Halpern 1989a; Stein 1995, sensitive to burning) *Salix scouleriana* (Halpern 1989a) *Sambucus cerulea* (Dyrness 1973, *Sambucus* spp., Stein 1995) Saprobic *Morchella* spp. (Weber et al. 1996)
Stand thinning	Susceptible to thinning and/or common in unthinned stands: Bryophytes (Thysell and Carey 2000) *Chimaphila menziesii* (Thomas et al. 1999) Ectomycorrhizal fungi such as *Cantharellus* spp. (Pilz et al. 1999a) *Goodyera oblongifolia* (Thysell and Carey 2000) *Kindbergia oregana* (He and Barclay 2000) *Polystichum munitum* (Bailey and Tappeiner 1998) *Taxus brevifolia* (Thomas et al. 1999) *Vaccinium parvifolium* (Thysell and Carey 2000) Increases: *Acer circinatum* (Tappeiner and Zasada 1993, O'Dea et al.1995) *Anaphalis margaritacea* (Bailey et al. 1998) *Dryopteris austriaca* (Alaback and Herman 1988) *Dryopteris expansa* (Deal 2001) *Gaultheria shallon* (Bailey and Tappeiner 1998; Huffman et al.1994a,b; Tappeiner and Zasada 1993; He and Barclay 2000; Thysell and Carey 2000) *Polystichum munitum* (Alaback and Hermann 1988, if moderate; Thomas et al. 1999; Thysell and Carey 2000) *Pteridium aquilinum* (Alaback and Herman 1988, if extreme; Bailey and Tappeiner 1998; Thysell and Carey 2000) *Rhamnus purshiana* (Thysell and Carey 2000) *Rubus spectabilis* (Alaback and Herman 1988, if extreme; Tappeiner and Zasada 1993) *Rubus ursinus* (Bailey and Tappeiner 1998, Thysell and Carey 2000) *Salix scouleriana* (Thysell and Carey 2000) *Sambucus racemosa* (Thysell and Carey 2000) Saprobic *Morchella* species (ephemeral fruiting, Weber et al. 1996) *Tricholoma magnivelare* (if thinned from below, removal of non-host trees, no soil compaction, Hosford et al. 1997) *Urtica dioica* (Thysell and Carey 2000) *Vaccinium alaskaense* (Alabeck and Herman 1988)
Fertilization	Decreases: *Berberis nervosa* (Thomas et al. 1999) Ectomycorrhizal fungi such as *Cantharellus* spp. (Arnolds 1991, Wallenda and Kottke 1998) *Gaultheria shallon* (Thomas et al.1999, He and Barclay 2000, Stanek et al. 1979, Prescott et al. 1993), especially if unthinned

Appendix 2. Common Pacific Northwest understory non-tree species with commercial, social, or cultural value, and associated response to forest management or forest development stage (continued).

Stand type/forest management	**Understory forest species response or association**
	Polystichum munitum (Prescott et al. 1993)
	Pteridium aquilinum (Stanek et al. 1979)
	Increases:
	Pteridium aquilinum (Prescott et al. 1993)
	Rubus spectabilis (Thomas et al. 1999)
	Sambucus racemosa (Thomas et al. 1999)
Maturation and diversification	Common associates:
	Acer circinatum (Halpern and Spies 1995, Bailey et al.1998)
	Chimaphila umbellata (Halpern and Spies 1995)
	Coptis laciniata (Halpern and Spies 1995)
	Coarse woody debris-saprobic fungi such as medicinal conks (Stamets and Yao 2002) and edible mushrooms (Stamets 2000)
	Epiphytes – *Lobaria*, *Usnea*, etc. (Alaback 1982, Halpern and Spies 1995)
	Goodyera oblongifolia (Halpern and Spies 1995, Bailey et al. 1998)
	Taxus brevifolia (Halpern and Spies 1995)
	Trillium ovatum (North et al. 1996)
	Vaccinium membranaceum (Halpern and Spies 1995)
	Vaccinium parvifolium (Halpern and Spies 1995)

Note: Species responses based information from experimental and retrospective studies presented in Appendix 1. Additional information regarding understory forest resources ecology can be found in Vance et al. (2001)

Appndix 3. Opportunities for compatible management of understory forest resources and timber across forest developmental stages.

Forest development stage:	**Opportunities for compatible management of understory forest resources and timber**
First: *Stand Initiation*	• Opportunities for some herbaceous, shrub, and exotic vascular species (e.g., *Hypericum perforatum*) (Appendix 2); depends on disturbance type, intensity, site preparation, amount of legacy retention, and subsequent vegetation control methods. • Alternative harvest and regeneration methods may provide more opportunities for understory species growth than conventional intensive timber management. • Retaining large shrubs and allowing some hardwood regeneration at this stage will provide more opportunities in later stages. • Nitrogen fertilization can increase conifer growth, accelerate canopy closure, and limit understory development, depending on competitive species interactions and species nutritional characteristics. Adding other nutrients can alter outcomes. • Berries: increased flowering and fruiting of species that survive and recover, better opportunities with retention of large shrubs. • Floral greenery species: desired characteristics for many products may be lacking in the short term. • Bryophytes and lichens: limited opportunities, legacy management may provide more opportunities. • Fungi: morels, CWD saprobes, and some parasites may be abundant; reforestation with appropriate host trees may be necessary for other species.

Second and Third: *Stand Closure and Exclusion*	• Forest structural and compositional factors can affect spatial and temporal patterns of understory species.
	• Densely stocked sites with canopy closure may have low herb and shrub abundance; ferns may be abundant.
	• Bryophytes and lichens: potential opportunities in dense stands, especially if hardwoods and large shrubs are present.
	• Fungi: potentially good opportunities, dependent on small-scale environments.
	• Canopy closure stage can be minimized through precommercial and commercial thinning and variable density thinning:
	    • Thinning can favor dominant understory species. Clonal, woody shrub, grass, sedge, nitrogen-fixing, and exotic species may increase flowering and fruiting. (Appendix 2).
	        • Outcomes depend on intensity, frequency, age and composition of stand and pre-thinning understory species composition.
	        • Uniform thinning may promote even species distributions.
	        • Shrub enhancement may require more intense thinning compared to herbs, but shrubs can be damaged during operations.
	        • Thinning may promote hardwood and conifer regeneration, especially in Alaska.
	        • Fungi:
	            • Productive patches can be mapped during sale planning and left uncut.
	            • Inoculations of leave trees with coarse woody debris-saprobes may improve harvesting opportunities.
	            • Too much slash can make forest stands hard to walk through and fungi difficult to find.
	• N-fertilization may accelerate density-dependant tree mortality; understory species tolerant of low N may decrease while others are enhanced. Fruiting of ectomycorrhizal fungi may be suppressed (Appendix 2).

Appendix 3. Opportunities for compatible management of understory forest resources and timber across forest developmental stages (continued).

Forest development stage:	**Opportunities for compatible management of understory forest resources and timber**
Fourth and Fifth: *Maturation and Diversification*	• Stages and management for these stages are more common on public lands. • Understory species can provide revenue in longer rotation frameworks. • Potential for compatible management exists. Heterogeneous resources create highly variable understory conditions with sites for understory regrowth, although conifer regeneration may be abundant. • Commercial thinning can be used to enhance herbs and shrubs and provide coarse woody debris for fungi. • Flowering and fruiting may increase in gaps. • Inefficiencies of variable species distribution, abundance, fruiting, and flowering may limit harvesting. • Bryophytes and lichens: conditions very favorable if abundant hardwoods and large shrubs are present. • Fungi: highly dependent on small-scale environments. • Conservation concerns regarding understory species harvesting on public lands may exist.

Note: Species-specific information for each forest management action is provided in Appendix 2.

Chapter 13

Managing Access to Nontimber Forest Products

Susan J. Alexander
USDA Forest Service, Pacific Northwest Research Station,
3200 SW Jefferson Way, Corvallis, OR 97331, USA
Roger D. Fight
USDA Forest Service, Pacific Northwest Research Station,
620 SW Main St., Suite 400, Portland, OR 97205, USA

1. Introduction

Nontimber forest products (NTFP) have received increasing attention from harvesters, forest managers, researchers, and the public in the 1990s. Nontimber forest products, referred to as special forest products by the U.S. Department of Agriculture, Forest Service, include floral greens, boughs, craft materials, wild edibles, medicinals, transplants, and other vegetative materials. Increased interest in NTFP is reflected in many publications, including two books on NTFP in the United States (Emery and McLain 2001, Jones et al. 2002). Several authors have described wild edible mushroom management in the Pacific Northwest (e.g., Liegel 1998, Pilz and Molina 1998, Pilz et al. 1999). Economic values for NTFP are not well known, but some research has explored prices and economic valuation methods for various species harvested in the Pacific Northwest (western Oregon and Washington, coastal British Columbia, and southeastern Alaska) (e.g., Blatner and Alexander 1998, Pilz et al. 1998, Alexander et al. 2002). Issues about NTFP harvest and sustainability are addressed in the U.S. National Report on Sustainable Forests (USDA FS 2003) in response to an international set of forest sustainability indicators and criteria referred to as the Montreal Process. Although forest management in the United States has not focused on the production of NTFP, they are abundant in forest ecosystems, and NTFP harvest is an important activity to many people.

Nontimber forest products play important roles in American culture and commerce. Over 200 species of NTFP are harvested from public and private

R.A. Monserud, R.W. Haynes and A.C. Johnson (eds.)
Compatible Forest Management, 383-400

lands in the Pacific Northwest for commercial, personal, and traditional purposes (de Geus 1995). Demand for NTFP has driven some of these resources into conditions of economic scarcity (Pearce 1992), creating a need for management. Although many NTFP are becoming scarce in an economic sense, their value generally has not been regarded sufficient to manage them for increased production. They are generally regarded as byproducts of forest management; therefore, the primary focus of NTFP management in the United States has been managing and regulating access to NTFP. Access management includes controlling both the physical ability to get to a place (e.g., road closures and gates), and legal access (e.g., permits, contracts, treaties, and regulation).

To the extent that NTFP are regarded as byproducts of forest management, they are considered compatible with forest production, but not worthy of joint management to produce an optimal mix of both outputs. As the value of NTFP increases relative to the value of timber, more opportunities are created to consider joint management. Although joint production of timber and NTFP is a promising area for study, there are few examples of joint management that produces an optimal mix of timber and nontimber products. Hence, the focus of the chapter is the mechanisms that managers use to manage access to NTFP.

1.1. Property Rights: Exclusion and Rivalness

Property rights in the United States are fairly explicit and are based on notions of exclusivity—that is, a landowner can determine who has access to his or her land. When talking about exclusion in economic terms, it means that the producer (the landowner in this case) can restrict access to the product (a tree or NTFP), limiting it to those who are willing to pay. Land ownership and the rights of owners regarding its use are well defined for both public (such as federal and state lands) and private land. It is ideal to define who has rights to a good, such as land, timber, or other forest resources, and agree on this definition before the resource is used, although this does not always happen. Disagreements about property rights are a significant source of conflict in natural resource use.

Timber and many other forest products are private goods irrespective of management. A private good is both rival and exclusive. Rivalness means one person's consumption of the resource reduces the quantity available to others. For instance, a harvested mushroom is rival in consumption; my eating it precludes you from eating it. Conversely, a sunset is nonrival in consumption; I can look at it all I want without reducing your ability to look at it all you want.

Timber has not always been a private good in the United States. In early settlement days, timber was regarded as a nonrival and nonexclusive good, and was even considered an impediment for agricultural development. In a sense, timber displayed nonrival characteristics because harvest by one person did not preclude another from harvesting some of the resource. Randall (1988) notes that pure nonrival goods are rare. He states that congestible goods characterize substantial sectors of the economy. When use of a congestible good is much less than capacity, nonrivarly results in that additional users impose small costs to the system. As the capacity constraint is approached, congestion sets in and additional users impose increasing costs on the system. It became clear to many people as settlement proceeded in the Unites States, that timber was becoming a congestible good and could become scarce. Consequently, land ownership became more closely defined and access to timber is regulated on all ownerships through harvest contracts, sales mechanisms, and pricing structures.

Most nontimber forest products, such as understory shrubs, berries, and mushrooms, were regarded as nonrival and nonexclusive goods long after timber became a private good, particularly those growing on public lands. Restricting access can be difficult (Tedder et al. 2002) or considered undesirable. Many NTFP are still regarded as nonrival and nonexclusive goods. These lingering social attitudes about NTFP have resulted in conflicts because some of these resources are now congestible and land managers are trying to establish exclusivity. The transition to congestion and scarcity and the resultant efforts to allocate harvest rights to NTFP challenge forest managers of both public and private lands.

1.2. NTFP Management History and Issues

Gathering of NTFP has historically been managed lightly or not at all in the United States. There are exceptions, such as areas on national forests in the temperate forests of the northwestern and northeastern United States that are managed for berry picking (especially, *Vaccinium* spp.). Many central and eastern states have addressed the harvest of select medicinal plants regarded as species of concern, such as American ginseng (*Panax quinquefolius* L.), in state legislation. Harvesting of NTFP in the United States is often rooted in traditions that have persisted despite decreasing availability, loss of knowledge, and the dominance of management for timber production, cattle grazing, and recreational activities on forestlands. Because many forest managers did not know or understand the cultural use patterns of NTFP, historically there has been a lack of NTFP management and policy, or even formal scientific study of NTFP ecology (Jones and Lynch 2002). In several regions of the United States, particularly the Pacific Northwest and the Northeast, NTFP industries such as the wild-harvested floral greens market and wild-harvested medicinal markets have been well-established for more than a century (Heckman 1951, Emery 1998). For most of this period, harvesters have gathered understory plant material and fungi on private and public lands with few restrictions.

Since the 1980s, however, forest management organizations have gradually increased funding levels for NTFP management and research. This shift in focus stems from several factors. There has been an increase in commercial demand for NTFP locally and globally. At the same time, scientists and managers have focused on increasing their understanding of the critical role that species and structural diversity have in sustainable forest management (Alexander and McLain 2001). Other factors include the decline of the timber industry in the Pacific Northwest and the loss of timber-related employment in many rural areas (von Hagen and Fight 1999; Lynch and McLain, in press). A sharp increase in Southeast Asian and Latino immigrants to Oregon and Washington and other regions has created a large pool of laborers with limited employment opportunities (Richards and Creasy 1996, Hansis 2002). Many rural people turn to NTFP harvesting when other job opportunities are lacking (Fisher 1997, Emery 1998). Concern about impacts on resource sustainability, impacts on roads, trespass, vandalism, and theft rose among public and private land managers. Over the past 15 years, the number of regulations, laws, and policy guidelines pertaining to the harvest of NTFP on state and federal lands increased greatly. This legislation applied not only to harvesting for personal use or sale, but also to first-level purchasing of NTFP, and the transport of NTFP on public roads (USDI BLM 1992, von Hagen and Fight 1999, McLain 2000).

Concerns common to harvesters and land managers include fairness and theft. Fairness is a goal often stated by managers and users, and encompasses social justice and equity issues. A primary function of access management is to assign property rights where such rights do not already exist. This is done through physical and legal mechanisms that, ideally, are effective, efficient, and equitable, while providing resource sustainability, some form of rent to the landowner, and income to the harvester. Lingering notions that NTFP are nonexclusive, along with shifting attitudes regarding who has rights to the resource, can make the efficient and fair allocation of NTFP problematic.

2. Managing Access

Forest managers on public and private lands in the United States have an array of choices to make when deciding how to allocate (formally or informally) harvest rights for NTFP. Forest managers may be aware of and support informal access rights. In these cases, gathering takes place without explicitly written rules, laws, or policies. Informal access rights may be public, or just a personal sense of ownership. Formal access rights, such as contracts and permits to harvest NTFP, are becoming more common in the United States.

This evolution in managing access has led to concerns that policymakers, land managers, and researchers need to acknowledge the existence of often long-standing customary claims to NTFP, including claims of non-native

Americans, on federal lands (Goodman 2002). Goodman's review of the international literature on customary claims and legal pluralism also strongly suggests that recognizing and embracing some elements of informal legal systems may allow viable approaches to sustainable NTFP management to develop.

Nontimber forest products have been gathered and used by people in North America for thousands of years. Native Americans and other groups have used forest resources to sustain themselves physically, culturally, and spiritually. Commercial trade in plants and animals, and plant and animal products has occurred for thousands of years, within and between aboriginal groups in North America. Highly mobile groups of Native Americans had a structured set of informal rules and traditions that dictated where, when, and who harvested NTFP, such as berries and salmon (*Oncorhynchus* spp.) (Fisher 1997, McLain and Jones 1997, Turner and Loewen 1998, Turner and Cocksedge 2001).

Today's managers make similar decisions when they formally allocate rights to gather NTFP. Managers use formal mechanisms such as policies, regulations, permits, and contracts to determine who can harvest, where they harvest, and when they harvest. For example, those harvesting for personal use may be sent to different places than commercial harvesters. Groups with specific formal rights, such as rights allocated through treaties, may have priority use in designated areas or for specific resources. Managers must then decide whether or not entry is limited and the duration of the access right. Will the harvester be given access rights for a season, a year, or several years? Finally, the manager selling NTFP must decide how the prices will be set and how the payment will be made. Prices may be set with an administrative decision or with a bid. Payment can be a lump sum or based on measurement of the amount harvested (scaled). We will discuss the decisions about harvest priority and location by using exclusive versus nonexclusive access as a framework.

2.1. Who and Where

The first two questions address who has harvest priority and where harvesting will occur. The first issue is allocating harvest rights that are either exclusive or nonexclusive. Exclusive access rights to a product or area for NTFP often take the form of area-based policies, contracts, and leases. The degree of exclusiveness depends on how the policy, contract, or lease is administered, in that the agreement may be transferable to subcontractors (harvesters), or may be with a specific individual or group. Exclusive access rights are more common with resources that grow in predictable ways from one year to the next, such as shrubs (floral greens) and boughs; with products that are harvested in large amounts, such as salal (*Gaultheria shallon* Pursh) or beargrass (*Xerophyllum tenax* (Pursh) Nutt.); with resources that are highly commercialized and not commonly harvested for personal or cultural use, such as salal and evergreen boughs; with products with high value, such as some medicinal plants; and where exclusion and enforcement of rules is possible and benefits to the landowner offset the costs. Exclusive access rights for NTFP are more common on private lands, or on public lands such as trust lands managed by a state where income generation is mandated. Exclusive access rights may also be granted through policy to one group of users, such as personal use harvesters. Exclusivity can be granted formally or may be part of an informal system.

A nonexclusive access right can take the form of a permit for gathering in a specified area, issued to all who apply. In this case there may be no restrictions on how many people pick there or where they can go in the area. Nonexclusive access rights are commonly granted for products with low predictability, such as wild edible fungi; resources that are harvested in small amounts, such as some medicinal plants; products that have a long tradition of personal and cultural use, such as berries; or with low value, such as some fungi; in places where exclusion would be costly and the benefits low, such as places with open physical access and low-value products; or on ownerships with a policy of nonexclusion, such as many federal lands. Nonexclusive rights are common on public lands; harvesters often get permits for personal or commercial use in designated areas where NTFP are regarded as nonexclusive goods, and may be congestible. The granting of non-exclusive rights may be a policy decision on the part of the landowner or manager for various reasons. For example, the landowner or manager may want to ensure some protection and support for more vulnerable members of society, or be unable to enforce exclusive rights.

Access rights to NTFP, whether they are exclusive or nonexclusive for a resource, are often granted in places where other uses are taking place. Property rights can be given for specific resource uses, such as commercial NTFP harvesting, and overlapped with other resource rights, such as personal use harvesting, or fishing. These partial interests are common in public lands (Tedder et al. 2002). The formalization and transferability of partial interests to resources in the United States has provided an alternative to trading or establishing full ownership rights, and have been used as policy tools to deal with environmental externalities. While the ability to transfer partial rights provides more explicit value, and the landowner can capture more of the rent, the web of partial rights also requires a well-functioning infrastructure for monitoring and enforcement (Weibe and Meinzen-Dick 1998). Partial interests in resources can accommodate a wider range of society's values and resource stewardship approaches at, perhaps, less cost than an outright shift to private property rights that have a higher potential to displace existing informal arrangements and potentially higher institutional costs (e.g., enforcement and cooperation) (Tedder et al. 2002). Partial interest in NTFP is often held through policy and treaty rights by Native Americans on lands outside reservations.

2.2. Entry and Time Limitations

Entry to an area to gather may be unlimited, as in a year-round lease with no seasonal restrictions, or limited, as in seasonal restrictions that limit the number of people that gather in an area and the quantity gathered. Restrictions can also limit gathering to certain uses (such as personal or traditional use). Certain groups, such as Native Americans with treaty rights, may have greater access to areas for traditional and personal-use gathering than other groups. Unlimited access rights still exist for a few NTFP in some places. Examples include resources that are harvested in small amounts for personal use, such as some craft materials, or wild foods.

Limitations on access can take many forms. There may be restrictions on the resource attributes, such as the stage of development or size of the plant or fungus, what proportion of plant material can be taken or must remain, and how the fungus or plant material is harvested. Limited access has many gradations. Lightly limited access may limit the area a resource can be harvested from, or control what kind of use can take place (e.g., personal use or cultural and traditional harvest only), but not limit the number of people in an area. Such lightly limited access may be used in many of the same situations where nonexclusive access rights are granted: for resources with low predictability; resources that are harvested in small amounts; resources with a long tradition of personal and cultural use; resources with low value, in places where exclusion is costly or difficult, or in places where relatively open access is a social equity issue. Slight limitations on access are also more likely for resources that are used in minor quantities or not perceived to be adversely impacted by harvesting.

Strict limited access mechanisms may be used where resources are predictable, harvested in large amounts for commercial markets, have high per-unit value, or where exclusion and monitoring are feasible. Limited access may be imposed for resources that are considered more sensitive to harvest pressures because of growth rates, harvest methods (such as taking roots as opposed to leaves), abundance issues, or other sustainability concerns. Strict limitations on access may also be used on private property, or for social, economic, or policy reasons.

The duration of a gathering right is an important decision that can affect both how resource rights are allocated, and how the resource is impacted by harvesting. Access may be granted for a season or shorter specified time (a short-term access right) or longer duration, perhaps years. The effect of duration on access rights will also be affected by exclusivity. Long-term exclusive access rights are increasingly being used for NTFP that have the potential to be over-harvested or that are not ephemeral (such as coniferous boughs, or shrubs). Long-term nonexclusive access rights may not afford much protection to the resource or the harvester. A harvester may know he or she can return year after year for a mushroom harvesting permit on federal land, but if the mushrooms are still a nonexclusive resource, this is no different than a short-term contract. The harvester, in this case, may have less incentive to practice stewardship, the value of the resource to the landowner may be difficult to capture, and the costs and difficulty of cooperation and enforcement can be higher. However, mushrooms are not very predictable, so long-term exclusive access rights may not be practical. Issues about stewardship, costs, and return to the landowner may not be solved by making access more exclusive or by lengthening contract duration. In addition, social equity issues may be more important than capturing all the value of the resource to the landowner or reducing enforcement costs.

2.3. Pricing

In the case of commercial sales, managers must decide whether to set prices through administrative decision or by competitive bid. Administered pricing mechanisms are generally easier and less costly to set up and administer than bid price contracts. Bid price mechanisms, while often more costly to set up and administer, offer the advantage of using economic markets to determine fair returns to the landowner.

Administered prices are set by the owner or manager of the resource with or without negotiation with buyers. This type of pricing may or may not take into account variations in experience, site conditions, and other factors that may make situations unique. Generally applied administered pricing mechanisms that are not situation-specific are frequently used for products with high costs associated with enforcement and cooperation, such as mushrooms and floral greens. This type of administered pricing is common on federal lands, where social equity issues may be more significant than financial returns. Site-specific administered prices that are adjusted for specific situations are common for products whose yield can be estimated, and where the harvesters will have exclusive rights, such as noble fir bough sales.

A bid price is set by the interaction of buyers and sellers in a market. Bid prices reflect what a willing buyer and seller consider a fair return to both the cost of providing the resource and providing a product for final sale to consumers. A bid price is for a specific place and situation, and is often defined in an auction type of process. Bid-price mechanisms for NTFP are more frequently used on private and state lands, but are becoming increasingly common on all ownerships. Bid pricing is easiest to use with NTFP that persist from year to year and can be consistently inventoried, such as posts and poles, or coniferous boughs.

Payment may be lump sum or scaled. Lump-sum payment is a set amount that is not necessarily directly related to the quantity harvested, whereas a scaled sale is based on measuring and paying a price per unit of weight or other measure. Paying for a product on the basis of scale means that a buyer will pay for the actual amount harvested. Unlike for timber, there are no formalized third party organizations that provide scaling information for NTFP. Reporting systems for scaled NTFP sales are arranged between the buyer and the seller. Lump sum payments are typically made in advance and can be based on bid or administered prices and an estimate of what will be harvested.

2.4. Summary

These management choices are a function of the goals of the land managers, and the regulations, laws, and policies that govern each landowner. The rights to harvest can vary from specific rights mentioned in treaties or law, which limits management choices, to entirely unregulated uses. Private landowners choose whether to allow access and if so, to whom. They also determine methods of sales from their property. In many states, legislation addresses how much volume or weight of NTFP a person on public roads may have in possession before they need a signed statement or permit from the landowner. County or municipal ordinances often determine where NTFP buyers can set up their temporary or permanent buying sheds, which can affect where harvesters gather. Public lands management of NTFP is determined by enabling legislation, such as the 1897 Organic Act for the U.S. Department of Agriculture, Forest Service, or the Federal Land Policy and Management Act of 1976 for the U.S. Department of the Interior, Bureau of Land Management. Regulations that follow legislation for public lands enable the determination of harvest permit or contract structures.

The degree of use, from incidental harvest to large-scale agricultural production, also affects management choices. Resource profitability affects regulation and management; a resource that is quite valuable per unit will have more harvest pressure, and conservation concerns will be significant determinants of management choices. The physical characteristics of the resource will affect how it is managed. For example, boughs are a product whose yield can be estimated, whereas mushrooms are resources that appear in less predictable ways and for short periods of time.

3. Case Study Examples

In this section, three case studies illustrate the choices managers make when formalizing access rights; they must decide who can harvest and where, for how long, and how to structure commercial sales. Each case study outlines the issues and management of a common nontimber forest product harvested in the Pacific Northwest, but with different characteristics and uses. Salal, a perennial shrub, is harvested heavily for commercial use in the floral market. Coniferous boughs are harvested commercially for the Christmas greens markets and for personal use, fundraising, and gifts by many groups and individuals. Wild edible mushrooms have a history of both commercial and personal-use harvest, and have become more widely sought for commercial markets in the past two decades.

3.1. Salal in Western Oregon and Washington

Salal harvest in western Washington and Oregon demonstrates many facets of informal and formal access management (Kerns et al. 2003). Salal is a NTFP that is almost entirely commercial; there is little harvest for personal use. For much of the past century, harvesters cut salal branches on many forest ownerships without restriction under informal agreements primarily centered in local communities. About 20 years ago, the floral market experienced a shift from an emphasis on evergreen huckleberry (*Vaccinium ovatum* Pursh), primarily sold domestically, to salal, primarily exported to Europe (Kerns et al., in press). Changes in the labor force, improvements in shipping technology, and investments in shipping companies' infrastructure all led to increased pressure on the resource. Landowners and managers began to impose formal access restrictions in the form of permits and contracts. On federal lands, nonexclusive seasonal permits are sold to harvesters without assigning rights to a specific area. The permits are generally for a set amount of salal and have a time limit. Prices are administratively set and paid lump sum. It is likely that the initial goal of the permits was to mirror informal rights that had existed for some time. Pressure on the resource has mounted steadily as markets expanded and labor structure shifted. With many people harvesting without permits on all ownerships, open road systems, and few funds available to enforce regulations, the difficulty and costs of enforcing rules and property rights are high for landowners in the area. Despite these issues, salal is a significant product in the western Oregon, Washington, and British Columbia floral industry. Some forest managers are examining the conditions under which salal productivity is enhanced, but silvicultural management generally has not focused on salal as a significant economic forest product that can be enhanced by comanagement with timber (Heidi Ballard, University of California Berkeley, personal communication, February 2003; Wendy Cocksedge, Royal Roads University, Victoria, British Columbia, personal communication, February 2003).

Several large private landowners, the Washington Department of Natural Resources, and the state of Oregon issue exclusive area-based leases for salal and other floral greens for short and long terms. The lessees, in turn, may issue permits to harvesters. In some places, these ownerships also issue 1-year nonexclusive, area-based permits. One private landowner with a number of salal contracts helps fund a law enforcement officer for the area. A group of landowners, managers, and law enforcement officers in northwestern

Washington meet regularly to discuss concerns (Mark Savage, Washington Department of Natural Resources, personal communication, February 2003). As landowners cooperate with one another, communicate through meetings, and form relationships with harvesters, the costs and difficulty of allocation and enforcing rules and property rights decrease. There is still some unpermitted harvesting (theft) and undocumented labor in the area. A formal management regime will exclude some of these current harvesters; some groups may regard this reallocation of property rights as unfair until the larger social group comes to terms with how landowners define property rights.

3.2. Coniferous Boughs

Coniferous boughs are widely harvested in the United States and Canada, particularly in the Pacific Northwest and the North Central regions. Commercial markets for coniferous boughs and products made from boughs, such as wreaths, are particularly active during the winter holiday season. Individuals and groups also harvest boughs for personal use. Individuals enjoy harvesting boughs for seasonal home decoration. Nonprofit groups often harvest boughs and make wreaths and swags as a fundraising activity; these groups may be granted an exclusive seasonal harvest right on forest lands by a private landowner (Pete Mastenbroek, Mason Bruce and Girard Consulting Foresters, personal communication, April 2003). Commercial large-scale bough harvesting has changed in ways similar to salal harvesting because of changes in the labor force available for forest work, improvements in technology, and investments in infrastructure. Landowners have recognized boughs as a private good with commercial value for a long time. Boughs are a direct consequence and byproduct of reforestation choices, and experienced managers can estimate yields. Only recently has research been done that will provide a scientific basis for estimating bough yields (Blatner et al., in review).

Noble fir (*Abies procera* Rehd.) is one of the most significant commercial bough species in the Pacific Northwest. The Forest Service generally sells bough contracts through sealed bid auctions as lump-sum sales. Bids are based on an estimate by the Forest Service manager of the weight of boughs in the designated sale area. Bid prices are affected by how much material the bidder believes is harvestable in the sale area. Harvesting is guided by rules regarding how the boughs will be removed and what proportion of crown must remain. Contracts may vary in duration from one to several years, and the contractor has exclusive access rights to the sale area.

Some large private forest companies sell permits based on a semi-negotiated administratively set price per unit weight, and payment is based on actual scaled weight. Although the contract length is one season, contractors operate under the assumption that they can renegotiate the contract unless terms of the agreement are violated. The contractor has exclusive rights to harvest in the sale area, which may be quite extensive.

A concern common to land managers and the bough industry is a projected decline of noble fir bough availability in the Pacific Northwest, particularly on federal lands. With less high-elevation harvest and a continued emphasis on wood production instead of bough production, total areas available for bough harvest appear to be declining. The Washington Department of Natural Resources is responding with proposals to plant noble fir at lower elevations than traditionally done. Noble fir bough production is an opportunity for forest managers to diversify income.

3.3. Wild Edible Mushrooms

Wild edible mushrooms are harvested by many people for personal use and for commercial sale. The Pacific Northwest has the largest and most active commercial wild edible mushroom industry in the United States. Wild edible mushrooms are sold fresh to restaurants and shipped fresh to Japan. Dried mushrooms are sold both within the United States and exported to Europe. The wild edible mushroom industry was relatively small until the mid-1980s; it has expanded rapidly because of increasing demand for exports, changes in the labor force, and improvements in shipping technology (Pilz et al. 1998, Alexander et al. 2002). State laws regarding transportation, possession, and buying of NTFP were enacted primarily because recreational mushroom harvesters were concerned about mushroom congestion and possible scarcity (McLain et al. 1998). Regulations and policies governing permits on federal lands have also become more structured.

On federal lands in the Pacific Northwest, personal-use mushroom harvesters are often required to obtain a free-use permit. The permit generally allows harvest in large specified areas for a specified time, and the harvester is limited in how much they can have in possession. Even when no personal-use permit is required, state law may restrict possession amounts for personal use. Permits for commercial harvest of wild edible fungi on federal lands are usually sold lump sum with an administratively set price. The permit specifies the area for commercial harvesting and the time period. Commercial mushroom permits do not generally limit the amount the harvester can have in possession.

Commercial harvesting permits on private lands, if available, are similar to those sold by federal agencies. Wild edible mushrooms can be valuable but they are difficult to treat as an exclusive good. Production varies from year to year, as do prices and markets for particular species. Yields are difficult, if not impossible, to estimate. Many private landowners do not issue permits to harvest mushrooms because of concerns about trespass, liability, and vandalism. They are not generally concerned about mushroom scarcity. Because of the unpredictable fruiting of edible fungi and the fact that they are ephemeral, mushrooms have more nonrival and nonexclusive good characteristics than products such as boughs and floral greens. Wild edible fungi contribute millions of dollars annually to rural economies in the Pacific Northwest, but comanagement of timber and mushrooms is poorly understood and has been very rarely attempted (e.g., Pilz and Molina 1998).

4. Discussion

Any access right will be some combination of management choices that determine who can harvest, where they can go, how they harvest, and for how long. If the product is sold, prices and payment must be determined. These choices have important implications for the economic value of the NTFP industries and for social values of interest to natural resource stakeholders. A commercial contract to harvest coniferous boughs on private land may be formal, exclusive, long-term, and limited, with bid prices based on scale. A commercial permit to gather mushrooms on public lands may be formal, nonexclusive, short term, and unlimited, with an administratively set lump-sum price. The rights to gather medicinals for personal use in a traditional harvesting area may be granted formally through legislation but still practiced as informal, exclusive, and long term. Management choices regarding access to NTFP depend on the goals of management and the degree of nonrivalry and nonexclusiveness of the particular product. Managers may be directed to maximize revenue or efficiency and so may treat the resource more as a private good and restrict access to a few harvesters. If the goal is to provide access to as many people as possible while sustaining the resource, maximizing revenue and clear assignment of property rights will be less significant.

There are many instances in which total exclusion is chosen. Complete exclusion of NTFP harvesters from an area can be based on a number of management concerns. The landowner or manager may be concerned about liability or resource sustainability. The landowner or manager may decide that the cost of law enforcement for an area outweigh the benefits. The landowner may be concerned with vandalism, theft, or ecoterrorism and exclude all users from the land base. Total exclusion is becoming increasingly common, particularly on private lands, and access to forests has been decreasing in the United States (Emery and Schroeder 2002). If people enter closed lands to harvest, even if they have a long tradition of informal access rights, that becomes trespassing and theft in the eyes of the landowner.

Management and regulation have struggled with the often conflicting goals of economic efficiency and social welfare issues, as exemplified by the changes in forest policy in the past few decades. Nontimber forest products emerged on forest policy agendas in the late 1980s and early 1990s when Oregon and Washington was embroiled in a struggle over how to manage the region's forests, and concerns about management and resource rights have since been expressed throughout the United States and Canada. Long studied in developing countries, NTFP have more recently become a subject of studies and management in developed countries, including Canada and the United States (Mohammed 1999, Jones et al. 2002, Tedder et al. 2002). The U.S. federal government adopted ecosystem management as the guiding policy for federal resource management in 1994. Resource management agencies, including the USDA Forest Service, USDI Bureau of Land Management, and the USDI Fish and Wildlife Service, have begun including NTFP in the planning process as required under the National Environmental Policy Act (Antypas et al. 2002). In 1999, the U.S. Congress included a rider on the FY2000 Appropriations Act (P.L. 106-113), known as Section 339. The rider is entitled "Pilot Program of Charges and Fees for Harvest of Forest Botanical Products." It requires that the Forest Service charge "fair market value" for NTFP harvesting permits, recover administrative costs, and ensure that harvesting levels of NTFP on national forests are sustainable. The rider allows certain funds from permit sales to be retained by the local National Forest. In Oregon and Washington, the Forest Service has responded by developing a standardized appraisal system for "special forest products." The Bureau of Land Management in Oregon and Washington has had a standardized appraisal system in place for several years.

Over the past 15 years, the number of regulations, laws, and policy guidelines pertaining to the harvest of NTFP on state and federal lands has expanded greatly (USDI BLM 1992, von Hagen and Fight 1999, McLain 2000). In the mushroom sector where regulatory expansion has been particularly noticeable, amateur mycologists or formally organized groups of amateur scientists interested in the collection, study, and consumption of fungi have been among the most politically vocal and visible stakeholders (McLain et al. 1998). Love et al. (1998) state that class and ethnic animosity have found their way into state legislation regulating commercial mushroom harvesting. They believe that current state mushroom regulatory policies and laws are based on anecdotal stories rather than research knowledge and facts, and are largely unenforceable. Love et al. (1998) argue that these policies and laws tend to discredit local knowledge and criminalize formerly legitimate commercial harvesting practices.

Common to the issues of regulation and inclusion in policymaking is the problem of defining the users and their purpose for harvesting. Carroll et al. (in press) found that differences expressed over the labeling of various types of huckleberry use in northeast Washington are reflective of a larger tension over the "real purpose" of national forests, and the competing claims that various stakeholders at differing levels make about the forests. Carroll et al. (in press) state that the social ecology of huckleberries (and probably all other nontimber forest resources) is at least as complicated as the biological complexity. The differing goals of harvest for personal use, commercial sale, subsistence, and cultural use often overlap. Some studies argue that by dividing harvesters into mutually exclusive categories of commercial and noncommercial use, it is more likely that policy solutions will have unintended, and possibly negative, social and ecological consequences (Love et al. 1998; Alexander et al. 1999; Jones and Lynch 2002; Carroll et al., in press).

Any management mechanism for NTFP, from a legal mandate for income generation to a system of permits and contracts based on enabling legislation, economic efficiency goals, or social welfare goals, must be flexible enough to fit the extremely variable products and the people that use them. Tedder et al. (2002) caution that any management mechanism must respect traditional and other personal uses by including the appropriate rights and responsibilities to ensure noncommercial property rights. If a new management regime is proposed, such as requiring permits where none were required before, the benefits of change should clearly exceed the costs of the change and the enforcement of the new rules, otherwise users will not willingly adopt the new management mechanism. Managing the resource stock and managing the industry that benefits from the resource have different requirements, and it cannot be assumed that addressing the issues of one will solve the issues of the other. Because of our general lack of information about NTFP and their management, an effective regime must be adaptive.

Perhaps the most important reason for carefully considering access mechanisms is that the mechanisms used affect the financial incentives faced by harvesters. We know little about how harvesters respond to changes in opportunities and incentives, but to the extent that they respond in ways that will increase their income we can draw conclusions about the nature of those effects on resource conservation and productivity. For example, under an exclusive contract to harvest mushrooms, a harvester has an incentive to leave immature mushrooms to be picked a few days later when they are larger; doing so will increase the overall production of the area. Under a nonexclusive contract, the harvester who leaves immature mushrooms to grow risks losing them altogether to another harvester. Another example is that under a long-term exclusive contract, a harvester of boughs or floral greens has an incentive to defer harvest of young plants and to fertilize them to increase the harvest in future years. Under a short-term contract, deferred harvest or investment in increasing growth is a gift to whichever harvester gets there first in future years. We have little understanding of the extent to which harvesters change their behavior in response to these and other incentives created by access mechanisms, but it is apparent that the choice of access management mechanisms could have significant long-term effects on the productivity and sustainability of nontimber forest resources.

5. References

Alexander, S.J., and McLain, R. 2001. An overview of non-timber forest products in the United States today. *Journal of Sustainable Forestry*, 13 (3/4): 59-66.

Alexander, S.J., McLain, R., Kim, Y.S., and Johnson, R. 1999. Recreational harvest of wild foods on the Gifford-Pinchot National Forest: resources and issues. Proceedings: Society of American Foresters 1999 National Convention SAF Publication 00-1. Bethesda, MD: Society of American Foresters.

Alexander, S.J., Pilz, D., Weber, N.S., Brown, E., and Rockwell, V.A. 2002. Mushrooms, trees and money: value estimates of commercial mushrooms and timber in the Pacific Northwest. *Environmental Management*, 30(1): 129-141.

Antypas, A., McLain, R.J., Gilden, J., and Dyson, G. 2002. Federal nontimber forest products policy and management. In: E.T. Jones, R.J. McLain, and J. Weigand (Eds.), Nontimber forest products in the United States (pp. 347-374). Lawrence, KS: University Press of Kansas.

Blatner, K.A., and Alexander, S. 1998. Recent price trends for non-timber forest products in the Pacific Northwest. *Forest Products Journal*, 48(1): 28-34.

Blatner, K.A., Fight, R.D., Vance, N., Savage, M., and Chapman, R. [In review]. A model to estimate noble fir bough weight. *Western Journal of Applied Forestry*.

Carroll, M.S., Blatner, K.A., and Cohn, P. [In press]. Somewhere between: social embeddedness of "recreational" and "commercial" huckleberry use. *Society and Natural Resources*.

de Geus, N. 1995. Botanical forest products in British Columbia: an overview. Victoria, BC: British Columbia Ministry of Forests, Integrated Resources Policy Branch.

Emery, M.R. 1998. Invisible livelihoods: non-timber forest products in Michigan's Upper Peninsula. New Brunswick, NJ: Department of Geography, Rutgers University. 291 p. Ph.D. dissertation.

Emery, M.R., and McLain, R.J. (Eds.). 2001. Non-timber forest products. Binghamton, NY: The Hayworth Press, Inc.

Emery, M.R., and Schroeder, R. 2002. Area and percent of land used for subsistence purposes. Criteria 6, Indicator 47, in National Report on Sustainable Forests—2003. Washington, DC: U.S. Department of Agriculture, Forest Service. Retrieved February 2003 from http://www.fs.fed.us/research/sustain/.

Fisher, A.H. 1997. The 1932 handshake agreement: Yakama Indian treaty rights and Forest Service policy in the Pacific Northwest. *Western Historical Quarterly*, 28: 187-217.

Goodman, E.C. 2002. Nontimber forest products customary claims. In: E.T. Jones, R.J. McLain, and J. Weigand (Eds.), Nontimber forest products in the United States (pp. 393-412). Lawrence, KS: University Press of Kansas.

Hansis, R. 2002. Workers in the woods: confronting rapid change. In: E.T. Jones, R.J. McLain, and J. Weigand (Eds.), Nontimber forest products in the United States (pp. 52-56). Lawrence, KS: University Press of Kansas.

Heckman, H. 1951. The happy brush pickers of the high Cascades. *Saturday Evening Post*, 4: 35-40.

Jones, E.T., and Lynch, K. 2002. The relevance of sociocultural variables to nontimber forest product research, policy, and management. In: E.T. Jones, R.J. McLain, and J. Weigand (Eds.), Nontimber forest products in the United States (pp. 26-51). Lawrence, KS: University Press of Kansas.

Jones, E.T., McLain, R.J., and Weigand, J. (Eds.). 2002. Nontimber forest products in the United States. Lawrence, KS: University Press of Kansas.

Kerns, B.K., Alexander, S.J., and Bailey, J.D. [In press]. Nontimber forest products and Douglas-fir stand conditions in western Oregon: ecology and economics of huckleberry species. *Economic Botany*.

Kerns, B., Pilz, D., Ballard, H., and Alexander, S.J. 2003. Compatible management of understory forest resources and timber. Chapter 12. In: R.A. Monserud, R.W. Haynes, and A.C. Johnson (Eds.), Compatible forest management. Dordrecht, The Netherlands: Kluwer Academic Publishers.

Liegel, L. (Comp.). 1998. The biological, socioeconomic, and managerial aspects of chanterelle mushroom harvesting: the Olympic Peninsula, Washington State, U.S.A. *Ambio,* Special Report No. 9. Stockholm, Sweden: Royal Swedish Academy of Sciences.

Love, T., Jones, E., and Liegel, L. 1998. Valuing the temperate rainforest: wild mushrooming on the Olympic Peninsula biosphere reserve. *Ambio*, Special Report No. 9: 16-25.

Lynch, K.A., and McLain, R.J. [In press]. Access, labor and wild floral greens management in western Washington's forests. General Technical Report. Portland, OR: U.S. Department of Agriculture, Forest Service, Pacific Northwest Research Station.

McLain, R.J. 2000. Controlling the forest understory: wild mushroom politics in Central Oregon. Seattle, WA: University of Washington. 331 p. Ph.D. dissertation.

McLain, R.J., Christensen, H.H., and Shannon, M.A. 1998. When amateurs are the experts: amateur mycologists and wild mushroom politics in the Pacific Northwest, USA. *Society and Natural Resources*, 11: 615-626.

McLain, R.J., and Jones, E. 1997. Challenging "community" definitions in sustainable natural resource management: the case of wild mushroom harvesting in the USA. Gatekeeper Series No. 68. International Institute for Environment and Development, Sustainable Agriculture Programme.

Mohammed, G.H. 1999. Non-timber forest products in Ontario: an overview. Forest Research Information Paper 145. Sault Ste. Marie, Ontario: Ontario Forest Research Institute, Ministry of Natural Resources.

Pearce, D.W. 1992. The MIT dictionary of modern economics. (4th ed.). Cambridge, MA: The MIT Press.

Pilz, D., Brodie, F.D., Alexander, S., and Molina, R. 1998. Relative value of chanterelles and timber as commercial forest products. *Ambio*, Special Report No. 9: 14-16.

Pilz, D., and Molina, R. 1998. A proposal for regional monitoring of edible forest mushrooms. Mushroom, *The Journal of Wild Mushrooming*, (16)3: 19-23.

Pilz, D., Smith, J., Amaranthus, M.P., Alexander, S., Molina, R., and Luoma, D. 1999. Mushrooms and timber: managing commercial harvesting in the Oregon Cascades. *Journal of Forestry*, 97(3): 4-11.

Randall, A. 1988. Market failure and the efficiency of irrigated agriculture. In: G. O'Mara (Ed.), Efficiency of irrigation: the conjunctive use of surface and ground water resources (pp. 12-21). Washington, DC: The World Bank.

Richards, R.T., and Creasy, M. 1996. Ethnic diversity, resource values, and ecosystem management: matsutake mushroom harvesting in the Klamath bioregion. *Society and Natural Resources*, 9: 359-374.

Tedder, S., Mitchell, D., and Hillyer, A. 2002. Property rights in the sustainable management of non-timber forest products. Victoria, BC: Renewal B.C; British Columbia. Economics and Trade Branch, British Columbia Ministry of Forestry.

Turner, N.J., and Cocksedge, W. 2001. Aboriginal use of non-timber forest products in northwestern North America: applications and issues. *Journal of Sustainable Forestry*, 13(3/4): 31-57.

Turner, N.J., and Loewen, D.C. 1998. The original "free trade": exchange of botanical products and associated plant knowledge in northeastern North America. *Anthropologica*, 40: 49-70.

U.S. Department of Agriculture, Forest Service [USDA FS]. 2003. National report on sustainable forests—2003. Washington, DC: U.S. Department of Agriculture, Forest Service. Retrieved March 2003 from http://www.fs.fed.us/research/sustain/.

U.S. Department of the Interior, Bureau of Land Management [USDI BLM]. 1992. BLM Task Force draft report: managing special forest products in Oregon and Washington. Portland, OR: U.S. Department of the Interior, Bureau of Land Management, Oregon State Office.

von Hagen, B., and Fight, R.D. 1999. Opportunities for conservation-based development of nontimber forest products in the Pacific Northwest. General Technical Report PNW-GTR-473. Portland, OR: U.S. Department of Agriculture, Forest Service, Pacific Northwest Research Station. 18 p.

Weibe, K.D., and Meinzen-Dick, R. 1998. Property rights as policy tools for sustainable development. *Land Use Policy*, 15(3): 203-215.

Chapter 14

Managing For Wildlife: A Key Component for Social Acceptance of Compatible Forest Management

Andrew B. Carey
USDA Forest Service, Pacific Northwest Research Station,
3625 93rd Ave., Olympia, WA 98512, USA

1. Introduction

Why manage for wildlife in U.S. forests? American society demands it. Which species should be favored? The social and cultural value of individual species continue to evolve. Large changes have taken place in less than 40 years; Kimmins (2002) states that changes in societal values have produced "future shock" in the forestry profession, with foresters and their institutions unable to adapt. Public demand for wildlife conservation has resulted in a long chain of legislation governing federal lands and supporting state and private wildlife conservation efforts (Hunter 1990). Nevertheless, controversies over forest management continue, and have led to a shift from active management for wildlife to establishment of large reserves off limits to active management (Hunter 1999). Now attention is shifting to second-growth forests where the public is concerned about biodiversity and ecosystem health (Hunter 1999, Lindenmayer and Franklin 2002, Shields et al. 2002).

Public scrutiny is no longer limited to federal lands. State agencies have recognized the need to address public concerns (Belcher 2001). Efforts to conserve wildlife on privately held forests are on the rise (Best and Wayburn 2001). Wood products companies find it necessary to develop compatible management approaches, including habitat conservation plans (Loehle et al. 2002). Public focus on wildlife conservation extends worldwide, even to coffee plantations (http://www.starbucks.com/aboutus/csr.asp). As public interest expands, it is prudent to ask what trends in values are relative to forest wildlife.

R.A. Monserud, R.W. Haynes, and A.C. Johnson (eds.)
Compatible Forest Management, 401-425

Surveys of American cultures (e.g., Ray 1996) found that sustainability—environmental, economic, and social—and other community-oriented values were on the rise. Sustainability in this sense is a broader concept than that outlined in the Montreal Process. It is more akin to the concept of compatibility Haynes et al. (2003) define as "...managing forests to produce wood and other uses in a socially acceptable manor without decreasing any other value," but with less emphasis on wood and more on human and ecological well-being (Lackey 2001). For example, in a survey of American forest values (Shields et al. 2002), the highest scoring values were (1) conserving and protecting watersheds (mean = 4.7 on a scale of 1 to 5); (2) protecting ecosystems and wildlife habitat (mean = 4.6); (3) placing as much importance on future generations as on the current in land use decisions (mean = 4.5); (5) acknowledging wildlife, plants, and humans have equal rights (mean = 4.3); and (6) preserving natural resources, even if some people must do without (mean = 4.1). Utilitarian scores averaged less than 3 (indicating disagreement). Still, the public believes in allowing for diverse uses (mean = 4.1).

In this chapter, I focus on forest management for wildlife outside reserves, and how such management pertains to compatible forest management. A core aspect of compatibility is social sustainability, or maintaining a civil society (Goodland 1995). Moving from adversarial to collaborative interactions is important to the public (Shields et al. 2002) and will help promote a civil society. Thus, I develop approaches to compatibility that are amenable to collaborative management (Wondolleck and Yaffee 2000, Behan 2001, Kemmis 2001). I use examples from Oregon and Washington where people are interested in conserving wildlife (from game to endangered species) and overall biological diversity, maintaining local communities, and obtaining useful goods and services from forests (e.g., open space, clean air and water, edible fungi and berries, floral greens, and wood products). These values are not limited to Oregon and Washington but are emerging throughout developed countries (e.g., Folke et al. 1996, Entwistle and Dunstone 2000, Holling 2001, Larson 2001).

I examine how wildlife management can address compatibility by formulating objectives for wildlife that can be (1) surrogates for diverse values related to wildlife and biodiversity in general, (2) modeled in planning exercises with a fair degree of confidence, (3) readily understood and evaluated by diverse groups of people, and (4) expressed in terms of measurable outputs. Managing forests requires managing multiple ecological processes over the long term. I summarize key processes, present a new classification of forest development based on processes, and provide an example of process-based management in simulation modeling. Finally, I discuss how well my predictions are holding up under experimental conditions.

2. Objectives for Wildlife Management

If compatibility is a goal of forest management, then it is necessary to develop mutually compatible objectives for the various components of forest ecosystems and the goods and services that forests provide. If economic objectives are paramount, then it may be useful to evaluate how environmental values are affected, to determine what some environmental opportunity costs are, and to answer the question: Is it likely that single focus management will be compatible with conservation of wildlife?

It is possible to formulate objectives for wildlife that can serve as surrogates for diverse values and that can be expressed in terms of measurable outputs. Maximizing one species of wildlife is rarely an environmental objective, and rarely will focus on just a few species meet broad objectives. In few situations will it be possible to achieve wildlife objectives in the short term. Instead, progress relative to initial conditions, management alternatives, and natural benchmarks is a realistic short-term objective. Common themes in the conservation literature can help formulate wildlife objectives, e.g., (1) keystone complexes, (2) flagship or charismatic species, (3) links among populations, communities, biodiversity, and biocomplexity, (4) ecological processes of development of forest biotic communities, and (5) forest development in dynamic landscapes.

2.1. Keystone Complexes

A keystone complex is relevant because it relates directly to ecosystem function by embodying trophic relationships at a hierarchy of spatial scales. In Oregon and Washington, the northern spotted owl (*Strix occidentalis caurina*) rests atop a complex, central, food web in natural, old forests (Figure 1). The primary prey of the spotted owl is the northern flying squirrel (*Glaucomys sabrinus*), which is also prey for American marten (*Martes americana*), long-tailed weasels (*Mustela frenata*), and other mammalian predators. The flying squirrel carries lichens and mosses to its nest, sometimes over long distances, and may be important in dissemination of those plants. The flying squirrel consumes sporocarps (truffles and mushrooms) and disseminates spores of fungi that are essential symbionts to the dominant tree species in lower elevation forests—Douglas-fir (*Pseudotsuga menziesii* (Mirb.) Franco)—and many other species of trees, both hardwoods and conifers. Fungi help trees take up water and nutrients and receive carbohydrates in return. The fungi use some of the carbohydrates; others are delivered to the soil and support a vast soil food web. This keystone complex can be expanded to a broader food web of predators-major prey species-primary production food bases even more representative of wildlife diversity and various ecosystem functions (Carey et al. 1999a,b; 2002).

Figure 1. A keystone complex: spotted owl-flying squirrel-truffle-Douglas-fir.

2.2. Flagship Species

Incorporating symbolic (flagship) species into forest management is essential because, by definition, people see these species as representative of highly valued ecosystems or as highly valued in their own right. People demand that these species be accounted for in management of natural resources. Some such species in the Pacific Northwest (western Oregon and Washington, coastal British Columbia, and southeastern Alaska) are the spotted owl, Roosevelt elk (a subspecies of *Cervus elaphus*), and Pacific salmon (seven species of *Oncorhynchus*). The spotted owl is symbolic of natural, old forests. The Roosevelt elk is valued by wildlife viewers because its impressive stature is easily seen from roads; it is also a prized game animal and a focus of subsistence hunting by Native Americans. Salmon are symbolic of historical Pacific Northwest culture, prized gastronomically, economically valuable in commercial and sport fisheries, and are a keystone species in many riparian ecosystems. Other groups of species are held in similar high esteem. Neotropical migratory birds and other forest birds are important to the American public. As Rachel Carson (1962) asked, What would a silent spring be like? What would a visit to a forest be without the songs of forest birds? In spring, bird songs are a major part of a forest experience (Wilson et al. in press).

2.3. Ecological Links

It is unlikely that scientists will determine the ecologies of most organisms in forests by the end of the 21st century. Forests, however, will be faced with increasing challenges from burgeoning human populations, globalization, introduction of exotic pests and diseases, and unavoidable natural catastrophes (Scheffer et al. 2001). Thus, many people want forest management to consider ecological links among species, biotic communities, overall biological diversity, biocomplexity and ecosystem function, resistance to disturbance, and resiliency in the face of disturbance (Holling 2001, Kinzig et al. 2002). Wildlife can be used to address these concerns, at least in part (Box 1). The occurrence of numerous wildlife species can be a measure of biological diversity because wildlife occupy such diverse niches, have complex trophic relationships with other organisms, and interact with the environment at different spatial scales (Loreau et al. 2001). For example, at the landscape scale, diverse wide-ranging species such as the spotted owl, Roosevelt elk, and salmon are indicative of total landscape function and the values that landscapes provide beyond those of local ecosystems. Recovery of the spotted owl is tied to recovery of naturalistic forest ecosystems with high prey biomass within landscapes; abundance of elk is tied to the biomass of low vegetation within landscapes (e.g., Carey et al. 1999a,b). Within ecosystems, complex above-ground food webs provide abundant and diverse prey and plant biomass, promote biodiversity, reflect ecosystem function, and contribute to the biocomplexity that underlies ecosystem health (resistance to disturbance and resilience in the face of disturbance) (Lawton 1994; Carey 2003a,b). Above-ground food webs are linked to below-ground food webs by wildlife that forage on and in the forest floor. For example, forest-floor mammals feed on arthropods, fungi, and seeds in the soil and litter; thus, the structure of the forest-floor mammal community is determined by the structure of the forest floor (Carey et al. 1999b, 2002, 2003a; Carey and Harrington 2001). The structure of the biotic communities in the soil depends on and influences multiple soil processes (e.g., Li and Strzelczyk 2000), and the biodiversity of soil organisms is an indicator of soil health (Pankhurst 1997). Thus, complete forest-floor small mammal communities suggest maintenance of the multiple forest-floor trophic pathways critical to ecosystem function and productivity.

2.4. Forest Ecosystem Development

Research has created a large database on wildlife in Oregon and Washington (e.g., Ruggiero et al. 1991). Wildlife populations have been measured, key habitat elements have been identified, biotic communities and their stages of development have been described, and relationships among species, communities, stages of development, and habitat elements have been

Box 1. Wildlife Objectives for Achieving Compatibility in Managed Forests in the Pacific Northwest and Ways to Measure Management Outcomes (Adapted from Carey et al. 1999b, 2003a)

Objective: Maintain capacity to support biological diversity

Method

(1) Use published wildlife-habitat relationship (WHR) models to evaluate the capacity of the landscape to support the full complement of forest wildlife indigenous to the area, or a diverse subset of indigenous wildlife (e.g., more than 125 species of markedly different life histories). Measure this capacity as a percentage of the capacity of the landscape if it were composed of natural old-growth forests.

(2) It is not practical to survey all wildlife; interested publics, however, can be involved in comparative and periodic surveys of some biotic communities. For example, stand- and landscape-level diversity of Neotropical migratory birds or forest bats can be compared to WHR model predictions for various alternative management scenarios.

Objective: Aid recovery of threatened forest species by providing habitat for them

Method

(1) The threatened spotted owl is the flagship species for natural forest ecosystems in the Pacific Northwest. Documented habitat requirements can be used to determine the management necessary to produce habitat (particularly foraging habitat), model that management, and track habitat development in alternative landscapes. Sufficient information exists to calculate the average area of habitat used by spotted owls within an average home range; thus, current and future landscapes can be evaluated for their capacity to support spotted owls.

(2) Spotted owls are easy to study; success in developing habitat can be determined by the actual use of managed, complex forest.

Objective: Provide for ecosystem productivity that supports complex food webs

Method

(1) A key trophic level can be used to evaluate food web complexity. For example, published WHR equations can be used to model changes in overall and individual biomass of northern flying squirrels, Douglas' squirrels, and Townsend's chipmunks (kg squirrels/ha). The biomass of these squirrels is a measure of the amount of primary productivity diverted to their foods, the reproductive outputs of fungi (truffles and mushrooms) and woody plant (fruits, seeds, nuts). Squirrel biomass measures the capacity of the ecosystem to support medium-size predators such as goshawks, spotted owls, long-tailed weasels, and American martens.

(2) Squirrels are expensive to study. Thus, over time, random sampling and periodic evaluation of a sample of stands could be used to evaluate models and compare capacities of different stand conditions.

Objective: Maintain long-term site productivity through forest-floor biocomplexity

Method

(1) Pacific Northwest old-growth forests have diverse forest-floor mammal communities that rely on diverse food webs based on invertebrates, fungi, lichens, and vascular plants. Forest-floor complexity also promotes diverse amphibian communities. Published WHR equations allow assessment of mammal community structure and comparison of the structure in future ecosystems to present conditions and to conditions in old-growth forests. Small mammals are prey for a variety of mammalian and avian predators. The integrity of the mammal community is an index to the diversity of food webs supported by forest-floor fauna and flora and, thus, the long-term outlook for continued ecological productivity.

(2) Forest-floor mammals are expensive to study; evaluation of management success is best done on a sampling basis.

Objective: Provide landscape productivity for wide-ranging wildlife

Method

(1) Roosevelt elk and black-tailed deer abundance is determined by amounts of forage (lichens and foliage of herbs, woody plants, and understory trees) in combination with shelter from weather and hiding cover. Existing models can predict their abundance in a forested landscape by forage production. Thus, alternative landscape conditions, current status, and projected future conditions can be compared. Cervid abundance determines landscape carrying capacity for (a) large wide-ranging predators such as mountain lions and gray wolves, (b) subsistence hunting by Native Americans, and (c) sport hunters.

summarized in regional volumes and wildlife-habitat relationships tables (e.g., Johnson and O'Neil 2001). These summaries provide valuable guidance to managers and the public and can be adapted to simulation models to compare management alternatives (e.g., Carey et al. 1999b). Additionally, narrowly defined wildlife communities (10 species) have been linked to stages of forest development, and the integrity of these communities, in turn, has been linked to plant species diversity and complexity in plant community structure (Carey et al. 1999a, Carey and Harrington 2001). Thus, biocomplexity—species diversity, variety in structure, and heterogeneity in spatial arrangements—seems to produce robust ecosystems.

The ecological succession of biotic communities, development of forest biocomplexity, and natural disturbances interact with forest management to produce dynamic landscapes composed of dynamic forest ecosystems.

Understanding the dynamics of disturbance and change seems essential to managing for wildlife, wood, and other values. Indeed small-, intermediate-, and large-scale disturbances that produce spatial and temporal heterogeneity are the basis for both biological diversity and biocomplexity (Whittaker 1975, Connell and Slatyer 1977, Bormann and Likens 1979, Oliver 1981, Canham et al. 1990, Carey et al. 1999a, Franklin et al. 2002). In conservation, too much emphasis has been placed on static future desired conditions. In timber management, too little emphasis has been placed on the consequences of shortening the duration and truncating the extent of dynamic processes on various values, including wood production (Curtis and Carey 1996, Carey 2003b). More emphasis is needed on managing ecological processes as opposed to focusing on static elements of structure, structural stages, or an end product. Improved classifications of forest development based on dynamic processes rather than static structures are necessary for managing processes. Key processes warrant clarification, especially those relevant to ecological stage setting, assembly of biotic communities, and maintenance of biodiversity. Wildlife can help identify those processes and evaluate the efficacy of management. A set of objectives and measures (Box 1) can be used to model:

(1) Existing conditions (baseline) and how the landscape is likely to change without management (no-management alternative)

(2) The landscape as composed of (a) old, natural forests (benchmark of potential) and (b) as representative of pre-settlement disturbance regimes (historical benchmark)

(3) The future landscape as it changes in response to management for compatibility

(4) The landscape as a shifting, steady-state mosaic of planned seral stages under continuous management for compatibility

(5) The landscape under alternative management regimes

3. Processes Underlying Forest Community Development

After a catastrophe destroys much or all of a forest canopy, the ecosystem reorganizes and begins to develop anew. The degree of retention of biological legacies from the preceding forest has profound influence on the site and the organisms available to the new ecosystem (Franklin et al. 2000, 2002). Legacy retention can range from a few live trees to a mixture of trees, shrubs, and coarse woody debris, and from intact forest floor to patches of intact forest. The more legacies retained, the more a forest-influenced environment is maintained and the greater the mycorrhizal networks, species and sizes of trees, degree of spatial heterogeneity, and available species pools. Some important legacies include seeds or seedlings of multiple species of conifers and hardwoods, ectomycorrhizal fungi, large coarse woody debris, and large live trees with epiphytic mosses and lichens. Legacies and the size and shape of the forest that was destroyed determine how distant any particular point in the reorganizing ecosystem is from sources of organisms that might colonize or recolonize a newly developing forest. The landscape context (biotic communities and seral stages) of the reorganizing ecosystem determines which species are available to recolonize a new forest. If a full complement of species is available, four basic processes determine how forests develop structurally in the Pacific Northwest and elsewhere (Carey et al. 1999a): crown-class differentiation, decadence, understory development, and canopy stratification. Each of these processes can be jumpstarted by legacies and hastened by active management and intermediate-scale disturbances. As basic structuring processes interact, two subsequent (higher order) processes determine the diversity, composition, and species structure of the biotic community: development of habitat breadth (Carey 1999a) and preinteractive niche diversification (Hutchinson 1978).

3.1. Crown-Class Differentiation

After trees have fully occupied the site, a tree canopy forms. Initially, the canopy may be dense and uniform, but over time some trees must become dominant, others codominant, subordinate, or suppressed for development to proceed. Crown-class differentiation is important for producing large trees, small dead and dying trees, and various other crown and canopy characteristics that develop habitat for a variety of forest wildlife. Differentiation occurs through natural and artificial disturbances that create gaps in the existing canopy.

Dense stocking, reliance on self-thinning, and tardy, light, evenly spaced thinning, however, can forestall differentiation, decrease biocomplexity, and lead to instability (Wilson and Oliver 2000). Structures and events that produce spatial heterogeneity can hasten the development of a complex community. Although crown-class differentiation can take place at small scales (one to a few trees), it affects the entire stand at larger scales (40 to 400 ha or more). Intermediate scale heterogeneity (0.1 to 0.5 ha), however, is necessary for development of biocomplexity. In addition to crown class differentiation, intermediate-scale legacy retention, natural disturbances, and management promote development of habitat breadth and niche diversification.

3.2. Decadence

Decadence is a complex process essential to biodiversity. Decadent trees (live trees with heart rot, standing dead trees, and fallen, decaying trees) can be retained during harvest operations. They can develop naturally through self-thinning, suppression, disease, insect attack, damage by falling trees, and weather-related events (lightning, windstorms, ice storms, and snowstorms). Or they can be created by intentional wounding, infecting, or killing. Decadent trees provide substrate for (1) a large variety of cavity-, hollow-, and crevice-nesting wildlife, (2) pecking and tearing foraging by insectivorous birds, (3) physical partitioning of the forest floor, which reduces competition between deer and elk and among small mammals, (4) cover for small mammals and salamanders, (5) support of invertebrate communites that are prey for insectivorous small mammals and amphibians, (6) foraging sites for mycophagous small mammals, (7) travel ways, and (8) entryways to subnivean environments (see Harmon et al. 1986, Bunnell et al. 1999, and Johnson and O'Neil 2001 for reviews). Typically, forest management for wildlife emphasizes two elements of decadence: large, dead, moderately decayed conifers and large, fallen, moderately decayed conifers. The former are the trees most commonly used by cavity-excavating birds and the latter provide important shelter for terrestrial amphibians and certain small mammals. However, to focus on these structures without considering the entire process of decadence and how the process varies with seral stage is a mistake. For example, a conifer with a broken top may continue to grow, develop a new top or "basket" top, develop top rot and provide perch, roost, and nest sites for hawks, owls, eagles, ospreys, woodpeckers, squirrels, bats, and a variety of other wildlife over a long period before and after the tree dies. In younger conifer forests, deciduous trees such as red alder (*Alnus rubra* Bong.), willow (*Salix* L.), and, aspen (*Populus tremuloides* Michx.), provide valuable cavity trees despite their relatively small size (Carey et al. 1997, Bunnell et al. 1999). For example, red-breasted sapsuckers (*Sphyrapicus ruber*) will nest in small decadent willows (20 cm diameter at breast height (dbh)) in second-growth forests but use only large snags averaging more than 1 m dbh in old-growth forests. Sapsuckers (and other woodpeckers (Picidae)) are often double keystone species in keystone species complexes. For example, in Rocky Mountain forests, red-naped sapsuckers (*S. nuchalis*) excavate cavities in fungus-infected aspens that are required as nest sites by two species of swallows (*Tachycineta* spp.) and drill sap wells into willows that provide nourishment for themselves, hummingbirds (Trochilidae), orange-crowned warblers (*Vermivora celata*), chipmunks (*Tamias* spp.), and an array of other sap feeders (Daily et al. 1993). Thus, additional emphasis is warranted for providing numerous live deciduous trees subject to eventual suppression or infection with top rot early in forest development, and both conifers and hardwoods with cavities or other evidence of top rot late in forest development to provide various sizes of cavity trees. Pileated woodpeckers (*Dryocopus pileatus*) play a cavity-creation keystone role in Pacific Northwest forests (Aubry and Raley 2002) and throughout many forests in North America. These large birds are capable of excavating nest cavities, entrances to hollow trees, and entrances to insect galleries in the interior of large, moderately decayed trees that later are used by a wide variety of birds and mammals.

3.3. Understory Development

Seeds and plants retained on site germinate, regrow if damaged, or continue to grow as light, water, and nutrients become available through canopy gaps. Canopy closure can extirpate many of the retained species and, if long enough, even eliminate seeds from the soil seed bank. If instead of full canopy closure (limited stocking or management action such as precommercial thinning) there is crown-class differentiation, gap formation, or commercial thinning, the understory develops in stature and composition. With sufficient light, the understory increases in foliage volume and fruit production, providing wildlife with a variety of food and cover. For example, foliage of deciduous shrubs provides forage for larval moths (*Lepidoptera*) that are important food to insectivorous birds and mammals (Muir et al. 2002). Maple seeds (*Acer circinatum* Pursh and *A. macrophyllum* Pursh) and hazelnuts (*Corylus cornuta* Marsh) are especially valuable to squirrels in coniferous forest where their staple foods are produced sporadically (conifer seed) or are of low nutritive value (truffles) (Carey et al. 1999a, 2002).

3.4. Canopy Stratification

With retention or recruitment of shade-tolerant conifers and hardwoods and continued gap formation through natural mortality or silvicultural thinning, the forest begins to develop various strata of vegetation—low herbs, short shrubs, tall shrubs, and a midstory of deciduous and coniferous trees. Increased botanical diversity may be accompanied by horizontal and vertical heterogeneity in composition and foliage volume. Thus, a large variety of trophic relationships develop, and the overall habitat space begins to differentiate into diverse niches that support an enhanced variety of plants and wildlife.

3.5. Development of Habitat Breadth

With legacy retention and following a long period of gap development or management such as variable-density thinning, the forest develops patchy overstory, midstory, shrub, and herb layers. The result is a fine-scale mosaic of 0.1- to 0.5-ha patches of 10 to 30 types with each type composed of a different mix of species with different growth habits. For example, one patch may have an understory of moss with a dense midstory of a shade-tolerant conifer under a relatively open overstory; another patch may exhibit a continuous column of foliage from different plants from the forest floor to the overstory. Note that the resulting structure is quite different than the development of ladder fuels in interior ponderosa pine (*Pinus ponderosa* Dougl. ex Laws.) forests where flammable Douglas-fir provides ladders for ground fires to reach canopies. The patchy structure resulting from development of habitat breadth includes less flammable species in the understory than in the overstory and provides biological and physical gaps in both understory and overstory that impede the spread of potentially catastrophic disturbances such as fire and disease. Thus, intermediate-scale heterogeneity has been recommended for wet to moist coastal forests (Carey et al. 1999a) and moist to dry interior forests (Reynolds et al. 1992, Graham et al. 1999, Harrod et al. 1999). The process of developing habitat breadth (the full range of small-scale vegetation site types or patches characteristic of old, natural forests in the region) seems essential to maintenance of biodiversity in two ways. First there is an overall increase in habitat space—the volume of space the forest occupies, the overall surface area of plants within that volume, and the architectural niches formed by various life forms. Second, the diversity of life forms provides a variety of substrates and foods (foliage, seeds, fruits, nuts, and carbohydrates in root exudates) for use by other plants, fungi, invertebrate animals, and wildlife at a scale suitable for exploitation by organisms of low to moderate mobility, resulting in niche diversification.

3.6. Preinteractive Niche Diversification

When legacies have been retained or large structures (large live trees, dead trees, and fallen trees) have developed and the four basic stage-setting processes have gone on to produce habitat breath, the phenomenon of preinteractive niche diversification (Hutchinson 1978) may take place. Simply put, the forest has sufficient variation in structure, plant composition, and patchiness so many species that typically compete in simpler environments can coexist, even in large numbers. For example, resident cavity-using birds overlap in space and resource use in young, simple forests, and occupancy of these forests by some of these species may be variable or sporadic. In old, complex forests there are more cavity-using birds than in young forests; each species tends to forage on different substrates (e.g., bole and branch sizes). Similarly, decaying plants (wood and foliage) host various sizes of saprophytic invertebrates, each primarily consumed by a different insectivorous mammal (e.g., shrew (Soricidae)). Diversity in foliage cover provides hunting perches for spotted owls and protective cover for their prey. A simple forest may provide the owl with one species of prey but a complex forest provides several species (Carey et al. 1992). Thus, a complex forest provides a more stable resource for the owl and less predator pressure on any single prey species. Niche diversification operates at various trophic levels, providing for diverse forest-floor invertebrate fauna, fungi, and vascular plants; diverse insectivorous, mycophagous, granivorous, and herbivorous mammals; and diverse predators at the top of food webs (Carey 2003a).

4. Stages of Development of Managed Forest Ecosystems

Considerable effort has gone into compiling wildlife-habitat relationships (WHR) databases across the United States. The weakest link in WHR databases may be their forest development models, which are for even-aged stands. These models were derived from timber type classifications that serve their intended purpose well. But they were not developed to account for the diverse elements of wildlife habitat; therefore they serve WHR purposes poorly unless they are augmented by numerous other habitat variables. For example, the most current WHR database for Oregon and Washington (Johnson and O'Neill 2001) cross-tabulates forest-dwelling species by 20 forest structural conditions, 9 habitat types, and almost 100 habitat elements and subcategories (see Box 2). The cross-tabulated WHR models are data-rich and useful, but cumbersome (see Appendix 1, Chapter 1 for a comparison with other classifications). Their complexity exceeds the cognitive limits of people engaging in discussions of field conditions or in collaborative management. Less complex, but still holistic models of forest development are more useful for heuristic modeling exercises and designing silvicultural prescriptions.

The simplest tree-focused model has four stages that follow large-scale disturbance: stand initiation, stem exclusion, understory development, and old growth (Oliver 1981; Appendix 1, Chapter 1). This model has proven too reductionistic for modeling WHR in simulations of Oregon and Washington landscapes (Carey et al. 1999b). Carey and Curtis (1996), building on Bormann and Likens (1979), went beyond structural classes to a set of eight developmental stages based on the processes taking place within the forest that influence development of the greater biotic (plant, fungal, and animal) community. Franklin et al. (2002) followed with an expanded classification of natural stand development. Naturally developing forests may go through as many as eight major sequential stand conditions, differing in duration from less than 10 years to more than 500 years (see Appendix 1, Chapter 1). However, management can truncate forest development, eliminate entire stages, speed up or slow down transition between stages, and produce stand conditions not found in natural forests (Carey et al. 1999b). Development may be limited to as little as 40 years in Douglas-fir forests, yet complex forests may require 70 years or more to develop, and forests producing a full array of values may require rotations of 125 to 250 years or longer (Carey and Curtis

Table 1. Characteristics of stages of development in managed forests without (simple) and with (complex) legacy retention, spatial heterogeneity in the canopy, and maintenance of decadence processes.

Stage	Biomass allocations	Decadence	Spatial complexity	Niche divergence[1]	Resistance to change[2]
Reorganization					
Simple	trees, herbs, shrubs	none	some	some	low
Complex	trees, shrubs, herbs	much in legacies	much	much	low
Exclusion					
Simple	trees	small trees, deciduous trees	some	none	high
Complex	trees	legacy logs and snags, small trees	some	some	high
Accumulation					
Simple	trees	well-decayed small trees	some	some	high
Complex	trees	large legacy structures; small trees	some	some	high
Understory reinitiation					
Simple	trees, shrubs	possible insect/disease mortality	some	some	moderate
Complex	trees, shrubs	various[3]	much	much	moderate
Understory development					
Simple	trees, shrubs, herbs	possible insect/disease mortality	some	some	moderate
Complex	various life forms[4]	various[3]	much[5]	much	moderate
Niche diversification	various life forms[4]	various[3]	much[5]	much	dynamic[6]
Gap dynamics	various life forms[4]	various[3]	much[5]	much	dynamic[6]

[1] Niche divergence refers to preinteractive niche diversification in which a variety of ecological niches are provided by disturbance reducing competition and promoting complexity.
[2] Resistance to change indicates stages in which management intervention is particularly important to foster continuing development.
[3] Large legacy live trees with decadence, snags, fallen trees, litter, and deep humus; created cavity trees and snags; felled trees, insect/disease mortality.
[4] Conifers of different species and age cohorts, deciduous trees and shrubs, evergreen shrubs, ferns, forbs, lichens, and bryophytes arrayed in patches that differ in composition and structure.
[5] Varying in vertical, horizontal, and temporal dimensions.
[6] These stages have internal dynamics of change in patches over space in time; still, they are highly resistant to disturbance.

1996). Thus, modeling active management and its effects on forest development requires a different type of classification—a non-sequential classification that can be subdivided into decadal periods. Furthermore, management for compatibility requires that models incorporate more than just the development of the tree community. Carey et al. (1999b), for example, modeled the Carey-Curtis 8 stages of biotic community development with 25 total substages. Field tests of the Carey-Curtis classification in formal experiments, retrospective comparisons of forests managed for various objectives, and in retrospective comparisons of managed and natural forests, suggested it was necessary to revise and expand the classification to account for the great diversity of conditions being produced in managed forests in Oregon and Washington (Box 2). This new classification incorporates seven stages, five of which are cross-classified as simple or complex in structure and composition. The stages differ in how resources (light, water, nutrients, and space) are allocated to plants, degree of decadence, spatial complexity, niche divergence, and resistance to change (Table 1). The latter is an important consideration in management because it indicates stages in which a relatively stable state alternative to late seral forest is likely to develop in the absence of external disturbance. If such a state does develop and persists for a relatively long time, it may be resistant to management efforts aimed at developing a complex, biologically diverse forest (Carey 2003a). In timber management, development often is limited to simple subsets of two stages, ecosystem reorganization and either competitive exclusion or biomass accumulation, with cycles (rotations) of 40 to 70 years, with few or with a variety of silvicultural manipulations (Table 2). In management for biodiversity, emphasis is on the complex subsets of ecosystem reorganization, understory reinitiation, understory development, and niche diversification with rotations of 130 years or more or, with the addition of gap dynamics, on very long rotations (350 years or more). A variety of silvicultural manipulations are used, as well as direct wildlife habitat improvements (Table 2). Despite the complexity of this classification, it, like any classification, is an artificial construct and is best applied when augmented with site-specific knowledge of environmental conditions (climate, weather, microclimate, and natural disturbance regimes), site fertility and productivity, plant community development, and special landscape elements (e.g., wetlands, riparian zones, talus slopes, and cliffs).

5. Modeling Two Approaches to Forest Management

In western Washington, one favored approach to managing industrial forests was maximizing net present value of wood (NPV management) by clearcutting on 40-year rotations (following forest practices rules for green-tree retention and riparian area management prior to 1995), preparing the site, and allowing natural regeneration of western hemlock (*Tsuga heterophylla*

Table 2. Effects of silviculture on ecosystem complexity.

Silvicultural treatment	More complexity	Less complexity
Regeneration harvest		
Legacy-retention harvest	X	
Clearcutting		X
Planting		
With natural regeneration	X	
Without natural regeneration		X
Weeding		
Retains some hardwoods & shrubs	X	
Removes hardwoods & shrubs		X
Precommercial thinning		
Clumped multi-species retention	X	
Systematic single species retention		X
Commercial thinning		
Variable density plus[1]	X	
Systematic minus[2]		X

[1] Variable-density thinning that mixes closed, moderately open, and open canopies on a 0.1- to 0.5-ha scale. It is designed to enhance woody plant diversity, maintain deciduous trees, promote recruitment of shade-tolerant trees with underplanting, and augment cavity trees and coarse woody debris when necessary.
[2] Light to moderate thinning with even spacing to favor one species.

(Raf.) Sarg). In a modeling exercise to develop practical, but holistic, management systems, an alternative emerged: manage for biodiversity, including wildlife, wood, water, and clean air benefits by (1) clearcutting with legacy retention, (2) planting Douglas-fir with natural regeneration of other conifers and hardwoods, (3) doing precommercial thinning to promote growth and maintain biodiversity, and (4) doing up to three variable-density thinnings to

Box 2. Stand Development in Second-Growth Douglas-fir Forests
Potential stages of development in managed, second-growth Douglas-fir forests (adapted from Carey and Curtis 1996) with approximate correspondence to stages of stand development and structural conditions.[1] Characteristics of stages of development in managed forests are subdivided based on the lack (simple) and presence (complex) of legacy retention, spatial heterogeneity in the canopy, and maintenance of decadence processes.

Stage of development	Management and subsequent ecological processes
Ecosystem reorganization	Management removes most overstory trees, with minor to major retention of biological components (future legacies) of the forest and planting or seeding of trees; vagile forest species and native and exotic non-forest species invade

Box 2. Stand Development in Second-Growth Douglas-fir Forests (continued)

Stage of development	Management and subsequent ecological processes	
	and grow; plant communities establish on bare ground initially as grass-forb-herb, then shrub-tree, and finally tree-dominated assemblages.	
Referents	Oliver (1981)	Stand initiation
	Johnson and O'Neil (2001)	6 conditions, grass/forb-open to sapling/pole-moderate
Simple	Clearcutting variants, site preparation, planting, vegetation control, and stocking control produce an even-aged monoculture or a forest dominated by one to a few tree species.	
Complex	Variable-retention harvest systems with 5 to 30% retention of legacies, reduced and variable site preparation, and planting with natural regeneration likely establishes multiple species of trees. Planting or seeding restores lost native plant diversity if necessary; limited vegetation control ensures full stocking and vegetative diversity; and precommercial thinning promotes multiple tree and shrub species and forestalls competitive exclusion of numerous species.	
Competitive exclusion	No management occurs beyond ensuring full stocking, selecting desired species, and controlling competitors.	
Referents	Oliver (1981)	Stem exclusion
	Johnson and O'Neil (2001)	Sapling/pole-closed, small tree-single story-closed, medium tree-single story-closed, large tree-single story-closed
Simple	Trees fully occupy the site and compete intensely for light, water, nutrients, and space by growing tall quickly. Some eventually overtop and suppress shorter life forms and slower growing species. Self-thinning produces even spacing and can reduce inter-tree competition. Failure to self-thin leads to loss of crown depth and spindly trees. Competitive exclusion may extend to most vascular plants, invertebrates, and wildlife.	
Complex	Similar to simple competitive exclusion, but legacies from the preceding stand (e.g., fallen trees, stumps, or patches of intact forest) provide refugia within the forest for a wide variety of species. Refugia maintain some spatial heterogeneity and species diversity. Suppressed trees provide substrate for cavity-excavating wildlife and foraging for saprophytic insects and insectivorous birds and mammals.	
Biomass accumulation	Management includes conventional thinning to moderately high relative densities, selection for desirable timber species, and removal of decadent, defective, and competing trees.	

Box 2. Stand Development in Second-Growth Douglas-fir Forests (continued)

Stage of development	Management and subsequent ecological processes	
Referents	Oliver (1981)	No equivalent
	Johnson and O'Neil (2001)	Sapling/pole-moderate, small tree-single story-moderate, medium tree-single story-moderate, large tree-single story-moderate
Simple	Trees fully occupy the site with moderate inter-tree competition; even-aged codominants grow and accumulate wood biomass rapidly while providing limited resources to other life forms.	
Complex	Low to moderate diversity and biomass of common plants species are maintained; growth of dominant trees contributes to homogeneity; legacies maintain some heterogeneity and diversity.	
Understory reinitiation	Silvicultural thinning, self-thinning, and natural growth and mortality promote dominance by some trees, death of others, and releases understory plants from overstory competition.	
Referents	Oliver (1981)	Understory reinitiation
	Johnson and O'Neil (2001)	Sapling/pole-open, small tree-single story-open, medium tree-single story-open, perhaps large tree-single story-open, depending on stand history
Simple	A uniform canopy of evenly spaced trees leads to uniform understory conditions with dominance by a few species such as a shade-tolerant conifer or the native clonal species salal (*Gaultheria shallon* Pursh) and swordfern (*Polystichum munitum* (Kaulfuss) K. Presl.).	
Complex	Variable-density thinning produces canopy heterogeneity that leads to variable light and moisture regimes in the understory, which, with legacies, produce a heterogeneous understory, generally low in stature. Underplanting augments the understory with key deciduous and conifer species.	
Understory development	Thinning or other mortality reduces competition among trees, increases growth of trees, and releases understory from competition; understory develops in stature, abundance, and species diversity.	
Referents	Oliver (1981)	No equivalent
	Johnson and O'Neil (2001)	6 small- to medium tree-multistory conditions, depending on legacies and decadence

Box 2. Stand Development in Second-Growth Douglas-fir Forests

Stage of development	Management and subsequent ecological processes (continued)	
Simple	A homogeneous overstory with moderate to low crown closure produces an understory that is botanically diverse but still dominated by a few species and lacking a distinctive patchy pattern; layers may develop, but lack of legacies precludes fully developed, complex biotic communities.	
Complex	Variable-density thinning produces canopy heterogeneity with high to moderate to low crown closure by dominants and codominants. Heterogeneity produces crown differentiation, including ingrowth by hardwoods and shade-tolerant conifers. Variable understory environmental conditions produce understory patches of differing composition. Cavity-tree creation and coarse woody debris augmentation during thinning and legacies from the preceding forest further compound the heterogeneity. The resulting complex structure provides a diversity of niches for species within various life forms, including fungi, mosses, lichens, achlorophyllous mycotrophs, grasses, forbs, evergreen shrubs, deciduous shrubs, deciduous trees, and shade-tolerant conifers.	
Niche diversification (Complex only)	Additional variable-density thinning or group selection may help develop high biocomplexity, including species diversity, structural diversity (live, dead, and fallen trees of various sizes; patchy understory; patchy midstory; canopy gaps; and gaps filled by shade-tolerant trees growing into the overstory), and vertical and horizontal spatial heterogeneity, but the forest still has not developed the giant structures characteristic of old-growth forests. When needed, cavity-tree creation and coarse woody debris augmentation during thinning maintain the decadence process.	
Referents	Oliver (1981)	No equivalent
	Johnson and O'Neil (2001)	6 medium to large tree-multi story conditions, depending on numerous variables
Gap dynamics (Complex only)	These managed forests more than 125 years old have high niche diversification and giant trees and tree-based structures, including legacies and the current dominant cohort. They are either passively managed for late-seral forest values along streams, on mass-wasting areas, in watersheds, or in reserves, on long rotations in shifting steady-state mosaic landscapes, or with group selection for uneven-aged management.	
Referents	Oliver (1981)	Old growth (but not natural)
	Johnson and O'Neil (2001)	Giant tree-multistory, depending on numerous variables

[1] See Appendix 1, Chapter 1 for cross-referencing to other classifications.

Table 3. Measures of wildlife outcomes linked to compatibility in a 6828-ha landscape in western Washington, in the last 100 years of 300-year simulations maximizing net present value of timber and conserving biodiversity to produce ecological services and economic goods (adapted from Carey et al. 1999b).

Ecological measure	Timber management	Biodiversity management
Habitat for spotted owls	No	Yes
Numbers of deer and elk[1]	423/134	401/200
Vertebrate diversity (% of maximum possible)[2]	64	100
Forest floor function (% of maximum possible)[3]	12	100
Ecological productivity (% of maximum possible)[4]	19	94
Landscape health (mean of the above %)[5]	32	98

[1] *Odocoileus hemionus* and *Cervus elaphus*.
[2] Based on 130 species of wildlife.
[3] Based on the integrity of the forest-floor small-mammal community of nine species.
[4] Based on the biomass of three squirrels (*Glaucomys sabrinus*, *Tamias townsendii*, and *Tamiasciurus douglasii*).
[5] Mean of vertebrate diversity, forest floor function, and ecological productivity.

remove wood products, add decadence (cavity trees), and accelerate the development of habitat breadth and niche diversification on rotations that alternated between 70 years and 130 years (Carey et al. 1999b). Biodiversity management included expanded protection of riparian and mass-wasting areas. Remarkably, the modeled loss of net present value between the two alternatives was slight—18% under old forest practices rules, 4% using comparable riparian protection developed for the biodiversity pathway, but a 13% gain with new Washington forest practices rules for riparian protection (Carey 2003b). Sustainable decadal revenues, however, increased 150% and forest-based employment increased fourfold once the landscape achieved a shifting, steady-state mosaic with a relatively even flow of wood products (Carey et al. 1999b). Of course, all economic outcomes are market dependent and may vary over time. The shifting, steady-state keeps the proportions of seral stages in the landscape constant while allowing individual sites in the landscape to undergo a progression of forest development in response to management. Thus, the spatial distribution of the various stages shifts in time while the proportions remain constant and a steady flow of wood products is produced. The differences in ecological performance of the managed landscapes were also marked (Table 3). Management for biodiversity was more than three times more effective than NPV management.

The reasons for the differences were clear. No species of wildlife were unique to the early stages in forests managed for NPV, but 14 species were unique to complex, late-seral stages and 11 to healthy riparian areas (Carey et al. 1999b). NPV management produced no habitat for spotted owls. Graphs for ecosystem productivity (Figure 2) and forest-floor productivity were similar to each other and showed that complex late seral stages supported much more

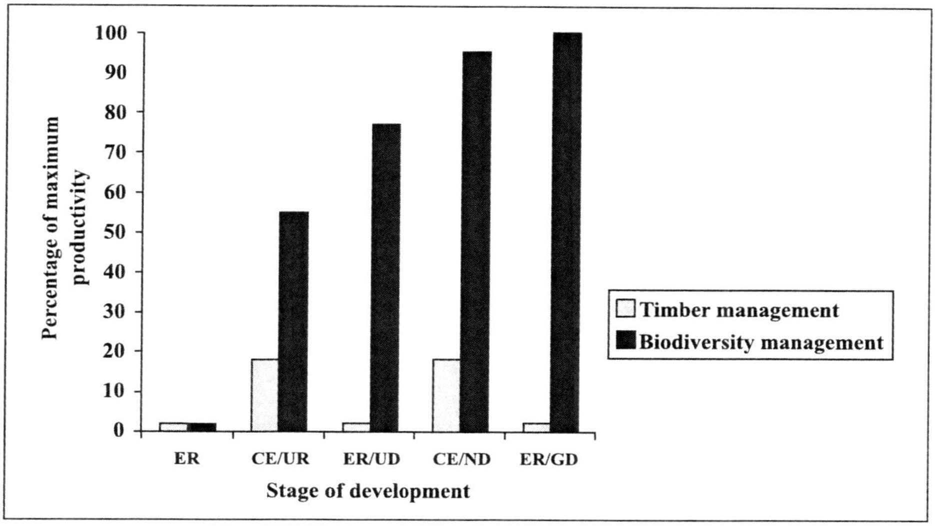

Figure 2. Effects of stage of development on ecological productivity. Ecological productivity is measured as biomass of squirrels as determined by production of truffles, mushroom, fruits, seeds, and nuts, (squirrel biomass determines carrying capacity for medium size predators). Timber management for maximum net present value alternates ecosystem reorganization and competitive exclusion on a 40-year rotation. Biodiversity management for wildlife, wood, water, and other values emphasizes five stages of development by alternating 70-year rotations with rotations of more than 130 years.
Note: ER = ecosystem reorganization; CE = competitive exclusion; UR = understory reinitiation; UD = understory development; ND = niche diversification; GD = gap dynamics.

diverse and abundant squirrel and forest-floor mammal communities than simple, early stages of forest development. Differences in production of deer and elk were slight. In forests managed for NPV, the large mammals find large concentrations of food in early clearcuts. In landscapes managed for biodiversity, smaller amounts of higher quality food are found throughout the landscape in all seral stages and in conjunction with shelter from the weather and hiding cover (Carey et al. 1999b). With management for biodiversity, more than 50% of the steady-state landscape was late-seral forest with less than 15% in early-seral stages, producing a landscape hospitable to late-seral wildlife and facilitative of dispersal and colonization processes. On the other hand, the NPV management landscape was universally inhospitable to late-seral wildlife.

6. Conclusion

Results from prospective experimental manipulation of second-growth forests to produce biocomplexity have proven successful. The results suggest active management for joint, efficient production of a wide variety of forest values is not only possible but also desirable from a forest health view (Haveri and Carey 2000; Carey 2001, 2002, 2003a; Carey and Wilson 2001; Carey et al. 2002; Wilson et al., in press). Retrospective comparisons of managed and natural forests (e.g., Carey et al. 1992, 1999a; Bunnell et al. 1999; Carey and Harrington 2001; Muir et al. 2002; Carey 2003a) suggest the same. Widespread recognition is emerging for the need to manage for structural, compositional, and spatial complexity at multiple scales when production of multiple values is the goal (Reynolds et al. 1992, Kaufmann et al. 1994, Carey and Curtis 1996, Curtis et al. 1998, Bunnell et al. 1999, Hunter 1999, Franklin et al. 2002, Lindenmayer and Franklin 2002, Muir et al. 2002, Carey 2003a,b). Habitat elements commonly cited as important are legacies, cavity trees, deciduous trees and shrubs, coarse woody debris, and spatial heterogeneity in the overstory and understory. Management at multiple spatial scales for dynamic ecosystems and landscapes will help meet (1) goals of biological reserves; (2) human needs for wood products, clean air, clean water, open space, and nature-based experiences; and (3) worldwide demands to maintain environmental health.

7. Acknowledgments

I thank Frank Davis (Olympic National Forest), Pete Holmberg (Washington Department of Natural Resources), Cathy Raley (Pacific Northwest Research Station), and Paula Sweden (Washington Department of Wildlife) for numerous discussions and their constructive reviews of an earlier draft of this manuscript. Annette Bristol (Pacific Northwest Research Station) provided helpful editorial review. I thank Jerry Franklin (University of Washington) for long-term, continuing discussions of forest ecology and management in the Pacific Northwest. I thank Bob Curtis, Dean DeBell, and Connie Harrington (Pacific Northwest Research Station), Bruce Lippke (University of Washington), and John Sessions (Oregon State University) for helpful discussions on the silviculture, economics, and engineering, respectively, of forest management.

8. References

Aubry, K., and Raley, C.M. 2002. The pileated woodpecker as a keystone habitat modifier in the Pacific Northwest. In: W.F. Laudenslayer, Jr., P.J. Shea, B.E. Valentine, [and others], Proceedings of the symposium on the ecology and management of dead wood in western forests (pp. 257-274). General Technical Report PSW-GTR-181. Albany, CA: U.S. Department of Agriculture, Forest Service, Pacific Southwest Research Station.

Behan, R.W. 2001. Plundered promise: capitalism, politics, and the fate of federal lands. Washington, DC: Island Press. 240 p.

Belcher, J.M. 2001. Turning the ship around: changing the policies and culture of a government agency to make ecosystem management work. *Conservation Biology in Practice*, 2(4): 17-23.

Best, C., and Wayburn, L.A. 2001. America's private forests: status and stewardship. Washington, DC: Island Press. 269 p.

Bormann, F.H., and Likens, G.E. 1979. Pattern and process in a forested ecosystem. New York: Springer-Verlag. 253 p.

Bunnell, F.L., Kremsater, L.L., and Wind, E. 1999. Managing to sustain vertebrate richness in forests of the Pacific Northwest: relationships within stands. *Environmental Review*, 7: 97-146.

Canham, D.D., Denslow, J.S., Platt, W.J. [and others]. 1990. Light regimes beneath closed canopies and tree fall gaps in temperate and tropical forests. *Canadian Journal of Forest Research*, 20: 620-631.

Carey, A.B. 2001. Experimental manipulation of spatial heterogeneity in Douglas-fir forests: effects on squirrels. *Forest Ecology and Management*, 152: 13-30.

Carey, A.B. 2002. Globalization of flora: inviting worldwide ecosystem disaster. *Renewable Resources Journal*, 20: 13-17.

Carey, A.B. 2003a. Biocomplexity and restoration of biodiversity in temperate coniferous forest. *Forestry*, 76(2): 131-140.

Carey, A.B. 2003b. Restoration of landscape function: Reserves or active management? *Forestry*, 76(2): 225-234.

Carey, A.B., Colgan, W., III, Trappe, J.M., and Molina, R. 2002. Effects of forest management on truffle abundance and squirrel diets. *Northwest Science*, 76: 148-157.

Carey, A.B., and Curtis, R.O. 1996. Conservation of biodiversity: a useful paradigm for forest ecosystem management. *Wildlife Society Bulletin*, 24: 610-620.

Carey, A.B., and Harrington, C.A. 2001. Small mammals in young forests: implications for management for sustainability. *Forest Ecology and Management*, 154: 289-309.

Carey, A.B., Horton, S.P., and Biswell, B.L. 1992. Northern spotted owls: influence of prey base and landscape character. *Ecological Monographs*, 62: 223-250.

Carey, A.B., Kershner, J., Biswell, B., and De Toledo, L.D. 1999a. Ecological scale and forest development: squirrels, dietary fungi, and vascular plants in managed and unmanaged forests. *Wildlife Monographs*, 142: 1-71.

Carey, A.B., Lippke, B.R., and Sessions, J. 1999b. Intentional systems management: managing forests for biodiversity. *Journal of Sustainable Forestry*, 9(3/4): 83-125.

Carey, A.B., and Wilson, S.M. 2001. Induced spatial heterogeneity in forest canopies: responses of small mammals. *Journal of Wildlife Management*, 65: 1014-1027.

Carey, A.B., Wilson, T.M., Maguire, C.C., and Biswell, B.L. 1997. Dens of northern flying squirrels in the Pacific Northwest. *Journal of Wildlife Management*, 61: 684-699.

Carson, R. 1962. Silent spring. Boston, MA: Houghton-Mifflin. 368 p.

Connell, J.H., and Slatyer, R.O. 1977. Mechanisms of succession in natural communities and their role in community stability and organization. *American Naturalist*, 111: 1119-1144.

Curtis, R.O., and Carey, A.B. 1996. Timber supply in the Pacific Northwest: managing for economic and ecological values. *Journal of Forestry*, 94(9): 4-7, 35-37.

Curtis, R.O., DeBell, D.S., Harrington, C.A., Lavender, D.P., St.Clair, J.B., Tappenier, J.C., and Walstad, J.D. 1998. Silviculture for multiple objectives in the Douglas-fir region. General Technical Report PNW-GTR-435. Portland, OR: U.S. Department of Agriculture, Forest Service, Pacific Northwest Research Station. 123 p.

Daily, G.C., Ehrlich, P.R., and Haddad, N.M. 1993. Double keystone bird in a keystone species complex. *Proceedings of the National Academy of Sciences*, 90: 592-594.

Entwistle, A., and Dunstone, N. 2000. Priorities for the conservation of mammalian diversity: Has the panda had its day? Cambridge, UK: Cambridge University Press.

Folke, C., Holling, C.S., and Perrings, C. 1996. Biological diversity, ecosystems, and the human scale. *Ecological Applications*, 6: 1018-1024.

Franklin, J.F., Lindemayer, D., MacMahon, J.A. [and others]. 2000. Threads of continuity. *Conservation Biology in Practice*, 1: 9-16.

Franklin, J.F., Spies, T.A., Van Pelt, R., Carey, A., Thornburgh, D.A., Berg, R., Lindenmayer, D.B., Harmon, M.E., Keeton, W.S., Shaw, D.C., Bible, K., and Chen, J. 2002. Disturbances and structural development of natural forest ecosystems with silvicultural implications, using Douglas-fir as an example. *Forest Ecology and Management*, 155: 399-423.

Goodland, R. 1995. The concept of environmental sustainability. *Annual Review of Ecology and Systematics*, 26: 1-24.

Graham, R.T., Harvey, A.E., Jain, T.B., and Tonn, J.R. 1999. The effects of thinning and similar stand treatments on fire behavior in western forests. General Technical Report PNW-GTR-463. Portland, OR: U.S. Department of Agriculture, Forest Service, Pacific Northwest Research Station. 27 p.

Harmon, M.E., Franklin, J.F., Swanson, F.J. [and others]. 1986. Ecology of coarse woody debris in temperate ecosystems. *Advances in Ecological Research*, 15: 133-302.

Harrod, R.J., McRae, B.H., and Hartl, W.E. 1999. Historical stand reconstruction in ponderosa pine forests to guide silvicultural prescriptions. *Forest Ecology and Management*, 114: 433-446.

Haveri, B.A., and Carey, A.B. 2000. Forest management strategy, spatial heterogeneity, and winter birds in Washington. *Wildlife Society Bulletin*, 28: 643-652.

Haynes, R.W., Monserud, R.A., and Johnson, A.C. 2003. Compatible forest management: background and context. Chapter 1. In: R.A. Monserud, R.W. Haynes, and A.C. Johnson (Eds.), Compatible forest management. Dordrecht, The Netherlands: Kluwer Academic Publishers.

Holling, C.S. 2001. Understanding the complexity of economic, ecological, and social systems. *Ecosystems*, 4: 390-405.

Hunter, M.L. 1990. Wildlife, forests, and forestry: principles of managing forests for biological diversity. Englewood Cliffs, NJ: Regents/Prentice Hall. 370 p.

Hunter, M.L. 1999. Maintaining biodiversity in forest ecosystems. Cambridge, UK: Cambridge University Press. 530 p.

Hutchinson, G.E. 1978. An introduction to population ecology. New Haven, CT: Yale University Press. 260 p.

Johnson, D.H., and O'Neil, T.A. 2001. Wildlife-habitat relationships in Oregon and Washington. Corvallis, OR: Oregon State University Press. 736 p.

Kaufmann, M.R., Graham, R.T., Boyce, D.A., Jr., [and others]. 1994. An ecological basis for ecosystem management. General Technical Report RM-246. Fort Collins, CO: U.S. Department of Agriculture, Forest Service, Rocky Mountain Forest and Range Experiment Station. 22 p.

Kemmis, D. 2001. This sovereign land: a new vision for governing the West. Washington, DC: Island Press. 263 p.

Kimmins, J.P. 2002. Future shock in forestry: Where have we come from; where are we going; is there a "right way" to manage forests? Lessons from Thoreau, Leopold, Toffler, Botkin, and nature. *Forestry Chronicle*, 78(2): 263-271.

Kinzig, A.P., Pacala, S.W., and Tilman, D. 2002. The functional consequences of biodiversity: empirical progress and theoretical extensions. *Monographs in Population Biology*, 33: 1-365.

Lackey, R.T. 2001. Values, policy, and ecosystem health. *BioScience*, 51: 437-443.

Larson, T.B. 2001. Biodiversity evaluation tools for European forests. *Ecological Bulletin*, 50: 1-237.

Lawton, J.H. 1994. What do species do in ecosystems? *Oikos*, 71: 367-374.

Li, C.Y., and Strzelczyk, E. 2000. Belowground microbial processes underpin forest productivity. *Phyton*, 40(4): 129-134.

Lindemayer, D.B., and Franklin, J.F. 2002. Conserving forest biodiversity: a comprehensive multiscale approach. Washington, DC: Island Press. 351 p.

Loehle, C., MacCracken, J.G., Runde, D., and Hicks, L. 2002. Forest management at landscape scales: solving the problems. *Journal of Forestry*, 100(6): 25-32.

Loreau, M., Naeem, S., Inchausti, P. [and others]. 2001. Biodiversity and ecosystem functioning: current knowledge and future challenges. *Science*, 294: 804-808.

Muir, P.S., Mattingly, R.L., Tappeiner, J.C., II [and others]. 2002. Managing for biodiversity in young Douglas-fir forests of western Oregon. Biological Science Report USGS/BRD/BSR--2002-0006. Corvallis, OR: U.S. Geological Survey, Forest and Rangeland Ecosystem Science Center. 76 p.

Oliver, C.D. 1981. Forest development in North America following major disturbances. *Forest Ecology and Management*, 3: 153-168.

Pankhurst, C.E. 1997. Biodiversity of soil organisms as an indicator of soil health. In: C.E. Pankhurst, B.M. Doube, V.V.S.R. Gupta (Eds.), Biological indicators of soil health (297-324). Cambridge, MA: CAB International.

Ray, P. 1996. The integral culture survey: a study of the emergence of transformational values in America. Research Paper 96-A. Sausalito, CA: Institute of Noetic Sciences. 160 p.

Reynolds, R.T., Graham, R.T., Reiser, M.H. [and others]. 1992. Management recommendations for the northern goshawk in the southwestern United States. General Technical Report RM-217. Fort Collins, CO: U.S. Department of Agriculture, Forest Service, Rocky Mountain Forest and Range Experiment Station. 90 p.

Ruggiero, L.F., Aubry, K.B., Carey, A.B., and Huff, M.H. (Tech. coords.). 1991. Wildlife and vegetation of unmanaged Douglas-fir forests. General Technical Report PNW-GTR-285. Portland, OR: U.S. Department of Agriculture, Forest Service, Pacific Northwest Research Station. 533 p.

Scheffer, M., Carpenter, S., Foley, A. [and others]. 2001. Catastrophic shifts in ecosystems. *Nature*, 413: 591-596.

Shields, D.J., Martin, I.M., Martin, W.E., and Haefele, M.A. 2002. Survey results of the American public's values, objectives, beliefs, and attitudes regarding forests and grasslands: a technical document supporting the 2000 USDA Forest Service RPA Assessment. General Technical Report RMRS-GTR-95. Fort Collins, CO: U.S. Department of Agriculture, Forest Service, Rocky Mountain Research Station. 111 p.

Whittaker, R.H. 1975. Communities and ecosystems. New York: MacMillan. 385 p.

Wilson, J.S., and Oliver, C.D. 2000. Stability and density management in Douglas-fir plantations. *Canadian Journal of Forest Research*, 30: 910-920.

Wilson, T.M., Carey, A.B., and Haveri, B.A. [In press]. Spring bird survey and social perceptions. In: R.O. Curtis [and others] (Eds.), Silvicultural options for young growth Douglas-fir: the Capitol Forest Study—establishment and first results. General Technical Report. Portland, OR: U.S. Department of Agriculture, Forest Service, Pacific Northwest Research Station.

Wondolleck, J.M., and Yaffee, S.L. 2000. Making collaboration work: lessons from innovation in natural resource management. Washington, DC: Island Press. 278 p.

Section 5—Social Aspects

Compatibility is basically about values. Every author in this book has looked at values and tradeoffs in one sense or another (the values of healthy trees, different stand structures, wood quality, wildlife, water quality, fish, and nontimber forest products, among others). But some values have been mentioned only briefly, and the concept of values has not been looked at directly.

In Chapter 15, Donoghue clarifies the concept of values and discusses the role of social values in compatible forest management. She also examines how information about social values is incorporated into discussions about compatible management. Over the past two centuries, ownership and management of federal land in the United States have been characterized by six overlapping eras that reflected the dominant social values of the time. The "new resource management paradigm" is characterized by ideas about sustainable development, harmony with nature, skepticism, and strong emphasis on public involvement in decisionmaking.

Values come in many forms. In the first four sections of this book, the focus has been mainly on silvicultural, ecological, and economic considerations, and less on the broader range of non-economic cultural, spiritual, and some recreational values. Social values shape how resources are managed and consumed. Shifting and expanding social values will continue to challenge our ability to achieve compatible forest management. Establishing values through market prices and non-market valuations may not reflect the most relevant social values pertaining to a particular issue.

Integrating information about values into decisionmaking is all the more difficult when values are conflicting, changing, or poorly understood. In Chapter 16, the authors review the social acceptability concept, including how judgments are revealed, evolution of the concept, and its limitations. The concept of social acceptability evolved from sociological theory first developed in the 1960s. Social acceptability is a judgmental process in which comparisons are involved, and knowledge and awareness shape the judgments. People's judgments are almost always conditional. Context (space, place, and time) is critical, and judgments may change as context changes. The failure to address this issue adequately has created an "either-or" mentality apparent in many forest management debates. People then take their issues to legal and political venues, and the result is too often management by legal and political edicts.

The social conflicts confounding resource management will not be resolved by better answers to the technical and scientific questions addressed in the first four sections of this book. To be accepted, forest management requires an approach that recognizes the legitimacy of many social values, especially those emphasizing historical, cultural, wildlife, and recreational values. The authors describe ways to engage people in mutual learning, trust building, and collaboration, and ways to build public support for new management practices; they also discuss some tools that have proved useful, such as the limits of acceptable change model. The social aspects of compatible management can be as, or more challenging than the technical and scientific challenges.

Given the diversity of often-conflicting and ever-changing values related to forests and forest resources, the challenge of achieving compatible forest management is formidable. Through better understanding of values, people may be able to expand the range of forest management options and thus be more likely to find an option that achieves compatible forest management.

Chapter 15

Social Values and Compatible Forest Management

Ellen M. Donoghue
USDA Forest Service, Pacific Northwest Research Station,
620 SW Main St., Suite 400, Portland, OR 97205, USA

1. Introduction

Although compatible forest management is often described as the simultaneous production of wood and other goods and services from forests (Haynes et al. 2003), it is fundamentally about values. In particular, it is about assigning values to forest goods and services and making judgments about how management actions affect those values. Values related to forests and forest resources come in many forms, including commodity, amenity, environmental quality, ecological, public use, and spiritual values (Stankey and Clark 1992). Some values conflict and others are not well understood. This poses a dilemma for forest management strategies, such as compatible forest management, that are not intended to negatively impact other values for the sake of wood production. If we do not understand a value and its relation to other values, how do we know if management actions negatively affect it?

Even though the concept of compatibility spans biological and social dimensions of forest management, the focus has tended toward silvicultural, ecological, and economic considerations and less on the broader range of noneconomic social values. This is due to several factors. One is that there is considerable ambiguity surrounding the concept of values. The ambiguity relates not only to what is meant by values, but also to how values are measured. Another factor is the lack of understanding about how to incorporate information on values into forest management decisionmaking. Some, such as economic values, have long been incorporated into forest management decisionmaking. However, they often reflect a limited range of social values related to forests. For instance, economic values do not reflect cultural, spiritual, and some recreational values.

R.A. Monserud, R.W. Haynes and A.C. Johnson (eds.)
Compatible Forest Management, 429-452

What do we mean by values? From soil science to social science, the term "values" has different meanings depending on the disciplinary use of the term. Sometimes various types of values are grouped together under terms such as "forest values." Such terms may have limited usefulness and may increase ambiguity in discussions about forest management. For instance, depending on a person's perspective, "forest values" may refer to ecological attributes, such as wildlife habitat, or social attributes, such as cultural amenities. Fortunately, most values can be defined as either ecosystem function values (Andrews and Waits 1978, Bengston 1994) or social values. Ecosystem function values are values, such as habitat and water quality, that exist whether or not humans are aware of them. Ecosystem function values exist independently of humans. However, because humans are concerned with resource consumption, conservation, and preservation, public and private institutions support biophysical sciences directed at measuring and assessing ecosystem function values. Public and private resource institutions also manage lands in support of ecosystem function values.

Social values related to forests and forest resources are derived directly from humans. Social values range from economic values determined through markets; to recreational values determined through visitor use, preferences, and impacts; to spiritual values determined through perceptions and opinions. Social values shape how people relate to forests through what they consume, use, observe, admire, and ritualize. Thus, social values lead to direct and indirect impacts on forest resources. Direct impacts are fairly obvious, such as impacts owing to a person's behavior hiking on a forest trail. Indirect impacts include the influence of social values on political and policymaking processes, as well as processes that are more localized, such as relations between neighboring landowners. Thus, social values shape how resources are managed and consumed. Social values also change over time, albeit some change more rapidly than others.

Improved understanding about values is relevant to private and public land managers attempting to engage in compatible forest management. Public land managers are charged with managing forests for diverse uses. Understanding values is important because values shape preferences, behaviors, and people's interaction with public forestlands.

Many private land managers also are concerned about social values largely because social values shape consumer preferences, influence regulations, and set agendas for environmental groups. For instance, societal preferences for certified wood may induce private landowners to change forest management practices in response to consumer preferences and expanding markets. In Oregon, private industry supported the creation of the nation's first Forest Practices Act in 1971. The forest practice rules apply to all private lands in the state, whether owned by individuals or large corporations, and to state-owned lands. The rules have been changed over the decades in response to new scientific information and changing public concerns (Oregon Department of Forestry 2003). And, in British Columbia, the timber company Weyerhaeuser has continued to implement changes in forest management induced by pressure from environmental groups that began under the former ownership and license of MacMillan Bloedel. These changes include an increased percentage of variable retention on the company's Crown and private lands (Beese et al. 2001). As public forest managers are increasingly concerned about managing for diverse values, private industries are increasingly concerned with maintaining their social license to operate.

The challenge for private and public forest managers, however, is that claiming compatibility has been achieved implies that knowledge about ecosystem function and social values, not only exists but also has been integrated into forest management decisionmaking. This implies that scientists and managers understand the fundamental differences in how people value forests within the context of a specific resource management problem. And, that managers understand these values sufficiently enough to make judgments about what is compatible and what is not. These may be lofty assumptions, given that there is considerable ambiguity about how scientists and managers define, measure, and integrate values in forest management. Therefore, this chapter responds to the need to clarify the concept of values and increase understanding about the role of social values in compatible forest management. I focus on the definition, measures, and characterizations of social values related to forests and forest resources. I also examine the integration of information about social values into discussions about compatible forest management.

2. The Role of Values in Shaping Change

The social and political processes that have shaped natural resource ownership and management throughout history reflect underlying sets of common social values. Values have been considered the lens of a worldview through which all is considered and acted upon (Dunlap and Van Liere 1984, Wildavsky and Dake 1990). In the United States, the ownership and management of federal land have been characterized by six overlapping eras that reflect the dominant social values of the times: acquisition, disposal, reservation, custodial management, intensive management, and consultation and conflict (Clawson 1983). In the era of disposal, occurring between the post-revolution years through the late 1800s, the notion of federal trusteeship prevailed. Land was cheap and seemingly plentiful. In the absence of a dominant social value for keeping land in the public domain, and in response to strong desires to settle the land and build the federal treasury, land was disposed to states, corporations, and individuals (Adams 1993). The intensive management era, generally depicted as the period from the 1950s through

1960s, is characterized by large increases in the volume of federal timber sales, the number of oil and gas leases, and in receipts for grazing, mineral, and recreation activities (Adams 1993). The current era, referred to as consultation and conflict (Clawson 1983), or extensive preservation (Culhane 1981), emerged out of the 1960s and is characterized by growing environmental awareness, a proliferation of environmental legislation, and greater public involvement in the policymaking processes related to the environment and natural resources. The seeds for the next era are already planted in international processes, such as the Montreal Process (Montreal Process Working Group 1998), that emphasize balance among social justice, environmentally sound practices, and economic benefits in the context of sustainable forest management.

The transition from one era to the next reflects shifting values. These are also characterized as shifting paradigms in resource management. Sociologists, political scientists, and others associate paradigm shifts with changing societal-environmental relations and changing social values toward the environment (Dunlap and Catton 1979, Koch and Kennedy 1991, Brown and Harris 1992, Bengston 1994, Kennedy et al. 1998, Shindler and Cramer 1999). Two commonly depicted paradigms are the "machine-model" and "organic-model" (Kennedy et al. 1998), which are similar to the "dominant resource management" and the "new resource management" paradigms (Bengston 1994, Shindler and Cramer 1999). Although depicted as dichotomies, suggesting a distinct shift from one to the other, the paradigms are generally viewed as representing long-evolving continuums in resource management.

Kennedy et al. (1998) say that in the early part of the 20th century, forestry agencies in many industrialized countries, including the United States, embraced a "machine-model" worldview. This worldview was based on thinking processes that were simple, segmented, and linear, and that reflected an illusion of control over an ordered world. In the United States, this type of thinking has been manifested in many ways, including the tendency to view forest ecosystems narrowly, to adopt intensive approaches to pest and fire management, to be fascinated with modeling, to create functional and hierarchical organizational structures and agency culture, and to believe that the agency knows what is best for the public. Like the "machine-model" view, the "dominant resource management" paradigm emphasizes the belief that natural resource management ought to be directed toward the production of goods and services that are beneficial to humans. Bengston (1994) describes the "dominant natural resource" paradigm as emphasizing economic growth, control of nature, faith in science and technology, ample reserves of natural resources, the substitutability of resources, and a dominant role for experts in decision-making.

Since the 1960s, there has been increasing recognition of the complex and integrated properties of natural resource issues. Indeed, it is suggested that scientific findings from as early as the 1960s have been forcing forestry professionals to rethink their approaches to forest management (Brooks and Grant 1992). The "organic-model" view is characterized by emphases on networks, processes, interactions, and dynamic properties of forest ecosystems. It has less emphasis on mechanistic approaches to forestry and greater emphasis on the people connected to forests. Similarly, the "new resource management" paradigm takes a relatively biocentric view that reflects a more environmentally holistic way of thinking about natural resources (Shindler and Cramer 1999). It is characterized by notions of sustainable development, harmony with nature, skepticism, and strong emphasis on public involvement in decisionmaking (Bengston 1994). These contemporary views of forest management share some similarities with sustainable forest management, a concept that is receiving greater emphasis in decisionmaking. Here, the focus is largely on the balance among economic, environmental, and social equity objectives in the context of forest management.

These paradigm shifts have been traced to shifts and expansions in environmental values in the United States, Canada, and other post-industrial nations (Koch and Kennedy 1991), as measured by public opinion polls, surveys, consumer trends, content analysis of news media, and other social science methodologies (Dunlap 1992, Shindler et al. 1993, Bengston 1994, Bengston and Xu 1996, Beckley et al. 1999, Shindler and Cramer 1999, Dunlap et al. 2000, Robson et al. 2000). Nonproduct benefits of forests, such as recreation, amenities, nature conservation, cultural heritage, and environmental protection have become more important to society since the 1970s (Koch and Kennedy 1991). These changing values helped shape environmental legislation since that time, for example the National Forest Management Act of 1976.

The power to communicate social values related to forests and the environment is now more widely and democratically distributed throughout an urban, post-industrial society (Koch and Kennedy 1991). At the same time, information about social values and their transitory nature comes in many forms, including the media and policymaking processes. Thus, it is not surprising that some forest managers feel unprepared for dealing with multiple, often conflicting values related to forests. Forestry agencies have had, and continue to have, difficulties making transitions from "machine-model" to the more inclusive and holistic "organic-model" of resource management (Kennedy et al. 1998). And yet, a better understanding of changing values is relevant for public and private land managers, whether they are trying to meet consumer needs, respond to pressure from environmental interest groups, or meet the needs of their clients.

For most post-industrial societies, values enter into forest management via two modes: populace and governance. In the populace mode, citizens, including direct and indirect users of forests and forest resources, reveal their values toward resource management through their choices, behaviors, attitudes, and opinions. Values are reflected in where they go, what they do when they get there, and how they do it. These behaviors and actions result in direct and indirect impacts to forest resources. Forest managers respond to these actions and behaviors in a variety of ways; for example, they may modify treatments to enhance wood quality to meet consumer preferences, or restore meadows that were degraded by heavy recreational use. In the governance mode, values enter into resource management through politics, policymaking, regulations, public participation in decisionmaking, legal appeals, political appointments, public agency contracts and collaborations, and other institutional means. These modes are not mutually exclusive; however, people may turn from one mode to the other if they do not see their values sufficiently reflected in forest management outcomes. This is occurring in the Pacific Northwest, defined as western Oregon and Washington, coastal British Columbia, and southeastern Alaska, where many public land management projects are held up in litigation.

Shindler and Cramer (1999) suggest that post-industrial society has altered individual value structures among citizens such that "higher order" needs (e.g., quality of life) have begun to supplant more fundamental subsistence needs as the motivation for societal behavior. They suggest that in the future, natural resource policy considerations will be determined by how well important political and social institutions, including land management agencies and constituent groups, respond to shifts in values.

3. Perspectives on Values

This section provides a more in-depth discussion on the definitions of values introduced earlier. Values and discussions about defining values have been at the center of academic debate throughout history. Although many debates remain within their respective disciplines, some debates transcend disciplinary distinctions. For example, discussions about values now occur among economists, philosophers, and sociologists because they recognize that complex environmental problems necessitate multi-disciplinary approaches (Bengston 1994, Brown 1984, Brown and Manfredo 1987).

Values are defined as preferences people share for certain types of outcomes in their lives, and for certain types of conduct (Ball-Rokeach and Loges 1992). Barbour (1980) defines values as general characteristics of a thing or circumstance that a person views with favor, believes is beneficial, and is inclined to promote. In another definition, a value is "an enduring belief that a specific mode of conduct or end-state of existence is personally or socially preferable to an opposite or converse mode of conduct or end-state of existence" (Rokeach 1973: 5). A distinction is often made in the literature between values *about* things and values *of* things (Brown and Manfredo 1987). For instance, definitions from philosophy emphasize that value is not just the worth of an object or goods, but also things that are right, obligatory, beautiful, or true (Edwards 1967). These definitions of values share a common emphasis on preferences for something positive that emanate from individuals and groups.

Sometimes the ambiguity surrounding the concept of values reflects the need to distinguish between individual and social values. Rokeach (1968) defines social values as preferential end-states of existence that include the larger society. The idea is that these desired end-states reflect a broader society, not just the individual. A more simplified definition is that social values are goods, services, or ideals that large groups of people will make sacrifices to achieve (Koch and Kennedy 1991).

Brown (1984) provides a typology that shows four ways that individuals and groups express values for individuals or groups:

	Source of value assignation	
Constituency	Individual	Group
Individual values	Case 1	Case 3
Group values	Case 2	Case 4

He describes that in case 1, the purchases of goods and responses to surveys are typically expressions of how an individual assigns value. Similarly, an individual can indicate the value he or she assigns to something on behalf of another individual; for example, a trustee administers a trust on behalf of a beneficiary. In case 2, individuals express opinions about the value they assign to things for groups of individuals. For instance, a person may state that maintaining an area as wilderness is of unmeasurable value to society. In case 3, group expressions of assigned values are derived from two or more individual expressions; for instance, parents advise a child how to responsibly make a campfire. In case 4, group expressions of assigned value on behalf of groups are typically reflected in market prices, election results, and survey results. Understanding the distinctions among the ways in which values are expressed by individuals and groups can clarify discussions about values.

Attitudes and values are closely related and often confused. An attitude is a learned predisposition to respond to a particular object in a generally favorable or unfavorable way (DeLamater 1992). Described in a slightly different way, attitude is a disposition, an antecedent, or determinant of behavior (Ajzen and Peterson 1988 in Taylor and Douglas 1999). It is a disposition to respond favorably or unfavorably to a commodity, condition, or event (Taylor and

Douglas 1999). Often we rely on measures of attitude as proxies for social values because values are difficult to define and measure, and can be misunderstood. This may be done through public opinion polls and other measures of attitudes. Values may have a temporary or transitory nature, as opposed to attitudes that may be more enduring. For instance, from an economist's perspective, a value exists only at one point in time—i.e., the value of lumber is today's price. Other social values, such as those related to culture and spirituality may be less transitory. However, even those values that appear more enduring are subject to change as context changes.

3.1. Three Properties of Values

Another way to understand the concept of values is to consider three properties that relate to the origination, relativity, and interaction of values. The first property is that values originate from human beings. Brown (1984: 233) states that "value is not an intrinsic quality of anything—rather, it emerges from the interaction between a subject and an object." People may state that a particular natural feature or animal has an intrinsic value, but that is a human value about that feature or animal. Similarly, values for which foresters manage forestland do not emerge from the land itself. Instead, public and private forestlands are managed for expressions of values that originate in the social system (Koch and Kennedy 1991). For example, commodity, wilderness, and endangered species values originate in and are assigned value by society, not by the land itself. Koch and Kennedy (1991) suggest that social values related to forests exist in interrelated sets of four major systems, represented by four interconnecting circles (Figure 1). Values related to forests are communicated by the economic, social, and political systems, but they originate in the social system. For instance, price of wood products and recreation expenditures are values communicated by the economic system. Legislation, taxes, and budget allocations reflect values communicated through the political system. However, values that individuals and groups assign to forest products and services originate from humans in the social system.

The second property of values is that they are relative. Values are not held in isolation; they are parts of a value system. Rokeach (1973: 5) states that a value system "is an enduring organization of beliefs concerning desirable modes…end-states…along a continuum of relative importance." Given that an individual may have a number of values as part of a system, it may be possible to rank values. However, ranking values can be difficult because there can be conflict among an individual's own values. Making tradeoffs as part of ranking may or may not be possible, depending on how absolute a value is. Context is also important for understanding an expressed value. As Brown (1984) suggests, a value is neither a concept held by the subject nor something attributed to the object, but merely that which arises from the preference of a

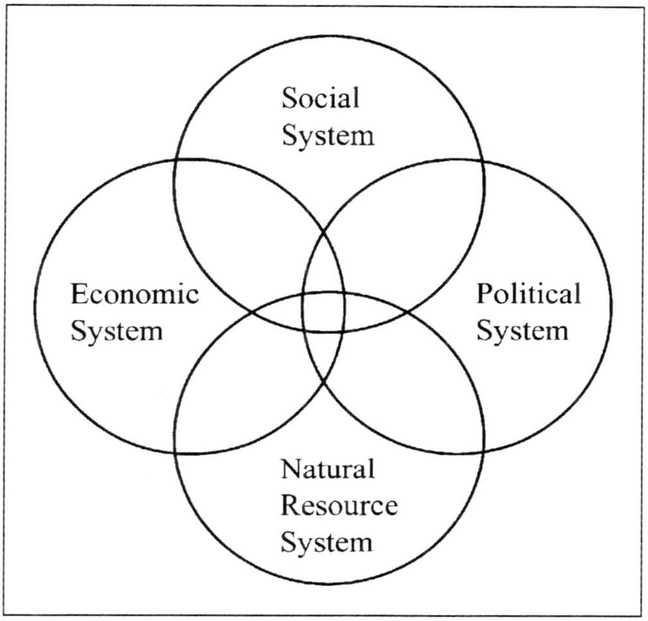

Figure 1. Systems that originate and communicate social values related to forests (Koch and Kennedy 1991).

subject for an object in a given context. Thus, a value that rises to the surface in a particular discussion about forests may be more important to an individual or society than other values that do not. However, understanding context is important for understanding the relative importance of values. For instance, one method of valuation may reveal an economic value that a group assigns to a forest resource, but spiritual or cultural values associated with that resource might be higher for the group at a different time, or in a different context. As values change with context, they also change with time. These contextual and temporal dimensions are important to understanding the relative nature of values.

And third, although the relativity of values suggests a relational quality, values do not harmoniously blend in ways that direct forest managers down a single, logical decisionmaking path. Instead, the pluralistic and conditionality of values implies that conflict is unavoidable. Some value conflicts are revealed in courts over management activities. Others are revealed by interactions among diverse stakeholder groups or neighboring landowners. Lack of understanding about the interactions and conflicts among values has prolonged debates in forest management (Daniels and Walker 1999). Individuals and groups are using an increasing range of mechanisms to express their values, such as markets, courts, public protests, and voting. Although there appear to be increasing opportunities for communicating values in political and social systems (Koch and Kennedy 1991), not all people or stakeholder groups are sufficiently engaged in those systems to express their values. And, social values that emerge through conflict and controversy surrounding forest management may not be held by the majority of society (Force and Fizzell 2000). Forest managers must contend with a range of often conflicting values from their clients, board members, stockholders, constituents, the public, and other stakeholders affected by the land management.

Inadequate consideration of the human origin, relative nature, and relational (potentially conflicting) properties of values likely contributes to controversies surrounding forest resource management. A better understanding of these three properties may help sharpen language and clarify assumptions related to forest management decisionmaking. For instance, an assessment of a proposed management action may reveal tradeoffs between providing recreation services and improving fish habitat in a particular location. Examining such tradeoffs while recognizing that the values associated with management options come from humans, are relative to other values, and likely conflict may clarify or broaden the range of management options.

3.2. Forms of Values

Values related to forests and forest resources can be divided into two categories: ecosystem function values and social values related to forests and forest resources (Table 1). The term "ecosystem function values" is used to describe the role of forest ecosystems in maintaining a healthy and resilient setting for all life (Bengston 1994). Also referred to as non-preference values, ecosystem function values are the biological and physical interactions among nonhuman entities that humans may not even know exist (Andrews and Waits 1978). One objective of biological and physical sciences is to understand ecosystem function values.

Social values, also referred to as preference values, are divided into held and assigned values. Held values are modes of conduct or behavior (e.g., bravery, loyalty), end-states (e.g., freedom, happiness), or qualities (e.g., beauty, symmetry) (Brown 1984). They are what economists may refer to as beliefs (Haynes and Weigand 1997) and what philosophers may consider ethical norms. Held values may have direct influence on attitudes, preferences, and beliefs (McFarlane and Boxall 2000).

Assigned values relate to the standing of the object relative to other objects. Thus, an assigned value reflects the perception and held values of the assigner(s) and the context of the valuation (Brown 1984). Assigned value can be measured in many ways and expressed in monetary and nonmonetary terms.

Table 1. Forms of values.

	Ecosystem function values (non-preference)	**Social values** (preference values)	
		Held values	Assigned values
Description	Biological and physical relations of one nonhuman entity to another that exist whether or not humans are aware of them.	A personal conception of the preferable that guides action and numerous modes of conduct, end-states, or qualities that might be part of that conception.	The preference relation between a person and object(s), given the person's held values, the relative importance to one or more other objects, and the context of the valuation.
Literature	Andrews and Waits (1978), Bengston (1994)	Brown (1984), Rokeach (1973)	Brown (1984)
Example	Cover for deer habitat.	Moral values (e.g., honesty), competence values (e.g., rationality), personal values (e.g., safety), and social values (e.g., camaraderie).	A person may assign subsistence, economic, or cultural values to a huckleberry field.
Methods	Applied biological and physical science methods.	Abstract and difficult to measure. May be easiest understood as a value system or ethical system. Closely related to human beliefs and norms.	Theoretically measurable through economic and noneconomic social science methods.

> For a given group of people, there are as many possible assigned values for something as there are valuation contexts and held value systems. Each assigned value is only an indication of value, and one may be more appropriate than another in a given decisionmaking situation (Brown 1984: 244).

Again, the three properties of values—that they are of human origin, are relative, and can conflict—are particularly apparent in assigned values.

The distinctions between ecosystem function values and social values can be confusing, but are important for achieving compatible forest management. For instance, studies on habitat for endangered species may be motivated by societal values for conservation or preservation, but such studies do not result in information about conservation or preservation values. Instead, the studies produce information about ecosystem function values, which in turn may shape social values related to conservation. Similarly, information about social values, particularly assigned values such as those related to recreational activities or wood commodities, are integral to understanding compatible forest management. Social values direct the choices people make, what they say, and what they do. These choices, words, and actions directly and indirectly affect the state of forests, as well as the decisionmakers and institutions that shape forest management and policy. Understanding ecosystem function values and social values are important to understanding how forests function, how people interact with forests, and what people expect from forest management.

3.3. Characterizing Values Related to Forests

Characterizations of values related to forests and forest resources have emerged out of philosophical and scientific discussions about what forests mean to people. Often, these characterizations combine ecological function values and social values. It is common to see them presented as descriptive categories of values that pertain to the environment and natural resources. For some authors, categorization implies mutual exclusivity across value categories (Xu et al. 1997). Others acknowledge an interwoven nature of values, and suggest that categories reflect dynamic and interconnected properties of values. Most characterizations share similarities; few claim to be exhaustive lists of all values related to forests, natural resources, and the environment. For instance, Rolston and Coufal's (1991) characterization of values includes 10 categories: life support, economic, scientific, recreation, aesthetic, wildlife, biotic diversity, natural history, spiritual, and intrinsic. Similarly, Stankey and Clark (1992) identify six forms of social values: commodity values (e.g., timber, range, minerals), amenity values (e.g., lifestyle, scenery, wildlife, nature), environmental quality values (e.g., air, water quality), ecological values (e.g., habitat conservation, sustainability, threatened and endangered species, biodiversity), public use values (e.g., subsistence, recreation, tourism), and spiritual values.

Characterizations of values serve several purposes. First, characterizations contribute to increased awareness of the diverse range of values that relate to forests and forest resources. Understanding this diversity is an important step in dealing with conflict and developing processes for mutual learning (Daniels and Walker 1999).

Second, value characterizations can be juxtaposed against characterizations of resource management activities, institutions, and other factors to identify knowledge gaps and develop research agendas (Luzadis et al. 1997). Rather than view values in isolation, it may be revealing to consider the interactions of values with other factors, such as ecosystem management principles and institutions.

Third, characterizations of values can be used as frameworks to structure discussions about specific values in the context of management. For instance, Bengston (1994) and Xu et al. (1997) distinguished four distinct ways in which people value forests and forest ecosystems: economic, life support, aesthetic, and moral value. These four categories became the basis for a trend analysis of values. They found that the shift and expansion of social values related to forestry suggests we are moving away from relatively easy-to-measure economic values toward values related to life support and morals that are much more difficult to measure and are poorly understood. Characterizations also can be used to help define the potential influences of values on particular management scenarios. For instance, decisionmaking pertaining to forest management activities, such as stand thinning, watershed restoration, road building, or revised policy on nontimber forest permits, can be examined across a selected range of values relevant to a particular issue. In this way, characterizations may guide policy evaluation by revealing areas where more information may be needed.

4. Assessing Social Values

Shifting and expanding social values will continue to challenge our ability to achieve compatible forest management. Continuous attention will be needed to understand how values change, how they are related, and how management actions affect them. If compatible forest management is a management goal, then the identification, development, and use of methods for assessing values may be necessary to determine if compatibility has been achieved. Without such value assessments, compatibility remains an assumption, or an unintended outcome, rather than an intentional pursuit. The selection of methods used for assessing values depends on the particular resource management issues and objectives. Issues and objectives can range from broad, complex, and conflictual, to fairly narrow and involving few stakeholders. Some methods for assessing values will reveal a variety of expected and unexpected values. Others may focus on specific values, such as recreation or wildlife values. Social and biological sciences use an assortment of methods for measuring and assessing social and ecosystem function values related to forests and forest resources. This section presents a general overview of the social science methods for assessing social values related to forest resources.

Hall and Bigler-Cole (2001) identify several methods for understanding stakeholder values, including methods for generating information from public hearings, mail comments, informal public contacts, workshops, focus groups, advisory boards, questionnaires, spatial or visual techniques, place assessments, and secondary sources. Group settings, such as public hearings, workshops, focus groups, and advisory boards, tend to be unrepresentative of the general population, but are inexpensive and can result in good quality and quantity of information about values. Hall and Bigler-Cole (2001) say that scientifically designed questionnaires can be representative of a range of stakeholder values, although they can be costly to administer. Other methods, such as spatial techniques, secondary data, and qualitative place assessments, are variable in their representativeness, expense, and quality of information. For each type of approach, Hall and Bigler-Cole (2001) provide examples of resource management studies, most of which were conducted in the Pacific Northwest of the United States.

4.1. Economic Valuation

Distinctions are often made between economic and noneconomic methods of assessing values related to forests, although some tools such as surveys and questionnaires are used in both. The distinctions likely reflect conceptual differences between economics and other social sciences. For instance, economists view value as observable, whereas social psychologists view values as attitudes inferred through behaviors (Heberlein 1988). Economists measure value preference via assigned economic value. This is done through observation of actions (through assessing price and time commitments) and through expressions about what people are willing to pay, willing to accept in compensation, and what they value as important (Taylor and Douglas 1999). Stated preference methods are often criticized for the hypothetical nature of the questions posed to consumers. In some cases, however, people's stated preferences may be good predictors of how they will behave in marketplaces (Loureiro et al. 2003). Where markets for ecosystem products exist, transactions between buyers and sellers reveal economic values.

The challenge is that no markets exist for many ecosystem goods and services (Haynes and Weigand 1997). In these situations, economists rely on nonmarket valuation techniques, such as contingent valuation, and methods such as travel cost and hedonic models. These methods adopt psychological and sociological approaches to valuation through the use of surveys and direct questioning of users, stakeholders, and the public at large. For instance, surveys can be used for developing economic measures, such as willingness to pay for recreation or fishery resource protection (Taylor and Douglas 1999). However, the focus of nonmarket valuation is on assigning a monetary value to an object or experience.

Economists generally recognize that monetary measures are convenient and easily communicated ways for assigning values, but may be weak approximations of social values (Cummings et al. 1986). Many forest managers, however, are trained to respond to economic values revealed through markets. Understanding such values is important to designing management objectives that consider profits, the time value of money, and stockholders. Economic values, however, only represent a small set of social values related to forest resources. Most public forest managers and some private forest managers consider a broader range of social values.

There is some concern that establishing values through market prices and nonmarket valuations may not reflect the most relevant social values pertaining to a particular issue. This is echoed in the concern that the economic marketplace model and the rational scientific model continue to be the dominant approaches for resource management decision support (Stankey et al. 1992).

4.2. Noneconomic Valuation

Social sciences have an assortment of qualitative and quantitative methods, such as surveys, in-depth interviewing, and observation, that are available for measuring noneconomic values. Questionnaires and interviews can be used to measure attitudes and assess value orientations of forest users and stakeholder groups (Steel et al. 1994, McFarlane and Boxall 2000). Sometimes qualitative methods, such as in-depth interviewing and ethnographic work, can complement quantitative approaches, such as charting market prices. For instance, data from interviews with gatherers of nontimber forest products (NTFPs) about why they collect NTFPs can be combined with data on permits sold to harvesters and the economic benefits from selling NTFPs that were collected under a permit. The combined information may provide a more complete picture of the value of NTFP gathering. Hansis (1995) also used qualitative methods to understand values in his study on the acceptability of clearcutting to rural and urban residents in the Pacific Northwest. Likewise, Fredrickson and Kerr (1999) applied qualitative methods in their study of spirituality and other meanings that people derive from, and attach to landscapes. As with economic valuation, noneconomic valuation uses secondary data in value research. For instance, content analysis is a research technique for making inferences from text by systematically identifying and analyzing words, phrases, and content of text. Xu et al. (1997) used newspaper text pertaining to national forests to assess trends in values related to forests.

4.3. Surveys

In survey research, scientists examine values associated with features of, activities on, and goods and services from forests by asking respondents about the importance of an attribute or the satisfaction they experience from forests and forest resources. A survey is a system for collecting information to describe, compare, or explain knowledge, attitude, and behavior (Fink 1995). Several examples are provided below to illustrate how surveys are used as decision-support tools for developing measures of public perceptions, attitudes, beliefs, and values about the environment and forest resources. In the Pacific Northwest, surveys have been a common tool for assessing environmental values. For instance, researchers conducted a survey of residents in Washington state to understand the role of environmental values on the relation between participation in outdoor recreation and pro-environmental behavior (Bright and Barro 2000). Mail surveys to residents of Oregon and the

United States about timber harvests, wildlife management, and economic development was designed to assess policy preferences (Shindler et al. 1993). Information from surveys can be integrated with data from economic valuations. For instance, a study about management of the Trinity River in northern California integrated willingness to pay and satisfaction-importance measures to assess economic and social values (Taylor and Douglas 1999). In all survey research, the research design should guard against threats to reliability and validity of data. A strong sampling scheme is one way to avoid bias.

4.4. Opinion Polls

Public opinion polls are another type of survey that typically ask people about current issues, policies, and institutions. Although they can be used to assess preferences and attitudes about the environment and natural resource management, it is questionable whether public opinion polls serve as good proxies for many social values related to natural resources. Some researchers have noted a sizeable discrepancy between social values related to forests as measured by public opinion polls and social values exhibited by people's behavior. The opinion polls indicate increasing public support for environmental protection whereas public behavior indicates an increasing demand for forest products (Shindler and Cramer 1999, Force and Fizzell 2000). People may say they disapprove of cutting trees, but they are purchasing larger homes that require more wood products. Some argue that public opinion is not useful as a decisionmaking tool if the public does not take responsibility for the consequences of its views. Public judgment is considered a more relevant form of public opinion about complex resource management problems (Yankelovich 1991, Force and Fizzell 2000). It includes more thoughtfulness, more weighing of alternatives, genuine engagement with issues, and emphasis on the normative, valuing, and ethical side of issues. Public judgment is achieved less through measures from opinion polls and more through communication and mutual learning between resource managers and the public. A better understanding about how people form judgments and accept ideas, decisions, and outcomes (Shindler et al. 2002) may help fill some of the gaps between public opinion, public values, and public behavior.

5. Participatory Approaches and Compatible Forest Management

Although the previous section describes an assortment of commonly used methods for producing information about social values, questions remain about the suitability and applicability of such information to many forest management issues. Is the information about values relevant to specific management issues? And, how, if at all, is information about values being integrated into management processes? In many situations, it may no longer be sufficient to use broad-scale, often noncontextual, information about social values as yardsticks for gauging public support for resource management decisions (Shulman and Penman 1994). These concerns reflect a growing awareness of the need for more upfront integration of values into research and management processes, particularly at the local level (Fight et al. 2002). They are also echoed in calls for alternative approaches to social science research for certain management situations (Kruger and Sturtevant 2003).

Participatory approaches to forest management have their roots in international fields related to community development, agriculture, and rural resource management where they have been used since the 1980s to address the needs of marginalized, rural, resource-based populations, and resource management agencies (Korten 1980; Chambers 1983, 1984; Cernea 1991). "Participatory approaches" is a general term used to describe an assortment of related methods, including participatory action research, rapid rural appraisal, participatory rural appraisal, and rapid assessment procedures. All of these use applied social science research methods and share the following characteristics: (1) a local focus, with an orientation toward explicitly acknowledged problems by local people and officials; (2) an emphasis on learning and action, with minimal time gaps between data collection, analysis, and feedback, and direct translation of knowledge into action; (3) a concern about process, including a communication strategy and clarity about roles and responsibilities (Found 1995, Barton et al. 1997); and (4) a minimization of distance between researcher and the people and issues being researched.

Participatory approaches to forest management and research involve local people in a cyclical process of understanding, adaptation, and action. Because participatory approaches are highly contextual and focus on participant involvement and mutual learning, they are promising methods for integrating values into forest management, particularly at the local level. Participatory approaches involve stakeholders in identifying issues, defining questions, designing and implementing research or management actions, collecting and analyzing data, monitoring and evaluating results, and using information to meet particular goals or objectives. Local participants function not as respondents to surveys or questionnaires, but as active participants who share knowledge, appreciations, and ideas.

Unlike valuation methods that are largely descriptive and broad scale (e.g., opinion polls), participatory approaches engage stakeholders in specific judgments about tough forest management choices. That is not to say that stakeholders make final decisions about forest management on public or private lands. Instead, by participating throughout management or research processes, stakeholders engage forest managers and researchers in mutual learning that can heighten appreciations for possible intended and unintended outcomes. Through these interactions, participants express their values in the context of a specific resource management issue, thereby increasing the likelihood of achieving compatible forest management. A common tenet of participatory approaches is that stakeholders who are involved throughout the science and management processes are more likely to develop a sense of commitment to the processes and ownership in the outcomes.

Identifying stakeholder interests and values continues to be an important theme in the literature on participatory rural development and resource management (Beckley et al. 1999, Bass 2001). Stakeholders are not limited to resource users, but may include community members, agency representatives, practitioners, and other stakeholders who may contribute knowledge. Bass (2001) describes four reasons why the inclusion of stakeholders is important to balancing economic, environmental, and social objectives related to forest management: (1) to ensure forestry standards and objectives are credible, (2) to make use of a broad range of ideas, skills, and inputs, (3) to evaluate trade-offs given current conditions and knowledge about an particular issue, and (4) to build a stronger foundation of stakeholder trust, accountability, and commitment.

In the Pacific Northwest, stakeholder groups related to public forestlands are numerous, owing to the relatively large percentage of publicly owned land in the region. In Oregon and Washington, for example, 57% of forestlands are managed by public agencies (Haynes 2003). Thus, most residents in the region engage in activities related to, or receive benefits from, public forestlands. Stakeholder groups value an increasing range of goods and services from forestlands, including clean water, clean air, wood and nonwood products, wildlife, scenic vistas, recreational opportunities, and other amenities.

Because public land managers are charged with managing for diverse uses, most participatory approaches will have more relevance to public land managers than private landowners. Issues with high stakeholder interest that may be appropriate for participatory research and management on public lands include nontimber forest products management, habitat restoration, trail maintenance and other recreation management issues, and issues surrounding American Indian treaty reserved rights. For instance, harvesters of nontimber forest products often have unique, locally specific, knowledge of forest resources and harvesting practices. They represent diverse stakeholders, including American Indian tribes, residents of rural communities, migrant workers, and recreationists. Nontimber forest product management has received relatively little institutional support from federal and state agencies and maintains high levels of management uncertainty. These features make it an appropriate topic for jointly framing questions, incorporating various forms of knowledge, and engaging in the type of mutual learning that occurs through participatory approaches.

Participatory approaches may not be suitable everywhere. They may be most appropriate for issues that have high levels of local context where there is institutional support to engage stakeholders in interactive, dynamic, and nontraditional processes. Participatory approaches will be less applicable for highly conflicting, broad-scale issues. Although participatory approaches may appear less relevant to private and industrial landowners, private landowners are increasingly interacting with neighbors and local interest groups in response to growing public concern about forests. Elements of participatory approaches, such as processes for mutual learning, may be relevant to private landowners who are interested in dealing with conflict. Participatory approaches on public lands will be increasingly challenged to demonstrate how lessons learned from participatory processes at the local level relate to management issues at regional levels. There may also be a need to relate lessons to other regions where private and industrial land ownership predominate, but where similar social values related to forests exist.

6. Discussion

This chapter suggests that understanding social values related to forests and integrating such knowledge into forest management are necessary for determining if compatible forest management is achievable. Indeed, without evidence of adequate attention to social values, statements about compatibility may be unjustified. Opportunities for greater integration of social values into forest management decisionmaking exist. However, several challenges remain for public and private forest management and research.

First, social values related to forests and forest resources may be best understood in the context of a specific resource management issue or objective, particularly where compatible forest management is concerned. To identify or develop appropriate methods for assessing social values, there needs to be clear understanding of the nature and complexity of a forest management issue. People hold and assign numerous values that change in relative importance, given differing circumstances. Thus, only by studying values in context will we be able to understand the relativity of values. Some methods, such as participatory approaches to management and research, may be particularly well suited for public land management issues with high local context and minimal or moderate conflict. By explicitly integrating values into problem-solving and action, participatory approaches may be appropriate for some forms of compatible forest management.

Second, social values do not come from forest resources or animals; they come from humans, and humans constantly change. Thus, values are not only relative but dynamic. For land managers to understand social acceptability of forest management actions they must consider public engagement to be a continuous process (Stankey et al. 2003). This is also true for understanding the dynamic nature of values related to forests and forest resources.

Some social values are enduring, however. For instance, indigenous groups, such as American Indian tribes in the Pacific Northwest, have enduring cultural values pertaining to the gathering, processing, and use of non-timber forest products such as huckleberry (*Vaccinium* spp.) and bear-grass (*Xerophyllum tenax* (Pursh) Nutt.). By better understanding these values, federal land managers are increasingly collaborating with tribal groups to develop and implement forest management actions to enhance the vitality of some forest resources. The existence of both enduring and dynamic social values related to forests and forest resources suggests that we treat compatible forest management as an adaptive and dynamic process.

And third, integrating information about values into decisionmaking is all the more difficult when values are conflicting, changing, or poorly understood. Improved understanding about the determinants of change will help set the context for understanding values and evaluating management options. By understanding determinants of change in a broader context, we may be better equipped to apply lessons learned from the interface of values and forest management at the local level to broad-scale issues in forest management.

Given the diversity of often-conflicting and ever-changing values related to forests and forest resources, the challenge of achieving compatible forest management is formidable. What will it take to develop management actions that do not negatively impact any value? Through better understanding of values—where they come from, how they change, and who assigns them—scientists, managers, and the public may be able to expand the range of forest management options. By expanding the pool from which forest management options are selected, we may be more likely to find an option that achieves compatible forest management. Therefore, where compatible forest management is a goal, the identification, development, and use of methods for assessing social values in the context of specific management issues may be necessary to achieve compatibility.

7. References

Adams, D.A. 1993. Renewable resource policy: the legal-institutional foundations. Washington, DC: Island Press. 557 p.

Andrews, R.N.L., and Waits, M.J. 1978. Environmental values in public decisions: a research agenda. Ann Arbor, MI: School of Natural Resources, University of Michigan. 90 p.

Ball-Rokeach, S.J., and Loges, W.E. 1992. Value theory and research. In: E.F. Borgatta, and M.L. Borgatta (Eds.), Encyclopedia of sociology. (Vol. 4, pp. 2222-2228). New York: MacMillan Publishing Co.

Barbour, I.G. 1980. Technology, environment, and human values. New York: Praeger. 331 p.

Barton, T., Borrini-Feyerabend, G., de Sherbinin, A., and Warren, P. 1997. Our people, our resources. Gland, Switzerland: IUCN, World Conservation Union. [irregular pagination]. Retrieved December 19, 2002 from http://www.iucn.org/themes/spg/opor.

Bass, S. 2001. Working for forest stakeholders. In: J. Evans (Ed.), The forests handbook. (Vol. 2, pp. 221-232). Mailden, MA: Blackwell Science.

Beese, W.J., Dunsworth, G., and Perry, J. 2001. The forest project: three-year review and update. *Ecoforestry*, 16(4): 10-17.

Beckley, T.M., Boxall, P.C., Just, L.K., and Wellstead, A.M. 1999. Forest stakeholder attitudes and values: selected social-science contributions. Information Report NOR-X-362. Edmonton, Alberta: Northern Forestry Centre. 24 p.

Bengston, D.N. 1994. Changing forest values and ecosystem management. *Society and Natural Resources*, 7(5): 515-533.

Bengston, D.N., and Xu, Z. 1996. Shifting and expanding forest values: the case of the U.S. national forests. *George Wright Forum*, 13(2): 10-19.

Bright, A.D., and Barro, S.C. 2000. The mediating effects of values on the relationship between outdoor recreation participation and pro-environmental behavior. In: I.E. Schneider, D. Chavez, B. Borrie, and K. James (Eds.), Diverse challenges of our times: people, products, places: third symposium on social aspects and recreation research (pp. 39-42). Tucson, AZ: Arizona State University.

Brooks, D.J., and Grant, G.E. 1992. New approaches to forest management: background, science issues, and research agenda. Part 1 of 2 parts. *Journal of Forestry*, 90(1): 25-28.

Brown, G., and Harris, C.C. 1992. The USDA Forest Service: toward the new resource management paradigm? *Society and Natural Resources*, 5(3): 231-245.

Brown, P.J., and Manfredo, M.J. 1987. Social values defined. In: D.J. Decker, and G.R. Goff (Eds.), Valuing wildlife: economic and social perspectives (pp. 12-23). Boulder, CO: Westview Press.

Brown, T.C. 1984. The concept of value in resource allocation. *Land Economics*, 60(3): 231-246.

Cernea, M.M. (Ed.). 1991. Putting people first: sociological variables in rural development. (2nd ed.). New York: Oxford University Press.

Chambers, R. 1983. Rural development: putting the last first. London: Longman House.

Chambers, R. 1994. The origins and practice of participatory rural appraisal. *World Development*, 22(7): 953-969.

Clawson, M. 1983. The federal lands revisited. Baltimore, MD: Johns Hopkins University Press. 302 p.

Culhane, P.J. 1981. Public lands policy: interest group influence on the Forest Service and the Bureau of Land Management. Baltimore, MD: Johns Hopkins University Press. 398 p.

Cummings, R.G., Brookshire, D.S., and Schulze, W.D. 1986. Valuing public goods: an assessment of the contingent valuation methods. Totowa, NJ: Rowman and Allanheld. 270 p.

Daniels, S.E., and Walker, G.B. 1999. Rethinking public participation in natural resource management: concepts from pluralism and five emerging approaches. In: J. Eldan, and J. Anderson (Eds.), Pluralism and sustainable forestry and rural development: proceedings of an international workshop (pp. 29-48). Rome: Food and Agriculture Organization.

DeLamater, J.D. 1992. Attitudes. In: E.F. Borgatta, and M.L. Borgatta (Eds.), Encyclopedia of sociology (Vol. 1, pp. 117-124). New York: MacMillan Publishing Co.

Dunlap, R. 1992. Trends in public opinion toward environmental issues: 1965-1990. In: R. Dunlap, and M. Mertig (Eds.), American environmentalism: the U.S. environmental movement, 1970-1990 (pp. 89-116). Philadelphia, PA: Taylor and Francis.

Dunlap, R.E., and Catton, Jr., W.R. 1979. Environmental sociology. *Annual Review of Sociology*, 5: 243-73.

Dunlap, R.E., and Van Liere, K.D. 1984. Commitment to the dominant social paradigm and concern for environmental quality. *Social Science Quarterly*, 65(40): 1013-1027.

Dunlap, R.E., Van Liere, K.D., Mertig, A.G., and Jones, R.E. 2000. Measuring endorsement of the new ecological paradigm: a revised NEP scale. *Journal of Social Issues*, 56(3): 425-442.

Edwards, P. (Ed.). 1967. The encyclopedia of philosophy. Vol. 8. New York: Macmillan Publishing.

Fight, R.D., Donoghue, E.M., and Christensen, H.H. 2002. Involving stakeholder communities in research on nontimber forest products. In: A.C. Johnson, R.W. Haynes, and R.A. Monserud (Eds.), Congruent management of multiple resources: proceedings from the wood compatibility initiative workshop (pp. 209-212). General Technical Report PNW-GTR-563. Portland, OR: U.S. Department of Agriculture, Forest Service, Pacific Northwest Research Station.

Fink, A. 1995. The survey handbook. Thousand Oaks, CA: Sage Publications. 129 p.

Force, J.E., and Fizzell, G. 2000. How social values have affected forest policy. In: Proceedings of the Society of American Foresters 1999 national convention (pp. 16-22). Bethesda, MD: Society of American Foresters.

Found, W.C. 1995. Participatory research and development: an assessment of IDRC's experience and prospects. A report to the International Development Research Centre. Toronto: York University. 63 p. Retrieved December 19, 2002 from http://www.iddrc.ca/evaluation/documents/participatory.

Fredrickson, L.M., and Kerr, W. 1999. Spiritual values: Can they be incorporated into forest management and planning? In: H.G. Vogelsong (Ed.), Proceedings of the 1998 northeastern recreation research symposium (pp. 239-245). General Technical Report NE-255. Newton Square, PA: U.S. Department of Agriculture, Forest Service, Northeastern Research Station.

Hall, T.E., and Bigler-Cole, H. 2001. Sociocultural factors and forest health management. *Northwest Science*, 75: 208-233.

Hansis, R. 1995. The social acceptability of clearcutting in the Pacific Northwest. *Human Organization*, 54: 95-101.

Haynes, R.W. (Tech. coord.). 2003. An analysis of the timber situation in the United States: 1952 to 2050. A technical document supporting the 2000 USDA Forest Service RPA assessment. General Technical Report PNW-GTR-560. Portland, OR: U.S. Department of Agriculture, Forest Service, Pacific Northwest Research Station. 254 p.

Haynes, R.W., Monserud, R.A., and Johnson, A.C. 2003. Compatible forest management: background and context. Chapter 1. In: R.A. Monserud, R.W. Haynes, and A.C. Johnson (Eds.), Compatible forest management. Dordrecht, The Netherlands: Kluwer Academic Publishers.

Haynes, R.W., and Weigand, J.F. 1997. The context for forest economics in the 21st century. In: K.A. Kohm, and J.F. Franklin (Eds.), Creating a forestry for the 21st century (pp. 285-301). Washington, DC: Island Press.

Heberlein, T.A. 1988. Economics and social psychology in amenity valuation. In: G.L. Peterson, B.L. Driver, and R. Gregory (Eds.), Amenity resource valuation: integrating economics with other disciplines (pp. 235-244). State College, PA: Venture.

Kennedy, J.J., Dombeck, M.P., and Koch, N.E. 1998. Values, beliefs and management of public forest in the Western World at the close of the 20th century. In: P. Schmidt [and others] (Eds.), New requirements for university education in forestry: proceedings of a workshop (pp. 15-34). DEMETER Series No. 1. Korbeek-Dijle (Bertem), Belgium: Drukkerij de Weide.

Koch, N.E., and Kennedy, J.J. 1991. Multiple-use forestry for social values. *Ambio*, 20(7): 330-333.

Korten, D.C. 1980. Community organization and rural development: a learning process approach. *Public Administration Review*, 40: 480-511.

Kruger, L.E., and Sturtevant, V.E. 2003. Divergent paradigms for community inquiry: an argument for including participatory action research. In: L. Kruger (Tech. ed.), Understanding community-forest relations (pp. 23-42). General Technical Report PNW-GTR-566. Portland, OR: U.S. Department of Agriculture, Forest Service, Pacific Northwest Research Station.

Loureiro, M.L., McCluskey, J.J., and Mittelhammer, R.C. 2003. Are stated preferences good predictors of market behavior? *Land Economics*, 79(1): 44-55.

Luzadis, V.A., Floyd, D.W., and Goergen, M.T. 1997. Values, institutions and ecosystem management: a research agenda for policy analysis. In: H.K. Cordell (Ed.), Integrating social science and ecosystem management: a national challenge. Proceedings: Conference on integrating social sciences and ecosystem management (pp. 177-183). General Technical Report SRS-17. Asheville, NC: U.S. Department of Agriculture, Forest Service, Southern Research Station.

McFarlane, B.L., and Boxall, P.C. 2000. Factors influencing forest values and attitudes of two stakeholder groups: the case of the Foothills Model Forest, Alberta, Canada. *Society and Natural Resources*, 13(7): 649-661.

Montreal Process Working Group. 1998. The Montreal Process. Retrieved January 9, 2002 from http://www.mpci.org.

Oregon Department of Forestry. 2003. Oregon Forest Practices Act. Retrieved March 29. 2003 from http://www.odf.state.or.us.

Robson, M., Hawley, A., and Robinson, D. 2000. Comparing the social values of forest-dependent, provincial and national publics for socially sustainable forest management. *The Forestry Chronicle*, 76(4): 615-622.

Rokeach, M. 1968. Beliefs, attitudes, and values: a theory of organization and change. San Francisco: Jossey-Bass. 214 p.

Rokeach, M. 1973. The nature of human values. New York: Free Press. 438 p.

Rolston, H., and Coufal, J. 1991. A forest ethic and multivalue forest management. *Journal of Forestry*, 89(4): 35-40.

Shindler, B.A., Brunson, M., and Stankey, G.H. 2002. Social acceptability of forest conditions and management practices: a problem analysis. General Technical Report PNW-GTR-537. Portland, OR: U.S. Department of Agriculture, Forest Service, Pacific Northwest Research Station. 68 p.

Shindler, B., and Cramer, L.A. 1999. Shifting public values for forest management: making sense of wicked problems. *Western Journal of Applied Forestry*, 14(1): 28-34.

Shindler, B., List, P., and Steel, B. 1993. Managing federal forests: public attitudes in Oregon and nationwide. *Journal of Forestry*, 94(6): 4-12.

Shulman, A.D., and Penman, R. 1994. Why study rangeland values? Some practices that scientists have much to answer for. *Rangeland Journal*, 16(2): 265-272.

Stankey, G.H., Brown, P.J., and Clark, R.N. 1992. Allocating and managing for diverse values of forests: the market place and beyond. In: N.E. Koch, and N.A. Moiseev (Comps.), Proceedings IUFRO international conference integrated sustainable multiple-use forest management under the market system (pp. 257-271). Copenhagen: Danish Forest and Landscape Research Institute.

Stankey, G.H., and Clark, R.N. 1992. Social aspects of new perspectives in forestry: a problem analysis. Milford, PA: Grey Towers Press.

Stankey, G.H., Clark, R.N., and Bliss, J. 2003. Fostering compatible forest resource management: the conditional nature of social acceptability. Chapter 16. In: R.A. Monserud, R.W. Haynes, and A.C. Johnson (Eds.), Compatible forest management. Dordrecht, The Netherlands: Kluwer Academic Publishers.

Steel, B.S., List, P., and Shindler, B. 1994. Conflicting values about federal forests: a comparison of national and Oregon publics. *Society and Natural Resources*, 7: 137-153.

Taylor, J.G., and Douglas, A.J. 1999. Diversifying natural resources value measurements: the Trinity River study. *Society and Natural Resources*, 12(4): 315-336.

Wildavsky, A., and Dake, K. 1990. Theories of risk perception: Who fears what and why? *Daedalus*, 119(4): 41-60.

Xu, Z., Bengston, D.N., and Fan, D. 1997. Identifying changes in forest values: a computerized content analysis. In: H.K. Cordell (Ed.) Integrating social science and ecosystem management: a national challenge: Proceedings, Conference on integrating social sciences and ecosystem management (pp. 142-149). General Technical Report SRS-GTR-17. Athens, GA: U.S. Department of Agriculture, Forest Service, Southern Research Station.

Yankelovich, D. 1991. Coming to public judgment: making democracy work in a complex world. Syracuse, NY: Syracuse University Press. 290 p.

Chapter 16

Fostering Compatible Forest Resource Management: The Conditional Nature of Social Acceptability

George H. Stankey
USDA Forest Service, Pacific Northwest Research Station,
3200 SW Jefferson Way, Corvallis, OR 97331, USA
Roger N. Clark
USDA Forest Service, Pacific Northwest Research Station,
400 N 34th St. Suite 201, Seattle, WA 98103, USA
John Bliss
Department of Forest Resources, College of Forestry, Oregon State University, Corvallis, OR 97331, USA

1. Introduction

For years, public and private resource management organizations have sought to implement sustainable policies and programs that achieve various forest management goals; yet success has proven elusive. Although such efforts have been marked by significant investments in assessing the biophysical and economic aspects of proposals, public opposition often has stymied implementation. One example—and the focus of this book—is reflected in the search for forest management policies and programs that increase the compatibility between wood production and other forest uses and values. Efforts to use timber harvesting as part of a forest management strategy that incorporates commodity production with a host of other purposes, from fire control to biodiversity enhancment, have been engulfed in controversy. At their core, such conflicts reveal the importance of social acceptability, which implies a sufficient level of public understanding and accord to permit policy implementation. Such judgments ultimately affect all forestland management, irrespective of tenure, although they are particularly salient at present with regard to public lands.

R.A. Monserud, R.W. Haynes and A.C. Johnson (eds.)
Compatible Forest Management, 453-480

Two aspects of citizen judgments regarding proposed forest management options contribute to this saliency. First, along with the complex biophysical and economic issues linked to any forest management policy, citizen judgments regarding the acceptability of those policies must be taken into account. Second, such judgments are often conditional. "It depends" accurately depicts the public sense on many issues and it reflects more than a simple reluctance or inability of people to make up their minds. These components of citizen judgments give rise to a host of questions that challenge resource managers. What does the idea of social acceptability embrace? How are judgments formed and altered? How do these judgments relate to biophysical and economic aspects of the decisionmaking process? What will people do when they are faced with situations they judge are unacceptable? What are the risks of operating in the face of adverse judgments? And in keeping with the theme of this book, how do social acceptability judgments relate to compatibility?

In this chapter we review the social acceptability concept, including how judgments are revealed, limitations to the concept, and its evolution. We discuss why the issue has gained increased attention in recent years and how the failure to address it adequately has created an "either-or" mentality apparent in much of the forest management debate. Such failures can drive the evolution of adverse judgments into legal and political edicts. We review studies investigating diverse topics such as fire management, dispersed recreation, and silviculture that reveal the key factors influencing social acceptability judgments. Although examples are drawn largely from Oregon, Washington, and Alaska, social acceptability is an international issue.

We outline examples of programs and strategies that reflect public concerns and are consistent with biophysical and economic considerations; these types of programs are needed to foster informed and supportive public judgments. We close with thoughts on the continuing challenges facing policymakers and scientists, and suggest steps to better accommodate social acceptability judgments in decisionmaking.

2. Social Acceptability: Origins, Meanings, and Function

The concept of social acceptability has roots in the work of sociologist Walter Firey (1960). Firey questioned why certain resource management practices and prescriptions persisted while others did not. He sought to identify the "mechanisms operating in every social order that forever sift and sort the resource processes...possible in a given habitat or...conceivable in a given culture" (Firey 1960, p. 19-20). His analysis suggests sustainable policies are:

- Physically possible; practices are consistent with sound scientific principles
- Economically feasible; practices generate benefits in excess of costs

- Culturally adoptable; practices are consistent with prevailing social customs, beliefs, and norms

Clawson (1974) produced an expanded but similar set of criteria for sustainable policies: (1) biological and physical feasibility, (2) economic efficiency, (3) economic welfare or equity, (4) social or cultural acceptability, and (5) operational or administrative practicality. Both frameworks acknowledge two premises. First, successful policy implementation is dependent on a variety of factors. Second, the lack of satisfactory performance in any criterion can stall implementation; each criterion is necessary but not sufficient to ensure sustained implementation (Stankey 1996).

Although considerable attention has been devoted to biophysical and economic analyses in support of forest management policies, less has focused on the issue of social acceptability. Stankey and Clark (1992, p. 23) address this imbalance, calling for research to "identify specific characteristics of stand and landscape level treatments and general social processes which enhance positive effects and mitigate negative effects on people and places." Since then, the body of social acceptability research has grown (Shindler et al. 2002), but theoretical and empirical understanding of the dynamics involved in the acceptability judgment process, and the role such judgments play in decision-making remain sketchy.

2.1. Challenges to Implementing Social Acceptability

Three challenges confront efforts to improve understanding and use of social acceptability. First, the concept lacks precision. Although social acceptability is a recurrent idea in resource management literature, the term lacks the rigorous definition and theoretical foundation that underlie social science concepts such as values and attitudes. As a result, various terms are used to describe public views regarding management practices (e.g., preferred, expected, appropriate, tolerable, and compatible). Yet these terms carry distinctive meanings and using them as synonyms for acceptability is inaccurate and confusing. For example, Clark and Gibbons (1991, p. 462-3) note that preferences are "the favored alternatives for activities, experiences, or setting attributes" while expectations are "conditions presumed to be encountered…based on previous knowledge and experience." In an effort to provide a standard definition, Brunson (1996, p. 9) proposed:

> A condition that results from a judgmental process by which individuals (1) compare the perceived reality with its known alternatives; and (2) decide whether the "real" condition is superior, or sufficiently similar, to the most favorable alternative condition. If the existing condition is not judged to be sufficient, the individual will initiate behavior—often, but not always, with-

in a constituency group—that is believed likely to shift conditions toward a more favorable alternative.

Whether this is a satisfactory definition is arguable, but it does highlight key aspects of the concept. Social acceptability

- Is a judgmental process
- Involves comparisons
- Includes knowledge and awareness that shape the judgments
- Has a behavioral component that can be individual, group, or both

A second concern is the extent to which social acceptability reflects the sum of individual judgments or some larger, collective assessment. Brunson's definition focuses on the processes through which individuals form judgments, and is the focus of most empirical studies. However, he also argues that individuals behave in ways that are both individual and collective; "Acceptability is characterized as a product of individual judgments, but it is subject to group influences and provides an impetus for group behaviors..." (1996, p. 9). Yet, limited understanding exists regarding the processes and dynamics through which assessments made by individuals become known, adopted, and expressed at the societal level. For example, as judgments move from individuals to a larger aggregate, do they take on emergent qualities not previously revealed?

Third, the ways in which social acceptability judgments are expressed, as well as how they evolve are not always clear. Virtually all policies have detractors, but many objections are little more than ill thoughts. What triggers thoughts into action? What prompts individuals to engage in collective action or to create groups supportive of their judgments (e.g., the growth of "friends of..." groups, reflecting a coalescence of interests)? With action, either individual or collective, judgments become public and can potentially mobilize political or legal action.

For example, as people assess current conditions, evaluate proposed options, and form judgments, they make implicit or explicit tradeoffs. What is judged acceptable depends on the costs and benefits individuals or groups might incur. These analyses are the basis for their response. Behavioral expressions regarding social acceptability embrace various forms: from silent acceptance, letter writing, public testimony, and forming advocacy groups, to demonstrations and civil disobedience. Perhaps the most powerful behavioral expressions are individual and collective political activities that press for legislative redress and reform. The failure to recognize and respond to public concern regarding a practice or policy early on (e.g., a letter to the ranger or a comment at a public meeting) can evolve into codified legal expressions that impose sanctions or directions that ultimately minimize or eliminate management discretion.

The factors that influence judgments on social acceptability suggest that because public expressions take various forms and occur at various scales, it is necessary to be sensitive and aware of them all. When social acceptability judgments trigger behavior, they can significantly impact organizations; thus, processes for monitoring public concerns are critical to avoiding major conflicts. Conversely, inadequate attention to individual dissatisfactions can evolve into a potent political or legal expression that dominates the decision-making process. For example, the social assessment by the Forest Ecosystem Management Assessment Team (FEMAT 1993) noted that many issues underlying public discontent in the 1990s were revealed in agency public involvement records spanning 25 years. Agency failure to adequately respond to those concerns contributed to public convictions that legal and political remedies were necessary. This failure is not limited to federal lands; similar issues underlie the evolution of various state forest practices to regulate activities on state and private lands.

2.2. Citizen Involvement in the Decisionmaking Process

Although various legislation (e.g., National Environmental Policy Act, 1969 and National Forest Management Act, 1976) ensure citizens have a visible and viable role in environmental decisionmaking, the nature of this role has long stirred debate. One argument supports a limited, consultative role for citizens because the technical complexity characteristic of natural resource issues marginalizes the value of citizen input. Because of this complexity, the argument goes, considerable technical expertise is required. Yaffee (1994, p. 294) notes, "while democratic norms suggest that each individual has the right and obligation to be represented in collective choices, as issues have become more technically complex, we have tended to elevate the role of the technical expert." Pierce et al. (1992, p. 14) describe this clash between democratic traditions and scientific complexity as the quandary facing modern industrialized societies; they ask, "How can the democratic idea of public control be made consistent with the realities of a society dominated by technically complex policy questions?"

In the face of such complexity, an implicit argument has evolved: the search for effective solutions must adhere to sound science and good economics. Public concerns are often treated as unfounded, inaccurate, or irrelevant. Yet, as discussed earlier, Firey's three criteria for enduring, sustainable policies include a social aspect. Sound science and good economics are important, but alone they are not sufficient. Judgments of social acceptability are critical and when absent, place any policy in jeopardy.

2.3. Four Arguments Support the Need for Socially Acceptable Decisions

First, few management decisions are simply a product of objective science that determines the "best" practice. Certainly, sound technical information is critical to describing alternatives, consequences, and implications. But ultimately, decisions reflect a value judgment embracing considerations deemed important by decisionmakers; science cannot resolve disputes regarding the merits of alternative value judgments. If decisions do not reflect wider public sentiment, political and legal feedback will so inform the decisionmaking institution. To garner support, decisionmakers need to treat public concerns, values, and interests as important and relevant input.

Second, accounting for public judgments in decisionmaking reflects a normative perspective deeply ingrained in American governance; i.e., the public has an inherent right of access to decisions about public resources. Wengert (1976, p. 25) argues, "increasing citizen participation is simply a matter of sound and desirable policy…this is a normative conclusion...(e.g.,) the person on whose land manmade rain falls has a right to be consulted."

Third, social acceptability judgments can influence an organization's survival. In a democracy, the absence of public understanding and support makes it difficult, if not impossible, to implement decisions. In the American political system, ultimate political power rests in the body politic. This means that the public holds veto power over public policies. Whether this power is recognized or exercised are separate issues (Friedmann 1987). Moreover, the legal authority held by agencies, such as the U.S. Department of Agriculture, Forest Service or U.S. Department of the Interior, Bureau of Land Management derive from the political power held by the body politic and delegated through the political process—and what is given can be taken back. This can occur irrespective of the validity of public concerns. In extreme cases, actions deemed wholly inconsistent with public values can trigger an agency's demise (Clarke and McCool 1996).

The previous points portray social acceptability as constraining decisionmaking. A fourth point, however, is that addressing social acceptability presents an opportunity to inform, influence, and educate all players—citizens, scientists, resource managers, policymakers—in short, to promote mutual learning (Michael 1995).

Nonetheless, Vining and Ebreo (1991) warn that the notion of "middle ground" can be confounding. Resource management agencies may define "the middle" differently than the public. Also, the oft-cited idea that agency plans must "be about right, because everyone is against them" is a bizarre, groundless perspective. Plans and policies lacking any constituency are doomed.

Nonetheless, efforts to achieve socially acceptable decisions create opportunity for constructing a positive relationship between agency and citizen. Working to achieve social acceptability presents an opportunity to discuss,

***Box 1.* Middle Ground Holds Opportunities for Compatibility**

An "either-or" mentality regarding natural resource management often results from protagonists at opposite ends of the preserve-develop spectrum dominating the policy agenda. Such polar positions are fed by differing ideologies and strategic concerns that act to suppress accommodation and compromise. However, diverse opinions and experiences exist between the poles: some inclined in one direction regarding forest policies, some in another, and many uninvolved, disinterested, or unaware. Here lie opportunities for moving beyond the "either-or" mentality, and where exploring opportunities for compatibility among forest uses and values are greatest.

debate, and learn about the complex issues at hand, as well as the concerns and perspectives of other interests (Yankelovich 1991, Wondolleck and Yaffee 2000).

2.4. Compatibility and Its Link to Social Acceptability

A traditional perspective is that conflict over natural resource management arises from inadequacies or disagreements regarding the technical or scientific grounds for policies and programs. Accordingly, development and dissemination of science-based information is seen as key: problems framed as technical or scientific presuppose technical or scientific solutions. And, indeed, review of the forest resource challenges facing the Pacific Northwest (western Oregon, and Washington, coastal British Columbia, and southeastern Alaska) reveals numerous questions for which we lack scientifically robust answers.

However, although improved understanding of technical and scientific questions is necessary, it will not resolve the social conflicts confounding modern resource management. In our view, framing natural resource conflicts as technical or scientific problems amounts to problem misspecification. Such misspecification appears common among natural resource organizations and within the scientific community. It has come to public attention with increasing frequency as scientific experts have been drawn into public policy debates (Jasanoff 1990). In public hearings, newspaper editorials, campaign speeches, and activist rallies, one can observe the technical rationality of the experts collide head-on with the cultural rationality of the citizenry. Whether one of these is more objective than the other is debatable. But we believe that there is a demonstrable need for greater understanding of ecosystem structure and function, *and* the interactions of humans with those ecosystems.

Political scientist Frank Fischer argues that citizens and scientists bring different rationalities to public policy debates. Following the work of Plough and Krimsky (1987), Fischer contrasts "technical rationality" (with its reverence for empirical evidence, the scientific method, and expert judgment) with "cultural rationality," which "is geared to—or at least gives equal weight to

***Box 2.* Public Preference in Natural Resource Management Remains a Contentious Issue**

The forest practice of clearcutting exemplifies how the rational of forestry professionals contrasts with that of the surveyed public. An article in the *Journal of Forestry* in 2000 summarized a decade of public opinion polling on the acceptability of clearcutting (Bliss 2000). It presented evidence that most Americans find the practice unacceptable, especially on public land. This was found to be true in several regions of the country, and among forest owners as well as non-owners. Reasons given for this overwhelming rejection included perceived environmental damage, negative aesthetic impacts, and lack of trust in the forestry profession. The author recommended that the profession incorporate social research results, such as those reported, into practice and policy.

The *Journal of Forestry* article received considerable written response from forestry professionals. The notion of allowing public judgment to influence the practice of forestry was condemned by many as detrimental to the profession. One forester concluded that encouraging the forestry profession to seek greater understanding of public values "accomplishes little more than retreat from science-based forestry...a dumbing down of forestry." Another wrote, "The Society of American Foresters professes to serve society as professionals based on sound science. [The author] seems to contradict this notion by suggesting that the profession buy into public perception and back away from sound science rather than trying to influence public perception." To these writers, only scientific arguments regarding the biophysical impacts of clearcutting were valid. They reflect confusion between science-based and science-informed decisions. The suggestion that public preference or cultural rationality has an essential role in natural resource policy development clearly is a contentious issue within the forestry community.

—personal and familiar experiences rather than depersonalized technical calculations" (2000, p. 132). Cultural rationality is, in Fischer's view, "the rationality of the social lifeworld" (p. 133); that is, the rationality that commonly drives public understanding and debate.

Cultural rationality does not reject technical arguments, but assesses them within a broader context of peer group values, social relationships, and the particulars of specific places and situations. Cultural rationality emphasizes the importance of process as well as substance in policy development; it focuses on the method for developing policy alternatives, who participates, how decisions are made and by whom, and whose science is brought to bear.

Having witnessed conflicting scientific evidence presented in support of opposing claims, Fischer argues that citizens dismiss the notion of science as being value-free and objective, and accordingly consider it as just another source of information. "Reliance on cultural rationality is especially strong when there is reason to believe in the possibility of deception or manipulation, which has often proved to be the case in environmental politics" (Fischer 2000, p. 137).

If public judgment regarding appropriate management of public forests relies heavily on cultural rationality and a dialogue that legitimizes the pluralistic nature of democracy, then it is likely that policies that depend exclusively on technical rationality and scientific evidence will not succeed politically. If this is the case, what should be done?

The link between cultural rationality and social acceptability is clear. Cultural rationality can be seen as a component of social acceptability judgments that includes assessments of risk, trustworthiness of agents, personal experience, and community values. Although the concept of compatibility is often portrayed in technical terms (e.g., what are impacts of alternative silvicultural treatments on soil erosion or wildlife habitat?), it is also a function of judgments that reflect cultural rational. For example, to what extent is the silvicultural treatment appropriate to the types of recreation experiences desired, or to the meaning assigned to places by long-term residents? In short, the compatibility concept transcends definitions grounded solely in technical or cultural rationality.

The compatibility concept holds great potential and is an important, legitimate focus for inquiry. However, this is only true if

- Compatibility represents a substantive effort to understand the nature of interactions and relations among forest uses and values, rather than a rhetorical slogan to justify timber harvesting.
- Compatibility is a key element in pursuing joint production of a host of forest uses and values and expanding options for these uses and values beyond the current emphasis in either-or solutions.
- Compatibility and the factors that shape it are recognized as essential for implementing integrative resource management.
- Compatibility is not conceived as an inherent aspect of any given activity, value, or specific kind of management approach, but rather the result of a comprehensive evaluative process.

In 1974, Clawson discussed possibilities for enhanced compatibility among various forest uses. Other than an absolute incompatibility between wilderness and both intensive recreation and timber harvesting, he argued a wide range of possible compatibilities among other forest uses existed. He noted, "it is absurd to pretend that all uses are possible on the same tract of forest land, but it is equally mistaken to assume that incompatibilities cannot be reduced or that compromises for different uses of different tracts are impossible" (Clawson 1974, p. 116).

Despite Clawson's analysis and his clear sense of possibilities for expanded compatibility among forest uses, little progress has occurred. The battle over the allocation of public lands and arguments about the role of private forestlands typically remain imbedded in "preserve or develop" terms. As a consequence, the goal of providing diverse landscapes to serve diverse uses and values has proven elusive. Despite the intuitive soundness of resource management programs that emphasize their compatibility with one another, the idea has remained little more than rhetoric. This is partially due to systemic flaws in identifying problems that resource management policies and programs intended to resolve.

3. Formative Factors in Social Acceptability Judgments

As local planners review comments from the public about proposed resource management actions, several aspects usually become apparent. Some concerns address technical elements of the proposal (e.g., competing science, and knowledge of interactions with other resources). Some express aesthetic concerns whereas others raise economic aspects. Some raise ethical issues related to exporting environmental impacts. Some reflect distrust of staff and the motives of public and private organizations. Others respondents express concern about specific details of the proposed action (e.g., size of affected area, location, and how it will affect favorite places). Finally, some respondents simply reject any change.

There are also patterns in responses based on who responds and why they feel as they do. Some responses are from long-term residents whereas others are from recent arrivals. Some respondents have a technical background whereas others ground their comments in personal experiences. Some respond from the perspective of recreationists who use the area whereas others comment as passer-bys who see the area from the highway. Some assert the need to husband resources by using our best knowledge to derive a variety of goods and services, whereas others argue that it is best to leave natural systems alone because we will only disrupt them.

Such comments highlight the many factors that contribute to the creation of a resource management policy agenda (dominated by extremes and a lack of capacity or interest in compromise) as well as an increasing inability of resource management institutions (public and private) to deal with socio-political ambiguity and scientific complexity. Private timberland owners often express concern about the impacts of public comments on their ability to generate economic returns from their lands. The constant controversy has resulted in lost energy, leadership, and innovation among these organizations and a decline in their ability to formulate progressive policies and programs that address the needs and concerns of citizens. What steps are needed to activate a more proactive approach?

Although there are many answers to this, we believe an important element is recognizing and acknowledging the role of the citizenry in resource management decisionmaking processes. It is also important to recognize that the process of forming social acceptability judgments is influenced and shaped by multiple factors, and that the role of any individual factor varies across space and time. Understanding the contextual setting is crucial (e.g., who is involved and history of the situation).

Attention needs to focus on the processes through which public *judgments*, as opposed to *opinions*, are derived and how these influence behavior. As used here, judgments reflect a position derived from a deliberative, reflective assessment of alternatives and tradeoffs. Yankelovich (1991, p. 47) contrasts judgments with opinions, noting that opinions stand in opposition to knowledge; i.e., "we fall back on opinion when knowledge is lacking." As used by Yankelovich, knowledge is broader than the conventional conception of a set of empirical facts; it is composed of various factors, including science, but also embraces an awareness of other values, contextual concerns, and personal experiences. Although any forest management policy likely will evoke an opinion from someone, there is no guarantee he or she will hold the same opinion over time. Judgments, like opinions, represent a disposition toward something but are characterized as (1) grounded in an understanding of the consequences (tradeoffs) associated with that view, (2) relatively firmly held (show low levels of volatility), and (3) consistent with other views and values held by the individual.

Thus, a judgment of acceptability about a natural resource policy reflects an understanding of the implications and consequences associated with that position. On which factors, then, do people base their judgment? And, how do changes in those factors lead to changes in judgments?

3.1. The Role of Context—Space, Place, and Time

Management that is acceptable in one situation may not be in another. This is true even when the management problem appears the same. The context of each situation involves a unique history and relationship with people and communities (Shindler 2000). Brunson (1993, p. 9) observed, "It is clear from even a cursory examination that the acceptability of a forest condition or a forestry practice cannot be considered apart from its context" and cites literature demonstrating how this can influence judgments. For example, Hodgson and Thayer (1980) found that forest photographs were deemed more scenic when labeled "forest growth" than when described as a "tree farm." Taylor and Daniel (1984) reported that while education about the beneficial effects of fire produced more favorable attitudes overall, it did not lead to acceptability of prescribed fire in recreation areas.

People consider many factors when forming judgments. These factors are interactive, and new information (e.g., what other people know) can trigger additional questions or concerns. Furthermore, as population shifts occur in communities, the balance of opinion about what is or is not acceptable can change significantly.

Thus, a fundamental challenge is understanding the context surrounding the social acceptability of any activity at a particular place. Despite limitations and problems regarding the formation of social acceptability judgments, failure to account for it has contributed to a situation where almost everything proposed is unacceptable, often for divergent reasons. The debate about compatibility and acceptability occurs in a world involving high levels of technical and scientific complexity and social-political-cultural ambiguity. As we think about socially acceptable ways to undertake wood production, we must place the discussion in its historical context. Reactions to a particular action often are as dependent on prior experiences as on the merits of the proposal.

Spatial context brings both biophysical and social concerns into consideration. Swanson and Sparks (1990) argue that unless a site's spatial context (e.g., its location within a region) is taken into account, the significance of management programs and scientific findings is difficult to interpret. Similarly, spatial context affects the social judgment process. Citizens often have difficulty identifying with larger landscape issues because their orientation is to places that have meaning for them, their family, or their community (Stankey and Shindler 1997). The focus on large scales by ecological scientists and small scales by citizens sets the stage for conflicts over acceptability (Lee and Stankey 1992).

Land ownership is a key component of spatial context. Although much of the public discussion over forest resource use focuses on public lands, over half of the nation's forestland is privately owned (Birch 1996). Some 10 million individuals and families hold 58% of the privately owned forestland, while another 27% is in corporate ownership. In Oregon and Washington where public forests dominate the landscape, the ecological significance of private forests is heightened because they are concentrated at lower elevations and along riparian corridors (Stanfield et al. 2002).

Numerous studies conclude that land tenure is a salient issue in forming public judgments of acceptability of forest practices and policies. Public opinion surveys consistently report that citizens hold different categories of forest owners to different standards (Bliss et al. 1997, Edwards and Bliss 2003). In general, citizens expect the highest level of environmental protection on public lands, but they also expect industrial forestlands to comply with environmental protection laws even though they accept timber production as a dominant activity on those lands. Survey respondents seem most lenient with their expectations of forest management behavior for individual and family forest owners.

Context also bears on the debate over compatibility among different forest uses as related to the interface between recreation and timber harvests. Clawson (1974) noted that recreation and timber harvests could be compatible, as long as the harvests were undertaken carefully. However, the relation appears more complex and conditional. Clark et al. (1984) reported that while motorized dispersed recreationists in the Cascade Range in Washington were strongly opposed to large clearcuts, they were supportive of small clearcuts, as long as their favorite campsites were protected. Brunson and Shelby (1992) found evaluations among forest users toward "new forestry" harvests differed if they were sightseers "looking in" or recreationists within the forests "looking out." For instance, partial cutting diminished the scenic quality of stands for passersby but the resulting increase in light and temperature made stands more pleasant to visit at certain times.

The idea of NIMBY (not in my backyard) often arises in debates regarding management practices, and is sometimes interpreted as a selfish, shortsighted perspective. However, such expressions can reflect dissatisfaction with decisionmaking processes that are seen as unresponsive to local concerns. The NIMBY phenomenon also can reflect place attachments, and influence whether people deem actions appropriate in that place (Clark et al. 1984, Williams and Patterson 1996). Treating forests as a collection of trees denies their contextual relevance to people. Knowledge about the spatial characteristics of unique places, and of the people who form attachments to them can be useful for resource professionals who seek to implement policies and actions (Kruger 2001). Place-specific interactions generate their own set of shared meanings among citizens and these can stand in stark contrast to the broad "placeless" planning efforts that people associate with some current landscape-level planning approaches. When this discrepancy occurs, it is likely that opposition will develop.

Context is critical for people judging if an option is acceptable, yet many public surveys lack this element. Surveys usually are focused on a discrete place and action to make analysis easier. This is fine for an academic study but not for formulating public or private forest management policy.

In sum, several levels of context come into play when people judge the merits of a project. First is the specific context in which the proposal rests (e.g., how many other areas have been harvested nearby?). Second is the context in which such proposals are developed and implemented (e.g., are local managers trusted?). Third is the context in which bureaucracies function to deal with issues that concern people (i.e., is the planning process responsive to the values and uses people expect from forests?). And there is even a larger context in which all of this fits—our society's struggle to understand the role of forests (both public and private) as well as other natural resources in an increasingly diverse culture.

3.2. The Role of Knowledge

The lack of public support for resource policies often is attributed to a lack of knowledge about the scientific basis for those policies. In short, there is a belief that "if they only understood the facts" they would support (or at least not oppose) the policy. However, confusing the provision of more facts with building public understanding and support is a mistake. As suggested earlier, although quality information is necessary in building and supporting sound decisions, it is not sufficient.

In recent years, resource professionals have expanded outreach efforts, such as public meetings and field trips. Unfortunately the flow of information in these efforts is often unidirectional; experts (the informed) provide citizens (the uninformed) with the technical knowledge needed to understand a policy. These approaches rest on the assumption that a lack of knowledge explains the opposition, which can be overcome by providing more facts.

Unfortunately, this rarely happens. As Jamieson (1994, p. 26) notes, "Generally programs that provide information are not very successful in improving understanding or changing behavior." For example, initial public responses to a proposed policy often are derived from visual and aesthetic images shaped by prior personal experiences and by prior experience in dealing with the organization (public and private). This does not mean that scientific input is not useful to the individual's judgment process, but it does suggest that how, why, and by whom information is presented is crucial. For example, research (Shindler et al. 1994, Yaffee and Wondolleck 1997) indicates that new management practices are more likely to be accepted by the public when

- The underlying rationale is clear.
- They have been genuinely engaged in a questioning process (preferably before implementation).
- They trust the information source (Moore 1995).
- They have an opportunity to consider the potential outcomes of the practice (which provides an opportunity to reflect on the implications and consequences of the practice) as well as possible alternatives, a key element in the judgment formation process (Yankelovich 1991).

Research indicates that people use scientific information as well as personal knowledge (knowledge based on experiences and observations) in forming judgments. However, the way people process scientific information and reconcile it with existing knowledge is not well understood. Cortner et al. (1998) and Shindler and Neburka (1997) report that people do not respond well to traditional agency information sharing programs or to scoping meetings, which are often used as part of the process mandated by the National Environmental Policy Act (1969). Providing citizens with an opportunity to evaluate the range of related information instead of selected bits of information is also important (Brunson and Reiter 1996). People show capacity for responding to information about tradeoffs, including positive and negative consequences, and often welcome the chance to do so. When given a rational set of choices, however limited or imperfect, citizens often will choose the lesser of two evils and accept it (Ehrenhaldt 1994). Thus, the ability of resource professionals to specify the context and communicate the nature of the options is as essential as the provision of objective, unbiased information.

How mutable are judgments of acceptability? Again, the short answer is "it depends." If concerns are ideologically-based, there is little likelihood they will change easily. If, however, concerns arise from incomplete understanding, then new knowledge can cause people to reevaluate the situation. One also must be aware of confusion over means and ends. For example, when people dismiss timber harvesting, they may be concerned about the means (clear-cutting versus selective harvest) or whether trees should be cut at all. New information and demonstrations likely would affect the former but not the latter.

At the same time, there is evidence that improved understanding of technical information can result in favorable social acceptability judgments. Stankey (1976) and McCool and Stankey (1986) found that higher levels of knowledge about the role of fire in forest ecosystems (as measured by a true-false test) was predictive of support among wilderness visitors for modified fire suppression policies in wilderness.

Shindler and Toman (2002) used a similar test to gauge public knowledge in the Blue Mountains region of northeastern Oregon regarding the role of prescribed burning and mechanical thinning as management strategies to reduce fire hazards. Overall, they found that citizens generally were knowledgeable about both practices (average score of 70%). Across the surveyed population, there was strong support for both prescribed fire and mechanized thinning to reduce fuel loads.

Shindler and Wright (2000) found that although citizen interest and concern about watershed management was high, few were well informed about the technical aspects of the issue. They also found that respondents held information providers in high trust and that the information provided was useful. They conclude that what people know and believe to be important, and how people feel about the information providers, the information, and the processes used to communicate it have equally important effects on trust building.

3.3. The Role of Aesthetics

A common presumption is that social acceptability judgments often are grounded in aesthetics more than in information regarding their biophysical or economic impacts. This can lead to a belief that public opposition is simply a reaction to how things look. Moreover, a common belief is that opposition will dissipate if either the "ugliness" can be replaced with something more attractive (e.g., eliminating straight edges in clearcuts) or by hiding these features from sight (e.g., leaving buffer strips between highways and clearcuts).

There is some support for these views. A series of research studies by Ribe (1989, 2002, in press) and Ribe et al. (2002) examine the aesthetic impacts associated with forest management prescriptions and provide evidence that the overall visual appearance of forest conditions provides an important initial basis for public judgments of acceptability. In one investigation (Ribe 2002), research focused on the extent to which perceptions of scenic beauty were equivalent to judgments of social acceptability and how this relation varied among different study participants. Although all those surveyed reported that "very beautiful" scenes were acceptable, those favoring resource production objectives had lower standards for both beauty and acceptability, rated acceptability higher than beauty, and reported "ugly" scenes as acceptable. Among those favoring resource protection objectives, the opposite was true. Ribe concludes that while scenic beauty may serve as a proxy for acceptability among similar populations, this is not necessarily true across diverse constituencies.

In recent research, Ribe (in press) examines the aesthetic impacts of two specific prescriptions associated with the Northwest Forest Plan: green-tree retention levels (see Hummel 2003) and down wood levels. Increased scenic beauty is strongly related to higher levels of green-tree retention and lower levels of down wood. The percentage of area in green-tree retention and the distribution of trees also affect scenic beauty ratings. For example, at 15% retention levels, dispersed retention patterns produce higher scenic beauty scores than aggregated patterns, which differ little from ordinary clearcuts. At higher retention levels—40 to 75%—distribution is less important.

Down wood from timber harvesting has been reported as a major source of adverse aesthetic ratings, particularly in near-view situations (Brown and Daniel 1986). To reduce such impacts, one prescription calls for removal. However, from a biodiversity perspective, removal of down wood could adversely affect biological functions and processes such as soil productivity or groundcover regeneration. As Gobster (1996) notes, actions taken to make timber harvesting compatible with visual quality objectives could lead to conditions incompatible with ecosystem and biodiversity objectives.

It is unlikely that information alone will offset adverse aesthetic effects associated with some management policies. Visual images convey powerful (if not necessarily accurate) messages about management practices. For example, programs designed to mimic or restore natural disturbance patterns might face considerable public resistance because of the visual impacts that disturbances such as fire and flooding produce. Convincing citizens of the long-term benefits of this approach may prove difficult.

However, research on the aesthetic aspects of forest management also reveals that adverse visual impacts often are interpreted as evidence of other impacts. Public controversy in the 1960s over timber harvesting in the Bitterroot National Forest in Montana initially was triggered by adverse aesthetic impacts associated with the practice of terracing, but broadened to include technical issues such as erosion and long-term site productivity. Surveys by Schuh (1995) and Bliss et al. (1994) reveal that respondents oppose intensive timber harvesting because they believe this practice leads to unacceptable resource degradation (erosion, stream sedimentation) in addition to aesthetic effects.

Research by Shindler and Collson (1998) provides further evidence that the acceptability question extends beyond aesthetics. For example, citizens were concerned about the relative priority of economic versus environmental issues; protecting traditional subsistence uses (e.g., hunting and gathering); having their opinions included in the planning process; and seeing that decisions reflected objective, credible information, rather than being dominated by interest group politics or by standardized decisions made in Washington, D.C. In Alaska, Burchfield et al. (in press) report that complex assessments underlie public evaluations of the acceptability of forest management practices; aesthetics are important, but so are concerns regarding the sustainability of practices for both resources and human communities.

Many studies have examined the acceptability of alternative silvicultural prescriptions. Some attempt to understand why people make certain judgments, but none reveal any management option that all subjects prefer or will accept. Because there always is a range of positive and negative responses, one must be wary of focusing on the central tendency—the average—when evaluating the acceptability of one strategy compared to others. Focusing on either averages or extremes when fashioning options misses the majority of people concerned about the project. Understanding why people judge proposals as they do is most important. Their reasons can offer clues about how to best interact and communicate with them. Although studies such as those described above are useful, understanding social acceptability is not simply about empirically determining how many favor a particular action.

3.4. The Role of Process

Discussion about social acceptability often centers on the specific merits or conditions associated with implementation of a policy. However, the processes through which those policies have been developed also can be the basis for acceptability judgments. For example, public involvement processes have attracted criticism (Blahna and Yonts-Shepard 1989) because of concerns that they constrain the content and substance of public input and lack a demonstrated link between input submitted and the subsequent decisions. There are also concerns that public involvement processes have done little to build understanding, sensitivity and tolerance, or mutual learning. Similar criticisms have been leveled at the underlying planning processes used by the natural resource agencies. As Wondolleck (1988) noted, the fundamental features of the technical-rational model (relying on the technical judgment of professionals) increasingly are at odds with the need to respond to the multi-party, multi-value disputes that typify resource management today; this results in processes that are seen as irrelevant and inappropriate.

Framing the problem is critical. Three distinctive components are involved—problem identification, problem clarification, and problem solving. Each step typically involves different people as well as different skills and knowledge. The key point is that the way a problem is framed greatly influences compatibility judgments. An inadequate or non-existent process for framing problems is unlikely to yield management actions that are compatible with other uses and values.

Failure to appropriately frame the problem perpetuates the problem. For example, consider how the course of inquiry and the likely strategies for resolution would change if we recast the question, "How do we develop socially acceptable wood production programs?" to, "Why is wood production frequently judged as socially unacceptable?" The first step is broadening the problem-framing process to include the underlying social-political-cultural "drivers." This has implications for determining who participates in framing the problem.

The procedures used to arrive at decisions often are "significant determinants of satisfaction separate from the effect of outcomes" (Lawrence et al. 1997, p. 579). In particular, there is a concern with determining the extent to which participants perceive processes as fair. Although little research has addressed this aspect of the social acceptability judgment process, the implications of research undertaken in other situations (e.g., arbitration awards and legal decisions) seem pertinent. In general, research suggests that when procedures are perceived as fair, outcomes (e.g., policies) tend to be evaluated as satisfactory, even when participants view outcomes as negative. In short, "even though the plan recommends actions that I don't agree with, I can accept them because the decision process was fair." Elements that define fair procedures include an opportunity to express one's views, even when it is known those views cannot or will not affect the eventual decision; the quality of justifications for the decision; and feedback to participants. There is evidence that nonparticipation often results from a perception that processes are unfair (e.g., the belief that one's views will not be listened to); such beliefs may explain why legal and political remedies are heavily relied upon (Lawrence et al. 1997).

3.5. The Role of Trust

Trust among individuals and between individuals and organizations influences social acceptability judgments. Moore (1995) found that personal trust derives from interpersonal attributes such as honesty and reciprocity whereas organizational trust stems from the perception that decision processes are open and fair. Generally, research reports the public is highly distrustful of federal forest and rangeland managers and therefore suspicious of efforts by federal managers to implement ecosystem-based management programs (Hoover et al. 1997). Ribe and Silvaggio (2002) report that nearly 40% of survey respondents agreed with the statement "I do not trust the Forest Service to manage public land." The problem is confounded because different stakeholders place greater trust in different parties in a controversy. Shindler (2000) found that while many citizens trust local resource managers, they are skeptical of the motives of the larger organization or its willingness to allow local managers flexibility and discretion. To the extent that a distrustful climate persists, the process of building the understanding and support required to achieve socially acceptable judgments is jeopardized.

Wondolleck and Yaffee (2000) report public dissatisfaction and distrust with the way planning processes incorporate its concerns in local, regional, and national issues. The public is dissatisfied with the balance of power, its access to the decisionmaking process, and the lack of effective venues to enhance mutual learning among managers and citizens. Public involvement processes need improvement so they will engage an increasingly diverse and distrustful public (Lang 1990) and foster mutual learning and understanding among resource managers, scientists, and citizens.

3.6. It Depends—Honest Answer or Just an Excuse?

Various factors affect the processes through which judgments of public acceptability are formed, communicated, and sustained or altered. Some judgments focus on the outcomes of a policy or action (e.g., how harvesting activities look or what happens to campsites). Others are grounded in process-related issues (e.g., fairness and trust).

With few exceptions, the answer to whether different resource management activities are compatible with one another is "it depends." At one level of analysis, this might seem an unsatisfactory answer or even convincing evidence of the inability or unwillingness of researchers and citizens to provide definitive answers to questions that land managers and policymakers cannot evade. In this sense, "it depends" becomes an excuse for failing to act.

However, if one considers the broader field of resource management, "it depends" is often the appropriate response to questions regarding relations among management activities, including biophysical aspects. For example, Clark and Gibbons (1991, p. 473) indicate that the impacts of recreational use on fisheries was "largely judgmental; seldom are definitive answers available to allow accurate prediction." Recreational effects on fisheries differed by region, river type, vegetation and soil conditions, season, and the nature and extent of recreational use.

Thus, the answer "it depends" is in no way code for "we don't know." In fact, we do know that details vary from place to place and that without context, equivocal answers are appropriate. For example, we have a substantial grasp of the formative factors from which judgments take form. What is less clear, and typically more situationally specific, is the relative importance of any specific factor from one situation relative to another, and the specific measures regarding that factor. In turn, while we cannot easily employ a linear, arithmetic process that calculates a mechanical index of acceptability, we can identify key factors that must be accounted for by managers and policy-makers concerning judgments held by the public and how those judgments have come about. For example, it is helpful for management authorities to know that perceived aesthetic impacts or a lack of trust trigger adverse public judgments so they can address these factors and ensure that proposed policies account for them in explicit, transparent terms. Kakoyannis et al. (2001) describe how a formal, knowledge-based decision model can help ensure identification and consideration of key factors affecting social acceptability judgments by making the various formative factors an explicit element of the model.

4. Concepts, Frameworks, and Tools

As our preceding discussion indicates, the social acceptability judgment process is complex. Judgments regarding a specific issue such as the compatibility between timber harvesting and other forest uses and values are formed in a contextual, situationally-specific setting. Moreover, these judgments are always conditional and subject to change as new information becomes apparent, and the situations and population changes.

The approaches taken to assess social acceptability judgments, and used by managers, citizens, and scientists to build sustainable and implementable decisions must have a capacity to deal with this complexity. Although public surveys provide a useful snapshot of public views on a resource management issue at a given time, surveys have limited utility for revealing the complex structure of those views or how those views change as context shifts. Survey information typically lacks three elements crucial to building an improved understanding of social acceptability: they seldom provide a contextual setting, there is limited information about the consequences of an individual's judgment, and the sentiment is solicited with little information about possible alternatives.

However, despite the difficulty in coming to grips with social acceptability, we do not start from scratch. A variety of tools have been developed and successfully used over the past 25 years. We provide three brief examples.

When efforts to assess compatibility identify place-based concerns, tools such as the Recreation Opportunity Spectrum (ROS) and concepts such as recreation habitat and site attributes can be helpful (Brown et al. 1978, Clark and Stankey 1979). The ROS helps identify ways to provide recreational opportunities ranging from primitive and unconfined to intensively developed. It recognizes the extent to which compatibility exists between recreation and other resource uses for part of the recreational spectrum. In some areas, for example, timber harvesting will preclude most remote and natural recreational opportunities while opening areas to people who do not mind evidence of roads and timber harvests. Elsewhere, opportunities for semiprimitive and undeveloped recreation might be possible in or near areas designated for logging or other resource uses if the scale of the project and its execution over time and space are modified to protect, enhance, or even create recreation sites.

When compatibility issues prompt questions and concerns that are knowledge-related, it is likely that collaborative and mutual learning models are needed (e.g., Stankey et al. 1985, Wondolleck and Yaffee 2000). Such approaches help build understanding of the social context in which the search for compatible approaches occurs. A specific framework for assessing impacts on recreational values is the Limits of Acceptable Change (LAC) model that attempts to determine acceptable levels of impact on various resources (Stankey et al. 1985). The LAC specifies physical and social conditions that must be maintained to protect recreational and other resource values. It is applicable across the recreational spectrum and should prove useful when attempting to determine the extent of roading, timber harvesting, or other resource management activities that are acceptable (compatible) with the provision of different recreational settings. Computer-aided knowledge helps synthesize and represent the knowledge from which social acceptability judgments derive, and for documenting assumptions and caveats that influence those judgments (Kakoyannis et al. 2001).

Concepts related to human habitats can help augment the analyses possible with the ROS. Many concepts from the wildlife habitat literature also apply to recreation and are useful when managing for recreation habitats. Understanding these concepts could help silviculturists and road engineers evaluate the potential effects of alternative prescriptions for recreational opportunities (Clark 1987, 1988). For example, people are "territorial" and often form strong attachments to favorite and frequently visited places that they do not want to see changed. It is important to identify these places before any on-the-ground management occurs so options for protecting or enhancing these places can be evaluated.

The nature and extent of change acceptable to recreationists varies. People have different expectations for "macro" versus "micro" sites (e.g., the watershed where one recreates versus the specific site where one camps). Microsites often are more susceptible to adverse change; that is, management activities acceptable in the general area (such as evidence of logging or roads) are considered intolerable at a campsite.

After problems associated with compatibilities have been appropriately framed, the applicability of tools such as ROS and LAC can be determined based on the specific details, location, and context of the problem.

5. Challenges for the Future

To conclude, we identify some challenges to implementing sustainable resource management practices in general and timber harvesting in particular. We believe that unless policymakers and managers successfully address these challenges, there is little hope that diverse segments of our society will accept the management programs, no matter how compatible and compelling policymakers think they are.

Resource management issues are complex. Although commonly acknowledged, few effective mechanisms and venues exist for helping citizens, scientists, and managers deal with this complexity. In spite of honest attempts by agencies to communicate the rationale underlying proposed actions, citizens receive most of their information through the media, where balanced perspectives and nuances are often poorly communicated. Generally, even "in-depth" media coverage is limited to several minutes. New ways to include the public are needed. Developing acceptable practices requires more than data comparing harvest treatments; we must deal with the multiple contexts in which judgments are made.

Competing paradigms exist for resource management. The reactions to Bliss (2000) (Box 2) illustrate that the paradigm for forest management has been professional foresters making rational scientific management decisions (Clark and Brown 1990). On the surface, this long-standing paradigm seems logical and right. Aspects of it, however, increasingly have been criticized (Wondolleck 1988). Although rational scientific management may best show how to *achieve* given ends, it is ill-suited to *choosing* ends. Other models, such as those used in political behavior are designed for that purpose and lead to consideration of the values inherent in forestry decisions. Forest management requires a new paradigm that recognizes the legitimacy of many social values, especially those emphasizing historical, cultural, wildlife, and recreational values. Professional foresters cannot be expected to represent all forest values and thus must learn from others who hold different values. The hope is that new approaches will reassert a middle ground in management prescriptions, thus lessening polarization.

Failure of organizations to follow through with their plans is a serious challenge. Citizens often complain that despite extensive planning, there are too few examples of implementation. Recent efforts to implement innovative management strategies on public lands, such as adaptive management and ecosystem management raised public expectations that quickly turned to skepticism and cynicism as on-the-ground applications and demonstrations failed to evolve in a timely manner. Planning on private lands around issues such as habitat conservation has been controversial as well. Increasing dependence on political and legal remedies often is caused by planning processes that fail to effectively address the concerns of citizens who care about forests. For example, citizens form groups based on common interests (e.g., "Friends of..." groups) because they feel their individual concerns and interests are not represented adequately in proposed policies and programs. People are increasingly demanding to be heard.

Resource management agencies have experienced a decline in their capacity to deal with the concerns of people they ultimately serve. Smaller budgets coupled with fewer employees have resulted in workforces without the skills, resources, and time to handle the multifaceted biophysical and social problems that they face. The search for routine, mechanistic approaches and answers to complex problems is a futile substitute for working with interested citizens to determine a solution (Yankelovich 1991). We need better ways to engage people with diverse interests and knowledge and focus on collaboration, mutual learning, and trust building.

Changes are inevitable. Developing transitional strategies that enable people and organizations to deal more effectively with emerging realities is critical (Bridges 1991). Successfully implementing and sustaining support requires more than a written and officially approved planning document to be revisited in 10 to 15 years. Adaptive management is one approach for dealing with uncertainty and change (Stankey et al. 2003). Unfortunately, evidence in the United States and elsewhere suggests that attempts to practice adaptive management have been largely rhetorical rather than substantive, further poisoning internal and external commitment and support.

Is it too late? Is it too hard? Can anything be done to better integrate public concerns into resource management? We increasingly hear such questions. We believe there are many positive opportunities. But if it were easy, we would likely be doing it. Ineffective leadership at many organizational levels is often described as a serious limiting factor in implementing and sustaining management programs. Failures to articulate a vision (both internal and external to organizations) at multiple levels, failures to follow through, and failures to create environments and provide resource for people to be effective often lead to lost commitment, creativity, and hope.

6. References

Birch, T.A. 1996. Private forest-land owners of the United States, 1994. Resource Bulletin NE-134. Radnor, PA: U.S. Department of Agriculture, Forest Service, Northeastern Forest Experiment Station. 183 p.

Blahna, D.J., and Yonts-Shepard, S. 1989. Public involvement in resource planning: toward bridging the gap between policy and implementation. *Society and Natural Resources*, 2: 209-227.

Bliss, J.C. 2000. Public perceptions of clearcutting. *Journal of Forestry*, 98(12): 4-10.

Bliss, J.C., Nepal, S.K., Brooks, R.T., Jr., and Larsen, M.D. 1994. Forestry community or granfalloon? *Journal of Forestry*, 92(9): 6-10.

Bliss, J.C., Nepal, S.K., Brooks, R.T., Jr., and Larsen, M.D. 1997. In the mainstream: environmental attitudes of mid-south NIPF owners. *Southern Journal of Applied Forestry*, 21(1): 37-42.

Bridges, W. 1991. Managing transitions: making the most of change. Reading, MA: Addison-Wesley Publishing Company, Inc. 130 p.

Brown, P.J., Driver, B.L., and McConnell, C. 1978. The opportunity spectrum concept and behavioral information in outdoor recreation resource supply inventories: background and application. In: G.H. Lund, V.J. LaBau, P.F. Ffolliott, and D.W. Robinson (Tech. coords.), Integrated inventories of renewable natural resources: proceedings of the workshop (pp. 73-84). General Technical Report RM-55. Fort Collins, CO: U.S. Department of Agriculture, Forest Service, Rocky Mountain Forest and Range Experiment Station.

Brown, T.C., and Daniel, T.C. 1986. Predicting scenic beauty of timber stands. *Forest Science*, 32: 471-487.

Brunson, M. 1993. "Socially acceptable" forestry: What does it imply for ecosystem management? *Western Journal of Applied Forestry*, 8(4): 116-119.

Brunson, M. 1996. A definition of "social acceptability" in ecosystem management. In: M.W. Brunson, L.E. Kruger, C.B. Tyler, and S.A. Schroeder (Eds.), Defining social acceptability in ecosystem management: workshop proceedings (pp. 7-16). General Technical Report PNW-GTR-369. Portland, OR: U.S. Department of Agriculture, Forest Service, Pacific Northwest Research Station.

Brunson, M.W., and Reiter, D.K. 1996. Effects of ecological information on judgments about scenic impacts of timber harvest. *Journal of Environmental Management*, 46: 31-41.

Brunson, M.W., and Shelby, B. 1992. Assessing recreational and scenic quality: How does "new forestry" rate? *Journal of Forestry*, 90(7): 37-41.

Burchfield, J.A., Miller, J.M., Allen, S., Schroeder, R.F., and Miller, T. [In press]. Social implications of alternatives to clearcutting on the Tongass National Forest: an exploratory study of residents' responses to alternative silvicultural treatments at Hanus Bay. General Technical Report. Portland, OR: U.S. Department of Agriculture, Forest Service, Pacific Northwest Research Station. 28 p.

Clark, R.N. 1987. Recreation management: a question of integration. *Western Wildlands*, 13(1): 20-23.

Clark, R.N. 1988. Enhancing recreation opportunities in silvicultural planning. In: W.C. Schmidt (Comp.), Proceedings, Future forests of the Mountain West: a stand culture symposium. (pp. 61-69). General Technical Report INT-243. Ogden, UT: U.S. Department of Agriculture, Forest Service, Intermountain Research Station.

Clark, R.N., and Brown, P.J. 1990. The emerging web of integrated resource management. Proceedings of the 1990 IUFRO XIX World Congress, Montreal, Canada. 6: 24-33.

Clark, R.N., and Gibbons, D.R. 1991. Recreation. In: W.R. Meehan (Ed.), Influences of forest and rangeland on salmonid fishes and their habitat. (Chapter 13 pp. 459-481), Special Publication 19. Bethesda, MD: American Fisheries Society.

Clark, R.N., Koch, R.W., Hogans, M.L., Christensen, H.H., and Hendee, J.C. 1984. The value of roaded, multiple-use areas as recreation sites in three national forests of the Pacific Northwest. Research Paper PNW-319. Portland, OR: U.S. Department of Agriculture, Forest Service, Pacific Northwest Research Station. 40 p.

Clark, R.N., and Stankey, G.H. 1979. The recreation opportunity spectrum: a framework for planning, management, and research. General Technical Report PNW-98. Portland, OR: U.S. Department of Agriculture, Forest Service, Pacific Northwest Forest and Range Experiment Station. 32 p.

Clarke, J.N., and McCool, D.C. 1996. Staking out the terrain: power and performance among natural resource agencies (2nd ed.). Albany, NY: State University of New York Press. 279 p.

Clawson, M. 1974. Conflicts, strategies, and possibilities for consensus in forest land use and management. In: M. Clawson (Ed.), Forest policy for the future: conflict, compromise, consensus (pp. 105-191). Papers and discussions from a forum on forest policy for the future. Washington, DC: Resources for the Future. 360 p.

Cortner, H.J., Wallace, M., Burke, S., and Moote, M.A. 1998. Institutions matter: the need to address the institutional challenges of ecosystem management. *Landscape and Urban Planning*, 40: 159-166.

Edwards, K.K., and Bliss, J.C. 2003. It's a neighborhood now: practicing forestry at the urban fringe. *Journal of Forestry*, 101(3): 6-11.

Ehrenhaldt, A. 1994. Let the people decide between spinach and broccoli. *Governing*, 7(10): 6-7.

Firey, W. 1960. Man, mind, and land: a theory of resource use. Glencoe, IL: The Free Press of Glencoe. 256 p.

Fischer, F. 2000. Citizens, experts, and the environment: the politics of local knowledge. Durham, NC: Duke University Press. 336 p.

Forest Ecosystem Management Assessment Team [FEMAT]. 1993. Forest ecosystem management: an ecological, economic, and social assessment. Portland, OR: U.S. Department of Agriculture; U.S. Department of the Interior [et al.]. [Irregular pagination].

Friedmann, J. 1987. Planning in the public domain: from knowledge to action. Princeton, NJ: Princeton University Press. 501 p.

Gobster, P.H. 1996. Forest aesthetics, biodiversity, and the perceived appropriateness of ecosystem management activities. In: M.W. Brunson, L.E. Kruger, C.B. Tyler, and S.A. Schroeder (Eds.), Defining social acceptability in ecosystem management: workshop proceedings (pp. 77-97). General Technical Report PNW-GTR-369. Portland, OR: U.S. Department of Agriculture, Forest Service, Pacific Northwest Research Station.

Hodgson, R.W., and Thayer, R.L., Jr. 1980. Implied human influence reduces landscape beauty. *Landscape Planning*, 7(2): 171-179.

Hoover, W.L., Mills, W.L., Jr., and Vasan, S. 1997. Nonindustrial private forest landowners in Indiana: are their objectives and attitudes consistent with ecosystem management? In: H.K. Cordell (Ed.), Integrating social science and ecosystem management: a national challenge: proceedings of the conference on integrating social sciences and ecosystem management (pp. 150-155). General Technical Report SRS-17. Asheville, NC: U.S. Department of Agriculture, Forest Service, Southern Research Station.

Hummel, S. 2003. Managing structural and compositional diversity with silviculture. Chapter 4. In: R.A. Monserud, R.W. Haynes, and A.C. Johnson (Eds.), Compatible forest management. Dordrecht, The Netherlands: Kluwer Academic Publishers.

Jamieson, D. 1994. Problems and prospects for a Forest Service program in the human dimensions of global change. In: K. Geyer, and B. Shindler (Eds.), Breaking the mold: global change, social responsibility, and natural resource management (pp. 23-28). Corvallis, OR: Oregon State University.

Jasanoff, S. 1990. The fifth branch: science advisers as policymakers. Cambridge, MA: Harvard University Press. 302 p.

Kakoyannis, C., Shindler, B., and Stankey, G. 2001. Understanding the social acceptability of natural resource decisionmaking processes by using a knowledge base modeling approach. General Technical Report PNW-GTR-518. Portland, OR: U.S. Department of Agriculture, Forest Service, Pacific Northwest Research Station. 40 p.

Kruger, L. 2001. What is essential is invisible to the eye: understanding the role of place and social learning in achieving sustainable landscapes. In: S.R.J. Sheppard, and H.W. Harshaw (Eds.), Forest and landscapes: linking ecology, sustainability and aesthetics (pp. 173-187). Wallingford, Oxon, United Kingdom: CABI Publishing.

Lang, R. 1990. Achieving integration in resource planning. In: R. Lang (Ed.), Integrated approaches to planning and management (pp. 27-50). Banff, AB: Banff School of Management.

Lawrence, R., Daniels, S.E., and Stankey, G. 1997. Procedural justice and public involvement in natural resources decision making. *Society and Natural Resources*, 10(6): 577-589.

Lee, R.G., and Stankey, G.H. 1992. Evaluating institutional arrangements for regulating large watersheds and river basins. In: P.W. Adams, and W.A. Atkinson (Comps.), Watershed resources: balancing environmental, social, political and economic factors in large basins (pp. 30-37). Corvallis, OR: Forest Engineering Department, Oregon State University.

McCool, S.F., and Stankey, G.H. 1986. Visitor attitudes toward wilderness fire management policy—1971-84. Research Paper INT-357. Ogden, UT: U.S. Department of Agriculture, Forest Service, Intermountain Forest and Range Experiment Station. 7 p.

Michael, D.N. 1995. Barriers and bridges to learning in a turbulent human ecology. In: L.H. Gunderson, C.S. Holling, and S.S. Light (Eds.), Barriers & bridges to the renewal of ecosystems and institutions (pp. 461-485). New York: Columbia University Press.

Moore, S.A. 1995. The role of trust in social networks: formation, function, and fragility. In: J. Craig, E.M. Mattiske, and D.A. Saunders (Eds.), Nature conservation 4: the role of networks (pp. 148-154). Chipping Norton, New South Wales, Australia: Surrey Beatty and Sons.

Pierce, J.C. Steger, M.A.E.; Steel, B.S., and Lovrich, N.P. 1992. Citizens, political communication, and interest groups: environmental organizations in Canada and the United States. Westport, CT: Praeger. 225 p.

Plough, A., and Krimsky, S. 1987. The emergence of risk communication studies: social and political context. *Science, Technology, and Human Values*, 12(3-4): 4-10.

Ribe, R., and Silvaggio, T. 2002. National forest management in timber and spotted owl country: a survey of interested people in western Oregon and Washington. Eugene, OR: Institute for a Sustainable Environment, University of Oregon. 104 p.

Ribe, R.G. 1989. The aesthetics of forestry: What has empirical preference research taught us? *Environmental Management*, 13: 55-74.

Ribe, R.G. 2002. Is scenic beauty a proxy for acceptable management? The influence of environmental attitudes on landscape perceptions. *Environment and Behavior*, 34(6): 757-780.

Ribe, R.G. [In press]. Aesthetic perceptions of the retention of green trees and down wood in timber harvests. Review draft. 36 p. *Ecological Applications*.

Ribe, R.G., Armstrong, E.T., and Gobster, P.H. 2002. Scenic vistas and the changing policy landscape: visualizing and testing the role of visual resources in ecosystem management. *Landscape Journal*, 21: 42-66.

Schuh, D. 1995. Managing esthetic values. *Journal of Forestry*, 93(2): 20-25.

Shindler, B. 2000. Landscape-level management: it's all about context. *Journal of Forestry*, 98(12): 10-14.

Shindler, B., Brunson, M., and Stankey, G.H. 2002. Social acceptability of forest conditions and management practices: a problem analysis. General Technical Report PNW-GTR-537. Portland, OR: U.S. Department of Agriculture, Forest Service, Pacific Northwest Research Station. 68 p.

Shindler, B., and Collson, P. 1998. Assessing public preferences for ecosystem management practices. In: D.L. Soden, B.L. Lamb, and J. Tennert (Eds.), Ecosystem management: a social science perspective (pp. 161-174). Dubuque, IA: Kendall-Hunt.

Shindler, B., and Neburka, J. 1997. Public participation in forest planning: eight attributes of success. *Journal of Forestry*, 91(7): 17-19.

Shindler, B., Peters, J., and Kruger, L. 1994. Social values and acceptability of alternative harvest practices on the Tongass National Forest. Corvallis, OR: Oregon State University. 95 p.

Shindler, B., and Toman, E. 2002. A longitudinal analysis of fuel reduction in the Blue Mountains: public perspectives on the use of prescribed fire and mechanized thinning. Corvallis, OR: Department of Forest Resources, Oregon State University. 76 p.

Shindler, B., and Wright, A. 2000. Watershed management in the central Cascades: a study of citizen knowledge and the value of information sources in the lower South Santiam basin. Corvallis, OR: Department of Forest Resources, Oregon State University. 90 p.

Stanfield, B.J., Bliss, J.C., and Spies, T.A. 2002. Land ownership and landscape structure: a spatial analysis of 66 Oregon Coast Range watersheds. *Landscape Ecology*, 17(8): 685-697.

Stankey, G.H. 1976. Wilderness fire policy: an investigation of visitor knowledge and beliefs. Research Paper INT-180. Ogden, UT: U.S. Department of Agriculture, Forest Service, Intermountain Forest and Range Experiment Station. 17 p.

Stankey, G.H. 1996. Defining the social acceptability of forest management practices and conditions: integrating science and social choice. In: M.W. Brunson, L.E. Kruger, C.B. Tyler, and S.A. Schroeder (Eds.), Defining social acceptability in ecosystem management: workshop proceedings (pp. 99-112). General Technical Report PNW-GTR-369. Portland, OR: U.S. Department of Agriculture, Forest Service, Pacific Northwest Research Station.

Stankey, G.H., Bormann, B.T., Ryan, C., Shindler, B., Sturtevant, V., Clark, R.N., and Philpot, C. 2003. Adaptive management and the Northwest Forest Plan: rhetoric and reality. *Journal of Forestry*, 101(1): 40-46.

Stankey, G.H., and Clark, R.N. 1992. Social aspects of new perspectives in forestry: a problem analysis. Milford, PA: Grey Towers Press. 33 p.

Stankey, G.H., Cole, D.N., Lucas, R.C., Petersen, M.E., and Frissell, S.S. 1985. The limits of acceptable change (LAC) system for wilderness planning. General Technical Report INT-176. Ogden, UT: U.S. Department of Agriculture, Forest Service, Intermountain Forest and Range Experiment Station. 37 p.

Stankey, G.H., and Shindler B. 1997. Adaptive management areas: achieving the promise, avoiding the peril. General Technical Report PNW-GTR-394. Portland, OR: U.S. Department of Agriculture, Forest Service, Pacific Northwest Research Station. 21 p.

Swanson, F.J., and Sparks, R.E. 1990. Long-term ecological research and the invisible place. *BioScience*, 40(7): 502-508.

Taylor, J.G., and Daniel, T.C. 1984. Prescribed fire: public education and perception. *Journal of Forestry*, 82: 361-365.

Vining, J., and Ebreo, A. 1991. Are you thinking what I think you are? A study of actual and estimated goal priorities and decision preferences of resource managers. *Society and Natural Resources*, 4(2): 177-196.

Wengert, N. 1976. Citizen participation: practice in search of a theory. In: A.E. Utton, W.R.D. Sewell, and T. O'Riordan (Eds.), Natural resources for a democratic society: public participation in decision-making (pp. 23-40). Boulder, CO: Westview Press.

Williams, D.R., and Patterson, M.E. 1996. Environmental meaning and ecosystem management: perspectives from environmental psychology and human geography. *Society and Natural Resources*, 9(5): 507-521.

Wondolleck, J.M. 1988. Public lands conflict and resolution: managing national forest disputes. New York: Plenum Press. 263 p.

Wondolleck, J.M., and Yaffee, S.L. 2000. Making collaboration work: lessons from innovation in natural resource management. Washington, DC: Island Press. 277 p.

Yaffee, S.L. 1994. The wisdom of the spotted owl: policy lessons for a new century. Washington, DC: Island Press. 430 p.

Yaffee, S.L., and Wondolleck, J.M. 1997. Building bridges across agency boundaries. In: J.F. Franklin, and K.A. Kohm, (Eds.), Creating a forestry for the 21st century (pp. 381-396). Washington, DC: Island Press.

Yankelovich, D. 1991. Coming to public judgment: making democracy work in a complex world. Syracuse, NY: Syracuse University Press. 290 p.

Section 6—Conclusion

This book began with a proposition about compatible forest management. In search of evidence and answers, the authors have explored the forests of the Pacific Northwest through research, experiences, historical context, and simulations predicting trends in the future. A number of themes have emerged. Spatial scale, for example, emerges as a major factor affecting compatibility. As in any complex story with many players, one or several themes have been prominent in some chapters and others less so; in different chapters, these other themes become the focus.

In this final chapter, the authors review the initial proposition about compatibility, discuss the eight themes (that have emerged from the reserach) and four reserach questions, weigh the evidence of these with respect to the proposition, and render a judgment of the propositon. Finally, the authors conclude with some thoughts on compatibility, sustainability, and land stewardship.

Chapter 17

The Search for Compatibility: What Have We Learned?

Robert A. Monserud
Richard W. Haynes
USDA Forest Service, Pacific Northwest Research Station,
620 SW Main St., Suite 400, Portland, OR 97205, USA
Adelaide C. Johnson
USDA Forest Service, Pacific Northwest Research Station,
2770 Sherwood Lane, Juneau, AK 99801, USA

1. Introduction

Forest management that seeks compatible production opportunities across several forest resources is the essence of compatible forest management. Most landowners and many land managers are keenly aware that their management actions produce a range of goods or services that contributes to both their personal prosperity and to societal well-being. It is this diversity of ownerships—each with different management objectives—that has a positive influence on compatibility when considered at the landscape scale.

Many of the examples in this book illustrate the inherent ambiguity in describing forest management. Forest prescriptions (plant, thin, harvest) may seem like relatively mechanistic strategies but it is often difficult to describe actual forest management because of the inherent uncertainty in various biophysical and socioeconomic processes and responses. In addition, the prescriptions must be site specific. This uncertainty lends a fuzzy—as opposed to definitive—nature to discussions of forest management. This lack of clarity is due to ambiguous treatment of scale, role of scientific information, specificity of local conditions, and selectivity in a strategic sense.

The experiences described in the various chapters emphasize that discussions of forest management should be addressed at the appropriate **scale**. For example, broad-scale discussions on the effects of forest management

R.A. Monserud, R.W. Haynes and A.C. Johnson (eds.)
Compatible Forest Management, 483-517

(e.g., the increased areal extent of softwood plantations) deal with the accumulation of numerous fine-scale effects and decisions by landowners. But these discussions at the broad scale set the context for decisions at the stand level. Much of the recent discussion (see Haynes et al. 2001) about the adoption of ecosystem perspectives for forest management deals with the necessity of shifting broad-scale management direction to specific management practices applied at the stand level.

Many chapters in this book implicitly assert that forest management should rest on the best **science** available. The first step is using the best available science to develop management practices. Scientists often play a key role in identifying and describing the utility of these practices. The best available science includes synthetic efforts that connect disparate research, including theoretical, experimental, and empirical approaches. Some chapters reveal how some scientific facts and, consequently, management practices have changed over time. This changing nature of science might raise questions about the confidence that should be placed in science-based management. But it is more productive to ask about the extent to which scientific knowledge is provisional. Everything that science uncovers is subject to reexamination as new information is revealed. The development of science is an incremental process that moves toward a more complete understanding of the observable world. At the same time, changes in economic and societal values modify the objectives of land management, altering the worth of different types of information. For example, the growing importance of western hemlock (*Tsuga heterophylla* (Raf.) Sarg) as a managed species creates the need for understanding the wood quality properties of managed hemlock.

The **specificity** of local conditions needs to be explicitly considered before management practices are applied. Too often we see a devolution into one-size-fits-all types of thinking in contemporary forest management. In the Pacific Northwest (western Oregon and Washington, coastal British Columbia, and southeastern Alaska) for example, we often see riparian buffer widths on public lands expressed as one tree height, regardless of the underlying biophysical conditions. Management prescriptions should depend on the site conditions and other local management conditions. General prescriptions are less likely to take advantage of localized differences, and, consequently, are less efficient in achieving the desired outcomes. Regrettably, these general prescriptions are defended by those who are adverse to risk (i.e., regulators wishing to avoid as much risk as possible), given their bias for cautionary practices. We need to develop management prescriptions on a site-by-site basis that consider variations in stand density, age, species composition, and history. We also need to consider that most of our research is confined to a relatively narrow range of conditions and stand origins, and its application for any stand will have to be partially tailored to local conditions.

Management should be **strategically selective**. Each landowner or manager approaches forest management with a different set of perspectives reflecting their objectives for managing the stand or forest in question. The requirements imposed by the regulatory environment (e.g., Forest Practices Code of British Columbia 1994, USDA and USDI 1994b) will influence the selection and application of different management practices. Although this regulatory environment is fixed at a given moment, it can be altered in response to changes in societal values and landowner actions. For example, by buffering roads or streams, landowners may achieve greater social acceptability for harvesting or thinning activities than by not buffering.

Perhaps the strongest example of management that is strategically selective is the adoption of variable retention in 1999 by Weyerhaeuser Co. Canada (Weyerhaeuser 2000). In 1998, the forest products company McMillan Bloedel publicly announced it had stopped clearcutting and adopted variable retention on private and tenured Crown forests in British Columbia. One year later the company was purchased by Weyerhaeuser. In effect, Weyerhaeuser purchased an unwritten social contract (no clearcutting) along with McMillan Bloedel's forestry assets. The lack of a well-developed scientific basis for alternatives to clearcutting did not allow this implicit contract to be suspended or broken. As a result, conversion to a variable-retention management system that addresses multiple forest values is being implemented in British Columbia and pilot research programs are in place (Weyerhaeuser 2000).

2. Examination of the Proposition from Chapter 1:

In Chapter 1, we stated a proposition regarding compatibility:

> *Commodity production (timber and nontimber forest products) and the other forest values (biodiversity, fish and wildlife habitat) can be simultaneously produced from the same area in a socially acceptable manner.*

We will examine the proposition by using an evidentiary approach to weigh information before rendering a final judgment. First, we will use direct evidence that includes results from empirical studies. Second, we will use syntheses of empirical data and expert judgments to develop an understanding of how management actions influence forest changes.

2.1. Major Themes

We begin by examining the evidence with respect to eight major themes.

(1) Multiple resource objectives. Perhaps the most distinctive feature of compatible forest management is the emphasis on joint production. Wood production was the dominant forest management objective in the Pacific Northwest for nearly a century. An ongoing debate on sustainability, ecological integrity, forest health, and social acceptability of forest management has been boiling for several decades. Our focus on the joint production of several forest resources is a direct response to this acrimonious debate, which is all too often characterized by dichotomous arguments to the extreme. Either-or choices and extreme arguments regarding a multiresource system are not fruitful.

We adapt the simple concept of Pareto optimality, which does not require a market. The idea is that a management action that increases the production of at least one desirable good without decreasing any other will increase social utility. Simply put, someone is better off and no one is worse off.

Economists expand this idea to the production possibility frontier, which Montgomery (Chapter 2) aptly applied to our problem of compatible forest management. She presents a hypothetical production possibility set for two outputs, wildlife and timber. The combinations of outputs represented by points inside the frontier are inefficient in the sense that one output can be increased at no cost to the other output. If the frontier accurately measures the productive capacity of the site and represents the outputs that matter to society, society will benefit by increasing at least one output while decreasing none. Other forest uses, besides wildlife and timber, may also be modeled (e.g., reductions in fire risk, recreation use, water yield, and aquatic species).

Our premise is that multiple resources can be managed simultaneously—at some scale—while producing wood. The difficulty enters when we add the caveat: in a socially acceptable manner. With it, forest values must be measured, or at least seriously examined, which is an inherently difficult task.

(2) Scale. Examine first the stand scale where management is actually implemented. As we learned after the Multiple-Use Sustained Yield Act of 1960, it is difficult to do everything in one small place. However, this does not mean that stands cannot be managed with multiple objectives. In Chapter 14, for example, Carey found that commercial thinning (wood production) could be used as an effective management tool for promoting wildlife habitat and increasing structural diversity. Also, Wipfli et al. (Chapter 3) found that even without thinning, there was greater understory species diversity in alder (*Alnus* spp.)-dominated second-growth stands than in adjacent second-growth spruce/hemlock (*Picea/Tsuga*) stands. Hence, alder provided increased structural diversity as well as promoting wildlife habitat. Monserud (Chapter 5) surveyed several stand management experiments that focus on joint production of wood and at least one other resource such as biodiversity. Each author is looking for alternatives to the intensive management regime (clearcutting followed by site preparation and planting) that has dominated managed Pacific

Northwest forests since the 1950s. Opportunities for compatible management of understory resources vary widely across forest developmental stages, management approaches, and land classes (Chapter 12). Opportunities will also vary due to site conditions, even across relatively small spatial scales, and that makes management of understory species particularly problematic.

At the watershed and landscape scale, which might consist of several thousand contiguous stands, it becomes easier to address several resource goals simultaneously. As Swanson et al. (Chapter 9) point out, it is important to consider the role of disturbance regimes in order to sustain native species and ecological processes. As we saw in the work of Cissel et al. (1999), an entire watershed (Blue River, Oregon) can be effectively managed with an overriding structural goal such as mimicking an historical disturbance regime (fire, in this case). The result increases the ecological integrity of the watershed from both wildlife and aquatic perspectives. Blocks of contiguous stands are treated as units, with the particular prescription depending on structural condition and disturbance risk. Wood is produced through a series of thinnings and harvests almost as a byproduct of the overall management goal. Fragmentation is considerably reduced compared to the Standards and Guides from the Northwest Forest Plan (USDA and USDI 1994a,b).

There is a growing recognition that aquatic ecosystems are dynamic in space and time and that periodic disturbances may contribute to their long-term productivity (see Chapter 9). Benda et al. (1998) model how aquatic systems respond to periodic disturbances such as wildfire, rain, and landslides in the Pacific Northwest. The model, which attempts to mimic the historical regime of fire, rain, and landslides, helps us understand how disturbance processes interact to introduce sediment and woody debris into channels with fish populations. Landslides are critical in bringing large wood into larger streams where complex habitat, preferred by salmonids, is created (Bisson et al. 1997, Benda et al. 2003, Reeves et al. in press). In fact, at least half of the wood in the fish-bearing portion of the channel network was delivered from smaller, nonfish-bearing tributaries by debris flows or other mechanisms (Reeves et al., in press). It is important to acknowledge large spatial scales, such as watersheds, and temporal scales of decades to centuries, to determine the pattern of natural and human-related disturbance on anadromous fish populations (Reeves et al. 1995, 2002)

Greenough et al. (1999a,b) offer another landscape demonstration (West Arm Demonstration Forest, British Columbia). They used a landscape model (Prognosis-EI) that employed detailed stand management simulations and tracked a broad array of environmental indicators. This allowed for a comparison of potential landscape prescriptions. Their "balanced" management regime succeeded in meeting five management objectives that are seemingly mutually exclusive (Greenough et al. 1999a,b). The management regime

(1) Produced a greater merchantable harvest than forecast for the baseline regimes
(2) Increased the average area of prime ungulate winter range
(3) Maintained a continuous supply of prime grizzly bear (*Ursus arctos*) summer foraging habitat
(4) Maintained a "retention" visual quality designation in all three designated viewsheds
(5) Reduced negative impacts on water quality by reducing the equivalent-clearcut-area in two out of three elevation bands in the watershed

These are simulation results, but they are based on the best available science. Results indicate that management objectives as disparate as wood, wildlife habitat, aesthetics, and water quality can be simultaneously met across a watershed with a management plan. This is compatible forest management.

At the subregional and regional scale, the possibilities for simultaneously producing numerous goods and services from the forest increase. Diversity in ownerships increases as scale increases. With a variety of ownerships comes a wide range of forest structures and conditions. Results from Spies and Johnson (Chapter 8), Haynes et al. (Chapter 10), and Montgomery (Chapter 2) all point to ownership itself as a surrogate driver of diverse management objectives and strategies.

However, long-term projections (100 years) from the Coastal Landscape Analysis and Modeling Study (CLAMS) (Chapter 8) indicate a coastal landscape headed for a condition dominated by two primary states: old, structurally diverse forest (federal land) and young, relatively uniform conifer plantations (private industrial land). Forests on nonindustrial or state lands may moderate this trend somewhat, but this dichotomy still appears likely to emerge over time under current policy. This trend will result in sharp edges at ownership boundaries, decline of intermediate-aged forests (e.g., 50 to 150 years old), and a decline in young, ecologically diverse forests characterized by hardwoods, shrubs, and patches of conifers.

A large landowner (e.g., the U.S. Forest Service) is unlikely to entertain diverse management objectives across the entire holding; organizations—be they businesses or governments—are guided by specific policy guidelines. In decades past, the U.S. Forest Service prided itself on decentralized management (Williams 2000). Recently, this decentralized management has been replaced by strong policy directives (e.g., the Northwest Forest Plan) that induce common management across a region. The result is less diversity in forest structures and conditions, especially in the future (Chapter 10). Uniform management is *de facto* land zoning. This works counter to the goals of compatible forest management.

At large scales, such as an entire region, it is possible that the sheer number of owners (with associated diverse management regimes) can produce a type of artificial compatibility. Assume that each owner has only one management objective, such as wood production to maximize net present value, or aquatic conservation, or the production of salal (*Gaultheria shallon* Pursh), or recreation. Joint production of forest resources does not occur on a single location, yet over the entire region a variety of forest products and services are produced. It is not clear if this multiple one-dimensional resource solution would ever be socially acceptable; if it were, we would have to admit that it met our definition of compatible management at the regional scale. This solution is not what we mean by joint production, although we acknowledge that not every resource can be produced on a given forest stand.

In Chapter 10, the evidence for implementing the various management regimes across the region show they differ both by ownerships and by the expectations of various owners about the financial feasibility of individual actions within the industry. These conditions are expected to remain stable, given the expected market and processing options. These conditions also offer opportunities for compatible management. We can sustain timber harvests while preserving habitat and the services derived from the various forest structural components. At the regional scale, the diversity of ownerships and management objectives helps achieve greater compatibility in forest management.

(3) Historical context. Across a region, we are presented with an array of forest conditions that exist for historical as well as ecological reasons. Treatises by Rajala (1998) and Williams (2000) describe some of the historical reasons for the current conditions of forests in the Pacific Northwest. Mathematicians refer to the current description of the forest as "initial conditions." They simply are, and we must deal with them as they are.

Some biologists refer to the present conditions as a contingency, something that cannot be ignored. Mayr (1983) points out that all biological phenomena have a proximate and an evolutionary cause. Historical context (the proximate cause) shaped the current forest conditions, the contingency that determines the basis for the future. Global vegetation modelers (e.g., Prentice et al. 1992) and global climate modelers refer to current conditions as a transient; the effect of the present on the path to a new equilibrium condition is a transient response. Only with time will its effect die out.

In Chapter 5, Monserud talks about the slow dynamics that characterize forest development. These time scales are long, in human terms. On the short end, an intensive industrial management regime may call for a 40-year rotation. At the other extreme is forest succession, which occurs on a scale close to a millennium in this temperate rainforest. The dominant forest species in western Oregon and Washington is Douglas-fir (*Pseudotsuga menziesii* (Mirb.) Franco), which is paradoxically, an early seral pioneer species that is capable of maintaining its dominance in a forest stand for 500 to 750 years. When we talk of "old growth," it is not a simple state that is quickly reached and maintained. As Franklin (1988) underscores, this rainforest is unique in that it has not only the tallest and largest trees in several important genera (e.g., *Abies*, *Picea*, *Pseudotsuga*, *Thuja*, *Tsuga*), but also the oldest.

Moving the region to a new forest condition is clearly a slow process. Simulation results from Haynes et al. (Chapter 10) and Spies and Johnson (Chapter 8) show that regional demographic changes among alternative policies can be great, but they nevertheless take 50 to 100 years for full effect. Swanson et al. (Chapter 9) model landscape change scenarios over a 600-year period to represent the broad range of stages in forest development. Policies, management, and disturbance regimes can alter the direction over time, but silvics and tree dynamics will determine the rate of growth.

In regard to productive fish habitat, long time scales, from decades to centuries are needed to take into account the full range of disturbances responsible for introduction of woody debris and sediment into stream systems (Reeves et al. 1995). Depending on characteristics of individual landscapes, woody debris and sediment are introduced through disturbance processes including landslides, bank erosion, catastrophic blowdown of trees, and tree mortality (Benda and Sias 1998, Martin and Benda 2001, Benda et al. in press, Reeves et al. in press). Such disturbance mechanisms influence the abundance of large wood in pools, an important element in fish habitat (Burnett 2001). Management policies and practices that are most similar in magnitude and frequency to the natural disturbance regimes will provide the best chances for compatible wood production with the maintenance of productive aquatic ecosystems.

(4) Social response. Social acceptability and many social values are *provisional* (Donoghue Chapter 15, Stankey et al. Chapter 16). The formation of social attitudes can be a fast process, compared to the slow dynamics of forest growth. This disparity between the rates of change for forest conditions and social attitudes results in a forest that is constantly lagging behind the current view of socially desirable conditions. As such, the forest is always viewed as being in a transient condition, with responsive forest managers trying to convert it to some new, more desirable state. Occasionally the problem is compounded by the very professionalism of foresters and their strong command of their subject matter. It is easy for professionals—including scientists—to fall into the trap of knowing the "best" management. The reality is that "best" is determined by a value system, which is inherently personal and provisional.

The provisional nature of social values and attitudes tempers any claims of attaining compatibility in an absolute sense. We can envision management scenarios where wood is produced economically in a manner that increases, or at least does not decrease a range of important environmental indicators. Other management scenarios may attempt to maintain cultural values rather than economics. Mangers on the Gifford Pinchot National Forest (Washington) are eliminating competing overstory vegetation to reach proper light levels to maintain some traditional huckleberry (*Vacccinium* spp.) areas. Nontimber forest products (NTFP) play important roles in American culture and commerce (Alexander and Fight, Chapter 13). In fact over 200 species of NTFP are harvested from public and private lands in the Pacific Northwest for commercial, personal, and traditional purposes (de Geus 1995).

As values change with context, they also change with time (Chapter 15). Although we can never be sure that the management regime will remain acceptable tomorrow, the manager can do much to ensure compatibility today. We can examine the aesthetics of a given compatible management scenario by using visualization tools such as McGaughey's (2002) EnVision to monitor the visual acceptability of the treatments (see Figure 1, Chapter 7). By involving local people in a cyclical process of understanding, adaptation, and action to evaluate potential forest management, confidence in the compatibility of the management can be increased. These participatory approaches are promising methods for better integrating values into forest management, particularly at the local level.

Donoghue (Chapter 15) argues that stating that compatibility has been achieved implies that knowledge about values, both ecosystem function and social values, exists and has been integrated into forest management decision-making. This implies that scientists and managers understand the fundamental differences in how people value forests within the context of a specific resource management problem. It also implies that managers understand these values sufficiently enough to make judgments about what is compatible and what is not. These are lofty assumptions, given the considerable ambiguity in how we define, measure, and integrate social values in forest management (Chapter 15).

(5) Site specificity. A fundamental feature of almost all forest management is that it is site specific. All the determinants of productivity that foresters refer to as site (soil, nutrient and moisture regimes, physiography, and climate) become factors affecting the success of a given forest management prescription. Genetics is also an important component of site productivity, as Monserud and Rehfeldt (1990) found. It is easy to think of silvicultural tools, such as thinning and site preparation, as procedures with generally uniform results, but they are applied in a management prescription for a specific location.

We are bound by the site-specific legacies of past land management. Rather than wish for a different past, we can look to examples such as in Chapter 14 and actively manage to preserve desirable stand legacies. Legacy retention can range from a few live trees to a mixture of trees, shrubs, and coarse woody debris, to entire patches of intact forest. The more legacies retained, the more a forest-influenced environment is maintained. Important legacies include seeds or seedlings of multiple species of conifers and hardwoods, ectomycorrhizal fungi, large coarse woody debris, and large live trees with epiphytic mosses and lichens (Chapter 14).

Although it is true that all site specificity, like politics, is local, there is a broader geographic and social context. We discuss compatible forest management in the context of the Pacific Northwest. We could provide examples of joint production in other regions of the world, but that is not where our research strength lies. Although we only examine the Pacific Northwest, we strongly feel that the ideas of compatible forest management could be applied to other regions.

(6) Science is incremental. Revolutions in science do occur (e.g., Kepler's Laws of Planetary Motion; natural selection; relativity; quantum mechanics; plate tectonics), but they are rare (Kuhn 1970). Science operates by attempting to reject hypotheses, a continual process of testing or refining the accepted paradigm in a given field. In the process, new knowledge is added in a seemingly haphazard fashion (Kuhn 1970). In the centuries between or following revolutions, most progress is made incrementally. As Swanson et al. note (Chapter 9), the history of forest policy has exhibited decades of relative stability punctuated by abrupt change, suggesting that surprising shifts in policy are an important component of long-term landscape management.

In Chapter 1 we gave a brief history of forest management in the Pacific Northwest. In it, we talked about the lack of a scientific basis (e.g., experiments) 50 years ago for the plantation management of Douglas-fir that became widespread throughout western Oregon and Washington. After a half century, the scientific basis has become quite solid (Smith et al. 1997). Nursery methods for efficiently raising healthy, superior planting stock are now common (Duryea and Dougherty 1991), including techniques for inoculating roots with mycorrhizal fungi to promote quick establishment and sustained growth (Castellano and Molina 1989). Average survival of seedlings has increased to 85% or better (Curtis et al. 1998). Effective methods have been developed for controlling competing shrub and nontimber vegetation, thus promoting rapid growth of established trees (Walstad and Kuch 1987). A range of harvesting systems has been developed to reduce problems such as soil compaction (Curtis et al. 1998). Knowledge was acquired incrementally, study by study. There was no grand experiment that clarified and explained all problems in one fell swoop.

In Chapter 5, Monserud discusses the difficulties and limitations of conducting long-term and large-scale forest management experiments. It is nearly impossible to use classic experimentation on landscapes larger than a watershed. No one would seriously propose a forest management experiment on forest-fire risk reduction, for example, where the treatment units were entire counties, much less states or provinces. No administrator would agree to allow his or her region to be the no-treatment control, even if it were randomly assigned. The spatial limitation in forestry experiments is paralleled by comparable problems arising from the enormous time periods needed to fully test management using long-lived forest trees. These considerations indicate a clear limitation to knowledge that can be gained from classic methods.

Nevertheless, science is still our ally. Fields such as geology and astrophysics are not able to experiment, yet they are well developed branches of science. Theory must be relied on when experimentation is not possible. In biology, our only strong theory (*sensu* Platt 1964) is Darwin's natural selection. Our next strong theory will be an understanding of the genome, but that is decades away.

Forestry and ecology have relied on a series of weak theories that are nevertheless useful. The oldest is the allometric property of self-self correlation, which captures the proportionality between the relative increments of two parts of a plant (Monserud and Marshall 1999). Allometric relations can be used to predict variables that are difficult to measure. For example, leaf area or crown biomass can be predicted from diameter at breast height (e.g., Monserud and Marshall 1999). Another useful theory is the –3/2 power law of self thinning (Yoda et al. 1963). Foresters have used the maximum size-density relationship inherent in self-thinning to construct stand management diagrams (Drew and Flewelling 1979, Farnden 1996). These are graphical representations of stand development which, in various formats, illustrate the interactions between density or some other measure of stocking, and various stand parameters such as mean diameter, top height, and volume (Farnden 1996). The stand management diagrams of Drew and Flewelling (1979) have long guided density control in Douglas-fir plantations in the Pacific Northwest.

Lacking strong theory (Platt 1964), forest scientists perforce have become adept at developing empirical relationships. A yield table is simply a concise summarization of observed stand development with respect to time. A stand growth simulation model is based on the same idea, but uses statistical methods to efficiently estimate the growth parameters (Chapter 6).

Even empiricism has its limits. Many of the landscape models described in Chapter 7 are based on heuristics (expert opinion and general rules that seem to work). SIMPPLLE (Chew 1995), for example, is a knowledge-based expert system. It was designed as a management tool to understand how vegetation and processes interact to affect landscape change (Barrett 2001). The processes modeled are succession, harvest, disease, insects, and fire. Analysis tools at the landscape-to-regional scale must rely strongly on data summarization methods (e.g., ATLAS by Mills and Kincaid 1992) and heuristics.

In the end, the weight of evidence becomes our guide, as it does in the legal system. We use science as much as we can to form decisions, but then we must rely on observation, experience, expert opinion, and heuristics. This is a fairly normal situation for most managers. As scientists, we sometimes would like to withhold judgment until the definitive experiment has been conducted. This is an unrealistic stance, and one that allows for hiding behind an impossibly high standard of knowledge. In the end, the weight of evidence combined with the best available science is a reasonable course of action. The alternatives-to-clearcutting treatments (ATC) study (McClellan et al. 2000) provides a case in point. With the ATC treatments recently established on only two blocks (in 1997 and 1999), managers are already responding to the existence of the treatments (see Figures 2 and 3 in Chapter 5). Most timber sales in southeastern Alaska now call for some form of alternatives-to-clearcutting treatments (M. McClellan, USDA, Forest Service, personal communication, 2002). Apparently, the demonstration that some new type of management is operationally feasible is sufficient for some managers to seriously consider it.

(7) Knowledge is fuzzy. Science is a powerful machine for discovering knowledge. We like to think of it as describing how the world works, in exact terms. Actually, uncertainty is a feature of almost all scientific knowledge. Heisenberg's Uncertainty Principle (we cannot simultaneously know an electron's momentum and location) shocked the scientific world in 1927 with its clear limitation on knowledge (Hawking 2001), just as Gödel's Incompleteness Theorem shocked mathematicians in 1931 by proving that not all true statements (theorems) were provable. The world has an essential indeterminacy; it can be pinned down only so far and no farther (Lightman 2000).

Statistical variation is a crucial feature of all experiments. A measure of variability coupled with statistical theory (e.g., Fisher's (1925) Analysis of Variance) allow for the separation of significant from insignificant effects. In experimentation, we know something is true only within the confidence bounds around the mean (some significant effect). Analyzing variation defines the field of statistics and is an essential feature in all observational science.

In forest management, variation and indeterminacy enter into decisions in many ways. All measurements of the forest resource contain measurement error. This is usually not a problem when estimating the dimensions of trees. However, it can be a serious problem when estimating population abundance for mobile, nocturnal, and elusive wildlife species such as the northern spotted owl (*Strix occidentalis caurina*) or its main prey, the northern flying squirrel (*Glaucomys sabrinus*). When managing for wildlife species, the uncertainty regarding population estimates makes it difficult to determine the effectiveness of prescribed management. Partly for this reason, the management focus is often on providing suitable habitat because habitat is more easily observed and measured.

As previously noted, certain social aspects of decisionmaking (social acceptance and values) are provisional and subject to change (Chapters 15 and 16). Donoghue (Chapter 15) points out that managers have several tools (participatory methods) that can help determine which forest values are important to various user groups. Cultural values are enduring, so, for example, it should not be too difficult to find out from local people which forest resources are most important to them. The problem is not how to manage for huckleberries (*Vacccinium* spp.) as much as realizing that it is important to manage for them (Fisher 2002).

Finally, decisionmaking is inherently indeterminant and fuzzy. The manager cannot ever know enough about management alternatives to make the ideal decision. Reynolds (2002) describes the general structure of a logic-based model for evaluating the sustainability of forests at regional and national levels, and illustrates how a logical formalism can be used to represent and evaluate compatibility among resource values and uses. Basically, the model evaluates a proposition. Fuzzy logic is used to weigh the evidence and arrive at a degree of truth for the proposition. Evidence can take many forms: empirical data, theoretical relations, and expert opinion. The intent of Reynolds'(2002) framework is to incorporate lexical uncertainty into the evaluation of a proposition. Even though most concepts that humans use are not precisely defined, people have little trouble making decisions in the course of everyday life.

Reynolds'(2002) tool can be used to compare across alternatives, with compatibility as a management goal (the proposition). The closer the "truth" of a particular proposition (a specific management regime) is to 1.0, the more compatible that regime will be. This is a powerful approach, but its utility relies on building a model of the proposition accurately describing compatible forest management.

(8) Markets provide incentives for management. The history of relentless change in the forest sector of the Douglas-fir region demonstrates the power of markets in altering forest resource conditions. In spite of arguments by advocates for forest management or environmental protection, a myriad of landowners with diverse land management objectives have changed forest resource conditions by responding to various market signals, mostly prices. For many years, changes in timber prices have provided the incentive for landowners (both public and private) to change the intensity and extent of forest management practices. Prices, which are also a function of wood quality, have also provided the incentive for the development (or contraction) of various forest products industries in the region (Chapters 10 and 11).

There is great angst among some that markets are not benign. Often, prices provide insufficient incentive for what advocates of forest management think are the necessary types and intensity of forest practices to ensure sustainable forest management. In the United States, we have frequently debated the role that prices play in private timberland management. In periods of weak stumpage markets (e.g., for small trees) there is concern that landowners and managers will not implement certain forest practices, such as thinning to reduce fire risk. In the case of private timberlands, relatively low financial returns may lead to changes in land use as landowners seek higher returns. These changes may include conversion to agriculture or residential developments (see Cubbage and Haynes 1988, USDA FS 1988, and Haynes 2003 for additional details).

These market-determined changes are not confined to timber markets. There is growing recognition of the values of other goods and services such as nontimber forest products (NTFP), and recreational use, and an emerging interest in valuing services such as carbon credits. Today's forests were determined by past consumer choices regarding their consumption of forest products, as well as the forest management choices made by landowners and managers.

Price plays a powerful role in both sets of choices. As shown in Figure 1 in Chapter 1, forest product prices are often characterized as highly volatile but increasing faster than the overall rate of inflation. The general history of rising prices suggests that timber is relatively scarce and, all else being equal, that there should be changes in various market factors to alleviate the price increases. After all, these are relatively free markets comprising numerous producers and consumers making decisions based on their observations of the market behavior. For example, consumers may substitute nonwood material in some uses such as residential construction (e.g., steel studs for framing) because the real prices of the substitute are more stable than wood prices. It follows that landowners and managers may increase the intensity or extent of land management to produce more timber (to the point where timber prices become stable).

The lack of general agreement about the effectiveness of free markets in increasing quantities of timber supplies (or other forest outputs) in periods of rising prices and perceived scarcity has caused concern among advocates of forest management. (see Cubbage and Haynes 1988 for a summary). Doubts about market effectiveness have often prompted the establishment of public and private programs designed to improve forest management with the intention of slowing the expected rises in timber prices.

Studies of the responsiveness of stumpage markets generally indicate that the price elasticities of supply and demand are unresponsive (see Adams and Haynes 1980, 1996). As economic theory dictates, shifts in the demand or supply curves will have substantial impacts on prices. All of this leads back to the observation that markets are relentless, leading to changes in forest resources that some favor and others strongly oppose.

Regardless of one's view, the Douglas-fir region is an example where rising prices for timber products led to the adoption of relatively systematic forest management regimes consisting of practices that tend to quicken development of well-stocked forests. This increases the economic returns to those (individuals, companies, and various publics) who own timberlands. Current and expected trends show private timberland owners continuing to invest in forest management, subject to increasing regulations of various forest practices. Public lands, on the other hand, are expected to be managed for diverse goals—many not involving the marketplace—reflecting increased recognition of the benefits of many forest goods and services.

2.2. Compatibility Questions

We will address four research questions regarding compatibility to help weigh evidence regarding the proposition:

(1) To what degree can wood production occur without impairing other forest values?
(2) How can the links between management actions and stand-level outputs of goods and services be developed?
(3) What are the methodological problems in developing broad-scale measures of ecosystem condition and performance?
(4) Can broad-scale measures illustrate compatibility or tradeoffs between biophysical and socioeconomic systems at national and ecoregional scales?

Early in the Wood Compatibility Initiative (see Chapter 1 and Haynes and Monserud 2002) we examined these research questions. They helped us assess the progress that was being made toward understanding the opportunities for compatible forest management. These questions focus on the broad issue of managing forests for greater compatibility. The preceding chapters describe numerous instances at several scales of analysis where excellent progress has been made toward understanding the nature of compatible wood production. Furthermore, examples in some chapters and other studies such as Cissel et al. (1999) and Greenough et al. (1999a) demonstrate that the practical needs of the land manager can be addressed and wood can be produced while designing management plans that increase the level of key environmental indicators and other forest values. The actual implementation of these concepts may be hindered by the attitudes and values of landowners toward the different types of approaches (market based or regulatory) taken to institute joint production of some ecosystem services (see Jacobson et al. 2000 for a summary).

(1) To what degree can wood production occur without impairing other forest values?

Forest management in the Douglas-fir region and the maritime forests of coastal British and southeastern Alaska has been evolving for over a century. Experience on even-aged silviculture and plantation management accumulated in the past half-century is vast (Walstad and Kuch 1987, Smith et al. 1997, Curtis et al. 1998, and Chapter 4).

For landowners whose primary goal was timber production, intensive plantation management evolved to take advantage of a highly productive ecosystem. As a result of a relatively predictable silvicultural system, the length of commercial rotations on high productivity lands decreased to as little as 40 to 50 years. Commercial thinning was often eschewed in favor of earlier harvests (Chapter 4).

For other landowners where timber production was one of several goals (e.g., on federal lands), silivicultural systems evolved that gave the manager considerable flexibility without appreciable loss of productivity. For example, federal land managers often used longer rotations (Curtis 1995). This strategy had several advantages: it reduced the area in the regeneration phase, and therefore, the up-front regeneration costs; larger trees with higher quality products were grown, with a continued flow of products from commercial thinning; opportunity to improve unbalanced regional age distributions increased; habitat for some wildlife was improved; carbon storage increased by delaying harvest; and opportunities were available to increase stand health and vigor through thinning.

The current decade has seen shifts in societal values that have led to major shifts in forest management practices in the Pacific Northwest. Greater emphasis, especially on public timberlands, has shifted toward managing for old-growth characteristics (see Appendix 1, Chapter 1), with related goals of protecting endangered species and fish habitat and promoting biodiversity. The popular management paradigm holds that the complex, multiple-story structure of typical old-growth stands derives from a stand development sequence that includes a dense closed-canopy stem-exclusion phase (Smith et al. 1997). Self-thinning following the stem-exclusion phase (second stage; Chapter 1, Appendix 1) then reduces stand density and allows understory regeneration of shade-tolerant tree species to form intermediate canopy layers. There is recent research by Tappeiner et al. (1997) suggesting a different successional approach. They found that regeneration on 10 old-growth sites in the Oregon Coast Range occurred over a prolonged period; trees grew at low density with little self-thinning and bypassed the dense stem-exclusion phase. These results suggest that thinning may be needed in dense young stands where the management objective is to speed development of old-growth characteristics (fourth and fifth stages; Chapter 1, Appendix 1). At the same time, a reexamination of a nearly forgotten experiment called "selective cutting" (similar to contemporary proposals for creation of gaps by using small [0.8-4.0 ha] clearcuts) in the Douglas-fir region in the 1930s would maintain the late-seral characteristics described in Chapter 1, Appendix 1 (Curtis et al. 1998). As Hummel (Chapter 4) points out, in order to measure biodiversity in a meaningful way, links need to be made with forest structure, species richness (a proxy for compositional diversity), and harvestable wood volume.

Examining the state of uneven-aged management in the Douglas-fir region, Emmingham (1998) concluded that regional silviculturists will need many decades to develop and maintain productive uneven-aged stands. Emmingham (1998) found that good natural models and reliable experience with uneven-aged stands are lacking. The lack of information about how to create and manage productive uneven-aged forests is a major impediment and threatens the ability of land managers to manage the late late-successional reserves according to the Record of Decision in the Northwest Forest Plan (USDA and USDI 1994b). Emingham notes that without further management, stands that have attained multilayer condition may return to single-canopy mature forests before they reach old-growth condition and thereby not attain the conditions desired for nontimber values (e.g., wildlife habitat), nor the flexibility desired for changing management objectives.

(2) How can the links between management actions and stand-levels outputs of goods and services be developed?

A relatively rich literature exists for selected products and species. For example, an enormous amount is known regarding almost all aspects of Douglas-fir management for wood production in the Pacific Northwest (e.g., Curtis et al. 1998 and Chapter 11). Considerable work also has been done on Sitka spruce (*Picea sitchensis* (Bong.) Carr) (Peterson et al. 1997). But in general, our knowledge is more limited for many other species (e.g., red alder (*Alnus rubra* (Bong.)), particularly for most nonwood values. A major impediment to effective management of nontimber forest products (e.g., mushrooms, salal, and huckleberry) is that almost all lack yield functions. Our knowledge is limited even for wood production of some key forest species, such as western hemlock (a climax species over most of the Pacific Northwest). In large part such limitations result from the singular success of intensive Douglas-fir plantation management for wood production in the past 50 years.

Limitations also result from a failure to bridge the gap between relatively abstract academic research and the high level of empiricism commonly employed to manage most forest stands. Although one can view a yield table or growth model projection as an abstraction, land managers invariably combine such forecasts with as much relevant site-specific information as they can marshal. The resulting management prescription is highly site-specific. Contrast this, for example, with work that attempts to explain ecological change based on changes in potential vegetation patterns (e.g., theoretical successional pathways) without any reference to actual current vegetation or site conditions. Another example is older growth and yield work based on fully stocked stands (McArdle and Meyer 1930) rather than the more common case of partially stocked stands.

The differences among various yield approaches, broader ranges of management regimes, and management and for a wider range of products are becoming clearer. In the past decade, for example, carbon storage has become increasingly important as a potential mitigating strategy to slow the rate of predicted global warming (Watson et al. 1996), and the development of various carbon accounting schemes has become part of inventory modeling research.

Progress is also being made in developing approaches to evaluate broad management policies for specific periods of time. Monserud (Chapters 6 and 7) discusses a wide range of stand and landscape management models. Few if any landscape models existed in the 1970s. Today, fire risk is one factor of many that can be examined in detail across a large landscape.

Spies et al. (2002a,b) have developed one of the most comprehensive large-scale management models. In the Coastal Landscape Analysis and Modeling Study (CLAMS), they have modeled the entire Coast Range of Oregon in a level of detail that allows for comprehensive examination of broad, regional policy. These simulated outcomes (Chapter 8) can be used as input into other analyses directed toward aquatic, terrestrial, and socioeconomic consequences. In the case of both aquatic and terrestrial wildlife species, simulated forest and range vegetation conditions are inputs to empirical or causal relations among factors that influence wildlife species viability.

However, less information exists on how and why public perceptions of acceptable forest management change and how to properly deal with them (Chapter 16). In general, knowledge about how the public makes and expresses choices about acceptability of forest management practices and the roles that different institutions play in these choices is poor. Related to this is a need to improve understanding of the social and economic acceptability of forest management, both on public and private timberlands. Furthermore, we need to understand how policy actions on one set of ownerships affects conditions on another set of ownerships (Chapters 8 and 10).

(3) What are the methodological problems in developing broad-scale measures of ecosystem condition and performance?

The tendency to focus on finer scale aspects of problems that are amenable to the scientific method has led to considerable resistance within the scientific community to developing composites of individual measures as broad-scale indicators. A National Research Council (2000) report describes many of the methodological issues. The forest science community, in particular, has been slow to embrace the development of broad-scale measures. This lack of enthusiasm partially stems from the experience of some scientists who find their issues become less significant—and even insignificant—when working at broader scales. Ultimately, management—like politics—is local. But policy is broad-scale, and policy drives management.

Another significant methodological problem is that work at broader scales has to rely more on simulation techniques and expert judgment models than on experimentation. Such models often rely on a mix of empirical and judgmental relations (Chapter 7). They are used to estimate how changes in input condition (especially those related to land management) result in changes in output measures of performance. Validation is difficult, and often is approximated by using sensitivity analyses to examine the soundness of the process relations and the robustness of projected outcomes.

Scientists' beliefs can be a problem. For example, some of the most contentious science issues involve understanding the extent and nature of tradeoffs that are assumed to occur among biophysical and socioeconomic components of large ecosystems. Much of the framework for this debate has been provided by scientists who often view themselves as advocates for a sustainable biosphere (Risser et al. 1991) and perceive limited opportunities for mutual gains in both biophysical and socioeconomic systems. The natural resource policy debate in the Pacific Northwest in the 1990s was often portrayed as involving direct tradeoffs between environmental changes and socioeconomic well-being. The implicit assumption is that this is a zero-sum game, which remains to be seen. This debate has been based on assertions or piecemeal collections of data. Composite measures can be developed to examine the direction and extent of tradeoffs between ecological and socioeconomic conditions as different management strategies are considered. The challenge is to develop composite measures that act as proxies for discussing these two dimensions, and for examining changes that may occur in ecological and socioeconomic conditions under different management alternatives.

(4) Can broad-scale measures illustrate compatibility or tradeoffs between biophysical and socioeconomic systems at the national and ecoregional scales?

There is a growing literature in forestry that illustrates the nature and extent of compatible wood production. Work on porcupine and owls described by Montgomery in Chapter 2 illustrates practical approaches. Much work also focuses on managing for multiple values (Lippke and Oliver 1993a) and on tradeoffs (Weyermann et al. 1991, Lippke and Oliver 1993b). The larger issue is what role science can play in the search for compatibility in terms of developing the methodological basis and dealing with issues that significantly affect broad-scale science.

First, broad-scale science poses significant challenges for the scientific community. An effective partnership among scientists, managers, and those

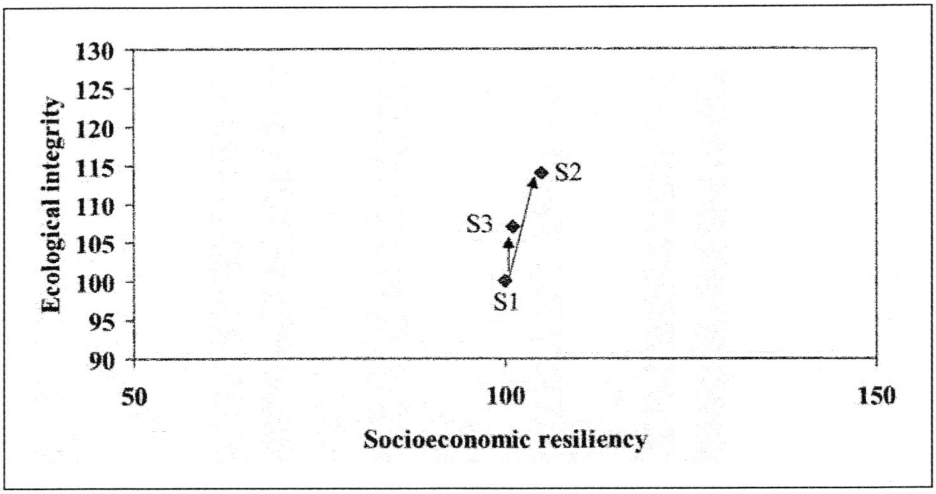

Figure 1–Tradeoff between ecological integrity and socioeconomic resiliency indices for management alternatives in the interior Columbia River basin assessment area for each management alternative (S1, S2, S3). Alternative S1 was used as a baseline comparison to index alternatives S2 and S3 (see Haynes and Quigley 2001).

engaged in the political tasks of governing is essential. The lack of clarity in the socioecological problems that lead to the need for comprehensive broad-scale strategies is frustrating for scientists. It becomes difficult to distinguish issues that reflect different policy preferences among the governing partnership from issues attributable to the lack of information. Furthermore, this lack of clarity surrounding the issue creates confusion about the appropriate spatial and temporal scale of response to various issues.

Second, the science and policy debate since the 1990s has been narrowed considerably by asserting ecological limitations but not acknowledging the full scope and complexity of deliberate societal choices. Evidence of this includes the relatively tight grouping of values for ecological integrity and for socioeconomic resiliency (Figure 1) for the three alternatives for managing almost 31 million hectares (ha) of federal land in the interior Columbia River basin (Haynes and Quigley 2001). These alternatives were designed with varying mixes of ecological restoration, maintenance of reserves, commodity production, and multiple-use management activities. While other alternatives might have been developed to reflect a broader spectrum of potential outcomes, public land managers reveal in their choices of alternatives that they have no incentive to consider a wider array of outcomes.

Third, the shift to managing ecosystems across relatively broad spatial extents stretches the limits of traditional science and of traditional management (Chapter 8). The key scientific tool, experimentation, (including the concepts of randomization, replication, and control) is essentially impossible

Table 1. Weighing the evidence for and against the proposition on compatibility.

		Testing the proposition		
Themes	**Weight, importance**	**For**	**Inconclusive**	**Against**
1. Multiple resource objectives	Moderate	X		
2. Scale	Strong	X		
3. Historical context	Weak	X		
4. Social response	Strong			X
5. Site specificity	Moderate	X		
6. Science is incremental	Strong	X		
7. Knowledge is fuzzy	Moderate		X	
8. Markets provide incentives	Strong		X	
Questions				
1. To what degree can wood production occur?	Strong	X		
2. How can links among management actions and outputs be developed?	Strong	X		
3. How can broad-scale measures be developed?	Weak		X	
4. Can broad-scale measures illustrate compatibility?	Moderate	X		

to use at the broad-scale level (Monserud 2002). As a result, "data" must be compiled and synthesized with less emphasis on the usual components of experimental science and little notion of its reliability.

Research into compatible forest management has better informed environmental policy discussion. But we need to acknowledge that these discussions can be characterized, since at least the early 1970s, as a constant tug-of-war among an unholy alliance of ignorance, ideology, and self-interest (see Blinder 1987 and Rolston 2000 for discussion on the roles scientists have played). Scientists can quickly be challenged beyond the limits of their scientific knowledge when asked to predict and interpret broad-scale social and biophysical consequences of management alternatives.

2.3. The Proposition: Conclusion

At the beginning of this section we stated a proposition and said we would offer evidence to assess whether we could accept or reject it. We considered direct evidence and synthetic information around four research questions that have been part of the Wood Compatibility Initiative. Our evaluation is summarized in Table 1.

2.3.1. The direct evidence—Considerable evidence indicates that compatible forest management is possible. The spatial scale is clearly the strongest factor affecting the proposition: as the scale of analysis (or management) increases, options are available to managers to increase overall regional compatibility between wood production and ecosystem functioning and integrity. Once the provincial or regional scale is reached, the sheer diversity of ownerships and management objectives ensure that multiple products and services are produced from the forest landbase.

At the stand scale, where management is actually implemented, we have evidence that multiple resources can be simultaneously produced in a mutually beneficial manner. Carey's (Chapter 14) wildlife management prescriptions call for increasing the diversity of forest structure and composition while removing some wood through thinning. By investigating the production possibility approach, Montgomery (Chapter 2) concludes that there appears to be substantial opportunity to improve current management. She provides strong evidence from the Oregon Coast Range that management for wood production and wildlife, or wood production and biodiversity, is not a zero-sum game. She demonstrates that current management is well below the production possibility frontier, and that increases in both resources can be made simultaneously. Opportunities for improvement do exist.

Working in southeastern Alaska forests largely dominated by only two species (western hemlock and Sitka spruce), Wipfli et al. (Chapter 3) found that the inclusion of red alder in young coniferous forests may mitigate some negative impacts of clearcutting on nontimber resources such as fish and wildlife, and provide acceptable levels of wood production from these young forests. Because of the short life span of red alder relative to its long-lived conifer associates, greater structural diversity was found in mixed alder-conifer forests than in those consisting nearly entirely of either alders or conifers. The greater diversity and abundance of vascular plants found with more alder increased forage production for deer and improved habitat for small mammals.

Both Kerns et al. (Chapter 12) and Alexander and Fight (Chapter 13) discuss the growing importance of managing for nontimber forest products, especially in the understory. These include a wide array of products, such as floral greens, boughs, craft materials, wild edibles, medicinals, transplants, and other materials. These species are also important for their ecological roles in forest communities. They contribute to biological diversity and long-term ecosystem productivity, underpin mammalian and avian abundance, and are important aesthetic components of forests (Kerns, Chapter 12). Often the management effort and decisions regarding NTFP are primarily related to managing people to control access to harvest NTFP (Chapter 13). Nevertheless, there is an increased focus on using a variety of forest management practices to enhance commercial understory products while maintaining an economically viable overstory tree crop (e.g., Weigand 1998). Kerns (Chapter 12) provides a detailed case study of salal management, including the use of overstory commercial thinning to control light levels for optimum salal production.

Perhaps the most promising scale to focus on is the watershed. Simulation results from Greenough et al. (1999a) in British Columbia and Cissel et al. (1999a) in Oregon both indicate that several environmental indicators can be increased while producing wood. The key is to manage the entire watershed simultaneously rather than in a piecemeal fashion. In Cissel's case, the organizing principle is duplicating the structure of the long-term disturbance (fire) regime by using thinnings and harvesting to match the desired structure across the watershed. This approach uses information on historical and current landscape conditions, disturbance history, and social goals to set objectives for future landscape structures that provide desired plant and wildlife habitat, watershed protection, timber, and other functions (Cissel et al. 1999). The result is a management plan that increases the ecological integrity of the entire watershed while still producing timber and other forest values.

Because clearcutting and plantation management have been the dominant management paradigm in the Pacific Northwest for over a half century, we do not have many demonstration areas to point to as examples of compatible forest management. Several operational-scale experiments in joint production are underway (see Chapter 5), but they are all less than a decade old. This situation forces us to rely on simulation models for analyses of compatibility. As Monserud demonstrates in Chapters 6 and 7, we have many strong tools available for both stand- and watershed-scale analyses. And the work of Spies et al. (2002a,b) on CLAMS (see Chapter 8) demonstrates that province-wide policy analyses can be conducted with enough detail to examine ecological integrity in relation to forest management. These numerous models are summaries of the best available science and knowledge on forest management. With them, we conclude that the possibilities for finding compatible forest management are great.

Of the eight themes reviewed in section 2.1., two (social response and markets) offer the least conclusive evidence (Table 1). Evidence from both of these is given strong weight because of their individual importance. In the case of the social response to compatible forest management, both Donoghue (Chapter 15) and Stankey et al. (Chapter 16) remind us that many social values are provisional. Consequently, the acceptability of forest practices depends on efforts by managers to engage the public and better integrate social values with forest management. Without this engagement, the available evidence suggests acceptance of compatible forest management is unlikely. In the case of markets, Haynes et al. (Chapter 10) describe how powerful markets have been agents of changes in forest management or resources. Because markets are not benign and exist only for a subset of goods and services, their impact on compatible forest management is inconclusive. Both cases suggest that there is great uncertainty about the possible implementation of compatible forest management when societal values, markets, and regulatory actions are all considered.

2.3.2. Synthetic information—We also considered the answers to the four research questions discussed earlier as evidence. The first two questions have the greatest weight importance (Table 1) and uniformly suggest that significant levels of compatible forest management already exist. The extensive application of multiple use management has given the Pacific Northwest a rich legacy of forests managed for a variety of goods and services. These forests demonstrate that there is a high degree of wood production taking place without impairing other forest values. There are examples where this takes place at the site or stand level as well as across broader landscapes. The relatively long history of deliberate forest management undertaken by many landowners with diverse objectives has contributed to this rich legacy. The application of forest practices to manipulate forest conditions has stimulated interest in both theoretical and applied research that has contributed to a rigorous scientific basis for management.

The third and fourth questions questions are important but carry less weight in this discussion (Table 1). The development of broad-scale measures is shrouded in methodological issues and as a result, this question is inconclusive in the context of judging compatibility. Measures that are developed raise interesting notions about unintended consequences, broad-scale science, or emphasize there are opportunities for mutual gains. These measures reveal the growing importance of expert judgments based on both experimental science and results derived from simulation models. There is not agreement about the development of broad-scale measures. Given the universal measures of economic, human health, and demographic factors used in today's society, it is only a matter of time before forest managers make statements about rating forest integrity or the socioeconomic resiliency of forest dependent communities.

2.3.3. Judging the proposition—Based on the weight of both the supporting direct and synthetic evidence (Table 1), we accept the proposition that we can produce commodity products and the other forest values simultaneously. Although both the scientific and empirical evidence suggests that this can be done at different scales, the provisional nature of social acceptance suggests caution in saying that this can be done at all scales in a socially acceptable fashion.

3. Closing

Before closing we return to several points. We evaluate the usefulness of the conceptual model that guided our efforts and summarize some of the limitations we have found. We also summarize how compatible forest management relates to sustainable forest management.

3.1. The Value of the Conceptual Model

Figure 2 lays out the conceptual model that has guided the development of this book. Some may ask if it has been a useful framework for our efforts and of its value to forest managers. Its greatest value to us, and to forest managers, has been contextual. It demonstrates both some of the components and the connections within the general management problem. It helped us define the relative roles of science and management, appreciate inherit differences in temporal or spatial scales, understand where the need for functional and integrated information exist, and realize that some aspects of forest management are so embedded in broader society that separate treatment is unnecessary.

As a framework, Figure 2 poses parallel challenges for both science and management because we often know more about what is in the boxes than we know about the lines (arrows) that connect them. From a systems analysis point of view, the boxes contain state variables and the lines connecting them are fluxes; it is much harder to model the fluxes. We have often thought about the five boxes as representing a set of processes or functions (e.g., How does stand composition change?). In this sense, the lines connecting the boxes represent the context setting and the cause and effect relations. But like most scientists, we struggle to consider the lines connecting the boxes as having equal stature and meriting equivalent levels of scholarly endeavor.

We have not treated the various boxes equally. The management actions component and the forest resource component have been discussed in many chapters. Much of the emphasis has been on the evolution of management strategies or regimes and the silvicultural practices employed in them. Descriptions of the regimes have changed over time (see Chapter 4 and Appendix 1, Chapter 1), and their application has become more selective (see Figure 3, Chapter 10) than often asserted. Highly robust silvicultural tools, such as thinning, are being applied in nontraditional ways to achieve an array of desired future conditions (see Chapters 12 to 14). Our treatment of the forest resource component has changed from relatively simple stand tables to more complex growth and yield models and on to inventory projection systems (see Chapters 6 and 7). While these have evolved to represent the forest resource, less progress has been made describing the nontimber outputs and conditions (Chapters 12 and 13). The same case can be said of the treat-

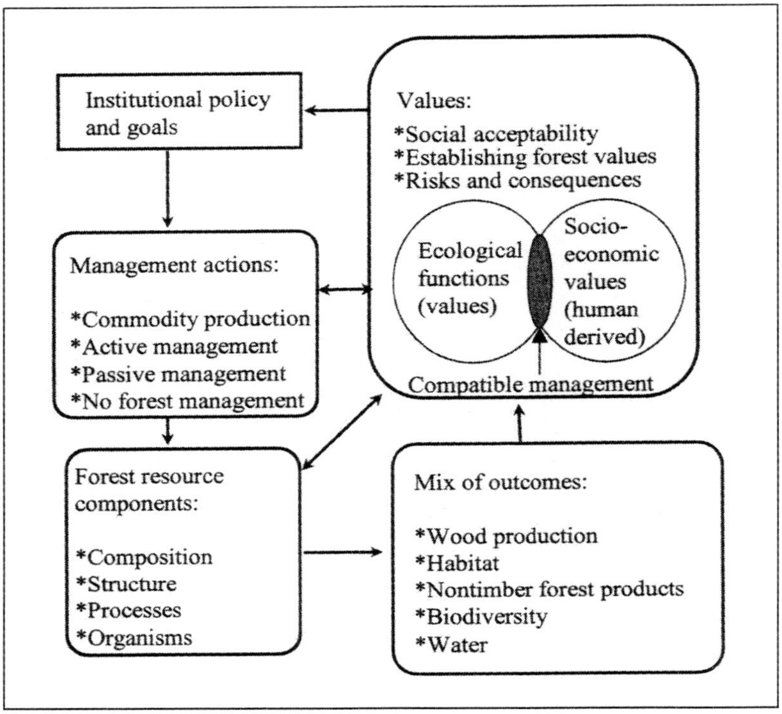

Figure 2. A general conceptual model for the Wood Compatibility Initiative.

ment or need for spatial resolution and the means that exist to aggregate across spatial scales.

Representing the values component focuses on two key issues: establishing notions of social values and social acceptability. These issues are indispensable in developing compatible forest management strategies. Raising the stature of these issues is a significant outcome of our efforts. We found that we did not need to explicitly consider the institutional policy and goals component because various institutional issues are already woven throughout contemporary forest management. This does not mean that foresters, landowners, and public are in universal agreement but, rather, there is an understating of the reality imposed on management debates by the existing instructional structures.

The connections between components are equally important but often neglected. Some of this reflects the way that contemporary research is structured; little attention is paid to cause and effect relations except in the case of some timber-related relations. Differences in spatial and temporal scales among the boxes add complexity to the lines that connect them. For example, the timber inventory resource is often described in terms of decadal age classes, harvests, and net growth that is applied annually, but societal values that influence social acceptability of forestry activities are in almost constant change. To some the volatility of social values seems chaotic and their coping strategy is to ignore social values. This often leads to gridlock among different interest groups. Acknowledging the volatility and working with it may lead to more successful forest management strategies.

Figure 2 also reveals areas where integrated and functional types of information are needed. A danger in contemporary forest management is thinking that every problem requires an integrated approach. As Figure 2 reveals, there is a need for both types of information. A land manager needs to know functional information, such as how trees grow or will respond to thinning. He or she will also need to know how to use that information to describe their contributions to broad management goals, such as sustainable forest management. The measure of success is addressing the right question at the right scale (either spatial or temporal) with the best information available.

Finally, Figure 2 suggests that the forest management decisions are hierarchical because they must recognize both the context set at higher scales and an understanding of specific lower scale processes. This allows managers to consider management actions at the level where they are to be implemented and where impacts will be observed. This gives managers a proactive way to deal with highly dynamic ecosystems and constantly changing social values.

3.2. Limitations to Achieving Greater Compatibility

There are a number of limitations to achieving greater compatibility in forest management. One of the greatest has been the focus on inputs of management strategies rather than the outputs. For example, the number of snag trees per hectare (input) is often managed for rather than the greater abundance of cavity nesting birds (output).

Another example is found in aquatics management. We have spent much of the last decade studying an aquatic conservation strategy that too often gets reduced to a discussion of buffer widths with scant mention of the results of different buffer widths in terms of habitat conditions or fish populations. This orientation has resulted from the focus of current land management around the implementation of different standards and guidelines imposed by regulatory or quasi-regulatory agencies (e.g., USDA and USDI 1994b). Hummel (Chapter 4) finds that there is insufficient data to support the supposition that silvicultural treatments used to create diverse forest structures conserve biodiversity better than those resulting in structural simplification.

It is not fully understood how selected management regimes affect the various components of the forest resource base. Generally we know how a management regime affects timber volumes but relatively less about how it affects other components, such as wildlife abundance or edible mushroom production. We tend to know more about the traditionally managed species such as Douglas-fir but less about other species types like western hemlock.

The growing importance of NTFPs in the Pacific Northwest (Chapters 12 and 13) is stymied by one fundamental shortcoming: yield functions do not exist for most of them. Yield functions for timber production are almost as old as forestry, with the first tables dating to the 18th century. In contrast, the expected yield of almost all NTFPs is uncertain and variable. Mycologists and experienced mushroom hunters have a wealth of experience that can direct them to likely locations (e.g., morels in the year following a forest fire), but actual yield functions or tables are speculative at best. Without a clear idea of productivity for a given product, harvest levels are rather arbitrary. Determing sustainable harvesting levels requires that the site potential be known. Unfortunately, all too often it is not.

Another limitation is the lack of background information on how social values influence forest management. While we understand that science-based information may be provisional because emerging information may alter our understanding, we are generally reluctant to acknowledge the provisional nature of social acceptance of management. Given the biases of foresters to provide sound stewardship of resources where change is measured in decades, we are often confused by the rapid dynamics of social change. The challenge is to better understand the linked nature of broad-scale societal changes and the goals for land management. After all, as land stewards we should strive to preserve as many options as possible for future generations who may value the forests for life support, economic, recreation, aesthetic, wildlife, biotic diversity, scientific, natural history, cultural, spiritual, and intrinsic values.

3.3. How Does Compatible Forest Management Relate to Sustainable Forest Management?

In this book we have examined forest management as a set of compatible production opportunities at the stand and landscape levels. Several examples have demonstrated that opportunities exist to simultaneously produce multiple forest goods and services, maintain the ecological integrity of the forest resource, and reduce social conflict regarding management. In this sense, compatible forest management is complementary to (and likely a component of) sustainable forest management.

The United States and Canada are signatories to the Montreal Process for Sustainable Forest Management. The Montreal Process is one of several frameworks that seeks to balance environment, economic, and equity considerations of forest management from a societal viewpoint (Mihajlovich 2001, Tittler et al. 2001). The criteria and indicators used in the debate about sustainable forest management reflect these three broad areas. Sustainable forest management often does not consider ownership differences other than the contributions of selected ownerships to broad-scale measures. Much of the debate about sustainable forest management is in the context of broader societal goals for sustainable growth (Mendoza and Prabhu 2000). These goals are described in terms of their contributions to economic prosperity that is socially just and environmentally sound.

One of the key differences in compatible forest management is its focus on compatible production opportunities at the stand and landscape levels from the perspectives of owners and managers. Here landowners or managers must decide how they can produce a range of goods and services as they practice forest management. This emphasis on relating management inputs and possible outcomes in an economic and financial context has long been recognized as the heart of the forest management question (see Baker 1950, Davis 1966, and Davis and Johnson 1987, Davis et al. 2001). It also implies that the benefits of compatible forest management need to accrue to those who pay the costs.

This focus on the financial aspects of management is a fundamental distinction between compatible forest management and sustainable forest management. Simply put, compatible forest management has to demonstrate that it can pay for itself on private timberlands. There are instances where forest regulations impose additional costs (e.g., riparian buffers). As long as these regulations are equitably applied, result in positive public benefits, and do not reduce the value of the land, they are generally not perceived as an onerous economic burden.

Compatible forest management on public timberlands often focuses on production of public goods, such as habitat. For example, much of the public land management in the Pacific Northwest is within the context of various habitat conservation strategies where the primary focus is on maintaining selected habitat while at the same time providing recreational opportunities, some timber harvest, gathering of nontimber forest products, and protection of culturally significant sites.

The spatial scales in compatible forest management are significantly different that those of sustainable forest management. Compatible forest management is often considered at the ownership and landscape scale. In the United States, we often ask about the management consequences across an ownership or across all owners at the regional level. The debate about sustainable forest management, however, is actually conducted at the broadest of scales: ecoregions and nations. As signatories to the Montreal Process, forest managers and planners in the United States and Canada have to assess the indicators of sustainability at the national level based on understanding processes at the next lowest spatial level (e.g., ecoregions).

Temporal scales are also different for compatible and sustainable forestry. In the current debate on sustainable forest management, the focus has been on assessing current progress toward sustainable forest management at a specific point in time. Discussions about compatible forest management tend to move back and forth along the time spectrum. Several chapters in this book trace the evolution of management strategies over the past 50 years whereas others consider the evolution of future trends.

Compatible forest management seeks production opportunities that feature mutual gains in selected sets of outputs. Some of the best examples are at finer scales where owners seek compatible forest management by focusing on different outputs. But on the broadest scale, the collective actions of many owners also contribute to both compatible and sustainable forest management.

Regardless of the distinction between compatible and sustainable forestry, sustainability of forest resources is an enduring issue in scientific and political discussion. Clearly, measures of sustainability become increasingly complex as the spatial scale increases from the stand to landscape to region and on to national and global scales. Although much of the work with criteria and indicators of sustainability are at the national scale, forest management is implemented at the stand scale where there has been an emphasis on sustained yield management.

4. Epilogue

The primary issue in land management is stewardship. As Solomon said some 2600 years ago, "One generation comes and another passes but the land remains." As land stewards, our task is to manage forests to contribute to economic prosperity with actions and goals that are socially just and environmentally sound. In one sense, compatible forest management merges our scientific understanding with changing societal values to shape the forest legacy for the next generation.

5. References

Adams, D.M., and Haynes, R.W. 1980. The softwood timber assessment market model: structure, projections, and policy simulations. *Forest Science Monograph*, No. 22. 62 p.

Adams, D.M., and Haynes, R.W. 1996. The 1993 timber assessment market model: structure, projections, and policy simulations. General Technical Report PNW-GTR-368. Portland, OR: U.S. Department of Agriculture, Forest Service, Pacific Northwest Research Station. 58 p.

Baker, F.S. 1950. The principles of silviculture. New York: McGraw-Hill.

Barrett, T.M. 2001. Models of vegetation change for landscape planning: a comparison of FETM, LANDSUM, SIMPPLLE, and VDDT. General Technical Report RMRS-GTR-76-WWW. Ogden, UT: U.S. Department of Agriculture, Forest Service, Rocky Mountain Research Station. 14 p.

Benda, L., Miller, D., Dunne, T., Agee, J., and Reeves, G. 1998. Dynamic landscape systems. In: R.J. Naiman, R.E. Bilby, and S. Kanter (Eds.), Stream ecosystems in the Pacific Northwest and coastal ecoregion (pp. 261-288). New York: Springer-Verlag.

Benda, L., and Sias, J. 1998. Landscape controls on wood abundance in streams. Seattle, WA: Earth Systems Institute Publication. 78 p.

Benda, L.E., Bigalow, P., and Worsley, W. 2003. Recruitment of in-stream large wood in old-growth and second-growth redwood forests, northern California, U.S.A. *Canadian Journal of Forest Research*, 32: 1460-1477.

Benda, L.E., D. Miller, D., Bigelow, P. , and Andras, K. [In press]. Effects of post-wildfire erosion on channel environments, Boise River, Idaho. *Forest Ecology and Management.*

Bisson, P.A., Reeves, G.H., Bilby, R.E., and Naiman, R.J. 1997. Watershed management and Pacific salmon: desired future conditions. In: D.J. Stouder, P.A. Bisson, and R.J. Naiman (Eds.), Pacific salmon and their ecosystems (pp. 447-474). New York: Chapman and Hall.

Blinder, A.S. 1987. Hard heads, soft hearts. Reading, MA: Addison-Wesley Publishing Company. 236 p.

Burnett, K.M. 2001. Relationships among juvenile anadromous salmonids, their freshwater habitat, and landscape characteristics over multiple years and spatial scales in the Elk River, Oregon. Ph.D. dissertation, Oregon State University, Corvallis, OR. 244 p.

Castellano, M.A., and Molina, R. 1989. Mycorrhizae. In: T.D. Landis, R.W. Tinus, S.E. McDonald, and J.P. Barnett (Eds.), The container tree nursery manual. Agricultural Handbook 674. (Vol. 5, pp. 101-167). Washington, DC: U.S. Department of Agriculture, Forest Service.

Chew, J.D. 1995. Development of a system for simulating vegetative patterns and processes at landscape scales. Missoula, MT: University of Montana. 182 p. Ph.D. dissertation.

Cissel, J.H., Swanson, F.J., and Weisberg, P.J. 1999. Landscape management using historical fire regimes: Blue River, Oregon. *Ecological Applications*, 9(4): 1217-1231.

Cubbage, F., and Haynes, R. 1988. Evaluation of the effectiveness of market responses to timber scarcity problems. Marketing Research Report 1149. Washington, DC: U.S. Department of Agriculture, Forest Service. 87 p.

Curtis, R.O. 1995. Extended rotations and culmination age of coast Douglas-fir: old studies speak to current issues. Research Paper PNW-RP-485. Portland, OR: U.S. Department of Agriculture, Forest Service, Pacific Northwest Research Station. 49 p.

Curtis, R.O., DeBell, D.S., Harrington, C.A., Lavender, D.P., St. Clair, J.B., Tappeiner, J.C., and Walstad, J.D. 1998. Silviculture for multiple objectives in the Douglas-fir region. General Technical Report PNW-GTR-435. Portland, OR: U.S. Department of Agriculture, Forest Service, Pacific Northwest Research Station. 123 p.

Davis, L.S. 1966. Forest management: regulation and valuation (2nd ed.). New York: McGraw-Hill. 498 p. plus appendices.

Davis, L.S., and Johnson, N.K. 1987. Forest management (3rd ed.) New York: McGraw-Hill. 790 p.

Davis, L., Johnson, K.N., Bettinger, P., and Howard, T. 2001. Forest management (4th ed.). New York: McGraw Hill. 804 p.

de Geus, N. 1995. Botanical forest products in British Columbia: an overview. Victoria, BC: British Columbia Ministry of Forests, Integrated Resources Policy Branch. 51 p.

Drew, T.J., and Flewelling, J.W. 1979. Stand density management: an alternative approach and its application to Douglas-fir plantations. *Forest Science*, 25: 518-532.

Duryea, M.L.; and Dougherty, P.M. 1991. Forest regeneration manual. Hingham, MA: Kluwer Academic Publishers. 433 p.

Emmingham, W. 1998. Uneven-aged management in the Pacific Northwest. *Journal of Forestry*, 96(7): 37-39.

Farnden, C. 1996. Stand density management diagrams for lodgepole pine, white spruce and interior Douglas-fir. Pacific Forestry Centre Information Report BC-X-360. Victoria, BC: Canadian Forest Service. 41 p.

Fisher, A.H. 2002. Making peace in the berry patch: The 1932 handshake agreement and the promise of cultural use zones. In: E.T. Jones, R.J. McLain, and J. Weigand. Nontimber forest products in the United States (pp. 293-299). Lawrence, KS: University Press of Kansas.

Fisher, R.A. 1925. Statistical methods for research workers. London: Oliver and Boyd. 318 p.

Forest Practices Code of British Columbia Act. 1994. Statutes of B.C., Bill 40. Victoria, BC: Queen's Printer. Retrieved April 4, 2003 from http://www.legis.gov.bc.ca/1994/3rd_read/gov40-3.htm.

Franklin, J.F. 1988. Pacific Northwest forests. In: M.G. Barbour, and W.D. Billings (Eds.), North American terrestrial vegetation (pp. 103-130). New York: Cambridge University Press.

Greenough, J.A., Robinson, D.C.E., Beukema, S.J., and Kurz, W.A. 1999a. Prognosis environmental indicators model: WADF case study methodology. Vancouver, BC: ESSA Technologies Ltd. 24 p.

Greenough, J.A., Robinson, D.C.E., Kurz, W.A., Beukema, S.J., Densmore, N., Winter, R., and Snowdon, B. 1999b. Use of the Prognosis-EI model in balancing timber and environmental values at the watershed-level. Vancouver, BC: ESSA Technologies Ltd. 30 p.

Hawking, S. 2001. The universe in a nutshell. New York: Bantam Books. 216 p.

Haynes, R.W. (Tech. Coordcoord.). 2003. An analysis of the timber situation in the United States: 1952 to 2050. A technical document supporting the 2000 USDA Forest Service RPA assessment. General Technical Report PNW-GTR-560. Portland, OR: U.S. Department of Agriculture, Forest Service, Pacific Northwest Research Station. 254 p.

Haynes, R.W., and Monserud, R.A. 2002. A basis for understanding compatibility among wood production and other forest values. General Technical Report PNW-GTR-529. Portland, OR: U.S. Department of Agriculture, Forest Service, Pacific Northwest Research Station. 46 p.

Haynes, R.W., and Quigley, T.M. 2001. Broad-scale consequences of land management: Columbia basin example. *Forest Ecology and Management*, 153(1-3): 179-188.

Haynes, R.W., Quigley, T.M., Clifford, J.L., and Gravenmier, R.A. 2001. Science and ecosystem management in the interior Columbia basin. *Forest Ecology and Management*, 153(1-3): 3-14.

Jacobson, M.G., Abt, R.C., and Carter, D.R. 2000. Attitudes toward joint forest planning among private landowners. *Journal of Sustainable Forestry*, 11(3): 95-112.

Kuhn, T.S. 1970. The structure of scientific revolutions (2nd ed.). Chicago: University of Chicago Press. 210 p.

Lightman, A.P. 2000. Great ideas in physics (3rd ed.). New York: McGraw-Hill. 300 p.

Lippke, B., and Oliver, C.D. 1993a. An economic tradeoff system for ecosystem management. In: M.E. Jensen, and P.S. Bourgeron (Tech eds.), Ecosystem management: principles and applications. (Vol. 2, pp. 337-345). General Technical Report PNW-GTR-318. Portland, OR: U.S. Department of Agriculture, National Forest SystemService, Forest ServicePacific Northwest Research Station.

Lippke, B., and Oliver, C.D. 1993b. Managing for multiple values. *Journal of Forestry*, 91(12): 14-18.

Martin, D., and Benda, L. 2001. Patterns of in-stream wood recruitment and transport at the watershed scale. *Transactions of American Fisheries Society*, 940-958.

Mayr, E. 1983. The growth of biological thought. Cambridge, MA: Harvard University Press. 974 p.

McArdle, R.E., and Meyer, W.H. 1930. The yield of Douglas-fir in the Pacific Northwest. Technical Bulletin 201. Washington, DC: U.S. Department of Agriculture. 64 p.

McClellan, M.H., Swanston, D.N., Hennon, P.E., Deal, R.L., De Santo, T.L., and Wipfli, M.S. 2000. Alternatives to clearcutting in the old growth forests of southeast Alaska: study plan and establishment report. General Technical Report PNW-GTR-494. Portland, OR: U.S. Department of Agriculture, Forest Service, Pacific Northwest Research Station. 40 p.

McGaughey, R.J. 2002. Creating visual simulations of fuel conditions predicted by the fire and fuels extension to the forest vegetation simulator. In: N.L. Crookston, and R.N. Havis (Comps.), Second Forest Vegetation Simulator conference proceedings (pp. 8-13). RMRS-P-25. Ogden, UT: U.S. Department of Agriculture, Forest Service, Rocky Mountain Research Station.

Mendoza, G.A., and Prabhu, R. 2000. Qualitative multi-criteria approaches to assessing indicators of sustainable forest resource management. *Forest Ecology and Management*, 131: 107-126.

Mihajlovich, M. 2001. Does forest certification assure sustainability? *Forestry Chronicle*, 77(6): 994-997.

Mills, J.R., and Kincaid, J.C. 1992. The aggregate timberland assessment system—ATLAS: a comprehensive timber projection model. General Technical Report PNW-GTR-281. Portland, OR: U.S. Department of Agriculture, Forest Service, Pacific Northwest Research Station. 160 p.

Monserud, R.A. 2002. Large-scale management experiments in the moist maritime forests of the Pacific Northwest. *Landscape and Urban Planning*, 59: 159-180.

Monserud, R.A., and Marshall, J.D. 1999. Allometric crown relations in three North Idaho conifer species. *Canadian Journal of Forest Research*, 29: 521-535.

Monserud, R.A., and Rehfeldt, G.E. 1990. Genetic and environmental components of variation of site index in inland Douglas-fir. *Forest Science*, 36(1): 1-9.

National Research Council. 2000. Ecological indicators for the Nation. Washington, DC: National Academy Press. 180 p.

Peterson, E.B., Peterson, N.M., Weetman, G.F., and Martin, P.J. 1997. Ecology and management of Sitka spruce, emphasizing its natural range. Vancouver, BC: University of British Columbia Press. 336 p.

Platt, J.I. 1964. Strong inference. *Science*, 146: 347-353.

Prentice, I.C., Cramer, W., Harrison, S.P., Leemans, R., Monserud, R.A., and Solomon, A.M. 1992. A global biome model based on plant physiology and dominance, soil properties and climate. *Journal of Biogeography*, 19: 117-134.

Rajala, R.A. 1998. Clearcutting the Pacific Rain Forest: production, science, and regulation. Vancouver, BC: University of British Columbia Press. 286 p.

Reeves, G.H., Benda, L.E., Burnett, K.M., Bisson, P.A., and Sedell, J.R. 1995. A disturbance-based ecosystem approach to maintaining and restoring freshwater habitats of evolutionarily significant units of anadromous salmonids in the Pacific Northwest. In: J.L. Nielsen (Ed.), Evolution and the aquatic ecosystem: defining unique units in population conservation. (pp. 334-349). Bethesda, MD: American Fisheries Society Symposium 17.

Reeves, G.H., Burnett, K.M., and Gregory, S.V. 2002. Fish and aquatic ecosystems in the Oregon Coast Range. In: S.D. Hobbs, J. P. Hayes, R.L. Johnson, G.H. Reeves, T.A. Spies, J. C. Tappeiner, II, and G.E. Wells (Eds.), Forest and stream management in the Oregon Coast Range (pp. 68-98). Corvallis, OR: Oregon State University Press.

Reeves, G.H., Burnett, K.M. and McGarry, E.V. [In press]. Sources of large wood in the mainstem of a fourth order watershed in coastal Oregon. *Canadian Journal of Forestry Research.*

Reynolds, K.M. 2002. Logic models as frameworks for thinking about compatibility. In: A.C. Johnson, R.W. Haynes, and R.A. Monserud (Eds.), Congruent management of multiple resources: proceedings of the wood compatibility initiative workshop. (pp. 215-224.) General Technical Report PNW-GTR-563. Portland, OR: U.S. Department of Agriculture, Forest Service, Pacific Northwest Research Station.

Risser, P.G.; Lubchenco, J.; and Levin, S.A. 1991. Biological research priorities–a sustain able biosphere. *Bioscience*, 41(9): 625-627.

Rolston, H. 2000. Environmental science and environmental ethics. Reflections: newsletter of the program for ethics, science, and the environment. (Vol. 4, pp. 2-3). Corvallis, OR: Oregon State University, Department of Philosophy.

Smith, J.P., Gresswell, R.E., and Hayes, J.P. 1997. A research problem analysis in support of the Cooperative Forest Ecosystem Research (CFER) Program. Corvallis, OR: Department of the Interior, United States Geological Service. 92 p.

Spies, T.A., Johnson, K.N., Reeves, G., Bettinger, P., McGrath, M.T., Pabst, R., Burnett, K. and Olsen, K. 2002a. An evaluation of tradeoffs between wood production and ecological integrity in the Oregon Coast Range. In: A.C. Johnson, R.W. Haynes, and R.A. Monserud (Eds.), Congruent management of multiple resources: proceedings from the wood compatibility workshop. (pp. 111-119.) General Technical Report PNW-GTR-563. Portland, OR: U.S. Department of Agriculture, Forest Service, Pacific Northwest Research Station.

Spies, T.A., Reeves, G.H., Burnett, K.M., McComb, W., Johnson, K.N., Grant, G., Ohmann, J., Garman, S., and Bettinger, P. 2002b. Assessing the ecological consequences of forest policies in a multi-ownership province in Oregon. In: J. Liu, and W.W. Taylor (Eds.), Integrating landscape ecology into natural resource management (Chapter 7, pp. 179-207). New York: Cambridge University Press.

Tappeiner, J.C., Lavender, D., Walstad, J., Curtis, R.O., and DeBell, D.S. 1997. Silvicultural systems and regeneration methods: current practices and new alternatives. In: K.A. Kohm, and J.F. Franklin (Eds.), Creating a forestry for the 21st century (pp. 151-164). Washington, DC: Island Press.

Tittler, R., Messier, C., and Burton, P.I. 2001. Hierarchical forest management planning and sustainable forest management in the boreal forest. *Forestry Chronicle*, 77(6): 998-1005.

U.S. Department of Agriculture, Forest Service. 1988. The South's fourth forest: alternatives for the future. Forest Resources Report 24. Washington, DC. 512 p.

U.S. Department of Agriculture, Forest Service, and U.S. Department of the Interior, Bureau of Land Management. 1994a. Final supplemental environmental impact statement on management of habitat for late-successional and old-growth forest related species within the range of the northern spotted owl. Vol. 1 and 2. Washington, DC: U.S. Department of Agriculture, Forest Service and U.S. Department of the Interior, Bureau of Land Management.

U.S. Department of Agriculture, Forest Service, and U.S. Department of the Interior, Bureau of Land Management. 1994b. Record of decision for the President's forest plan. Washington, DC: U.S. Department of Agriculture, Forest Service and U.S. Department of the Interior, Bureau of Land Management. 74 p. plus appendices.

Walstad, J.D., and Kuch, P.J. (Eds.) 1987. Forest vegetation management for conifer production. New York: John Wiley & Sons. 523 p.

Watson, R.T., Zinyowera, M.C., and Moss, R.H. (Eds.) 1996. Climate change 1995: impacts, adaptations and mitigation of climate change: scientific-technical analyses. Contribution of Working Group 2 to the second assessment report of the Intergovernmental Panel on Climate-Change. New York: Cambridge University Press. 800 p.

Weigand, J.F. 1998. Management experiments for high-elevation agroforestry systems jointly producing matsutake mushrooms and high-quality timber in the Cascade Range of southern Oregon. General Technical Report GTR-PNW-424. Portland, OR: U.S. Department of Agriculture, Forest Service, Pacific Northwest Research Station.

Weyerhaeuser. 2000. TFL 39 Enhanced forest management pilot project. EFMPP Report No. 1. 4 p. Retrieved April 4, 2003 from http://www.weyerhaeuser.com/ourproducts/timberland-sp/bc/new_efmpp/pdf/efm_brochure.pdf.

Weyermann, D.L., Fight, R.D., and Garrett, L.D. 1991. A users guide for SAMM: a proto type southeast Alaska multiresource model. General Technical Report PNW-GTR-274. Portland, OR: U.S. Department of Agriculture, Forest Service, Pacific Northwest Research Station. 50 p.

Williams, G.W. 2000. The USDA Forest Service—the first century. FS-650. Washington, DC: U.S. Department of Agriculture, Forest Service. 154 p.

Yoda, K., T. Kira, H. Ogawa, and Hozumi, K. 1963. Self-thinning in overcrowded pure stands under cultivated and natural conditions. *Journal of Biology of Osaka City University*, 14: 107-129.